Analysis of Engineering Structures and Material Behavior

Analysis of Engineering Structures and Material Behavior

Josip Brnić

University of Rijeka – Faculty of Engineering, Department of Engineering Mechanics, Rijeka, Croatia

WILEY

This edition first published 2018

Registered Office(s)
John Wiley & Sons, Inc., 111 River Street, Hoboken, NJ 07030, USA
John Wiley & Sons Ltd, The Atrium, Southern Gate, Chichester, West Sussex, PO19 8SQ, UK

Editorial Office
The Atrium, Southern Gate, Chichester, West Sussex, PO19 8SQ, UK

For details of our global editorial offices, customer services, and more information about Wiley products visit us at www.wiley.com.

Wiley also publishes its books in a variety of electronic formats and by print-on-demand. Some content that appears in standard print versions of this book may not be available in other formats.

Library of Congress Cataloging-in-Publication Data

Names: Brnić, Josip, 1951– author.
Title: Analysis of engineering structures and material behavior / by Josip Brnić University of Rijeka.
Description: First edition. | Hoboken, NJ : Wiley, 2018. | Includes bibliographical references and index. |
Identifiers: LCCN 2017042809 (print) | LCCN 2017048581 (ebook) | ISBN 9781119329107 (pdf) | ISBN 9781119329060 (epub) | ISBN 9781119329077 (cloth)
Subjects: LCSH: Structural analysis (Engineering)
Classification: LCC TA645 (ebook) | LCC TA645 .B659 2018 (print) | DDC 624.1/7–dc23
LC record available at https://lccn.loc.gov/2017042809

Cover Design: Wiley
Cover Image: © Photo ephemera/Gettyimages

Set in 10/12pt Warnock by SPi Global, Pondicherry, India

Printed and bound by CPI Group (UK) Ltd, Croydon, CR0 4YY

10 9 8 7 6 5 4 3 2 1

"Theory and the real process are referred to each other; theory teaches understanding of the process and the process testifies to the real event."

Let this book be dedicated to the memory of my parents, who instilled in me respect towards people.

Contents

Frequently Used Symbols and the Meaning of Symbols

Symbol	Meaning
A	Cross-sectional area
A_0	Initial cross-sectional area
A_1	Cross-sectional area after deformation
A_e	Finite element area
a	Crack length
a, b, c, d, e, t	Constants in stiffness matrix
$\mathbf{a}$, $[a]$	Polynomial matrix
$[a]$	Transformation matrix
a_0, (a_i)	Initial crack length
$\boldsymbol{a}^b$, $[a^b]$	Polynomial matrix at the boundary of finite element
a_{eff}	Effective crack length
a_f	Failure crack length
B	Strain-displacement matrix
B, N	Parameters
b	Width of rectangular
C	Constant, contour of considered curve, compliance
$\mathbf{C}$	Elasticity matrix (matrix of elastic constants), structural damping matrix
C, m	Constants in Paris equation("m" is strain hardening coefficient)
C, n, p, q	Experimentally derived constants in Forman-Newman –Koning equation
$\mathbf{C}_b$	Generalized elasticity matrix (bending of plate)
C_F, m_F	Constants in Forman equation
C_{ijkl}	Fourth-order tensor (elasticity tensor, elastic matrix or stiffness matrix)
CVN	Charpy impact energy(specimen with V-notch)
$\mathbf{C}_S$	Elasticity matrix refers to shear stresses

(*Continued*)

Symbol	Meaning
$\mathbf{c}^e$	Finite element damping matrix (local coord. System)
D	Diameter
D	Plate flexural rigidity
D, p, r	Parameters
$\mathbf{d}_\varepsilon$	Differential operator
dA_n	Differential area of an arbitrary sloping section (plane)
dA_x, dA_y, dA_z	Differential area on x, y, z direction
da	Increase in crack length (length of crack: a)
$d\lambda$	Coefficient
E	Modulus of elasticity
E_x, E_y, E_z	Young moduli for orthotropic materials
e	Position of shear center, distance between the centroid and the neutral axis, distance
e_i	Invariants of strain deviator
e_{ij}, $[e]$	Deviator strain tensor
F	Force (intensity)
F	Force, loading
$\mathbf{F}$, $[F]$	Force vector, concentrated force vector
F_{cr}	Critical force
F_i, M_i	Nodal forces
F_m	Known components of $\mathbf{F}_R$
$\mathbf{F}_R$	Vector of structure nodal forces.
${}^t\mathbf{F}_R$	Vector of externally applied nodal forces in the considered structure at time t
F_r	Unknown components of $\mathbf{F}_R$
$\mathbf{F}_V$	Vector of volume forces
${}^t\mathbf{F}_\sigma$	Vector of nodal forces that corresponds to the element stresses at the time t
f	Yield function, crack opening parameter (in Forman-Newman-de Koning Equation)
$\mathbf{f}^e$	Finite element nodal forces vector (local coord. system)
f_{ij}	Dimensionless function
f_v, f_x, f_y, f_z	Volume force vector and components forces(unit)
G	Shear modulus (modulus of rigidity)
G_{xy}, G_{yz}, G_{zx}	Shear moduli for orthotropic materials
h	Height of rectangular
I_{min}	Minimum moment of inertia
I_p	Polar moment of inertia

Symbol	Meaning
I_t	Torsion moment of inertia
I_x, I_y	Axial moment of inertia (area moment of inertia about an in-plane axis)
I_{xy}	Centrifugal/deviation moment of inertia(product of areas)
I_1, I_2	Principal (principal centroidal) moments of inertia
I_1, I_2, I_3	Stress tensor invariants
i_{min}	Minimum radius of inertia
J	J-integral (contour integral)
$\mathbf{J}$	Jacobi matrix
J_e	Elastic part of J
J_{Ic}	Fracture toughness
J_{pl}	Plastic part of J
J_1, J_2, J_3	Invariants of deviator stress tensor
K	Bulk modulus, kinetic energy, stress intensity factor
$\mathbf{K}$	Global stiffness matrix (structural matrix)
K/SIF, K_I, K_{II}, K_{III}	Stress intensity factor, stress intensity factors for three opening modes (I, II, III)
K^*	Cyclic strength coefficient
K_c, K_{Ic}, K_{IIc}, K_{IIIc}	Critical stress intensity factor
K_{eff}	Effective stress intensity factor
K_{Ic}	Fracture toughness (Plane strain fracture toughness)
K_{tot}	Total stress intensity factor (as effect of assembled load)
k	Constant
$\mathbf{k}^e$	Finite element stiffness matrix (local coord. system)
$\left[\bar{k}_c^e\right]$	Condensed stiffness matrix
$\left[\bar{k}^e\right]$, $\left[\bar{f}^e\right]$, $[\bar{u}^e]$	Finite element stiffness matrix, vector of nodal forces and vector of nodal displacements in global coordinate system
L	Length (of beam, element)
L_e	Effective (or free) buckling length
L_i (i = 1,2,3)	Natural coordinates
L_0(= G)	Gage length
L_1	Length of considered element after loading
l	Length
l, m, r	Direction cosines
$l_i(z)$, $l_i(L_i)$, $l_i(\xi)$, $l_i(\eta)$, $l_i(\zeta)$	Lagrange interpolating polynomials
$\mathbf{M}$	Structural mass matrix
M_f	Bending moment (flexural moment)

(Continued)

Symbol	Meaning
M_t	Torsion moment(torque)
M_x, M_y	Bending moment (flexural moment) about in-plane axis of cross-section of element (beam), moments in the plate related to the unit of the length of plate.
M_{xy}	Twisting moment in plate
$\mathbf{m}^e$	Finite element mass matrix (local coord. system)
N	Axial force, number of cycles, shape function
N	Axial force (internal)
N, $[N]$	Interpolation matrix(matrix of interpolation functions, shape functions matrix)
n	Strain hardening exponent (in Holloman-Ludwig equation), normal to the considered section (plane), degree of polynomial, number of nodes
n^*	Cyclic hardening exponent
P	Larson-Miller parameter
$P_n(x)$	Polynomial
$\vec{p}_a$	Vector of an average stress
$\vec{p}_n$, p_{nx}, p_{ny}, p_{nz}	Vector of total stress and its intensity components
R_p, r_p	Radius of plastic zone around crack tip
Q	Shear force, heat
Q_{xz}, Q_{yz}	Shear force in plate
q	Shear flow, distributed load intensity
$\boldsymbol{q}_i$	Eigenvectors (shape vectors)
$\mathbf{q}_v$	Body force
R	Radius, stress ratio
S_{ij}, $[S]$	Deviator stress tensor
S_{ijkl}	Fourth-order tensor(compliance tensor)
S_x, S_y	First (static) moment of the area with respect to axis x, y
S_1, S_2, S_3	Principal values of deviator tensor
T	Temperature
T_i	Traction vector
T_m	Melting temperature
t	Wall thickness, time, thickness
$\mathbf{t}_0$, $[t_0]$	Basic transformation matrix (rotational matrix)
$\mathbf{t}^e$	Finite element transformation matrix
U	Vector of structure nodal displacements
u, [u], $\{u\}$	Displacement vector

Symbol	Meaning
u, v, w	Displacements, on x, y, z
$\dot{\mathbf{U}}$	Structural velocity vector
$[\ddot{\mathbf{U}}]$	Structural acceleration vector
U_i, V_i, W_i, Φ_i	Nodal displacements
$\mathbf{U}_m$	Unknown components of **U**
$\mathbf{U}_r$	Known components of **U**
U_0	Strain energy density
U_{0D}	Distortional energy density
$\mathbf{u}^e$, $[u^e]$, $\{u^e\}$	Finite element nodal displacements vector (local coord. system)
V	Volume
V	Potential of external load
W	Work done by external forces, elastic strain energy density
W_p	Polar moment of resistance
W_t	Torsion moment of resistance
W_x, W_y	Section modulus
x, y, z	Cartesian coordinates
α	Angle of neutral axis, angle of principal stresses / strain
$\{\alpha\}$, $\boldsymbol{\alpha}$	Vector of constants(vector of generalized coordinates)
α, β, γ	Functions with respect to ratio h / b of rectangular
α_T (or α)	Coefficient of thermal dilatation
α_x, α_y	Rotation about x,y axes
β	Factor
$\mathcal{G}$	Strain energy release rate (*crack extension force* or *crack driving force*).
γ	Shear strain, material constant
$\gamma_{xy},\dots\gamma_{zx}$	Shear strain components
γ_R, γ_ρ	Angle
ΔK_{th}	Fatigue crack growth threshold
ΔK	Difference between K_{max} and K_{min}
ΔL, $\Delta \boldsymbol{l}$	Elongation
ΔT	Change in temperature
ΔV	Change in volume
ε	Strain
$\dot{\varepsilon}$	Strain rate
$\boldsymbol{\varepsilon}$, $[\varepsilon]$, ε_{ij}	Strain tensor

(*Continued*)

Symbol	Meaning
ε_i	Principal strains (dilatations)
ε_{ij}	Strain components
ε_{ij}^0, $[\varepsilon^0]$	Mean strain tensor
ε_{max}	Maximum principal strain (dilatation)
ε_V (ε_v)	Volumetric strain
$\varepsilon_x, \varepsilon_y, \varepsilon_z$	Normal strain components-dilatations (directions: x, y, z)
ε_0	Mean strain (average dilatation)
$\varepsilon_1, \varepsilon_2, \varepsilon_3$	Principal strains (dilatations)
η	Coefficient of viscosity
θ	Twisting angle per unit length
κ	Curvature
κ_x, κ_y	Curvature of the midsurface
λ	Slenderness, introduced substitute
λ, μ	Lame constants
μ	Coefficient of friction
ν	Poisson ratio
$\nu_{xy}, \nu_{xz}, \nu_{yx}, \nu_{yz}, \nu_{zx}, \nu_{zy}$	Generalized Poisson ratio (orthotropic materials)
ξ, η, ζ	Coordinates, dimensionless coordinates
Π	Total potential (total potential energy)
ρ	Radius
$\boldsymbol{\sigma}$, $[\sigma]$, σ_{ij}	Stress tensor
σ_a	Stress amplitude
σ_{all}	Normal stress allowable
σ_{cr}	Critical stress
σ_e	Equivalent stress
σ_{ij}	Stress components
σ_{ij}^m	Mean (spherical) stress tensor
σ_{ij}^0, $[\sigma^0]$	Mean (spherical) stress tensor
σ_m, *UTS*, *US*, σ_{US}	Ultimate tensile strength, mean stress
$\sigma_{max}, \sigma_{min}$	Stress maximum, minimum
σ_n	Normal stress on an arbitrary sloping section (plane)
σ^{oct}	Normal octahedral stress
$\sigma_x, \sigma_y, \sigma_z$	Normal stress components(directions: x, y, z)
σ_0	Mean normal stress, initial stress in rheological models
$\sigma_{0.2}$, *YS*, σ_{YS}, σ_Y	0.2 offset yield strength
$\sigma_1, \sigma_2, \sigma_3$	Principal stresses, stresses in rheological models
τ_{all}	Shear stress allowable

Symbol	Meaning
$\tau_{\mathrm{I}}, \tau_{\mathrm{II}}$	Extreme values of shear stresses at the directions I, II (plane stress state)
$\tau_{\max}$	Shear stress maximum
τ_n	Shear stress on an arbitrary sloping section (plane)
τ^{oct}	Shear octahedral stress
$\tau_{\mathrm{x}},\ldots..\tau_{\mathrm{zx}}$	Shear stress components
τ_y	Shear stress associated with yielding in uniaxial tension
τ_1, τ_2, τ_3 or $\tau_{\mathrm{I}}, \tau_{\mathrm{II}}, \tau_{\mathrm{III}}$	Extreme values of shear stresses at three –dimensional state of stress
Φ	Prandtl stress function
φ	Angle, twisting angle, angle of an arbitrary plane (section) with respect to perpendicular plane
$\varphi_x, \varphi_y, \varphi_z$	Rotational displacement
ψ	Reduction in area(contraction of cross-sectional area), radius ratio, S. Venant warping function
ω_i	Eigenvalues (free vibration frequencies)

Principal SI Units and the US Equivalents

Quantity	SI Unit	US Equivalent
length	meter (m)	39.370079 inch (in)
	meter (m)	3.2808399 feet (ft)
area	millimeter^2 (mm^2)	0.001549907 in^2
	meter^2 (m^2)	10.7639104 feet^2 (ft^2)
mass	kilogram (kg)	2.2045855 lb – mass
	kilogram (kg)	0.06852177 slug (lb s^2/ft)
volume (solid)	meter^3 (m^3)	35.3146667 feet^3 (ft^3)
	meter^3 (m^3)	61012.8 inch^3 (in^3)
volume (liquid)	liter (l)	0.03531566 feet^3 (ft^3)
	liter (l)	0.26417944 gallon (gal)
force	newton (N = kg m/s^2)	0.22480894 pound (lb, lbf)
	kilonewton (kN)	0.22480894 kilopound (kip)
stress/pressure	pascal (Pa = N/m^2)	0.855470208 pound/foot^2 (psf)
	kilopascal (kPa/m^2)	0.145037944 pound/inch^2 (psi)
	megapascal (MPa = N/mm^2)	0.145037944 kilopound/inch^2 (ksi)
moment of a force	newton meter (N · m)	8.8507456 pound inch (lb · in)
	newton meter (N · m)	0.73756215 pound · foot (lb ft)
energy/work	joule (newton meter)	0.73756215 pound · foot (lb ft)
velocity	meters per second (m/s)	39.37 inches per second (in/s)
	meters per second (m/s)	3.2808399 feet per second (ft/s)
density	kilograms per cubic meter (kg/m^3)	0.06238 pounds per cubic foot (lb/ft^3)
stress intensity factor *SIF* (K); fracture toughness (K_{Ic})	(MPa$\sqrt{m}$);... (K)	1.099 ksi$\sqrt{in}$...(*SIF*)

SI Prefixes, Basic Units, Physical Constants, the Greek Alphabet

SI Prefixes

Prefix	tera	giga	mega	kilo	centi	milli	micro	nano	pico
SI symbol	T	G	M	k	c	m	μ	n	p
Factor	10^{12}	10^{9}	10^{6}	10^{3}	10^{-2}	10^{-3}	10^{-6}	10^{-9}	10^{-12}

Basic Units

Quantity	SI		US	
length	meter (m)		foot (ft)	
mass	kilogram (kg)		slug (lb·s^2/ft)	
time	second (s)		second (sec)	
force	newton (N = kgm/s^2)		pound (lb)	
temperature	degree:	Kelvin (K) Celsius (°C)	degree:	Fahrenheit (°F) Rankine (°R)
	$T_K = T_{°C} + 273.15$		$T_{°F} = T_{°R} + 459.67$	

Physical Constants

Quantity	SI	US
acceleration of gravity(g)	9.80665 m/s^2	32.1740 ft/s^2
density (ρ), unit weight of water (at 4 °C = 39.2 °F)	1000 kg/m^3	62.43 pcf
Normal atmospheric pressure (at)	101,325 kPa 0.101325 MPa	14.6960 psi

The Greek Alphabet

α	Α	Alpha	ν	Ν	Nu
β	Β	Beta	ξ	Ξ	Xi
γ	Γ	Gamma	ο	Ο	Omicron
δ	Δ	Delta	π	Π	Pi
ε	Ε	Epsilon	ρ	Ρ	Rho
ζ	Ζ	Zeta	σ	Σ	Sigma
η	Η	Eta	τ	Τ	Tau
θ	Θ	Theta	υ	Υ	Upsilon
ι	Ι	Iota	φ	Φ	Phi
κ	Κ	Kappa	χ	Χ	Chi
λ	Λ	Lambda	ψ	Ψ	Psi
μ	Μ	Mu	ω	Ω	Omega

Important Notice Before Reading the Book

The author and publisher of this book have invested reasonable efforts in its preparation. The book presents material that is the subject of the author's research and lectures and covers a wide spectrum of engineering disciplines. It provides a concise written guide to theories and applications, showing the methods for solving particular problems. However, the book contains the author's interpretations of the above and not facts. The material in the book is provided as a study aid for the reader and not for business-related activity. Since the book may contain different types of error, the author and the publisher make no warranty of any kind with regards to any of the content of the book or its usage. The author and the publisher shall not be liable in any event for any damages in connection with the usage of any contents of this book. By proceeding to read this book, the reader agrees with the above. This note is a part of this book.

Preface

This book, in its concise but clear form, provides to students, engineers and researchers an insight into the analysis of possible stresses and strains on engineering components which can arise when these components are subjected to loads. It also provides an insight into experimental investigations of the properties of materials from which the components are made. As such, the book is organized in such a way as to encourage, in engineering students and engineers, development of the ability to analyze an existing problem and solve it in a simple but logical way.

The contents of this book refer to matters related to the strength of materials, the theory of stress and strain, the theory of elasticity, the testing of materials, fracture mechanics, creep and the finite element method. In other words, the book covers topics related to structural analysis and the selection of materials, in order to create optimal designs and products.

Engineering components are intended to carry load, to conduct heat, to be exposed to wear and/or corrosive environments and so on. Whatever their predicted service conditions, the components need to be shaped and, of course, manufactured. The designer and manufacturer of a structure should be familiar with all of the above requirements relating to the stress and strain analysis of a structure as well as with the behavior of materials under load.

This book is intended as a reference of lasting value. The author would be grateful to anyone who wishes to contribute suggestions through which to improve the content of the next edition of this book.

Josip Brnić
Rijeka, November 2017.

About the Author

Josip Brnić, DSc, is a Professor in the Department of Engineering Mechanics at the Faculty of Engineering, University of Rijeka, Croatia. He graduated in Mechanical Engineering at the Faculty of Engineering, University of Rijeka. He received his MSc in Mechanical Engineering from the Faculty of Mechanical Engineering, University of Ljubljana, Slovenia in 1983 and his DSc in Mechanical Engineering from the Faculty of Engineering, University of Rijeka in 1988. He received the title of Consulting Professor from the Harbin Institute of Technology, Harbin, China in 2012, and the title of Honorary Professor from the Henan Polytechnic University, Jiaozuo, China in 2011.

At the beginning of his career, he worked in parallel on the "Brodoprojekt", dealing with structural analysis and design, mainly of underwater objects, and at the Faculty of Engineering, dealing with research and teaching. After this period (1978–2000), he moved to permanent employment at the Faculty of Engineering. He was Vice Dean and Dean of the Faculty of Engineering of the University of Rijeka, Vice Rector and Rector of the University of Rijeka. He has also been a member of the National Council for Science of the Republic of Croatia.

In addition, Professor Brnić is an Associate member of the Croatian Academy of Sciences and Arts. He is the reviewer of papers for several prestigious international journals indexed in CC. His research, teaching and publications are in the areas of computational and experimental mechanics, finite element structural analysis and materials testing. He has authored more than 300 scholarly articles and ten books. For achievements in his research he has received several awards, including the Award of the Croatian Academy of Arts and Sciences in the field of Engineering in 2010 and a Lifetime Achievement award from the University of Rijeka in the field of Engineering Sciences in 2012.

Acknowledgements

During the years of my work in teaching, research, design and writing, I have met a lot of respectable people to whom I am grateful for everything that I have learned from them. We have shared beautiful but also hard times and temptations. My experience and knowledge I have tried to convey to others, and now, in writing this book, I am trying to record certain facts and results and leave them to future generations.

Among those whose contributions to my education are unforgettable are professors from the days of my primary, secondary and university education, professors from my masters and doctoral degree education, professors and friends from Croatian and foreign universities, many colleagues and friends from industry and many others.

A particular source of inspiration when choosing material and methods of presentation were my students, who always find a specific approach to problem solving.

In terms of encouragement and support in the preparation of this book and cooperation in research, my gratitude goes to my friends and colleagues with whom I have worked for many years. My special thanks go to Assistant Professor Sanjin Krscanski, DSc, for the creation and preparation of the figures in this book. I also thank those at Wiley who have been involved in the preparation, editing and production of this edition.

1

Introduction

1.1 The Task of Design and Manufacture

The design of structures and machines that need to be safe, reliable and economical, and the estimation of their service life at some point in time, has recently been based on iterative procedures, capacitive computers and numerical analysis. In recent times, a very powerful tool in the aforementioned analysis has been the finite element method.

Practically all fields of human activity are involved in design processes. Problems that need to be solved relate to transportation, buildings, water systems, communication systems, and so on [1]. The same goes for the production processes and metal-forming processes by which products are created.

The designer (manufacturer) of the product needs to be familiar with the mechanical behavior of materials, specifically with such topics as deformation, fatigue and fracture. Emphasis is placed on methods (numerical and predictive) that can be useful in avoiding any kind of failure. Such methods usually take a mechanics viewpoint. The resistance of materials to failure is expressed by material properties such as tensile strength, yield strength, fracture toughness, creep resistance, and so on. Of course, an understanding of the data related to the mentioned material properties is also required. The main task in designing a structure is to ensure that the final product is the optimal one; i.e. it must meet its purpose, it must be economical and it must comply with the prescribed durability and safety requirements.

1.2 Factors that Influence the Design of Engineering Structures

When we use the word structure, we are usually referring to the design of a construction. Therefore, in engineering design we often use the words "structure" and "construction" interchangeably. However, in human life, the word "structure" can have another meaning: the structure of a process or system. The words "structures," "machines" and "materials" are interrelated. Each machine is also a construction, but each structure does not have to be a machine. For example, a car has an engine. This engine is a kind of structure but it is not the primary structure (the body) of the car. Also, for example, the structure of

Analysis of Engineering Structures and Material Behavior, First Edition. Josip Brnić.

an aircraft is not an engine. Many factors are involved in the design, optimization and production of the structure, such as:

- Purpose
- Materials
- Lifetime
- Safety
- Manufacturing technology.

With respect to the spatial position of elements that make up a structure, they may be combined in such a way as to produce planar or spatial structures (see Figure 1.1).

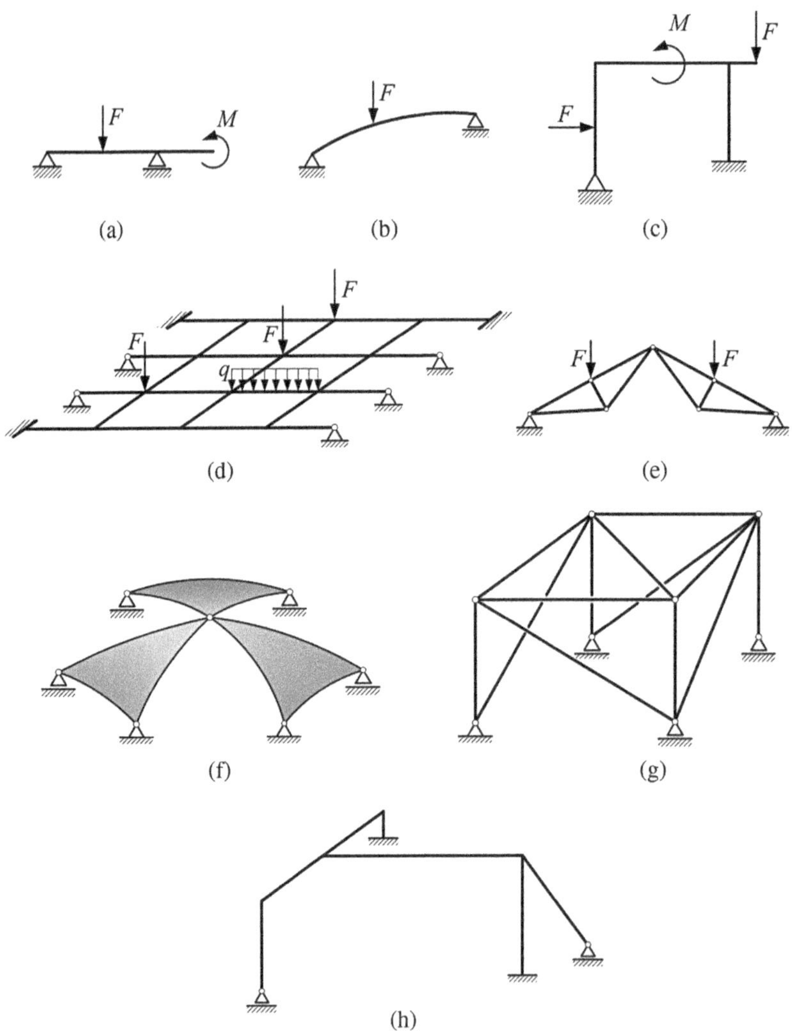

Figure 1.1 Types of structures. a–e) Planar structures. f–h) Spatial structures.

According to the relative proportions of dimensions of the structures (and their elements), structures may be categorized as:

- One-dimensional systems (rods, beams, frames, arches, grillages, etc.);
- Two-dimensional systems (plates, shells, disks);
- Three-dimensional systems.

1.3 The Importance of Optimization in the Process of Design and the Selection of Structural Materials

In recent times, the processes of design and manufacture of structures have involved the application of optimization procedures. Optimization may be treated as the act of obtaining the best product under given circumstances [2]. Three main classes of structural optimization can be described:

- Sizing optimization;
- Layout/topology optimization;
- Shape optimization.

Sometimes material optimization is added as a special type of optimization. In any case, the optimization procedure that yields the optimal product includes both optimal design and manufacturing processes.

Stress and strain analysis is a part of structural analysis. However, differences between the engineering analysis of a structure, the design process and the manufacturing process need to be explained [3]. While the analysis process is concerned with determining the behavior of an existing structure, the design process is intended to calculate sizes, shapes, topology and materials for a structure. Decreasing resources, both in terms of materials and production costs, require the optimization of these processes. In this sense, when choosing a material, its availability and convenience should be considered. Modern structural design is based on the criteria of optimal design, the use of finite element analysis of stress and strain and high-capacitive computers. Consequently, as a result of both design and manufacturing processes, the market should be offered the best product.

Many engineering applications will be subjected to high temperature conditions, or high levels of load, or are intended to be used in conditions with aggressive environments. The materials chosen for structures such as ships, pressure vessels, bridges and so on have to withstand and resist all of these different operating conditions for the required lifetime of the structure. In accordance with this, it is important to know the properties of the various materials that need to comply with the conditions of service life. Physical, chemical and mechanical properties determine the utility of the material. In this sense, designers must be familiar with different types of structures, different types of materials as well as their properties and possible manufacturing processes [4]. Materials science is concerned primarily with the basic knowledge of materials while materials engineering is concerned with the use of this knowledge, i.e. how to transform materials into products. It is known that chemical composition, the processing path and the resulting microstructure define the properties of a material [5]. The kind of properties that depend on the microstructure are called *structure-sensitive properties* (for example, yield strength and hardness). Processes such as cold rolling, hot rolling and so on provide the means to develop and control the microstructure. An engineering structure is designed, manufactured, maintained and controlled in order to guarantee that it does not fail and that it serves the purpose for which it was intended. Structure lifetime predictions and safety during service life are key indicators regarding a structure's quality and reliability.

1.4 Commonly Observed Failure Modes in Engineering Practice

A structure is designed and manufactured for a specific purpose, with a certain safety level and for a certain lifetime. The purpose of the designed structure may be, for example, to carry a load, to transport something, to store a liquid or gas or whatever. Any mechanical failure, which may be defined as any change in size, shape or material properties, can make the considered structure incapable of performing its intended function. Engineering analysis of failures, a newly established discipline in engineering practice, is of the utmost importance in engineering practice. It is an approach to determining why and how an engineering member (component) has failed, because the particular failure has a cause of origin and a form of manifestation. Understanding why and how a failure occurred can be helpful in establishing improvements in design, manufacture and the use of structures, and may provide a means to avoid similar situations in the future. Despite all the possible failures, the number of successfully designed structures may be considered satisfactory. Although in engineering practice many types of failure may occur, the structure will have been designed and manufactured with the assumption that it does not contain a fault. With regard to failures, we may consider:

- The time at which they arose;
- Their causes (*why* the component failed);
- The modes of their manifestation (*how* the component failed).

In general, two main categories of failure can be distinguished:

- Pre-existing failures/defects;
- Failures that arose in the design, manufacture, assembly, service life, maintenance, control processes, etc. [6–13].

Pre-existing defects can arise in a material and may be defects stemming from imperfections.

Common causes of failure that can arise in processes such as design, manufacture, service life, maintenance, and so on include misuse (or abuse), design errors, improper choice of material, manufacturing defects, assembly errors, unforeseen operating conditions, improper maintenance, the transition of temperature effect and inadequate control. Knowledge of these causes helps provide the answer as to why an engineering component failed. Misuse means that the structure was used in conditions for which it was not designed, for example, overload. Design errors/failures may involve the following items: the material used and its chemical composition, the size and shape of an engineering component and/or the material properties. Similar explanations may be given for other causes of failure.

Failure modes in engineering practice that arise in the aforementioned processes (design, manufacture, maintenance, and so on) can be listed as: force-induced elastic deformation, yielding, buckling, fatigue, fracture, corrosion, creep, thermal shock, and so on. In addition, three different fracture modes or mechanisms that occur especially in metals may arise: the sudden fracture of a brittle material, the fracture of flawed members or progressive fracture fatigue.

1.5 Structures and the Analysis of Structures

Types of structures, loads, stress states, the analysis of structures and in analysis used finite elements are interconnected. Each of the mentioned subjects is explained in more detail at various points in the book. Briefly, types of structures are mentioned in Sections 1.2 and 11.2.2; stresses and stress states are mentioned in Chapter 2; strains and strain states in Chapter 3; loads in Chapter 6; finite elements and their applications in Chapter 11.

With respect to the dimensionality of its elements, a structure may be one-dimensional, two-dimensional or three-dimensional. With respect to the spatial positions of its elements, it may be planar or spatial. Analysis of the structure depends on the used finite elements and can be one-dimensional (1-D), two-dimensional (2-D) or three-dimensional (3-D). The same division is valid for finite elements (1-D, 2-D and 3-D finite elements). A state of stress (strain) may be uniaxial, biaxial or multiaxial. In this case, the use of the terms one-, two- or three-dimensional is rarer.

In short, an engineering structure or an appropriate engineering element is designed for a specific purpose. When the engineering element is subjected to a load, it is, at the same time, subjected to stress. Consequently, its volume and/or shape may change. It is known that, depending on the level of the load, the response of the structure may belong to the elastic, plastic or elastic–plastic field. It is possible to analyze the type of load, state of stress and the strain state. The load can be considered basic (axial, shear, torsion, bending, twisting) or complex (a combination of loads). The state of stress can be uniaxial, biaxial or multiaxial. According to the type of stress, stress can be normal or shear. The state of strain can also be uniaxial, biaxial or multiaxial. According to the type of strain, strain can also be normal or shear. A geometrical manifestation of achange in volume and/or achange in shape is visible, for example, under axial load in the extension/shortening of the element; under torsion in the rotation of one cross-section of the element relative to another; under bending in the deflection of a considered point in the undeformed and the deformed state, etc. Relationships between applied loads and the development of stress in an engineering element that is subjected to load are considered in the appropriate chapters dealing with this matter. The design of an element cannot be completed until the level of stress, strain and any changes in volume and/or shape have been considered.

References

1 Boresi, A. P., Schmidt, R. J. and Sidebottom, O. M. (1993) *Advanced Mechanics of Materials*, John Wiley & Sons, New York.

2 Rao, S. S. (2009) *Engineering Optimization*, 4th edition, John Wiley & Sons, New Jersey.
3 Arora, J. S. (2004) *Introduction to Optimal Design*, 2nd edition, Elsevier Academic Press, San Diego.
4 Smith, W. F. and Hashemi, J. (2010) *Foundations of Materials Science and Engineering*, 5th edition, McGraw-Hill, New York.
5 Bramfitt, B. I. (1997) Effects of Composition, Processing and Structure on Properties of Irons and Steels, in *ASTM Handbook*, Vol. **20**, Materials Selection and Design, Dieter, G. E. (Volume Chair), USA, ASM International, pp. 357–382.
6 Brooks, C. R. and Choudhury, A. (2002) *Failure Analysis of Engineering Materials*, McGraw-Hill, USA.
7 Farahmand, B., Bockrath, G. and Glassco, J. (1997) *Fatigue and Fracture Mechanics of High Risk Parts*, Chapman & Hall, New York.
8 Collins, J. A. (1993) *Failure of Materials in Mechanical Design*, 2nd edition, John Wiley & Sons, New York.
9 Brnic, J., Turkalj, G., Krscanski, S., Lanc, D., Canadija, M. and Brcic, M. (2014) Information relevant for the design of structure – ferritic-heat resistant high chromium steel X10CrAlSi25. *Materials and Design*, **63**, 508–518.
10 Brnic, J., Turkalj, G., Canadija, M. and Niu, J. (2014) Experimental determination and prediction of the mechanical properties of steel 1.7225. *Materials Science and Engineering A*, **600**, 47–52.
11 Brnic, J., Turkalj, G., Lanc, D., Canadija, M., Brcic, M. and Vukelic, G. (2014) Comparison of material properties: Steel 20MnCr5 and similar steels. *Journal of Constructional Steel Research*, **95**, 81–89.
12 Brnic, J., Turkalj, G., Niu, J., Canadija, M. and Lanc, D. (2013) Analysis of experimental data on the behavior of steel S275JR – Reliability of modern design. *Materials & Design*, **47**, 497–504.
13 Brnic, J., Turkalj, G., Canadija, M. and Lanc, D. (2011) AISI 316Ti (1.4571) Steel – Mechanical, Creep and Fracture Properties versus Temperature. *Journal of Constructional Steel Research*, **67**(12), 1948–1952.

2

Stress

2.1 Definition of Average Stress and Stress at a Point

In structural analysis it is possible to distinguish the analysis of stress and/or strain of a structure's members in two scenarios:

- When loads and dimensions (of cross-sections) of the members are known;
- When loads are known or predicted and dimensions need to be determined in accordance with the allowable stress/strain.

In any case, a complete analysis can be done when the load, dimensions and material are known. However, it is necessary to define some terms. Consider a three-dimensional body (solid, continuous medium) in a state of equilibrium subjected to external forces acting on its surface, as shown in Figure 2.1 [1–6]. Let the body be divided/separated by a fictitious plane (R) into two parts, I and II. Consider part I; the forces acting through the surface (R) are maintained in order to preserve equilibrium. In this way, part I is subjected to external forces which belong to part I as well as to internal forces acting on the surface (R). The intensity of the internal forces per unit area is called *stress*. Further, consider the internal forces which belong to the small area (ΔA). These forces, when reduced in the point O, form a force-couple system (ΔM, ΔF). If the surface ΔA decreases to zero, then stress vector $\vec{p}_{\rm n}$ is obtained. This can be divided into two components: $\vec{\sigma}$ (or $\vec{\sigma}_{\rm n}$) and $\vec{\tau}$ or ($\vec{\tau}_{\rm n}$). In this case, "n" denotes the normal to the plane (R).

In accordance with Figure 2.1:

$$\vec{p}_{\rm a} = \frac{\Delta \vec{F}}{\Delta A};\ \vec{p}_{\rm n} = \lim_{\Delta A \to 0} \frac{\Delta \vec{F}}{\Delta A} = \frac{{\rm d}\vec{F}}{{\rm d}A};\ \lim_{\Delta A \to 0} \frac{\Delta \vec{M}}{\Delta A} = \vec{0};\ \vec{p}_{\rm n} = \vec{\sigma}_{\rm n} + \vec{\tau}_{\rm n}. \tag{2.1}$$

In addition to the vector notation shown in Equation (2.1), the stresses may also be written in so-called symbolic notation:

$$\mathbf{p}_{\rm a} = \frac{\Delta \mathbf{F}}{\Delta A},;\ \mathbf{p}_{\rm n} = \lim_{\Delta A \to 0} \frac{\Delta \mathbf{F}}{\Delta A} = \frac{{\rm d}\mathbf{F}}{{\rm d}A} \tag{a}$$

In Equation (2.1), $\vec{p}_{\rm a}$ is called *the vector of an average stress*, $\vec{p}_{\rm n}$ is called *the vector of total stress at a point*. The intensity of the force ΔF per area ΔA is called an *average stress* while the intensity of differential force, dF, per diffential area, dA, is called the *total stress*. Stress σ at a point O is called *perpendicular stress* or *direct stress* (*normal stress*), while stress τ is

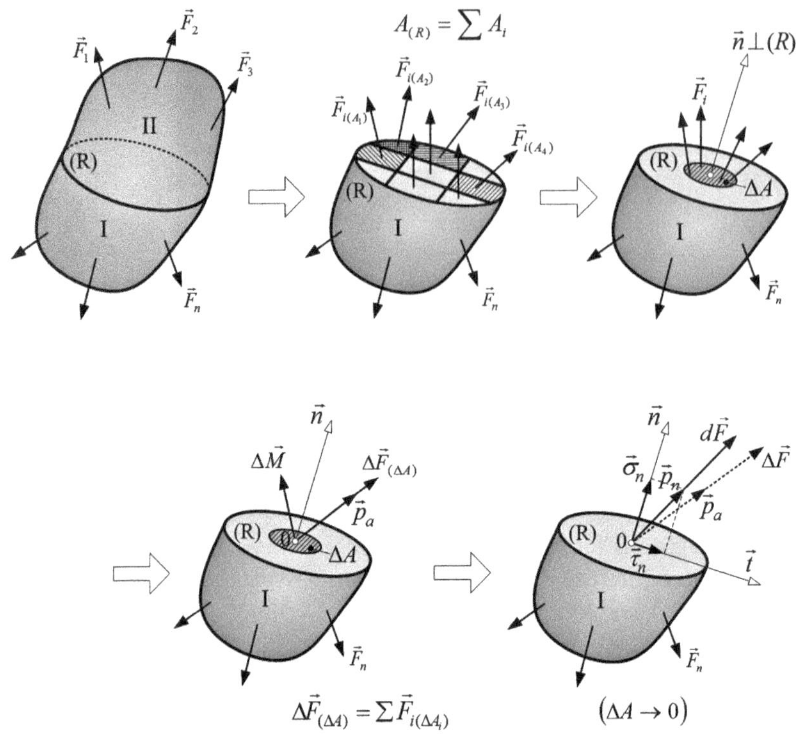

Figure 2.1 Definition of stress.

called *shear stress* or *tangential stress*. If, for example, a rod element is subjected to force F (Figure 6.2), then average stress ($p_a = \sigma$) acting on the cross-sectional area (A) of the element is $p_a = \sigma = (F/A) = E \cdot \varepsilon$, where E is the *modulus of elasticity* and ε is *normal strain*, as given by Hooke's Law, Equation (6.8).

2.2 Stress Components and Equilibrium Equations

The solid subjected to external load may change its shape and/or volume. However, after the deformation process has finished, the body is in an equilibrium (in its deformed state). To gain a better understanding of the body's behavior in this equilibrium as well as a better understanding of the transition of internal forces through the body, analyze an infinitesimal volume element (Figure 2.2), extracted from the stressed body shown in Figure 2.1 [7–9]. Stress components placed on the considered infinitesimal volume are presented in the Cartesian coordinate system.

2.2.1 Stress Components

The resultant stresses (forces) are attached to the center of the faces (planes, areas) of the infinitesimal element and the resultant moments are neglected. Consequently,

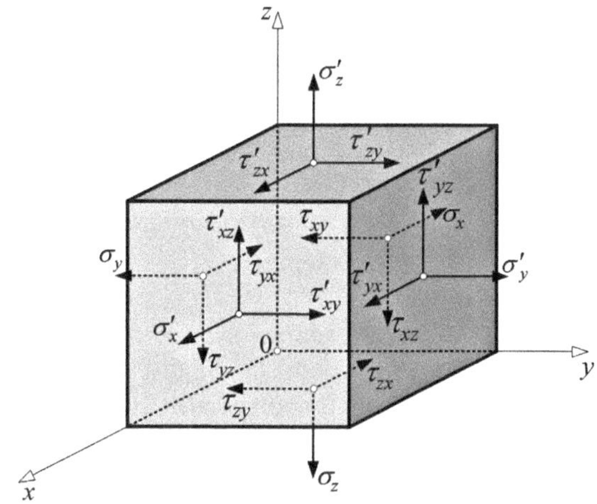

Figure 2.2 Stress components acting on an infinitesimal volume element – Cartesian coordinate system.

the stresses are distributed continuously through the body. According to Figure 2.2, the following stress components may be distinguished:

$$\sigma_x, \ldots ..\tau_{zx}, \quad \sigma'_x, \ldots \ldots \tau'_{zx}.$$

As can be seen, components designated with (′) refer to the planes that do not contain coordinate axes passing through the origin of the coordinate system (0, x, y, z) in Figure 2.2. Further, in accordance with the Taylor rule, the relationship between the stress components placed on the planes passing through the origin of the coordinate system and those placed on the planes that are at distances of dx, dy, and dz, can be written as:

$$\begin{aligned}
\sigma'_x &= \sigma_x + \frac{\partial \sigma_x}{\partial x}\mathrm{d}x + \ldots \\
\sigma'_y &= \sigma_y + \frac{\partial \sigma_y}{\partial y}\mathrm{d}y + \ldots \\
&\vdots \\
\tau'_{zy} &= \tau_{zy} + \frac{\partial \tau_{zy}}{\partial z}\mathrm{d}z + \ldots
\end{aligned} \tag{2.2}$$

Finally, it follows that only nine different stress components exist. If volumetric forces $\vec{f}_V = (f_x, f_y, f_z)$ are also considered, then equilibrium equations (*Louise Navier equations*) in index notation can be written as follows.

2.2.2 Equilibrium Equations

$$\frac{\partial \sigma_{ji}}{\partial x_j} + f_i = \sigma_{ji,j} + f_i = 0 \tag{2.3}$$

Or, written by components:

$$\frac{\partial \sigma_x}{\partial x} + \frac{\partial \tau_{yx}}{\partial y} + \frac{\partial \tau_{zx}}{\partial z} + f_x = 0$$
$$\frac{\partial \tau_{xy}}{\partial x} + \frac{\partial \sigma_y}{\partial y} + \frac{\partial \tau_{zy}}{\partial z} + f_y = 0. \tag{2.4}$$
$$\frac{\partial \tau_{xz}}{\partial x} + \frac{\partial \tau_{yz}}{\partial y} + \frac{\partial \sigma_z}{\partial z} + f_z = 0.$$

Imagine that the coordinate system is now placed in the center of the considered infinitesimal volume (point O_1, Figure 2.9), instead of the position shown in Figure 2.2. Only stress components acting on the visible sides of the infinitesimal volume are shown. Consider the equilibrium equations regarding the moments for its axes ($\Sigma M_i = 0$, i = considered axis). Now, the following equalities are valid:

$$\tau_{xy} = \tau_{yx}, \quad \tau_{yz} = \tau_{zy}, \quad \tau_{zx} = \tau_{xz} \tag{2.5}$$

In this case, moments of the stresses which are parallel or intersect the considered axes are equal to zero.

2.3 Stress Tensor

From the nine stress components (as nine scalars) which represent a three-dimensional state of stress, only six components are required to describe the state of stress at a point in an engineering component (see also the discussion in Section 2.4.3). It is known that an array of $3^2 = 9$ components (in this case stress components), whereby the components are transformed from one to another Cartesian coordinate system by a particular rule, is called a *second-order tensor*. Since the stress components behave in the same way, stress is a second-order tensor. So these nine stress components may be tabulated in array form as follows [3, 5, 8, 10–19]:

$$\boldsymbol{\sigma} = \sigma_{ij} = [\sigma] = \begin{bmatrix} \sigma_{xx} & \sigma_{xy} & \sigma_{xz} \\ \sigma_{yx} & \sigma_{yy} & \sigma_{yz} \\ \sigma_{zx} & \sigma_{zy} & \sigma_{zz} \end{bmatrix} = \begin{bmatrix} \sigma_x & \tau_{xy} & \tau_{xz} \\ \tau_{yx} & \sigma_y & \tau_{yz} \\ \tau_{zx} & \tau_{zy} & \sigma_z \end{bmatrix} = \begin{bmatrix} \sigma_1 & 0 & 0 \\ 0 & \sigma_2 & 0 \\ 0 & 0 & \sigma_3 \end{bmatrix} \tag{2.6}$$

Equation (2.6) represents the *stress tensor* ($\boldsymbol{\sigma}$ – symbolic notation; σ_{ij} – index record /notation; $[\sigma]$ – matrix form/notation). This stress tensor is also known as the *Cauchy stress tensor* since its stress components arise in Cauchy stress equations (Equations (2.25) or (2.36)).

2.3.1 Mean and Deviatoric Stress Tensors

The stress tensor can be decomposed into two parts. One is the *mean* (*spherical*) *stress tensor* and the other is the *deviatoric stress tensor*:

$$\sigma_{ij} = \sigma_{ij}^0 + S_{ij} \quad \text{or} \quad [\sigma] = [\sigma^0] + [S] \tag{2.7}$$

where σ_{ij} = the stress tensor; $\sigma_{ij}^{0} \equiv \sigma_{ij}^{\mathrm{m}}$ = the mean stress tensor; S_{ij} = the deviatoric stress tensor.

Presentation of the stress tensor using components as well as principal stresses, in accordance with Equation (2.7), may be given as follows:

$$\begin{bmatrix} \sigma_x & \tau_{xy} & \tau_{xz} \\ \tau_{yx} & \sigma_y & \tau_{yz} \\ \tau_{zx} & \tau_{zy} & \sigma_z \end{bmatrix} = \begin{bmatrix} \sigma_0 & 0 & 0 \\ 0 & \sigma_0 & 0 \\ 0 & 0 & \sigma_0 \end{bmatrix} + \begin{bmatrix} \sigma_x - \sigma_0 & \tau_{xy} & \tau_{xz} \\ \tau_{yx} & \sigma_y - \sigma_0 & \tau_{yz} \\ \tau_{zx} & \tau_{zy} & \sigma_z - \sigma_0 \end{bmatrix} \tag{2.7a}$$

$$\begin{bmatrix} \sigma_1 & 0 & 0 \\ 0 & \sigma_2 & 0 \\ 0 & 0 & \sigma_3 \end{bmatrix} = \begin{bmatrix} \sigma_0 & 0 & 0 \\ 0 & \sigma_0 & 0 \\ 0 & 0 & \sigma_0 \end{bmatrix} + \begin{bmatrix} \sigma_1 - \sigma_0 & 0 & 0 \\ 0 & \sigma_2 - \sigma_0 & 0 \\ 0 & 0 & \sigma_3 - \sigma_0 \end{bmatrix}. \tag{2.7b}$$

As can be seen, based on Equations (2.7), the mean (spherical) stress tensor is of the form:

$$\begin{aligned} \sigma_{ij}^{0} \equiv \sigma_{ij}^{\mathrm{m}} = \sigma_0 \delta_{ij} \equiv \sigma_{\mathrm{m}} \delta_{ij} &= \begin{bmatrix} \sigma_0 & 0 & 0 \\ 0 & \sigma_0 & 0 \\ 0 & 0 & \sigma_0 \end{bmatrix} \equiv \begin{bmatrix} \sigma_{\mathrm{m}} & 0 & 0 \\ 0 & \sigma_{\mathrm{m}} & 0 \\ 0 & 0 & \sigma_{\mathrm{m}} \end{bmatrix} \\ &= \frac{1}{3} \sigma_{kk} \delta_{ij} \end{aligned} \tag{2.8}$$

Based on Equations (2.7) and (2.8), it follows that:

$$\sigma_{ij} = \frac{1}{3} \sigma_{kk} \delta_{ij} + S_{ij} \tag{2.8a}$$

The deviatoric stress tensor (expressed by principal stresses) is:

$$\begin{aligned} S_{ij} = [S] &= \begin{bmatrix} \sigma_1 - \sigma_0 & 0 & 0 \\ 0 & \sigma_2 - \sigma_0 & 0 \\ 0 & 0 & \sigma_3 - \sigma_0 \end{bmatrix} \equiv \begin{bmatrix} \sigma_1 - \sigma_{\mathrm{m}} & 0 & 0 \\ 0 & \sigma_2 - \sigma_{\mathrm{m}} & 0 \\ 0 & 0 & \sigma_3 - \sigma_{\mathrm{m}} \end{bmatrix} \\ &= \begin{bmatrix} \frac{2\sigma_1 - \sigma_2 - \sigma_3}{3} & 0 & 0 \\ 0 & \frac{2\sigma_2 - \sigma_3 - \sigma_1}{3} & 0 \\ 0 & 0 & \frac{2\sigma_3 - \sigma_1 - \sigma_2}{3} \end{bmatrix} \end{aligned} \tag{2.9}$$

$$S_{kk} = 0 \tag{2.9a}$$

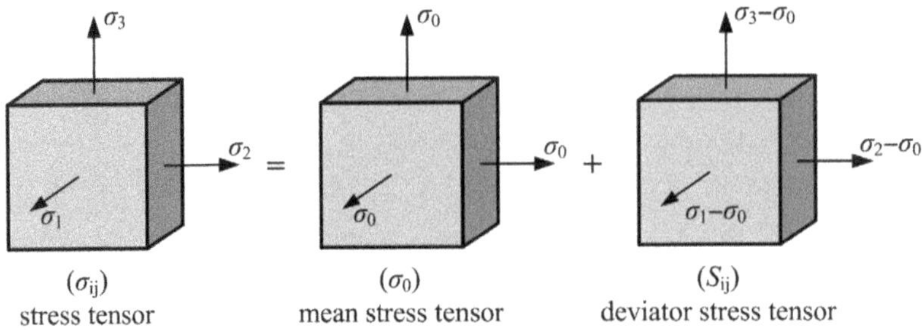

Figure 2.3 Graphical representation of the stress tensor.

The deviatoric stress tensor in Equation (2.9) can also be written in terms of stress components:

$$S_{ij} = \begin{bmatrix} \dfrac{2\sigma_x - \sigma_y - \sigma_z}{3} & \tau_{xy} & \tau_{xz} \\ \tau_{yx} & \dfrac{2\sigma_y - \sigma_z - \sigma_x}{3} & \tau_{yz} \\ \tau_{zx} & \tau_{zy} & \dfrac{2\sigma_z - \sigma_x - \sigma_y}{3} \end{bmatrix} \tag{2.9b}$$

Deviatoric stresses are of interest in the theory of plasticity, since they are accepted as being the most important factor influencing the yielding of ductile materials. Graphical representation of the meaning of the stress tensor, the mean stress tensor and the deviatoric stress tensor is given in Figure 2.3.

In Equations (2.8) and (2.9) above, *the mean normal stress* (the average normal stress or hydrostatic stress) is:

$$\sigma_0 \equiv \sigma_m = \frac{\sigma_x + \sigma_y + \sigma_z}{3} = \frac{\sigma_1 + \sigma_2 + \sigma_3}{3} = \frac{1}{3}\sigma_{kk} = \frac{1}{3}I_1 \tag{2.10}$$

Experiments carried out on a lot of metallic materials found that their yielding and plastic deformation were essentially independent of the applied mean normal stress. Hence, most theories of plasticity postulate that plastic behavior of materials is related primarily to that part of the stress tensor that is independent of mean normal stress.

2.4 States of Stress

2.4.1 Uniaxial State of Stress

Consider the effect of axial force on the one-dimensional element in Figure 2.4(a).

In the vicinity of the point N was cut an elementary volume (presented in 2-D in Figure 2.4(b)). On the basis of equilibrium of all forces acting on it, with respect to the axis n, it follows that:

$$\sum F_n = 0 \rightarrow \sigma_n \mathrm{d}A_n - \sigma_z \mathrm{d}A_z \cos\varphi = 0 \rightarrow \sigma_n = \sigma_z \cos^2\varphi, \tag{2.11}$$

where: $\mathrm{d}A_z = \mathrm{d}A_n \cos\varphi$, $\mathrm{d}A_y = \mathrm{d}A_n \sin\varphi$.

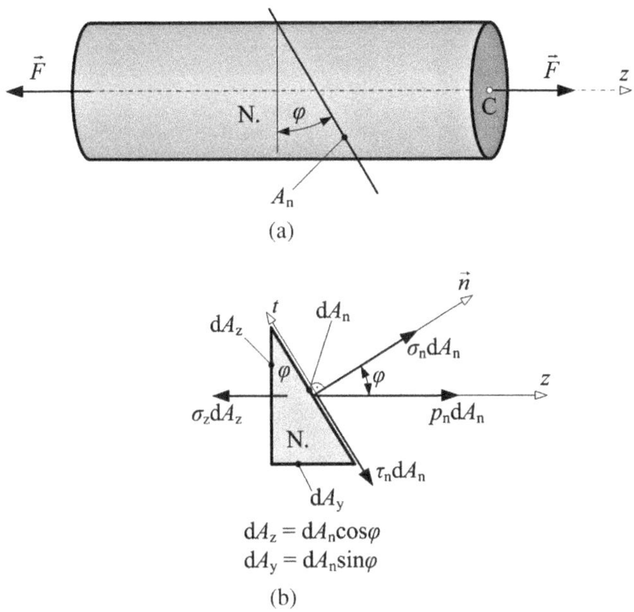

Figure 2.4 One-dimensional engineering element subjected to axial force. (a) One-dimensional element; (b) forces acting on the elementary volume presented in 2-D.

Search the extreme values of the function defined by Equation (2.11). It is found that they occur for $\varphi = 0$ and for $\varphi = \frac{\pi}{2}$ and they are of the level:

$$\begin{aligned} \varphi = 0 \quad &\rightarrow \quad \sigma_n = \sigma_z = \sigma_{\max} = \sigma_1 \\ \varphi = \frac{\pi}{2} \quad &\rightarrow \quad \sigma_n = 0 = \sigma_{\min} = \sigma_2. \end{aligned} \tag{2.12}$$

Thus, by applying the principle of determination to the extreme values of function (2.11), the angles φ_i are obtained by first derivative, and for these angles, the second derivative of Equation (2.11) gives a minimum and maximum.

On the basis of this consideration, it can be seen that uniaxial loading of an engineering element causes a uniaxial state of stress. As such, the extreme value of the normal stress is called the *principal stress*, and in this case it is σ_1. The direction of the line of action of the principal stress is called the *principal direction* and it is perpendicular to the principal plane on which the shear stress is zero. In the same manner, considering the equilibrium of all forces with respect to the t direction, we find the following:

$$\sum F_t = 0 \quad \rightarrow \quad -\tau_n \mathrm{d}A_n + \sigma_z \mathrm{d}A_z \sin\varphi = 0 \rightarrow \ \tau_n = \sigma_z \cos\varphi \sin\varphi. \tag{2.13}$$

The extreme values of shear stress are found as follows:

$$\begin{aligned} \varphi = -\frac{\pi}{4} \quad &\rightarrow \quad \tau_n = \frac{1}{2}\sigma_z = \tau_{\mathrm{I}} = \tau_{\max} \\ \varphi = \frac{\pi}{4} &\rightarrow \tau_n = -\frac{1}{2}\sigma_z = \tau_{\mathrm{II}} = \tau_{\min} \end{aligned} \tag{2.14}$$

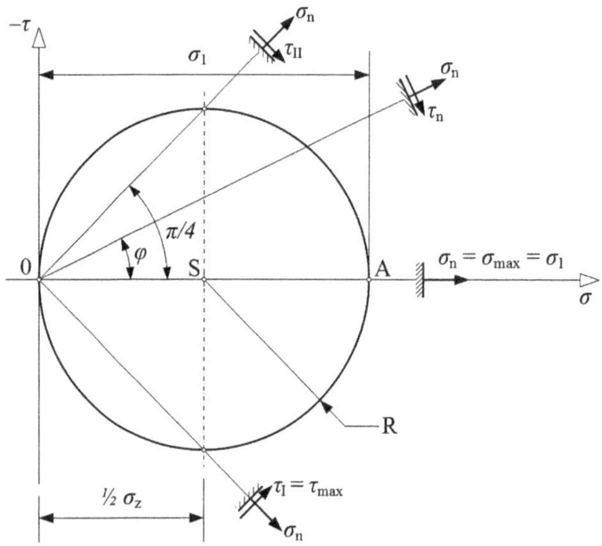

Figure 2.5 Uniaxial state of stress – Mohr's circle.

If the angle φ is eliminated from Equations (2.11) and (2.13), Mohr's circle equation follows:

$$\left(\sigma_n - \frac{1}{2}\sigma_z\right)^2 + \tau_n{}^2 = \left(\frac{1}{2}\sigma_z\right)^2. \tag{2.15}$$

Mohr's circle corresponding to a uniaxial state of stress is presented in Figure 2.5. The stress tensor is:

$$[\sigma] = [\sigma_z] = [\sigma_1] \tag{2.15a}$$

2.4.2 Two-dimensional State of Stress

Consider now a plate under plane stress conditions, i.e. on the sides of the plate, normal stresses and shear stresses act simultaneously, as shown in Figure 2.6.

Similar to the case of the uniaxial state of stress, in the vicinity of the point N from the plate is cut one elementary volume (presented in Figure 2.6(b)).

On the basis of the equilibrium equation with respect to the direction (axis) n, (Figure 2.6(b)), it follows that:

$$\sum F_n = 0 \rightarrow \sigma_n \mathrm{d}A_n - \sigma_x \mathrm{d}A_x \cos\varphi - \sigma_y \mathrm{d}A_y \sin\varphi - \tau_{xy}\mathrm{d}A_x \sin\varphi - \tau_{yx}\mathrm{d}A_y \cos\varphi = 0.$$
$$\rightarrow \sigma_n = \sigma_x \cos^2\varphi + \sigma_y \sin^2\varphi + \tau_{xy}\cos\varphi\sin\varphi + \tau_{yx}\sin\varphi\cos\varphi. \tag{2.16}$$

Applying certain mathematical transformations relating to trigonometric functions, such as:

$$\sin^2\varphi = \frac{1}{2}(1 - \cos 2\varphi);\ \cos^2\varphi = \frac{1}{2}(1 + \cos 2\varphi);\ \sin\varphi\cos\varphi = \frac{1}{2}\sin 2\varphi, \tag{2.16a}$$

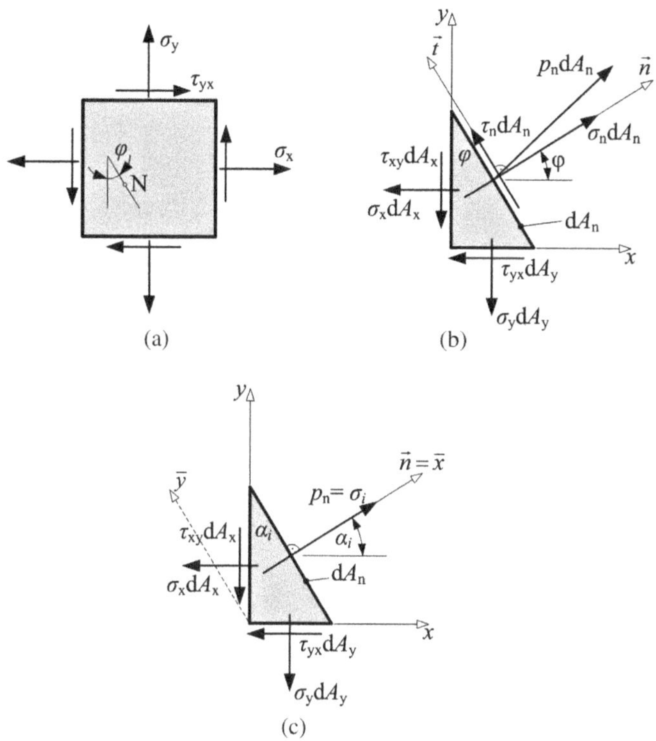

Figure 2.6 Body subjected to plane stress conditions. (a) Plate under plane stress conditions; (b) forces acting on an inclined plane (elementary volume); (c) elementary volume subjected to principal stresses.

yields:

$$\sigma_n = \frac{1}{2}(\sigma_x + \sigma_y) + \frac{1}{2}(\sigma_x + \sigma_y)\cos 2\varphi + \tau_{xy}\sin 2\varphi \tag{2.17}$$

If the equilibrium equation is considered with respect to the t axis, this yields:

$$\begin{aligned}&\sum F_t = 0 \rightarrow \tau_n dA_n + \sigma_x dA_x \sin\varphi - \sigma_y dA_y \cos\varphi - \tau_{xy} dA_x \cos\varphi + \tau_{yx} dA_y \sin\varphi = 0.\\&\rightarrow \tau_n = -(\sigma_x - \sigma_y)\sin\varphi\cos\varphi + \tau_{xy}\cos^2\varphi - \tau_{yx}\sin^2\varphi,\end{aligned} \tag{2.18}$$

and, after similar transformations, shear stress is:

$$\tau_n = -\frac{1}{2}(\sigma_x - \sigma_y)\sin 2\varphi + \tau_{xy}\cos 2\varphi. \tag{2.19}$$

Searching the extremes of the functions (2.17) and (2.19) yields the angle that defines the directions of principal stresses (principal directions) and the directions of the extreme shear stresses. The directions of the principal stresses are defined as:

$$\tan 2\alpha = \frac{2\tau_{xy}}{\sigma_x - \sigma_y}, \quad \alpha = \alpha_1, \quad \alpha_2 = \alpha_1 + \frac{\pi}{2} \tag{2.20}$$

and the *principal stresses* are:

$$\sigma_{1,2} = \frac{1}{2}\left(\sigma_x + \sigma_y\right) \pm \frac{1}{2}\sqrt{\left(\sigma_x - \sigma_y\right)^2 + 4\tau_{xy}^2}, \tag{2.21}$$

while the directions of the extreme shear stresses are defined by:

$$\tan 2\beta = -\frac{\sigma_x - \sigma_y}{2\tau_{xy}}; \ \beta_{\text{I, II}} = \alpha_1 \mp \frac{\pi}{4}, \tag{2.22}$$

and the *extreme values of the shear stresses* are:

$$\tau_{\text{I,II}} = \pm\frac{1}{2}\sqrt{\left(\sigma_x - \sigma_y\right)^2 + 4\tau_{xy}^2} = \pm\frac{1}{2}\left(\sigma_1 - \sigma_2\right). \tag{2.23}$$

Based on Equations (2.17) and (2.19), Mohr's circle is defined as:

$$\left[\sigma_n - \frac{1}{2}\left(\sigma_x + \sigma_y\right)\right]^2 + \tau_n{}^2 = \frac{1}{4}\left[\left(\sigma_x - \sigma_y\right)^2 + 4\tau_{xy}{}^2\right]. \tag{2.24}$$

All of the equations are derived respecting the constraint: $\sigma_x > \sigma_y$.

Mohr's circle for the plane stress state can be presented in the form shown in Figure 2.7.

The stress tensor is:

$$\boldsymbol{\sigma} = \sigma_{ij} = [\sigma] = \begin{bmatrix} \sigma_x & \tau_{xy} \\ \tau_{yx} & \sigma_y \end{bmatrix} = \begin{bmatrix} \sigma_1 & 0 \\ 0 & \sigma_2 \end{bmatrix}. \tag{2.24a}$$

The stress tensor in Equation (2.24a) corresponds to the biaxial (plane) stress state shown in Figure 2.6(a). The principal stresses (σ_1, σ_2) can be determined in two ways. One is mentioned above and relates to the application of the principle of extremes of the function on the normal stress given by Equation (2.17).

Another method of determining the principal stresses is as follows. In accordance with Figure 2.6(c), let the vector of total stress ($\vec{p}_n = \vec{\sigma}_i$), which consists of two components (p_{nx}, p_{ny}), lie on the normal $\vec{n} = \bar{x}$ that is defined by the angle α_i, where this angle is the angle of the principal planes (principal axes). Using the equilibrium equation for the forces on the directions x ($\Sigma F_x = 0$), and y $\left(\Sigma F_y = 0\right)$, we have:

$$\begin{aligned} p_{nx} &= \sigma_i \cos\alpha_i = \sigma_x \cos\alpha_i + \tau_{xy}\sin\alpha_i \\ p_{ny} &= \sigma_i \sin\alpha_i = \tau_{xy}\cos\alpha_i + \sigma_y \sin\alpha_i \end{aligned} \ \longrightarrow \ \begin{aligned} &(\sigma_x - \sigma_i)\cos\alpha_i + \tau_{xy}\sin\alpha_i = 0 \\ &\tau_{xy}\cos\alpha_i + \left(\sigma_y - \sigma_i\right)\sin\alpha_i = 0, \end{aligned} \tag{2.25}$$

The principal planes (and principal stresses) are obtained when the biaxial element in Figure 2.6(a) is rotated by the appropriate angle $\varphi = \alpha$, and then $i = 1 \text{ or } 2$. The same results as given by Equation (2.25) can be obtained based on Equation (2.38).

A nontrivial solution of Equations (2.25) exists if:

$$\begin{vmatrix} \sigma_x - \sigma_i & \tau_{xy} \\ \tau_{xy} & \sigma_y - \sigma_i \end{vmatrix} = 0. \tag{2.26}$$

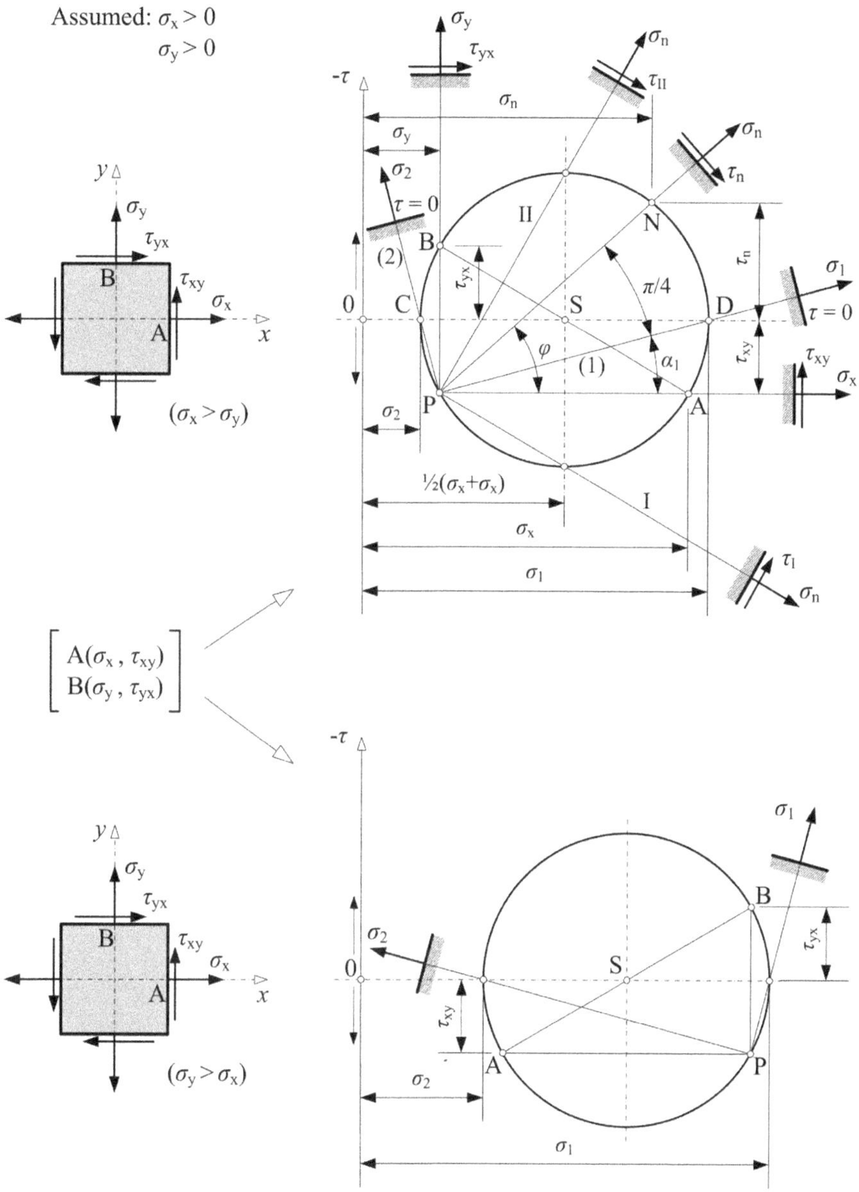

Figure 2.7 Mohr's circle for a plane stress state.

From Equation (2.26), the following quadratic equation can be obtained:

$$\sigma_i^2 - \sigma_i\left(\sigma_x + \sigma_y\right) + \left(\sigma_x\sigma_y - \tau_{xy}^2\right) = 0 \tag{2.27a}$$

Further, the solution to this equation is Equation (2.21):

$$\sigma_{1,2} = \frac{\sigma_x + \sigma_y}{2} \pm \frac{1}{2}\sqrt{\left(\sigma_x - \sigma_y\right)^2 + 4\tau_{xy}^2} \tag{a}$$

The solution for the determinant in Equation (2.26), except in the form of Equation (2.27a), may also be shown as:

$$\sigma_i^2 - I_1\sigma_i + I_2 = 0, \tag{2.27b}$$

where the two roots (σ_1, σ_2) are the two principal stresses or so-called *eigenvalues* of the stress tensor. In Equation (2.27b), I_i are *stress tensor invariants* (invariants of stress):

$$\begin{aligned} I_1 &= \sigma_x + \sigma_y = \bar{\sigma}_x + \bar{\sigma}_y = \sigma_1 + \sigma_2 \\ I_2 &= \sigma_x\sigma_y - \tau_{xy}^2 = \bar{\sigma}_x\bar{\sigma}_y - \bar{\tau}_{xy}^2 = \sigma_1\sigma_2. \end{aligned} \tag{2.28}$$

Stresses $(\bar{\sigma}_x, \bar{\sigma}_y)$ are the stresses for the rotated coordinate axes $(\bar{x}, \bar{y})$.

2.4.3 Three-dimensional State of Stress

Figure 2.8 shows an infinitesimal tetrahedron.

Figure 2.9 shows an infinitesimal parallelepiped with stresses acting on its surfaces. As stated previously, only stress components acting on the visible surfaces of the infinitesimal parallelepiped are shown.

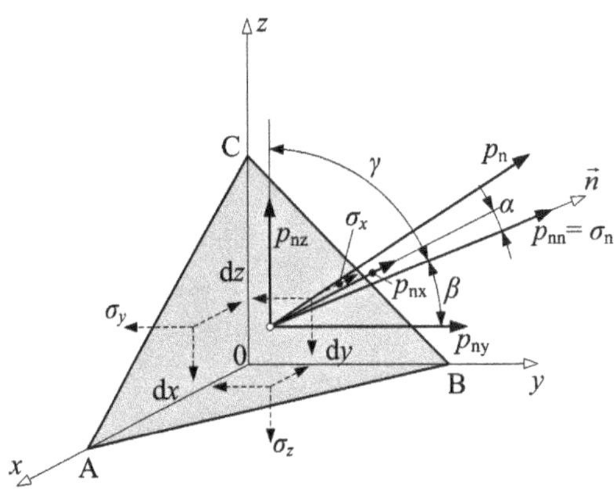

Figure 2.8 An infinitesimal tetrahedron.

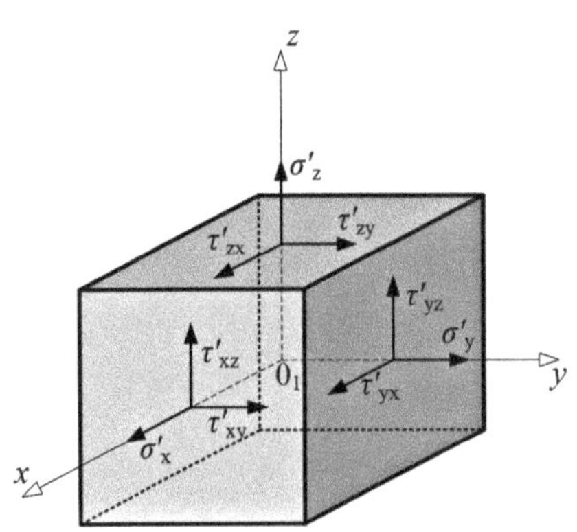

Figure 2.9 An infinitesimal parallelepiped in a three-dimensional state of stress.

Assuming that elastic deformation is under consideration, on the basis of superposition of deformation, Hooke's Law may be written in the following form:

$$\begin{aligned}
\varepsilon_x &= \varepsilon_x^x + \varepsilon_x^y + \varepsilon_x^z = \frac{1}{E}\left[\sigma_x - \nu\left(\sigma_y + \sigma_z\right)\right]; \ \gamma_{xy} = \frac{\tau_{xy}}{G} \\
\varepsilon_y &= \varepsilon_y^y + \varepsilon_y^x + \varepsilon_y^z = \frac{1}{E}\left[\sigma_y - \nu\left(\sigma_x + \sigma_z\right)\right]; \ \gamma_{yz} = \frac{\tau_{yz}}{G} \\
\varepsilon_z &= \varepsilon_z^z + \varepsilon_z^x + \varepsilon_z^y = \frac{1}{E}\left[\sigma_z - \nu\left(\sigma_x + \sigma_y\right)\right]; \ \gamma_{zx} = \frac{\tau_{zx}}{G}
\end{aligned} \tag{2.29}$$

or:

$$\left.\begin{aligned}
\sigma_x &= \frac{E}{(1+\nu)(1-2\nu)}\left[(1-\nu)\varepsilon_x + \nu(\varepsilon_x + \varepsilon_z)\right], \quad \tau_{xy} = G\gamma_{xy} \\
\sigma_y &= \frac{E}{(1+\nu)(1-2\nu)}\left[(1-\nu)\varepsilon_y + \nu(\varepsilon_x + \varepsilon_z)\right], \quad \tau_{yz} = G\gamma_{yz} \\
\sigma_z &= \frac{E}{(1+\nu)(1-2\nu)}\left[(1-\nu)\varepsilon_z + \nu(\varepsilon_x + \varepsilon_y)\right], \quad \tau_{zx} = G\gamma_{zx}
\end{aligned}\right\}, \tag{2.30}$$

where E is the modulus of elasticity and G is the *shear modulus* (see Section 6.1).

Let the parallelepiped axes x, y and z (Figure 2.9) be its principal axes. In this case, the magnitudes of the extreme values of the shear stresses are:

$$\begin{aligned}
\tau_1 &= \pm\frac{1}{2}(\sigma_2 - \sigma_3) \\
\tau_2 &= \pm\frac{1}{2}(\sigma_3 - \sigma_1). \\
\tau_3 &= \pm\frac{1}{2}(\sigma_1 - \sigma_2)
\end{aligned} \tag{2.31}$$

If it is necessary to write the maximal shearing strains in terms of the principal normal strains, i.e. $\gamma_i = f(\tau_i)$, then these values can be obtained if the values given by Equations (2.30), but written in the form of principal stresses, are inserted in the following equations:

$$\begin{aligned}
\gamma_1 &= \pm\frac{\tau_1}{G} = \pm\frac{\sigma_2 - \sigma_3}{2G} \\
\gamma_2 &= \pm\frac{\tau_2}{G} = \pm\frac{\sigma_3 - \sigma_1}{2G} \\
\gamma_3 &= \pm\frac{\tau_3}{G} = \pm\frac{\sigma_1 - \sigma_2}{2G}
\end{aligned} \tag{2.31a}$$

Finally, it follows that:

$$\gamma_1 = \varepsilon_2 - \varepsilon_3; \ \gamma_2 = \varepsilon_3 - \varepsilon_1; \ \gamma_3 = \varepsilon_1 - \varepsilon_2 \tag{2.31b}$$

The maximal shear stress is the largest value of τ_i (i = 1, 2, 3). Sometimes, instead of subscripts (1, 2, 3), subscripts (I, II, III) are used. The planes of extreme values of shear stresses are parallel with the directions of the principal stresses (σ_i), and they are diagonal planes of the parallelepiped.

2.4.3.1 Stress on an Arbitrary Plane

In engineering practice, it is of interest to design an engineering element, subjected to stresses, in an appropriate way. With this in mind, the concept of optimal design is usually used. However, independent of the concept used, one needs to know how many pieces of data are required to define the stress state at a point. From previously conducted analysis, it is clear that the stress at a point is determined by the vector of the total stress, $\vec{p}_n$. Since, through the point of the body, an infinite number of planes exist, and each of these planes has one total stress vector, this means that an infinite number of total stress vectors is required. In addition to this, a new setting can be considered. Consider an infinitesimal tetrahedron (dx, dy, dz), as in Figure 2.8, extracted from the stressed body. Its oblique plane, ABC, falls away from the origin of the coordinate system for dh. The stress vectors $(\boldsymbol{\sigma}_1, \boldsymbol{\sigma}_2, \boldsymbol{\sigma}_3)$ or $(\boldsymbol{\sigma}_x, \boldsymbol{\sigma}_y, \boldsymbol{\sigma}_z)$ acting on mutually perpendicular planes of the tetrahedron, are:

$$\begin{aligned} \boldsymbol{\sigma}_x &= \sigma_x\mathbf{i} + \tau_{xy}\mathbf{j} + \tau_{xz}\mathbf{k} \\ \boldsymbol{\sigma}_y &= \tau_{yx}\mathbf{i} + \sigma_y\mathbf{j} + \tau_{yz}\mathbf{k}, \\ \boldsymbol{\sigma}_z &= \tau_{zx}\mathbf{i} + \tau_{zy}\mathbf{j} + \sigma_z\mathbf{k} \end{aligned} \tag{2.32}$$

where **i**, **j**, **k** are unit vectors in the x, y, z directions. The oblique plane ABC of the tetrahedron is defined by unit normal vector **n** (through point O) whose cosine directions relative to x, y and z are l, m and r:

$$l = \cos\alpha, \quad m = \cos\beta, \quad r = \cos\gamma \tag{2.33}$$

$$\mathbf{n} = l\mathbf{i} + m\mathbf{j} + r\mathbf{k} = \mathbf{n_x} + \mathbf{n_y} + \mathbf{n_z} \tag{2.34}$$

The angles α, β and γ are the angles of the normal to the plane with respect to the axes of the coordinate system (x, y, z). The vector of the total stress on the oblique plane is $\mathbf{p}_n$; its components relating to the coordinate axes are $\mathbf{p}_{nx}$, $\mathbf{p}_{ny}$, $\mathbf{p}_{nz}$, while its normal and shear components (i.e., the normal and shear stresses on the oblique plane) are $\mathbf{p}_{nn}$ $(=\sigma_n)$ and $\mathbf{p}_{ns}$ $(=\tau_n)$, as shown in Figure 2.8. In equilibrium equations, body forces will be omitted. Stress vector $\mathbf{p}_n$ on an oblique plane ABC can be obtained by summation of the forces acting on the tetrahedron (equilibrium of the tetrahedron), taking into consideration the ratios of the areas of three perpendicular planes of the tetrahedron to the area of its oblique plane, i.e., l, m and r. Equilibrium equations of the forces in the directions of the coordinate axes are (shown only for the x direction):

$$\begin{gathered} \sum F_{ix} = 0 \rightarrow p_x \mathrm{d}A - \sigma_x \mathrm{d}A_x - \tau_{yx}\mathrm{d}A_y - \tau_{zx}\mathrm{d}A_z = 0 \\ \Sigma F_{iy} = 0; \ \Sigma F_{iz} = 0. \end{gathered} \tag{2.35}$$

Since:

$$\mathrm{d}A_x = \mathrm{d}A\cdot l; \ \ \mathrm{d}A_y = \mathrm{d}A\cdot m; \ \ \mathrm{d}A_z = \mathrm{d}A\cdot r \tag{2.35a}$$

on the basis of Equations (2.35), finally, the following equations can be established:

$$\begin{aligned} p_{nx} &= \sigma_x l + \tau_{yx} m + \tau_{zx} r \\ p_{ny} &= \tau_{xy} l + \sigma_y m + \tau_{zy} r. \\ p_{nz} &= \tau_{xz} l + \tau_{yz} m + \sigma_z r \end{aligned} \tag{2.36}$$

Equations (2.36) are known as the *Cauchy equations of stresses.* In index notation, Equation (2.36) becomes:

$$p_{ni} = \sigma_{ji}\, n_j \tag{2.37}$$

The normal and shear stress components on an oblique plane, considering Equations (2.32, 2.35 and 2.36), are:

$$\begin{aligned} p_{nn} = \sigma_n = \mathbf{p}_n\,\mathbf{n} = \sigma_x l^2 + \sigma_y m^2 + \sigma_z r^2 + 2\tau_{yz} mr + 2\tau_{xz} lr + 2\tau_{xy} lm \\ p_{ns} = \tau_n = \sqrt{p_n^2 - p_{nn}^2} \end{aligned} \tag{2.38}$$

Finally, the state of stress at a point is defined by the stress tensor, since it is possible to determine the stress at any oblique plane through the considered point if stresses on three mutually perpendicular planes that intersect at that point are known.

2.4.3.2 Stress on an Octahedral Plane

For simplicity, imagine that the coordinate axes x, y, z coincide with the principal axes (1, 2, 3), i.e. with the principal stresses ($\sigma_1, \sigma_2, \sigma_3$). The family of planes whose unit vectors $\mathbf{n}$ ($\mathbf{n} = l\mathbf{i} + m\mathbf{j} + r\mathbf{k}$) satisfy the relation $l^2 + m^2 + n^2 = 1$, i.e. $l = m = n = 1/\sqrt{3}$, are known as *octahedral planes* (eight planes). Here, l, m and n are also the direction cosines of unit vector $\mathbf{n}$, as given in Equation (2.33). The normal and shear stresses associated with these planes are called the *octahedral stresses.* The *octahedral normal stress* can be obtained in accordance with Equation (2.38), i.e. $p_{nn} = \sigma^{\text{oct}}$, taking into consideration that in the case of principal axes $\sigma_x = \sigma_1$, $\sigma_y = \sigma_2$, $\sigma_z = \sigma_3$, and shear stresses $\tau_{ij} = 0$. In this respect:

$$\sigma^{\text{oct}} = \frac{1}{3} I_1 = \frac{\sigma_1 + \sigma_2 + \sigma_3}{3} = \frac{\sigma_x + \sigma_y + \sigma_z}{3} = \frac{1}{3}\sigma_{kk}. \tag{2.39}$$

In accordance with Equation (2.10) and Equation (2.43) below,

$$\sigma_{kk} \equiv I_1 = \sigma_x + \sigma_y + \sigma_z = \sigma_1 + \sigma_2 + \sigma_3 \tag{2.39a}$$

Also, since I_i are invariants (as explained later), the quantities that relate to them can be taken as those given by Equations (2.43)–(2.45) below to arbitrary axes.

The *octahedral shear stress,* based on Equations (2.38), (2.35) and (2.32), is:

$$\tau^{\text{oct}} = \sqrt{\frac{2}{9} I_1^2 - \frac{2}{3} I_2} = \frac{1}{3}\sqrt{(\sigma_1 - \sigma_2)^2 + (\sigma_2 - \sigma_3)^2 + (\sigma_3 - \sigma_1)^2} \tag{2.40a}$$

This equation may also be written in the form:

$$\tau^{\text{oct}} = \frac{1}{3}\sqrt{(\sigma_x - \sigma_y)^2 + (\sigma_y - \sigma_z)^2 + (\sigma_z - \sigma_x)^2 + 6\left(\tau_{xy}^2 + \tau_{yz}^2 + \tau_{zx}^2\right)}. \tag{2.40b}$$

In the above equations, I_i are stress invariants.

2.4.3.3 Principal Stresses and Stress Invariants

Stress Tensor

When a three-dimensional state of stress is considered, then, analagous to Equation (2.26), it follows that [10, 17–18]:

$$\begin{vmatrix} \sigma_x - \sigma_i & \tau_{xy} & \tau_{xz} \\ \tau_{xy} & \sigma_y - \sigma_i & \tau_{yz} \\ \tau_{xz} & \tau_{yz} & \sigma_z - \sigma_i \end{vmatrix} = 0, \tag{2.41}$$

and this now yields a cubic equation – the so-called *characteristic equation*:

$$\sigma_i^3 - I_1\sigma_i^2 + I_2\sigma_i - I_3 = 0, \tag{2.42}$$

Where the *stress invariants* (stress tensor invariants) are [7, 12, 16]:

$$I_1 = \sigma_x + \sigma_y + \sigma_z = \sigma_1 + \sigma_2 + \sigma_3 \tag{2.43}$$

$$\begin{aligned} I_2 &= \begin{vmatrix} \sigma_x & \tau_{xy} \\ \tau_{yx} & \sigma_y \end{vmatrix} + \begin{vmatrix} \sigma_x & \tau_{xz} \\ \tau_{zx} & \sigma_z \end{vmatrix} + \begin{vmatrix} \sigma_y & \tau_{yz} \\ \tau_{zy} & \sigma_z \end{vmatrix} \\ &= \sigma_x\sigma_y + \sigma_y\sigma_z + \sigma_z\sigma_x - \tau_{yx}^2 - \tau_{zy}^2 - \tau_{xz}^2 = \sigma_1\sigma_2 + \sigma_2\sigma_3 + \sigma_3\sigma_1 \end{aligned} \tag{2.44}$$

$$I_3 = \begin{vmatrix} \sigma_x & \tau_{xy} & \tau_{xz} \\ \tau_{yx} & \sigma_y & \tau_{yz} \\ \tau_{zx} & \tau_{zy} & \sigma_z \end{vmatrix} = \sigma_x\sigma_y\sigma_z - \sigma_x\tau_{yz}^2 - \sigma_y\tau_{xz}^2 - \sigma_z\tau_{xy}^2 + 2\tau_{xy}\tau_{yz}\tau_{xz} = \sigma_1\sigma_2\sigma_3 \tag{2.45}$$

Sometimes the notation used for stress invariants may take the following form:

$$I_i \rightarrow I_{i\sigma};\ (i = 1,2,3).$$

If Equations (2.7), (2.8), (2.10), (2.39a) and (2.43)–(2.45) are considered, then the stress invariants for the mean stress tensor σ_{ij}^{m} in Equation (2.8) can be defined as:

$$\begin{aligned} I_1^{m} &= I_1 = 3\sigma_m \\ I_2^{m} &= \frac{1}{3}I_1^2 = 3\sigma_m^2 \\ I_3^{m} &= \frac{1}{27}I_1^3 = \sigma_m^3 \end{aligned} \tag{2.45a}$$

The three *principal stresses* (*eigenvalues, extreme values, characteristic values*) $\sigma_1, \sigma_2, \sigma_3$, where ($\sigma_1 > \sigma_2 > \sigma_3$) as the three roots of Equation (2.42) can be obtained. These three principal stresses, which are normal stresses whose directions (principal axes of stress) intersect at one point, form three principal planes of stress through the point which are mutually perpendicular and on which the shearing stresses are equal to zero (shearing stresses vanish). However, the principal stresses and their directions depend only on the applied load and do not depend on the choice of coordinate system (x, y, z). In this sense, I_i, given by Equations (2.43)–(2.45), are stress invariants and they are independent of the choice of coordinate system, i.e. their magnitudes remain constant regardless of the choice of coordinate system.

Deviatoric Stress Tensor

The deviatoric stress tensor is also a symmetrical tensor.

In accordance with Equations (2.7)–(2.10), the deviatoric stress tensor is:

$$S_{ij} = \sigma_{ij} - \sigma_{ij}^{0} \equiv \sigma_{ij} - \sigma_{ij}^{\mathrm{m}} = \sigma_{ij} - \frac{1}{3}\sigma_{kk}\delta_{ij}, \tag{2.46}$$

and the principal values (S_1, S_2, S_3) of the deviatoric tensor (deviator tensor) can be obtained:

$$\begin{aligned} S_1 &= \sigma_1 - \sigma_{\mathrm{m}} = \frac{2\sigma_1 - \sigma_2 - \sigma_3}{3} \\ S_2 &= \sigma_2 - \sigma_{\mathrm{m}} = \frac{2\sigma_2 - \sigma_1 - \sigma_3}{3} \\ S_3 &= \sigma_3 - \sigma_{\mathrm{m}} = \frac{2\sigma_3 - \sigma_2 - \sigma_1}{3} \end{aligned} \tag{2.47}$$

Invariants J_i(or I_{id} or I_{iS}) of the deviatoric stress tensor, S_{ij}, where S_i are the principal values of the deviatoric stress tensor, in accordance with Equations (2.9) and (2.43)–(2.45), may be written as follows.

Since, on the basis of Equations (2.7), (2.10) and (2.39a), we can write:

$$\sigma_{ij} = S_{ij} + \frac{1}{3}\sigma_{\mathrm{m}}\delta_{ij} = S_{ij} + \frac{1}{3}I_1\delta_{ij}, \tag{2.48}$$

it follows that any diagonal member may be written as:

$$\sigma_i = S_i + \frac{1}{3}I_1 \tag{2.48a}$$

If this quantity is inserted into Equation (2.42), it follows that:

$$\left(S_i + \frac{1}{3}I_1\right)^3 - I_1\left(S_i + \frac{1}{3}I_1\right)^2 + I_2\left(S_i + \frac{1}{3}I_1\right) - I_3 = 0$$

After mathematical calculation, it follows that:

$$S_i^3 + \frac{1}{27}I_1^3 + S_i^2 I_1 + \frac{1}{3}S_i I_1^2 - S_i^2 I_1 - \frac{2}{3}S_i I_1^2 - \frac{1}{9}I_1^3 + S_i I_2 + \frac{1}{3}I_1 I_2 - I_3 = 0,$$

and finally, the following equation can be written:

$$S_i^3 + J_2 S - J_3 = 0 \tag{2.49}$$

On the basis of Equation (2.49) and equation above it, the following result can be obtained.

The first invariant is:

$$J_1(\text{or}\, I_{1d}\, \text{or}\, I_{1S}) = S_1 + S_2 + S_3 = 0, \tag{2.50a}$$

as is evident from Equation (2.9), i.e. $S_{kk} = 0$.

The second invariant is (from Equation (2.49) and the equation above it):

$$J_2 = I_2 - \frac{1}{3}I_1^2 = (S_1S_2 + S_2S_3 + S_3S_1)$$
$$= \frac{2\sigma_1-\sigma_2-\sigma_3}{3}\cdot\frac{2\sigma_2-\sigma_1-\sigma_3}{3} + \frac{2\sigma_2-\sigma_1-\sigma_3}{3}\cdot\frac{2\sigma_3-\sigma_2-\sigma_1}{3}$$
$$+ \frac{2\sigma_3-\sigma_2-\sigma_1}{3}\cdot\frac{2\sigma_1-\sigma_2-\sigma_3}{3} = \frac{-3\,\sigma_1^2 - 3\,\sigma_2^2 - 3\,\sigma_3^2 + 3\sigma_1\sigma_2 + 3\sigma_2\sigma_3 + 3\sigma_3\sigma_1}{9}$$
$$= -\frac{1}{6}\left[(\sigma_1-\sigma_2)^2 + (\sigma_2-\sigma_3)^2 + (\sigma_3-\sigma_1)^2\right]. \tag{2.50b}$$

Finally, the third invariant is (from Equation (2.49) and the equation above it):

$$J_3 = I_3 - \frac{1}{3}I_1I_2 + \frac{2}{27}I_1^3 = S_1S_2S_3$$
$$= \frac{1}{27}(2\sigma_1-\sigma_2-\sigma_3)(2\sigma_2-\sigma_3-\sigma_1)(2\sigma_3-\sigma_1-\sigma_2) \tag{2.50c}$$

Mohr's circle can also be shown for a three-dimensional state of stress, and its construction can be found in the literature, for example, see [3, 5, 12].

2.5 Transformation of Stress Components

An example of stress component transformations from one Cartesian system to another is presented in Figure 2.10 [3].

Stress components belonging to an (x, y) system are:

$$\sigma_{ij} = [\sigma] = \begin{bmatrix} \sigma_x & \tau_{xy} \\ \tau_{yx} & \sigma_y \end{bmatrix} \tag{2.51}$$

Stress components in a plane coordinate system $(\bar{x},\bar{y})$ are:

$$\bar{\sigma}_{ij} = [\bar{\sigma}] = \begin{bmatrix} \bar{\sigma}_x & \bar{\tau}_{xy} \\ \bar{\tau}_{yx} & \bar{\sigma}_y \end{bmatrix}. \tag{2.52}$$

The following relationships may be established (see also Equations (2.16)–(2.19), [1–3, 13, 20–22]:

$$\begin{aligned} \bar{\sigma}_x &= \sigma_x\cos^2\varphi + \sigma_y\sin^2\varphi + 2\tau_{xy}\sin\varphi\cos\varphi \\ \bar{\sigma}_y &= \sigma_x\sin^2\varphi + \sigma_y\cos^2\varphi + 2\tau_{xy}\sin\varphi\cos\varphi \\ \bar{\tau}_{xy} &= -\left(\sigma_x-\sigma_y\right)\sin\varphi\cos\varphi + \tau_{xy}\left(\cos^2\varphi - \sin^2\varphi\right) \end{aligned} \tag{2.53a}$$

If the following equalities are used:

$$2\sin\varphi\cos\varphi = \sin 2\varphi;\ \cos^2\varphi - \sin^2\varphi = \cos 2\varphi;\ \sin^2\varphi = (1-\cos 2\varphi)/2;$$
$$\cos^2\varphi = (1+\cos 2\varphi)/2, \tag{a}$$

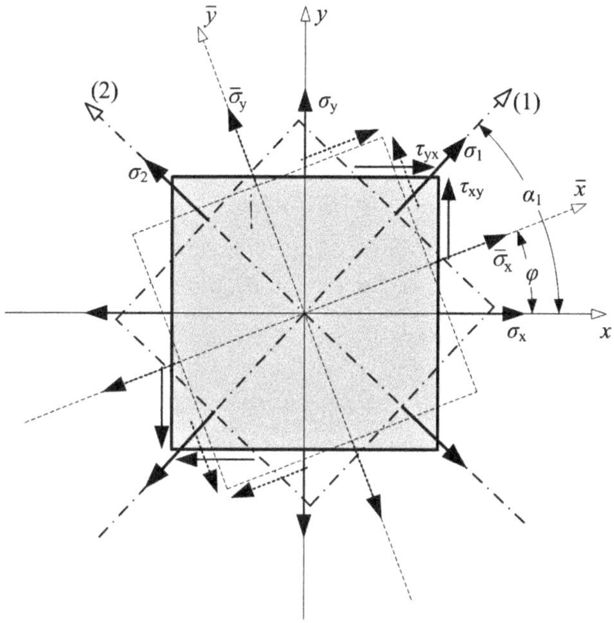

Figure 2.10 Plane stress component transformation.

then Equations (2.53a) can be written in the form:

$$\begin{aligned}\bar{\sigma}_x &= \frac{\sigma_x+\sigma_y}{2}+\frac{\sigma_x-\sigma_y}{2}\cos 2\varphi+\tau_{xy}\sin 2\varphi\\ \bar{\sigma}_y &= \frac{\sigma_x+\sigma_y}{2}-\frac{\sigma_x-\sigma_y}{2}\cos 2\varphi-\tau_{xy}\sin 2\varphi\\ \bar{\tau}_{xy} = \left(\bar{\tau}_{yx}\right) &= -\frac{\sigma_x-\sigma_y}{2}\sin 2\varphi+\tau_{xy}\cos 2\varphi\end{aligned} \tag{2.53b}$$

In matrix form, Equations (2.53) can be written as follows:

$$\begin{bmatrix}\bar{\sigma}_x & \bar{\tau}_{xy}\\ \bar{\tau}_{yx} & \bar{\sigma}_y\end{bmatrix}=\begin{bmatrix}\cos\varphi & \sin\varphi\\ -\sin\varphi & \cos\varphi\end{bmatrix}\begin{bmatrix}\sigma_x & \tau_{xy}\\ \tau_{yx} & \sigma_y\end{bmatrix}\begin{bmatrix}\cos\varphi & -\sin\varphi\\ \sin\varphi & \cos\varphi\end{bmatrix} \tag{2.54}$$

or:

$$[\bar{\sigma}]=[a][\sigma][a]^{\mathrm{T}}, \tag{2.55}$$

where $[a]$ is a transformation matrix. When rotated axes reach a certain angle ($\varphi=\alpha$), normal stresses reach their extreme values (principal stresses) σ_1 and σ_2 (Figure 2.10), and the stress tensor now has the following form:

$$\bar{\sigma}_{ij}=[\bar{\sigma}]=\begin{bmatrix}\sigma_1 & 0\\ 0 & \sigma_2\end{bmatrix}. \tag{2.56}$$

Example 2.1.
Consider a structural member that is subjected to plane stress, as shown in Figure 2.11(a). Determine the principal stresses.

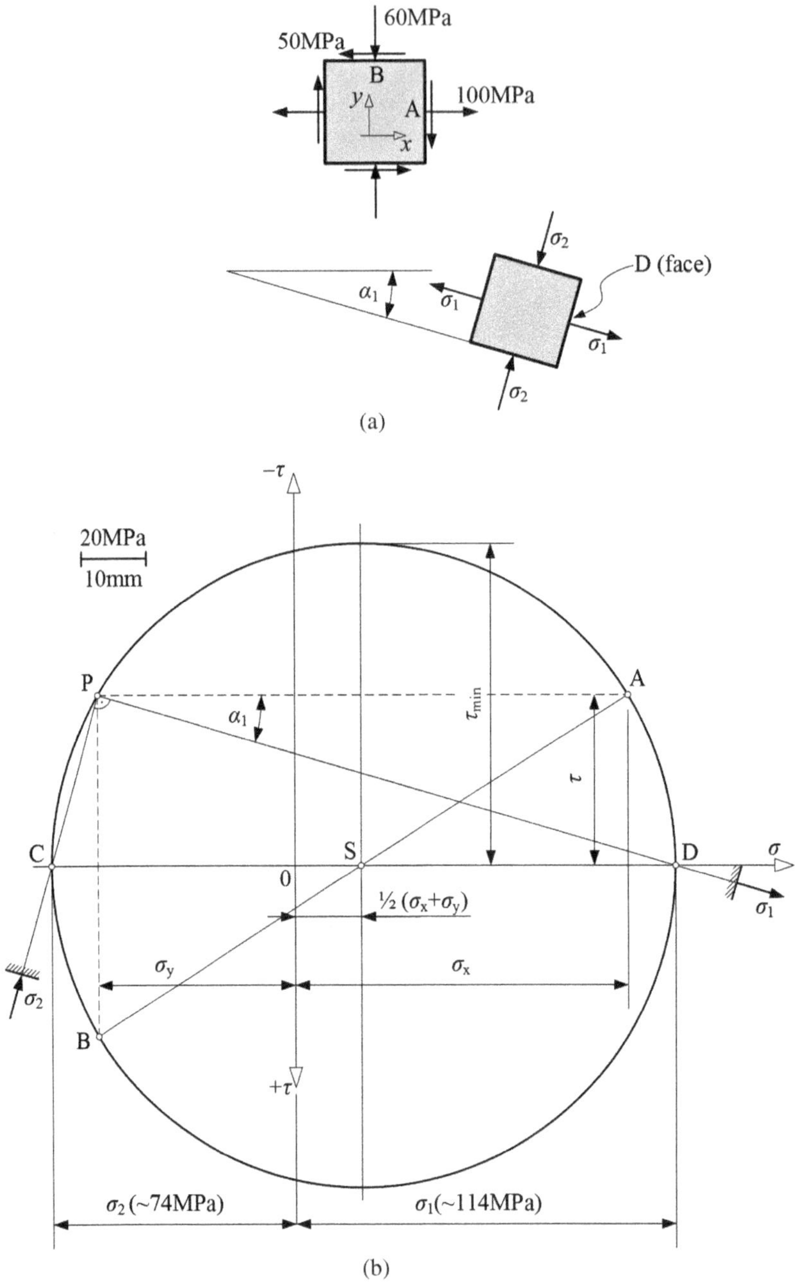

Figure 2.11 Mohr's circle (Example 2.1). (a) Given data; (b) solution to the problem.

Data and Solution:
The data and the solution can be seen in Figure 2.11.

Example 2.2.
Consider a point in a structural member that is subjected to plane stress, as shown in Figure 2.12. Determine the stresses at a point if the angle of the plane through the point is 60° with respect to the y axis, as shown.

Data:
The data are given in Figure 2.12.
Solution:
In accordance with Equation (2.17):

$$\sigma_n = \frac{1}{2}(\sigma_x + \sigma_y) + \frac{1}{2}(\sigma_x - \sigma_y)\cos 2\varphi + \tau_{xy}\sin 2\varphi$$
$$\sigma_n = \frac{1}{2}(30+70) + \frac{1}{2}(30-70)\cos(2\cdot 60°) + (-40)\sin(2\cdot 60°)$$
$$= 25.36\,\text{MPa}$$

And in accordance with Equation (2.18):

$$\tau_n = -(\sigma_x - \sigma_y)\sin\varphi\cos\varphi + \tau_{xy}\cos^2\varphi - \tau_{yx}\sin^2\varphi$$
$$\tau_n = -(30-70)\cos 60^{\circ}\sin 60^{\circ} + (-40)(\cos^2\varphi - \sin^2\varphi)$$
$$= 37.32\,\text{MPa}$$

In addition, if Equation (2.21) is applied, the principal stresses are:

$$\sigma_{1,2} = \frac{1}{2}(\sigma_x + \sigma_y) \pm \frac{1}{2}\sqrt{(\sigma_x - \sigma_y)^2 + 4\tau_{xy}^2}$$
$$\sigma_1 = \frac{1}{2}(30+70) + \frac{1}{2}\sqrt{(30+70)^2 + 4(-40)^2} = 114.03\,\text{MPa}$$
$$\sigma_2 = -14.03\,\text{MPa},$$

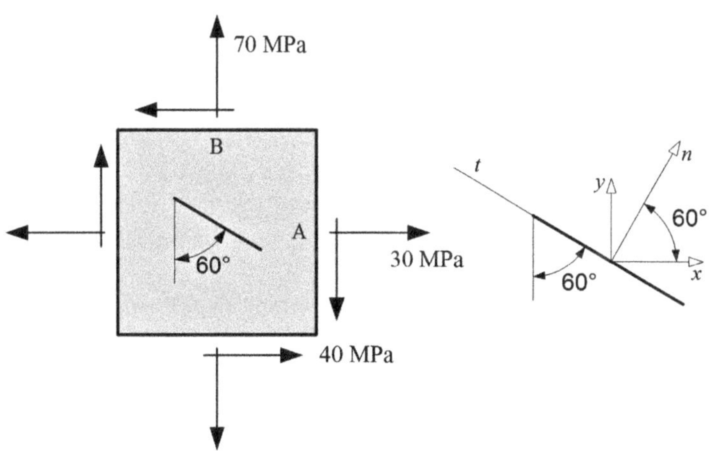

Figure 2.12 An engineering component in a plane stress state – the stresses at a point.

and, according to Equation (2.23), the extreme values of the shear stresses are:

$$\tau_{\mathrm{I,II}} = \pm\frac{1}{2}\sqrt{\left(\sigma_x-\sigma_y\right)^2+4\tau_{xy}^2} = \pm\frac{1}{2}(\sigma_1-\sigma_2)$$

$$\tau_{\mathrm{I}} = \frac{1}{2}(114.03-(-14.03)) = 64.03\,\mathrm{MPa}$$

$$\tau_{\mathrm{II}} = -64.02\,\mathrm{MPa}.$$

References

1 Alfirević, I. (1995) *Strength of Materials I* (in Croatian), ITP "Tehnička knjiga" D.D., Zagreb.
2 Alfirević, I. (2006) *Linear Structural Analysis*, Pretei d.o.o., Zagreb.
3 Brnic, J. and Turkalj, G. (2004) *Strength of Materials I* (in Croatian), Faculty of Engineering, University of Rijeka, Rijeka.
4 Seed, G. M. (2000) *Strength of Materials*, Saxe-Coburg Publications, Edinburgh.
5 Boresi, A. P., Schmidt, R. J. and Sidebottom, O. M. (1993) *Advanced Mechanics of Materials*, John Wiley & Sons, New York.
6 Beer, F. P., Johnston, E. R. Jr, DeWolf, J. T. and Mazurek, D. F. (2012) *Mechanics of Materials*, McGraw-Hill, USA.
7 Solecki, R. and Conant, R. J. (2003) *Advanced Mechanics of Materials*, Oxford University Press, New York.
8 Kačanov, L. M. (1969) *Basic Theory of Plasticity* (in Russian), Nauka, Moscow.
9 Sun, C. T. (2006) *Mechanics of Aircraft Structures*, John Wiley & Sons, New York.
10 Boresi, A. and Schmidt, R. J. (2003) *Advanced Mechanics of Materials*, John Wiley & Sons, New York.
11 Khan, A. S. and Huang, S. (1995) *Continuum Theory of Plasticity*, John Wiley & Sons, New York.
12 Boresi, A., Chong, J. D. and Lee, J. D. (2011) *Elasticity in Engineering Mechanics*, John Wiley & Sons, New Jersey.
13 Sun, C. T. (2006) *Mechanics of Aircraft Structures*, John Wiley & Sons, New York.
14 Liu, A. F. (2005) *Mechanics and Mechanisms of Fracture*, ASM International, USA.
15 Megson, T. H. G. (2005). *Structural and Stress Analysis*, 3rd edition, Butterworth-Heinemann, New York.
16 Bowman, K. (2004) *Mechanical Behavior of Materials*, John Wiley & Sons, New York.
17 Alfirević, I. (2003) Introduction to Tensors and Continuum Mechanics *(in Croatian)*, Golden Marketing, Zagreb.
18 Prelog, E. (1978) *Elasto and Plastomechanics* (in Slovenian), Faculty of Mechanical Engineering, University of Ljubljana, Ljubljana.
19 Štok, B. (1988) *Mechanics of Deformable Bodies* (in Slovenian), *part I*, Faculty of Mechanical Engineering, University of Ljubljana, Ljubljana, Slovenia.
20 Alfirević, I. (1975) *Advanced Strength of Materials* (in Croatian), *Faculty of Mechanical Engineering and Naval* Architecture, Zagreb.
21 Craig, R. R. (2011) *Mechanics of Materials*, 3rd edition, John Wiley & Sons, New Jersey.
22 Riley, W. F., Sturges, L. D. and Morris, D. H. (2007) *Mechanics of Materials*, John Wiley & Sons, USA.

3

Strain

3.1 Definition of Strain

As stated previously, after a body has been subjected to a load, because it is deformable, its volume and/or shape will be changed. It is said that the body *deforms*. Particles of material at certain points on the body change their positions relative to each other.

A body can undergo two types of strain: normal (engineering) strain, ε, and shear (tangential) strain, γ.

Provided that the elongation (ΔL) of the line element between two considered particles at two points remains small and that the body remains elastic throughout deformation, then the *engineering strain* (*normal strain*) of the line element (Figure 3.1(a)) is defined as the ratio of the change in length to the original length of a considered line element [1, 2]:

$$\varepsilon = \lim_{L\to 0}\frac{\Delta L}{L};\ \Delta L = L_1 - L, (\text{designation}\, L\, \text{or}\, l) \tag{3.1a}$$

where L is the initial length and L_1 is the length after elongation of the considered line element.

Strain is a dimensionless quantity. A strain at a fracture is considered a measure of ductility.

When considering two originally perpendicular lines in the vicinity of the point O, the *shear strain* is the change in angle between two considered, mutually originally perpendicular lines (Figure 3.1(b)) [1, 2, 3]:

$$\begin{aligned} &\gamma = \lim(\measuredangle 0AB - \measuredangle 0A'B'),\\ &A \to 0\\ &B \to 0 \end{aligned} \tag{3.1b}$$

where "$\measuredangle$" designates the angle.

However, for an isotropic material that obeys Hooke's Law, normal strain is caused by normal stress and shear strain is caused by shear stress. Normal strain is related to the extension (dilatation) of the element. Also, it can be said that normal strain is perpendicular to the face of the considered element while shear strain is parallel to it.

Analysis of Engineering Structures and Material Behavior, First Edition. Josip Brnić.

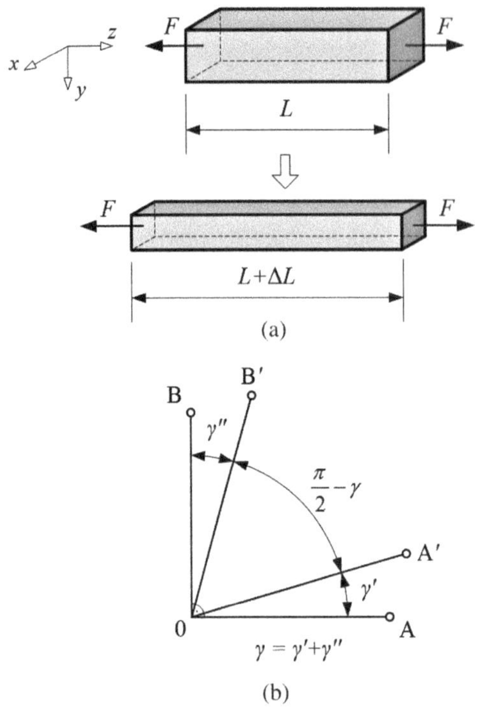

Figure 3.1 Engineering and shear strains. (a) Elongation (ΔL) and normal strain (ε); (b) shear strain (γ).

3.1.1 Some Properties of Materials Associated with Strain

3.1.1.1 Poisson's Ratio

When the elongation of a bar is considered (Figure 3.1(a)), then the axial extension induces a lateral contraction. The ratio of lateral strain (x direction or y direction) to longitudinal strain (z direction) is referred to as *Poisson's ratio*, ν [4, 5]:

$$\nu = -\frac{\text{lateral strain}}{\text{longitudinal strain}} = -\frac{\varepsilon_x}{\varepsilon_z} = -\frac{\varepsilon_y}{\varepsilon_z} \tag{3.2}$$

Properties such as *volumetric strain*, ε_V, and *bulk modulus*, *K*, may also be mentioned.

3.1.1.2 Volumetric Strain

Volume change is defined by volumetric strain or cubic dilatation (see Figure 3.2):

$$\varepsilon_V = \frac{\Delta V}{V} \tag{3.3}$$

In accordance with Figure 3.2, if an infinitesimal volume is subjected to normal strains, then the change in volume can be written as:

$$\Delta V = V_1 - V \tag{3.4a}$$

The volume of an undeformed (unstrained) solid element is:

$$\mathrm{d}V = \mathrm{d}x \cdot \mathrm{d}y \cdot \mathrm{d}z, \tag{3.4b}$$

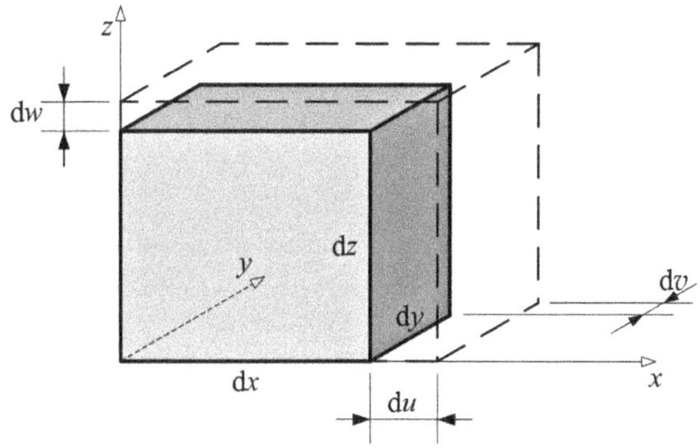

Figure 3.2 Infinitesimal volume.

while the volume of a deformed element is:

$$dV_1 = (dx + du)\cdot(dy + dv)\cdot(dz + dw). \tag{3.4c}$$

Volumetric strain then is:

$$\varepsilon_V = \frac{dV_1 - dV}{dV} = \frac{\Delta dV}{dV} = \frac{(dx + du)\cdot(dy + dv)\cdot(dz + dw) - dx\cdot dy\cdot dz}{dx\cdot dy\cdot dz},$$

or:

$$\varepsilon_V = (1 + \varepsilon_x)\cdot(1 + \varepsilon_y)\cdot(1 + \varepsilon_z) - 1. \tag{3.5a}$$

Equation (3.5a) also contains the terms representing the products of the infinitesimal values. In accordance with small-displacement theory, they are discarded. Equations (3.12)–(3.17) below also belong to small-deformation theory. In a general form, deformation theory developed as purely geometrical and terms of higher order arise within it. So, applying small-deformation theory, Equation (3.5a) for volumetric strain takes the form:

$$\varepsilon_V = \varepsilon_x + \varepsilon_y + \varepsilon_z = \varepsilon_1 + \varepsilon_2 + \varepsilon_3, \tag{3.5b}$$

where $\varepsilon_1, \varepsilon_2, \varepsilon_3$ are principal strains (principal dilatations).

3.1.1.3 Bulk Modulus

The *bulk modulus* (volume/volumetric modulus, modulus of compression) of a material refers to the ratio of volumetric stress and volumetric strain [4, 5]. It is a property of isotropic materials and it is a measure of the change in volume that a material element undergoes when exposed to hydrostatic stress (that is, it characterizes the change in volume of a material under hydrostatic pressure). If the material element under consideration is exposed to hydrostatic stress, then we have:

$$\varepsilon_x = \varepsilon_y = \varepsilon_z = \varepsilon, \text{ and } \sigma_x = \sigma_y = \sigma_z = \sigma$$

In accordance with the generalized Hooke's Law (Equation (7.29a)), the strain component in the x direction is:

$$\varepsilon_x = \frac{\sigma_x}{E} - \frac{\nu\sigma_y}{E} - \frac{\nu\sigma_z}{E} = \frac{1}{E}\left[\sigma_x - \nu\left(\sigma_y + \sigma_z\right)\right] = \frac{\sigma}{E}(1-2\nu) \tag{3.6}$$

For other directions, strains can be found in a similar way. Volumetric strain is:

$$\varepsilon_V = \varepsilon_x + \varepsilon_y + \varepsilon_z = 3\varepsilon = 3\frac{\sigma}{E}(1-2\nu). \tag{3.7}$$

At hydrostatic pressure it is: $p = -\sigma$, and then:

$$\varepsilon_V = -\frac{3}{E}(1-2\nu)p. \tag{3.8}$$

Finally, the bulk modulus is:

$$K = \frac{\text{volumetric stress}}{\text{volumetric strain}} = \frac{-p}{-\frac{3}{E}(1-2\nu)p}$$

$$K = \frac{E}{3(1-2\nu)}. \tag{3.9}$$

Observations indicate that material subjected to hydrostatic pressure decreases in volume, that is, dilatation is negative (Equation (3.8)), and from that it follows that the modulus K is a positive quantity. From Equation (3.9) it follows that $(1 - 2\nu) > 0$ or $\nu < ½$. Because the Poisson ratio is positive for engineering materials, it follows that $0 < \nu < ½$. Also, for $\nu = ½$, $K = \infty$ and the material will then be perfectly incompressible.

3.1.1.4 Modulus of Elasticity

The modulus of elasticity, E, is a property of materials that is mentioned in several chapters of this book – for example, in Chapters 6 and 7. Consider, for example, Figure 6.1(a). Let the extension of a rod of initial length $L_0 = L$ and initial cross-sectional area $A_0 = A$, and let these extensions be caused by a small force F such that deformation belongs to the elastic domain. The extension is $\Delta L = L_1 - L$. Since a small deformation is being considered, the change in the cross-sectional area $A_0 = A$ can be neglected. The modulus of elasticity is defined as:

$$E = \frac{\text{stress}(\sigma)}{\text{strain}(\varepsilon)} = \frac{F/A}{\Delta L/L} \tag{3.10}$$

3.1.1.5 Shear Modulus (Modulus of Rigidity)

Consider, for example, an element with a small volume ($dx \cdot dy \cdot dz$) which is subjected to the force $F_x = F$ acting in the x direction iņ its upper surface ($dA = dx \cdot dy$) and let this force cause the edge dx to deform (move) in the x direction by Δdx with respect to the base edge (dx). The base plane of this element is xy.

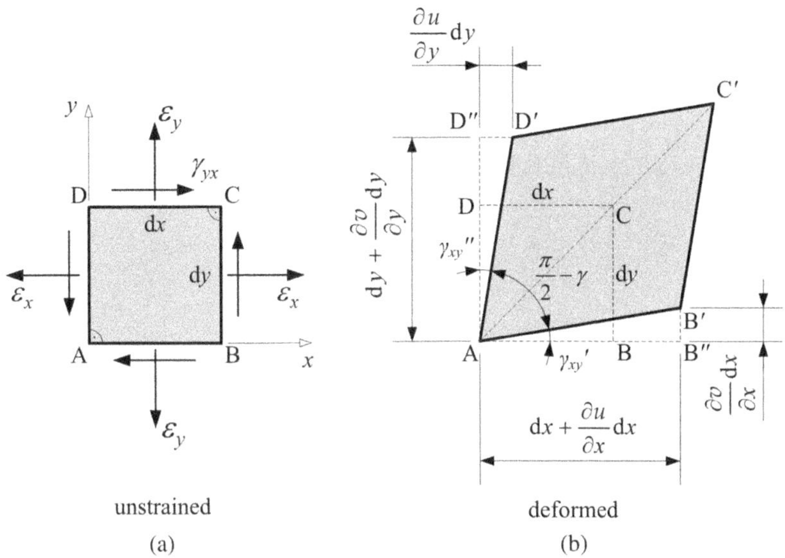

Figure 3.3 An infinitesimal element in its undeformed (unstrained) and deformed states. (a) Unstrained state; (b) deformed state.

In accordance with the above explanation, the shear modulus, G, is defined as:

$$G = \frac{\text{stress}(\tau)}{\text{strain}(\gamma)} = \frac{\tau_{zx}}{\gamma_{zx}} = \frac{F/A}{\Delta\mathrm{d}x/\mathrm{d}z} \tag{3.11}$$

3.2 Strain–Displacement Equations

The intention here is to develop expressions relating to the relationship between strain and displacement. For simplicity, consider a biaxial strain state problem (plane strain problem), as shown in Figure 3.3. Imagine that this infinitesimal element (Figure 3.3 (a)) has been cut out from an infinitesimal parallelepiped and that the x–y plane will be used as the reference plane; in other words, the length dz of the mentioned parallelepiped does not change, and strains associated with axis z are equal to zero ($\varepsilon_z = \gamma_{xz} = \gamma_{yz} = 0$). A considered element is subjected to a positive strain ε_x, ε_y, γ_{xy}.

Consider what happens when an infinitesimal element ABCD is subjected to deformation. Its new shape will be A′B′C′D′. In accordance with the situation presented, the following equations relating to normal and shear strains can be given:

$$\varepsilon_x = \lim_{\mathrm{B}\to\mathrm{A}} \frac{\overline{\mathrm{A'B'}} - \overline{\mathrm{AB}}}{\overline{\mathrm{AB}}}, \quad \varepsilon_y = \lim_{\mathrm{D}\to\mathrm{A}} \frac{\overline{\mathrm{A'D'}} - \overline{\mathrm{AD}}}{\overline{\mathrm{AD}}} \tag{3.12}$$

$$\gamma_{xy} = \gamma_{yx} = \lim_{\substack{\mathrm{B}\to\mathrm{A}\\ \mathrm{D}\to\mathrm{A}}} (\text{angle ABD} - \text{angle A'B'D'}). \tag{3.13}$$

Other normal and shear strains, for example related to a spatial problem, may be derived easily.

Based on Figure 3.3(b), and assuming that small strains are being considered, that is, the difference between A′B′ and A′B″ is small (i.e. the difference is neglected), and the same holds for the AD side, the following relationships can be derived [1, 6–9]:

$$\begin{aligned} \varepsilon_x &= \frac{\overline{A'B''} - \overline{AB}}{\overline{AB}} = \frac{dx + \frac{\partial u}{\partial x}dx - dx}{dx} = \frac{\partial u}{\partial x} \\ \varepsilon_y &= \frac{\overline{A'D''} - \overline{AD}}{\overline{AD}} = \frac{dy + \frac{\partial v}{\partial y}dy - dy}{dy} = \frac{\partial v}{\partial y} \end{aligned} \tag{3.14}$$

$$\gamma_{xy} = \gamma'_{xy} + \gamma''_{xy};$$

$$\begin{aligned} \tan\gamma'_{xy} &= \frac{\overline{B'B''}}{\overline{A'B''}} = \frac{\frac{\partial v}{\partial x}dx}{dx + \frac{\partial u}{\partial x}dx} = \frac{\frac{\partial v}{\partial x}}{1 + \frac{\partial u}{\partial x}} = \frac{\frac{\partial v}{\partial x}}{1 + \varepsilon_x} \approx \frac{\partial v}{\partial x} \\ \tan\gamma''_{xy} &= \frac{\overline{D'D''}}{\overline{A'D''}} = \frac{\frac{\partial u}{\partial y}dy}{dy + \frac{\partial v}{\partial y}dy} = \frac{\frac{\partial u}{\partial y}}{1 + \frac{\partial v}{\partial y}} = \frac{\frac{\partial u}{\partial y}}{1 + \varepsilon_y} \approx \frac{\partial u}{\partial y} \end{aligned} \tag{3.15}$$

Assumption: $(\varepsilon_x, \varepsilon_y)$ is much less than 1.

When considering a three-dimensional problem, the strain – displacement relations are:

$$\begin{aligned} \varepsilon_x &= \frac{\partial u}{\partial x}, \quad 2\varepsilon_{xy} = 2\varepsilon_{yx} = \gamma_{xy} = \frac{\partial u}{\partial y} + \frac{\partial v}{\partial x} = \gamma_{yx} \\ \varepsilon_y &= \frac{\partial v}{\partial y}, \quad 2\varepsilon_{yz} = 2\varepsilon_{zy} = \gamma_{yz} = \frac{\partial v}{\partial z} + \frac{\partial w}{\partial y} = \gamma_{zy} \\ \varepsilon_z &= \frac{\partial w}{\partial z}, \quad 2\varepsilon_{zx} = 2\varepsilon_{xz} = \gamma_{zx} = \frac{\partial u}{\partial z} + \frac{\partial w}{\partial x} = \gamma_{xz} \end{aligned} \tag{3.16}$$

These equations, known as infinitesimal strain-displacement relations, can be written in index notation as:

$$\varepsilon_{ij} = \frac{1}{2}\left(u_{i,j} + u_{j,i}\right). \tag{3.17}$$

The strain components shown in Equations (3.16) are also known as the *Cauchy equations of deformation* [9], and are often referred to as infinitesimal strain components because of the assumption concerning the displacement derivatives [7]. In the equations, shear strain components $\gamma_{xy}, \ldots$ are often referred to as *engineering shear strains*, while shear strain components $\varepsilon_{xy}, \ldots$ are usually referred to as *mathematical shear strains* and they are usually used in elasticity theory.

The relationship is:

$$\varepsilon_{xy} = \frac{1}{2}\gamma_{xy}. \tag{3.18}$$

However, the relationships in Equations (3.16) are the strain-displacement relations for small-displacement theory [10].

3.3 Strain Tensors

3.3.1 Small Strain Tensor

The meaning and values of the strain (deformation) tensor components are clear from Equations (3.12)–(3.16). The procedure for constructing the tensor is clear from Equations below. The strain tensor (for a three-dimensional state of strain) is usually written in the following form [2, 6–7, 9, 11–13]:

$$\varepsilon_{ij} = [\varepsilon] = \begin{bmatrix} \varepsilon_{xx} & \varepsilon_{xy} & \varepsilon_{xz} \\ \varepsilon_{yx} & \varepsilon_{yy} & \varepsilon_{yz} \\ \varepsilon_{zx} & \varepsilon_{zy} & \varepsilon_{zz} \end{bmatrix} = \begin{bmatrix} \varepsilon_x & \frac{1}{2}\gamma_{xy} & \frac{1}{2}\gamma_{xz} \\ \frac{1}{2}\gamma_{yx} & \varepsilon_y & \frac{1}{2}\gamma_{yz} \\ \frac{1}{2}\gamma_{zx} & \frac{1}{2}\gamma_{zy} & \varepsilon_z \end{bmatrix}, \tag{3.19}$$

and is usually referred to as the *Cauchy tensor of infinitesimal strain* [9], or the Cauchy tensor of small strain or the small strain tensor. In engineering practice, the matrix form of this tensor is written as shown on the right-hand side of Equation (3.19). Components of this tensor are derived under the assumption of infinitesimal strain, that is, squares of strains and mutual products of strains are neglected. More exact values will be obtained if simplifications are not performed, and in this situation we have under consideration the so-called *finite strain* (total strain) tensor [1, 9, 14, 15].

Let us consider the process of deformation of the solid (deformable) body shown in Figure 3.4, that is, the deformation behavior of this solid body subjected to forces. The deformation behavior may be considered in a small-deformation domain (small-displacement theory) or in a finite-deformation domain (finite- or large-deformation theory). Here, as stated at the beginning of this section, small strains are considered.

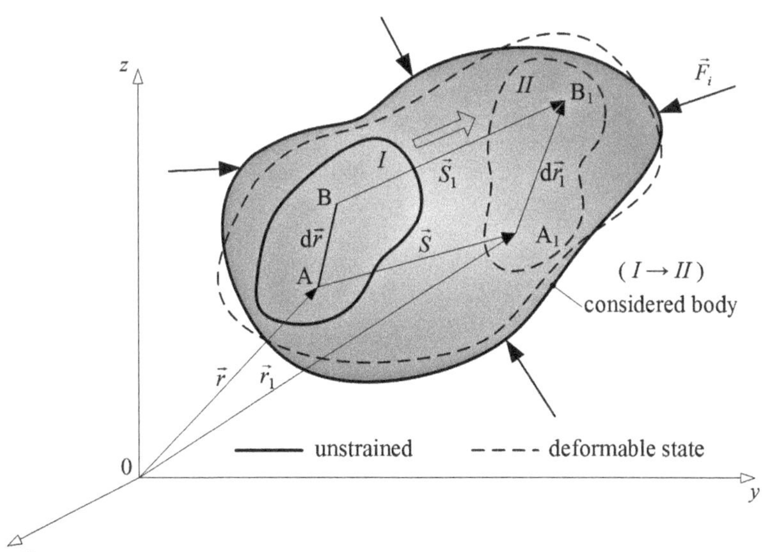

Figure 3.4 Displacements.

Consider a very small part, I, of the loaded body, which, after the deformation process is complete, takes the position II. Our interest is in considering what happens to the length $\overline{AB}$ of part I when it progresses to position $\overline{A_1B_1}$ of part II after the process of deformation due to forces is complete. Considering that the deformation is infinitesimal and continuous, displacements u_i, v_i and w_i of any point must be continuous, unambiguous and differentiable functions of the coordinates x, y, z. Suppose that u, v and w are displacements of the point A while u_1, v_1 and w_1 are displacements of the point B during the deformation process of $\overline{AB}$ to achieve the position $\overline{A_1B_1}$. In accordance with the above, and based on Figure 3.4, it follows that [9]:

$$\vec{s} = u\vec{i} + v\vec{j} + w\vec{k}; (u,v,w) = f(x,y,z) \tag{3.20a}$$

or, in symbolic notation:

$$\mathbf{s} = \mathbf{s}(\mathbf{r}) = u\,\mathbf{i} + v\,\mathbf{j} + w\,\mathbf{k}, \tag{3.20b}$$

Also, we have:

$$\vec{s}_1 = u_1\vec{i} + v_1\vec{j} + w_1\vec{k} \tag{3.21a}$$

For small strains, the first derivatives in the Taylor series are used, and since displacements are differentiable functions of the coordinates x, y, z, we have:

$$\vec{s}_1 = \vec{s} + \mathrm{d}\vec{s} + \ldots = \vec{s} + \frac{\partial \vec{s}}{\partial x}\mathrm{d}x + \frac{\partial \vec{s}}{\partial y}\mathrm{d}y + \frac{\partial \vec{s}}{\partial z}\mathrm{d}z \tag{3.21b}$$

or:

$$\begin{aligned} u_1 &= u + \mathrm{d}u = u + \frac{\partial u}{\partial x}\mathrm{d}x + \frac{\partial u}{\partial y}\mathrm{d}y + \frac{\partial u}{\partial z}\mathrm{d}z \\ v_1 &= v + \mathrm{d}v = \ldots \\ w_1 &= w + \mathrm{d}w = \ldots \end{aligned} \tag{3.22}$$

So, based on Equation (3.20a) and the above, we have:

$$\begin{aligned} \frac{\partial \mathbf{s}}{\partial x} &= \frac{\partial u}{\partial x}\mathbf{i} + \frac{\partial v}{\partial x}\mathbf{j} + \frac{\partial w}{\partial x}\mathbf{k}, \\ \frac{\partial \mathbf{s}}{\partial y} &= \frac{\partial u}{\partial y}\mathbf{i} + \frac{\partial v}{\partial y}\mathbf{j} + \frac{\partial w}{\partial y}\mathbf{k}, \\ \frac{\partial \mathbf{s}}{\partial z} &= \frac{\partial u}{\partial z}\mathbf{i} + \frac{\partial v}{\partial z}\mathbf{j} + \frac{\partial w}{\partial z}\mathbf{k}, \end{aligned} \tag{3.23}$$

Thus:

$$\begin{aligned} \mathrm{d}\mathbf{s} &= \frac{\partial \mathbf{s}}{\partial x}\mathrm{d}x + \frac{\partial \mathbf{s}}{\partial y}\mathrm{d}y + \frac{\partial \mathbf{s}}{\partial z}\mathrm{d}z \\ &= \left[\frac{\partial u}{\partial x}\mathrm{d}x + \frac{\partial u}{\partial y}\mathrm{d}y + \frac{\partial u}{\partial z}\mathrm{d}z\right]\mathbf{i} + \left[\frac{\partial v}{\partial x}\mathrm{d}x + \frac{\partial v}{\partial y}\mathrm{d}y + \frac{\partial v}{\partial z}\mathrm{d}z\right]\mathbf{j} \\ &+ \left[\frac{\partial w}{\partial x}\mathrm{d}x + \frac{\partial w}{\partial y}\mathrm{d}y + \frac{\partial w}{\partial z}\mathrm{d}z\right]\mathbf{k} \end{aligned} \tag{3.24}$$

We can edit Equation (3.22) as follows:

$$
\begin{aligned}
\frac{u_1-u}{\mathrm{d}r} &= \frac{\partial u}{\partial x}l+\frac{\partial u}{\partial y}m+\frac{\partial u}{\partial z}r\\
\frac{v_1-v}{\mathrm{d}r} &= \frac{\partial v}{\partial x}l+\frac{\partial v}{\partial y}m+\frac{\partial v}{\partial z}r\\
\frac{w_1-w}{\mathrm{d}r} &= \frac{\partial w}{\partial x}l+\frac{\partial w}{\partial y}m+\frac{\partial w}{\partial z}r
\end{aligned}
\tag{3.25}
$$

where:

$$l=\cos\alpha=\frac{\mathrm{d}x}{\mathrm{d}r},\ m=\cos\beta=\frac{\mathrm{d}y}{\mathrm{d}r},\ r=\cos\gamma=\frac{\mathrm{d}z}{\mathrm{d}r} \tag{3.26}$$

However, in matrix notation, Equation (3.24) is:

$$[\mathrm{d}s]=\begin{bmatrix}\mathrm{d}s_x\\ \mathrm{d}s_y\\ \mathrm{d}s_z\end{bmatrix}=\begin{bmatrix}\frac{\partial u}{\partial x}&\frac{\partial u}{\partial y}&\frac{\partial u}{\partial z}\\ \frac{\partial v}{\partial x}&\frac{\partial v}{\partial y}&\frac{\partial v}{\partial z}\\ \frac{\partial w}{\partial x}&\frac{\partial w}{\partial y}&\frac{\partial w}{\partial z}\end{bmatrix}\cdot\begin{Bmatrix}\mathrm{d}x\\ \mathrm{d}y\\ \mathrm{d}z\end{Bmatrix}=[S][\mathrm{d}r]=[\mathrm{d}r_1-\mathrm{d}r] \tag{3.27}$$

$$[\mathrm{d}r_1]=[\mathrm{d}r]+[S][\mathrm{d}r] \tag{3.28}$$

Also, the so-called specific strain, based on Equations (3.25) and (3.26) is:

$$[\varepsilon']=\frac{\vec{s}_1-\vec{s}}{\mathrm{d}r}=\frac{u_1-u}{\mathrm{d}r}\vec{i}+\frac{v_1-v}{\mathrm{d}r}\vec{j}+\frac{w_1-w}{\mathrm{d}r}\vec{k} \tag{3.29}$$

$$[\varepsilon']=[\tilde{\varepsilon}][n]=\begin{bmatrix}\frac{\partial u}{\partial x}&\frac{\partial u}{\partial y}&\frac{\partial u}{\partial z}\\ \frac{\partial v}{\partial x}&\frac{\partial v}{\partial y}&\frac{\partial v}{\partial z}\\ \frac{\partial w}{\partial x}&\frac{\partial w}{\partial y}&\frac{\partial w}{\partial z}\end{bmatrix}\begin{bmatrix}l\\ m\\ r\end{bmatrix} \tag{3.30}$$

Tensor $\tilde{\varepsilon}_{ij}$, in matrix notation $[\tilde{\varepsilon}]$, represents the gradient of the displacement vector; it belongs to the second-order tensor but is asymmetric. It can be decomposed into symmetric and antisymmetric parts:

$$[\tilde{\varepsilon}]=\left[\tilde{\varepsilon}^{\mathrm{S}}\right]+\left[\tilde{\varepsilon}^{\mathrm{AS}}\right] \tag{3.31}$$

$$\left[\tilde{\varepsilon}^{\mathrm{S}}\right]=\frac{1}{2}\left([\tilde{\varepsilon}]+[\tilde{\varepsilon}]^{\mathrm{T}}\right)=\begin{bmatrix}\frac{\partial u}{\partial x}&\frac{1}{2}\left(\frac{\partial u}{\partial y}+\frac{\partial v}{\partial x}\right)&\frac{1}{2}\left(\frac{\partial u}{\partial z}+\frac{\partial w}{\partial x}\right)\\ \frac{1}{2}\left(\frac{\partial v}{\partial x}+\frac{\partial u}{\partial y}\right)&\frac{\partial v}{\partial y}&\frac{1}{2}\left(\frac{\partial v}{\partial z}+\frac{\partial w}{\partial y}\right)\\ \frac{1}{2}\left(\frac{\partial w}{\partial x}+\frac{\partial u}{\partial z}\right)&\frac{1}{2}\left(\frac{\partial w}{\partial y}+\frac{\partial v}{\partial z}\right)&\frac{\partial w}{\partial z}\end{bmatrix} \tag{3.32}$$

$$[\tilde{\varepsilon}^{\mathrm{AS}}] = \frac{1}{2}\left([\tilde{\varepsilon}] - [\tilde{\varepsilon}]^{\mathrm{T}}\right) = \begin{bmatrix} 0 & \frac{1}{2}\left(\frac{\partial u}{\partial y} - \frac{\partial v}{\partial x}\right) & \frac{1}{2}\left(\frac{\partial u}{\partial z} - \frac{\partial w}{\partial x}\right) \\ \frac{1}{2}\left(\frac{\partial v}{\partial x} - \frac{\partial u}{\partial y}\right) & 0 & \frac{1}{2}\left(\frac{\partial v}{\partial z} - \frac{\partial w}{\partial y}\right) \\ \frac{1}{2}\left(\frac{\partial w}{\partial x} - \frac{\partial u}{\partial z}\right) & \frac{1}{2}\left(\frac{\partial w}{\partial y} - \frac{\partial v}{\partial z}\right) & 0 \end{bmatrix} \tag{3.33}$$

Equations (3.32) and (3.33) can be written in the form:

$$\varepsilon_{ij}^{\mathrm{S}} = \varepsilon_{ij} = \frac{1}{2}\left(u_{i,j} + u_{j,i}\right) \tag{3.34}$$

$$\varepsilon_{ij}^{\mathrm{AS}} = \frac{1}{2}\left(u_{i,j} - u_{j,i}\right) \tag{3.35}$$

As we know, the antisymmetric part of the tensor $\tilde{\varepsilon}_{ij}$ given in Equation (3.33) describes the rotation, so this is a tensor of rotation, while the symmetric part of the tensor $\tilde{\varepsilon}_{ij}$ from Equation (3.32) describes pure infinitesimal deformation, and this will be considered the strain tensor, ε_{ij} from Equation (3.19).

3.3.2 Finite Strain Tensor

In engineering practice, some deformation problems concerning structures or structural elements may be solved using small-displacement theory. This is the case when displacements u_i and their gradients are small and thus using the equations of small strains and the tensor of small strains is reasonable. On the other hand, some structures are subjected to forces that cause large (finite) strains. In deformation theory, it is always important to derive the expression of the so-called magnification factor. This defines the ratio between the extended length of the considered fiber and its starting infinitesimal length. As mentioned previously, in classical elasticity (and plasticity), approximations are made in such a way that consideration is based on the concept of small (infinitesimal) displacements, which leads to the linearization of deformation theory. The strain tensor and its components (from Equation (3.19)) were developed under the assumption that strains are small (infinitesimal), that is, it is possible to neglect the terms of higher order (quadratic terms and products of strains). More appropriate values can be obtained if such values are not discarded/neglected.

Attempts have been made to develop the theory of deformation without using the above-mentioned approximations. Two strain tensors in particular can be mentioned: the *Green–Saint Venant* finite strain tensor (sometimes also known as the *Lagrange–Green* finite strain tensor) and the *Almansi* finite strain tensor (sometimes also known as the *Euler–Almansi* finite strain tensor). The Green–Saint Venant strain tensor enables the determination of strains in deformed configurations when certain quantities are known at the time of the initial configurations. This procedure is based on *material description*, that is, on the *Lagrangian* (*material*) *coordinates* (y_i) where the displacement components are functions of material coordinates. On the other hand, the Almansi strain tensor enables the determination of strains in initial configurations when certain quantities are known in the deformed configurations. This procedure is based on *spatial description*, that is, on *Eulerian* (*spatial*) *coordinates* (x_i) where the displacement components are functions of the spatial coordinates.

Let's consider the procedure for obtaining the components of *finite strains* via *material description*; that is, using Lagrangian (material) coordinates. Consider, in accordance with Figure 3.4 [9, 12, 13, 16], an infinitesimal fiber (line element) $\overline{AB}$ of length dr, which, under deformation of the medium, becomes the line element $\overline{A_1B_1}$ of length dr_1. This means that $A(y_1, y_2, y_3)$ moves to $A_1(x_1, x_2, x_3)$, while $B(y_1 + \mathrm{d}y_1, y_2 + \mathrm{d}y_2, y_3 + \mathrm{d}y_3)$ moves to $B_1(x_1 + \mathrm{d}x_1,\ x_2 + \mathrm{d}x_2,\ x_3 + \mathrm{d}x_3)$. It can be seen that coordinates x, y, z in this consideration at time $t = 0$ are replaced by y_1, y_2, y_3; that is, with coordinate system y_i. This coordinate system is used to describe the undeformed state of the body. The deformed state of the body, that is, the deformed configuration, is described in the x_i coordinate system. Thus, the y_i coordinate system corresponds to the undeformed (unstrained) state of the configuration (the configuration at time $t = 0$) while the x_i coordinate system corresponds to the configuration at time $t > 0$. Consider the change in length of the mentioned line element due to the deformation process. For simplicity, let system x_i be designated y_α. The square of the length of the considered line element $(\overline{AB})$ before deformation is:

$$\mathrm{d}r^2 = \mathrm{d}y_1^2 + \mathrm{d}y_2^2 + \mathrm{d}y_3^2, \tag{3.36a}$$

or, in index noatation:

$$\mathrm{d}r^2 = \mathrm{d}y_i\,\mathrm{d}y_i, \quad i = 1, 2, 3. \tag{3.36b}$$

The length of the deformed element $\overline{A_1B_1}$, that is, of the considered element $(\overline{AB})$ after deformation, is:

$$\mathrm{d}r_1^2 = \mathrm{d}x_1^2 + \mathrm{d}x_2^2 + \mathrm{d}x_3^2 \equiv \mathrm{d}y_{\alpha 1}^2 + \mathrm{d}y_{\alpha 2}^2 + \mathrm{d}y_{\alpha 3}^2 \tag{3.37a}$$

or, in index notation:

$$\mathrm{d}r_1^2 = \mathrm{d}y_\alpha\,\mathrm{d}y_\alpha, \; \alpha = 1, 2, 3. \tag{3.37b}$$

The difference between the deformed $(\overline{A_1B_1})$ and undeformed $(\overline{AB})$ lengths of the line element is:

$$\mathrm{d}r_1^2 - \mathrm{d}r^2 = \mathrm{d}y_\alpha\,\mathrm{d}y_\alpha - \mathrm{d}y_i\,\mathrm{d}y_i \tag{3.38}$$

Since, in the material description, the displacement components are considered functions of the material coordinates, we express the coordinates y_α as functions of the coordinates y_i:

$$y_\alpha = y_\alpha(y_1, y_2, y_3) = y_\alpha(y_i) \tag{3.39}$$

Further,

$$\mathrm{d}y_\alpha = \frac{\partial y_\alpha}{\partial y_i}\mathrm{d}y_i, \tag{3.40}$$

and Equation (3.38) becomes:

$$\begin{aligned}\mathrm{d}r_1^2 - \mathrm{d}r^2 &= \frac{\partial y_\alpha}{\partial y_i}\mathrm{d}y_i \frac{\partial y_\alpha}{\partial y_j}\mathrm{d}y_j - \mathrm{d}y_i\,\mathrm{d}y_i = \left(\frac{\partial y_\alpha}{\partial y_i}\frac{\partial y_\alpha}{\partial y_j} - \delta_{ij}\right)\mathrm{d}y_i\,\mathrm{d}y_j \\ &= 2L_{ij}\mathrm{d}y_i\,\mathrm{d}y_j\end{aligned} \tag{3.41}$$

So, the Lagrange–Green (Green–Saint Venant) finite strain tensor is:

$$L_{ij} = \frac{1}{2}\left(\frac{\partial y_\alpha}{\partial y_i}\frac{\partial y_\alpha}{\partial y_j} - \delta_{ij}\right) \tag{3.42}$$

and, after further editing [12, 13, 16], is:

$$L_{ij} = \epsilon_{ij} = \frac{1}{2}\left(u_{i,j} + u_{j,i} + u_{k,i}\,u_{k,j}\right) \tag{3.43}$$

The first two terms in Equation (3.43) are identical to the value represented by Equation (3.17) (or Equation (3.34)) and represent the small strain tensor, while the other terms are nonlinear. The components of the finite strain tensor represented by Equation (3.43) are:

$$\begin{aligned}
\varepsilon_x &= \frac{\partial u}{\partial x} + \frac{1}{2}\left[\left(\frac{\partial u}{\partial x}\right)^2 + \left(\frac{\partial v}{\partial x}\right)^2 + \left(\frac{\partial w}{\partial x}\right)^2\right]\\
\varepsilon_y &= \frac{\partial v}{\partial y} + \frac{1}{2}\left[\left(\frac{\partial u}{\partial y}\right)^2 + \left(\frac{\partial v}{\partial y}\right)^2 + \left(\frac{\partial w}{\partial y}\right)^2\right]\\
\varepsilon_z &= \frac{\partial w}{\partial z} + \frac{1}{2}\left[\left(\frac{\partial u}{\partial z}\right)^2 + \left(\frac{\partial v}{\partial z}\right)^2 + \left(\frac{\partial w}{\partial z}\right)^2\right]\\
\varepsilon_{xy} &= \frac{1}{2}\gamma_{xy} = \frac{1}{2}\left(\frac{\partial u}{\partial y} + \frac{\partial v}{\partial x} + \frac{\partial u}{\partial x}\frac{\partial u}{\partial y} + \frac{\partial v}{\partial x}\frac{\partial v}{\partial y} + \frac{\partial w}{\partial x}\frac{\partial w}{\partial y}\right) = \frac{1}{2}\gamma_{yx}\\
\varepsilon_{yz} &= \frac{1}{2}\gamma_{yz} = \frac{1}{2}\left(\frac{\partial v}{\partial z} + \frac{\partial w}{\partial y} + \frac{\partial u}{\partial y}\frac{\partial u}{\partial z} + \frac{\partial v}{\partial y}\frac{\partial v}{\partial z} + \frac{\partial w}{\partial y}\frac{\partial w}{\partial z}\right) = \frac{1}{2}\gamma_{zy}\\
\varepsilon_{zx} &= \frac{1}{2}\gamma_{zx} = \frac{1}{2}\left(\frac{\partial u}{\partial z} + \frac{\partial w}{\partial x} + \frac{\partial u}{\partial x}\frac{\partial u}{\partial z} + \frac{\partial v}{\partial x}\frac{\partial v}{\partial z} + \frac{\partial w}{\partial x}\frac{\partial w}{\partial z}\right) = \frac{1}{2}\gamma_{xz}
\end{aligned} \tag{3.44}$$

The components defined by Equation (3.44) are known as Lagrange–Green (Green–Saint Venant) strain components (that is, they are valid for material coordinates). In addition, the Euler–Almansi finite strain tensor is:

$$E_{ij} = \mathtt{E_{ij}} = \frac{1}{2}\left(u_{i,j} + u_{j,i} - u_{k,i}\,u_{k,j}\right) \tag{3.45}$$

3.3.3 Mean and Deviatoric Strain Tensors

The previously described small strain (Cauchy) tensor from Equation (3.19) can also be decomposed into the *mean strain tensor* and the *deviatoric strain tensor*:

$$\varepsilon_{ij} = \varepsilon_{ij}^0 + e_{ij} \quad \text{or} \quad [\varepsilon] = \left[\varepsilon^0\right] + [e] \tag{3.46}$$

This decomposition can be displayed in such a way that the strain tensor is presented using strain components or using principal strains (dilatations). In Equation (3.46) we have: ε_{ij} = strain tensor; $\varepsilon_{ij}^0 \equiv \varepsilon_{ij}^{\mathrm{m}}$ = mean strain tensor; e_{ij} = deviatoric strain tensor.

Depending on the aforementioned display capabilities, the strain tensor decomposition in Equation (3.46) is:

$$\begin{bmatrix} \varepsilon_x & \frac{1}{2}\gamma_{xy} & \frac{1}{2}\gamma_{xz} \\ \frac{1}{2}\gamma_{yx} & \varepsilon_y & \frac{1}{2}\gamma_{yz} \\ \frac{1}{2}\gamma_{zx} & \frac{1}{2}\gamma_{zy} & \varepsilon_z \end{bmatrix} = \begin{bmatrix} \varepsilon_0 & 0 & 0 \\ 0 & \varepsilon_0 & 0 \\ 0 & 0 & \varepsilon_0 \end{bmatrix} + \begin{bmatrix} \varepsilon_x-\varepsilon_0 & \frac{1}{2}\gamma_{xy} & \frac{1}{2}\gamma_{xz} \\ \frac{1}{2}\gamma_{yx} & \varepsilon_y-\varepsilon_0 & \frac{1}{2}\gamma_{yz} \\ \frac{1}{2}\gamma_{zx} & \frac{1}{2}\gamma_{zy} & \varepsilon_z-\varepsilon_0 \end{bmatrix} \tag{3.47a}$$

or it can be written in the form:

$$\begin{bmatrix} \varepsilon_1 & 0 & 0 \\ 0 & \varepsilon_2 & 0 \\ 0 & 0 & \varepsilon_3 \end{bmatrix} = \begin{bmatrix} \varepsilon_0 & 0 & 0 \\ 0 & \varepsilon_0 & 0 \\ 0 & 0 & \varepsilon_0 \end{bmatrix} + \begin{bmatrix} \varepsilon_1-\varepsilon_0 & 0 & 0 \\ 0 & \varepsilon_2-\varepsilon_0 & 0 \\ 0 & 0 & \varepsilon_2-\varepsilon_0 \end{bmatrix} \tag{3.47b}$$

In accordance with Equation (3.46), the mean strain tensor is:

$$\begin{aligned} \varepsilon_{ij}^{0} \equiv \varepsilon_{ij}^{\mathrm{m}} = \varepsilon_0\,\delta_{ij} \equiv \varepsilon_{\mathrm{m}}\delta_{ij} &= \begin{bmatrix} \varepsilon_0 & 0 & 0 \\ 0 & \varepsilon_0 & 0 \\ 0 & 0 & \varepsilon_0 \end{bmatrix} \equiv \begin{bmatrix} \varepsilon_{\mathrm{m}} & 0 & 0 \\ 0 & \varepsilon_{\mathrm{m}} & 0 \\ 0 & 0 & \varepsilon_{\mathrm{m}} \end{bmatrix} \\ &= \frac{1}{3}\varepsilon_{kk}\delta_{ij} \end{aligned} \tag{3.47c}$$

Based on Equations (3.46) and (3.47c), it follows that:

$$\varepsilon_{ij} = \frac{1}{3}\varepsilon_{kk}\delta_{ij} + e_{ij} \tag{3.47d}$$

In Equation (3.47c), ε_{m} is the mean normal strain:

$$\varepsilon_{\mathrm{m}} = \varepsilon_0 = \frac{\varepsilon_1+\varepsilon_2+\varepsilon_3}{3} = \frac{\varepsilon_x+\varepsilon_y+\varepsilon_z}{3} = \frac{1}{3}\varepsilon_{kk} = \frac{1}{3}\varepsilon_v = \frac{1}{3}\bar{I}_1 \tag{3.48}$$

where $\bar{I}_1$ is the first invariant of the strain tensor, to which we will return later.

As already mentioned in Section 2.3.1, experiments performed on metallic materials indicate that yielding and plastic deformation are essentially independent of mean normal strain. Hence, plasticity theories postulate that plastic behavior of materials is related primarily to that part of the strain tensor that is independent of the mean normal strain.

3.3.4 Principal Strains and Strain Invariants

3.3.4.1 Strain Tensor

The principal strains and strain invariants can be calculated using a procedure analogous to that used to find the principal values and invariants in the case of stress (refer back to Equations (2.26) and (2.41)) [10, –14]. In this case we have:

$$\begin{vmatrix} \varepsilon_x - \varepsilon_i & \frac{1}{2}\gamma_{xy} & \frac{1}{2}\gamma_{xz} \\ \frac{1}{2}\gamma_{yx} & \varepsilon_y - \varepsilon_i & \frac{1}{2}\gamma_{yz} \\ \frac{1}{2}\gamma_{zx} & \frac{1}{2}\gamma_{xz} & \varepsilon_z - \varepsilon_i \end{vmatrix} = 0 \tag{3.49}$$

Equation (3.49) yields the cubic equation (characteristic equation):

$$\varepsilon_i^3 - \bar{I}_1 \varepsilon_i^2 + \bar{I}_2 \varepsilon_i - \bar{I}_3 = 0 \tag{3.50}$$

In Equation (3.50), ε_i are the three *principal strains* as the three roots of this equation. The invariants of the strain tensor, again analogous with those for stress in Equations (2.43)–(2.45), are:

$$\begin{aligned} \bar{I}_1 &= \varepsilon_x + \varepsilon_y + \varepsilon_z = \varepsilon_1 + \varepsilon_2 + \varepsilon_3 = \varepsilon_V \\ \bar{I}_2 &= \varepsilon_x\varepsilon_y + \varepsilon_y\varepsilon_z + \varepsilon_z\varepsilon_x - \frac{1}{4}\left(\gamma_{xy}^2 + \gamma_{yz}^2 + \gamma_{zx}^2\right) = \varepsilon_1\varepsilon_2 + \varepsilon_2\varepsilon_3 + \varepsilon_3\varepsilon_1 \\ \bar{I}_3 &= \varepsilon_x\varepsilon_y\varepsilon_z + \frac{1}{4}\gamma_{xy}\gamma_{yz}\gamma_{zx} - \frac{1}{4}\left(\varepsilon_x\gamma_{yz}^2 + \varepsilon_y\gamma_{zx}^2 + \varepsilon_z\gamma_{xy}^2\right) = \varepsilon_1\varepsilon_2\varepsilon_3 \end{aligned} \tag{3.51}$$

where ε_i are the *principal strains* (principal dilatations) and $\bar{I}_i$ the *strain invariants* [15]. Invariants of the mean strain tensor from Equation (3.47c) are:

$$\begin{aligned} \bar{I}_1^{\mathrm{m}} &= \bar{I}_1 = 3\varepsilon_{\mathrm{m}} \\ \bar{I}_2^{\mathrm{m}} &= \frac{1}{3}\bar{I}_1^2 = 3\varepsilon_{\mathrm{m}}^2 \\ \bar{I}_3^{\mathrm{m}} &= \frac{1}{27}\bar{I}_1^3 = \varepsilon_{\mathrm{m}}^3 \end{aligned} \tag{3.52}$$

where: $\bar{I}_i^{\mathrm{m}}$ are the invariants of the mean strain tensor; ε_{m} is the mean normal strain.

3.3.4.2 Deviatoric Strain Tensor

In a similar manner to the deviatoric stress tensor from Equation (2.9), the *deviatoric strain tensor* written in matrix form in Equation (3.47b) can also be written as:

$$\begin{aligned} [e] &= \begin{bmatrix} \varepsilon_1 - \varepsilon_{\mathrm{m}} & 0 & 0 \\ 0 & \varepsilon_2 - \varepsilon_{\mathrm{m}} & 0 \\ 0 & 0 & \varepsilon_3 - \varepsilon_{\mathrm{m}} \end{bmatrix} \\ &= \begin{bmatrix} \frac{2\varepsilon_1 - \varepsilon_2 - \varepsilon_3}{3} & 0 & 0 \\ 0 & \frac{2\varepsilon_2 - \varepsilon_3 - \varepsilon_1}{3} & 0 \\ 0 & 0 & \frac{2\varepsilon_3 - \varepsilon_1 - \varepsilon_2}{3} \end{bmatrix} \end{aligned} \tag{3.53}$$

Based on this equation, it can be seen that the first invariant, defined as in Equation (2.43), is equal to zero. So, the invariants of the deviatoric strain tensor are:

$$\bar{J}_1 = 0\ldots..(= \mathrm{tr}[e]) = e_1 + e_2 + e_3$$

$$\begin{aligned}\bar{J}_2 &= \bar{I}_2 - \frac{1}{3}\bar{I}_1^2 = e_1e_2 + e_2e_3 + e_3e_1 \\ &= -\frac{1}{6}\left[(\varepsilon_1-\varepsilon_2)^2 + (\varepsilon_2-\varepsilon_3)^2 + (\varepsilon_3-\varepsilon_1)^2\right] \\ \bar{J}_3 &= \det[e] = \bar{I}_3 - \frac{1}{3}\bar{I}_1\bar{I}_2 + \frac{2}{27}\bar{I}_1^3 = e_1e_2e_3 \\ &= \frac{1}{27}(2\varepsilon_1-\varepsilon_2-\varepsilon_3)(2\varepsilon_2-\varepsilon_3-\varepsilon_1)(2\varepsilon_3-\varepsilon_1-\varepsilon_2)\end{aligned} \tag{3.54}$$

3.4 Transformation of Strain Components

The transformation of strain components is shown in Figure 3.5. When, at the considered point, strain components u and v defined for coordinates x and y are known, then, using the transformation rule, we can determine components $\bar{u}$ and $\bar{v}$ for the rotated coordinates $\bar{x}$ and $\bar{y}$.

Whereas the displacement components $\bar{u}$ and $\bar{v}$ are functions of $\bar{x}$ and $\bar{y}$, and at the same time $\bar{x}$ and $\bar{y}$ are functions of x and y, to determine the value of $\bar{\varepsilon}_x$, for example, it is necessary to apply the chain rule derivation of complex functions of several variables. However, consider the determination of the mentioned strain, $\bar{\varepsilon}_x$. In accordance with Figure 3.5, we have [8, 17, 18]:

$$\begin{aligned}&\bar{\varepsilon}_x = \frac{\partial\bar{u}}{\partial\bar{x}};\ \bar{u} = u\cos\varphi + v\sin\varphi;\ \bar{\varepsilon}_x = \frac{\partial u}{\partial\bar{x}}\cos\varphi + \frac{\partial v}{\partial\bar{x}}\sin\varphi; \\ &\frac{\partial u}{\partial\bar{x}} = \frac{\partial u}{\partial x}\frac{\partial x}{\partial\bar{x}} + \frac{\partial u}{\partial y}\frac{\partial y}{\partial\bar{x}};\ \frac{\partial v}{\partial\bar{x}} = \frac{\partial v}{\partial x}\frac{\partial x}{\partial\bar{x}} + \frac{\partial v}{\partial y}\frac{\partial y}{\partial\bar{x}}. \\ &\frac{\mathrm{d}x}{\mathrm{d}\bar{x}} = \cos\varphi;\ \frac{\mathrm{d}y}{\mathrm{d}\bar{x}} = \sin\varphi\end{aligned} \tag{3.55a}$$

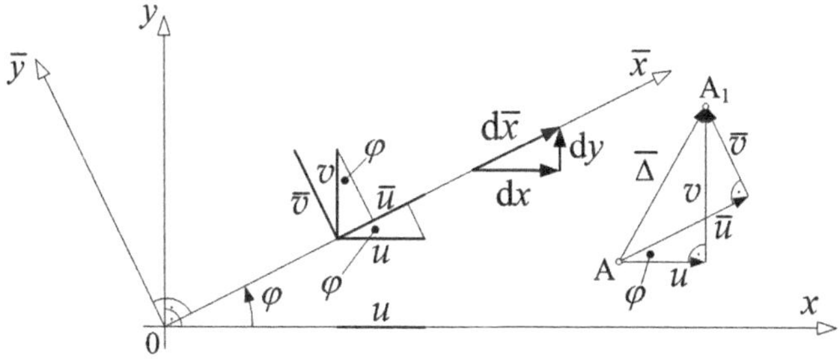

Figure 3.5 Transformation of strain components.

Finally, we have:

$$\bar{\varepsilon}_x = \varepsilon_x \cos^2\varphi + \varepsilon_y \sin^2\varphi + 2\varepsilon_{xy} \sin\varphi \cos\varphi$$
$$\bar{\varepsilon}_x = \varepsilon_x \cos^2\varphi + \varepsilon_y \sin^2\varphi + \gamma_{xy} \sin\varphi \cos\varphi \tag{3.55b}$$

Transformations for other components can be carried out in a similar way. On the basis of this consideration, it can be concluded that the transformation of the strain components is carried out in the same manner as the transformation of the stress components in Equations (2.53). As a result, the following transformations of the strain components for the planar model are known:

$$\begin{aligned}
\bar{\varepsilon}_x &= \varepsilon_x \cos^2\varphi + \varepsilon_y \sin^2\varphi + \gamma_{xy} \sin\varphi \cos\varphi \\
\bar{\varepsilon}_y &= \varepsilon_x \sin^2\varphi + \varepsilon_y \cos^2\varphi - \gamma_{xy} \sin\varphi \cos\varphi \\
\bar{\gamma}_{xy} &= -2\left(\varepsilon_x - \varepsilon_y\right) \cos\varphi \sin\varphi + \gamma_{xy}\left(\cos^2\varphi - \sin^2\varphi\right)
\end{aligned} \tag{3.55c}$$

If now the identities mentioned in Equations (a) in Section 2.5 are applied, the following equations can be written:

$$\begin{aligned}
\bar{\varepsilon}_x &= \frac{\varepsilon_x + \varepsilon_y}{2} + \frac{\varepsilon_x - \varepsilon_y}{2} \cos 2\varphi + \frac{1}{2}\gamma_{xy} \sin 2\varphi \\
\bar{\varepsilon}_y &= \frac{\varepsilon_x + \varepsilon_y}{2} - \frac{\varepsilon_x - \varepsilon_y}{2} \cos 2\varphi - \frac{1}{1}\gamma_{xy} \sin 2\varphi \\
\frac{1}{2}\bar{\gamma}_{xy} &= -\frac{\varepsilon_x - \varepsilon_y}{2} \sin 2\varphi + \frac{1}{2}\gamma_{xy} \cos 2\varphi
\end{aligned} \tag{3.55d}$$

3.4.1 Mohr's Circle

Mohr's circle relating to strain can be found in a manner analogous to that discussed in Chapter 2 on stress. We simply replace stress by strain: ($\sigma_x \to \varepsilon_x$; $\sigma_y \to \varepsilon_y$; $\tau_{xy} \to \frac{1}{2}\gamma_{xy}$).

We should add that principal strains (dilatations) may be determined in the same manner as principal stresses. The same is valid in the procedure for the determination of the direction of principal strains and the direction of the extreme values of shear strains.

3.5 Strain Measurement

In the previous chapter, when we discussed axial loading, it was shown that axial elongation, ΔL, can be obtained by measuring the length of the considered engineering element before and after a load is applied – that is, as the difference in length between the deformed and unstrained element. If this elongation is divided by the length of the element in its undeformed (unstrained) state, one can calculate the normal strain, ε. The question is how to determine the normal strain or shear strain at a given point on the free surface of the body with respect to an arbitrary direction, or how to determine the principal strain(s). A frequently used and very convenient method in the measurement of normal strains is the use of strain gages, as shown in Figure 3.6. In engineering practice, the arrangement of strain gages measuring three normal strains – ε_i $(i = 1,2,3)$ – is known as a *strain rosette*. Strain gages are taped (glued) to the element at the place where elongation needs to be determined. The wire of the strain gage increases in length (decreases in diameter) and causes an increase in the electrical resistance of the gage. As shown in the figure, the strain components $(\varepsilon_x, \varepsilon_y)$ can be determined by measuring along axes x and y. The same is valid for the ξ direction.

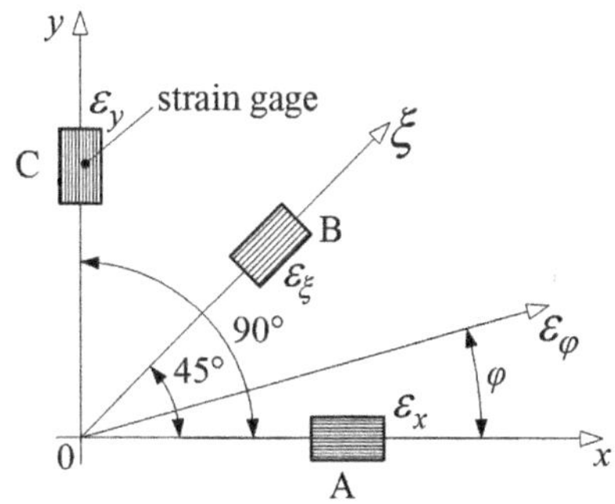

Figure 3.6 Strain measurement.

In accordance with Equation (3.55b), the normal strain related to any direction that is defined by angle φ can be calculated as follows:

$$\varepsilon(\varphi) = \varepsilon_x \cos^2\varphi + \varepsilon_y \sin^2\varphi + \gamma_{xy} \sin\varphi\cos\varphi. \tag{3.56}$$

For example, if $\varphi = 0$, it follows that $\varepsilon(\varphi) = \varepsilon_x$. In addition, if, for example, $\varphi = 45°$ (strain gage B, Figure 3.6), then the shear strain is:

$$\gamma_{xy} = 2\varepsilon_\xi - (\varepsilon_x + \varepsilon_y); (\varepsilon_\xi = \varepsilon_{\varphi = 45°}). \tag{3.57a}$$

So, if normal strains $\varepsilon_x, \varepsilon_y, \varepsilon_\xi$ have been measured, then the shear strain can be calculated. In general, strain components ε_x, ε_y and γ_{xy} can be determined at a given point if normal strains are measured along any three lines through that point. Given this, three equations can be written. In a general form, these equations may appear as follows (see Equation 3.55b):

$$\varepsilon_i = \varepsilon_x \cos^2\varphi_i + \varepsilon_y \sin^2\varphi_i + \gamma_{xy} \sin\varphi_i \cos\varphi_i \tag{3.57b}$$

where $i = 1, 2, 3$, while angle φ of each strain gage is measured from the x axis (as shown, for example, in Figure 3.9). By solving the three equations of the form shown in Equation (3.57b), the above-mentioned strain components ε_x, ε_y and γ_{xy} can be determined.

As previously mentioned, because of the analogy between stress and strain, all other elements, like the principal strains and the principal directions, may be determined in the same manner as for stresses. For example, for the same scheme of strain gages shown in Figure 3.6 (the rosette), the principal strains can be calculated (as for stress in Equation (2.21)) according to:

$$\varepsilon_{1,2} = \frac{1}{2}(\varepsilon_x + \varepsilon_y) \pm \frac{1}{2}\sqrt{(\varepsilon_x - \varepsilon_y)^2 + \gamma_{xy}^2}. \tag{3.58}$$

If the shear strain in Equation (3.58) is substituted by Equation (3.57), then the principal strains (dilatations) can be written in the following form [5]:

$$\varepsilon_{1,2} = \varepsilon_{\max,\min} = \frac{1}{2}\left(\varepsilon_x + \varepsilon_y\right) \pm \frac{1}{\sqrt{2}}\left[\left(\varepsilon_x - \varepsilon_\xi\right)^2 + \left(\varepsilon_\xi - \varepsilon_y\right)^2\right]^{1/2}. \tag{3.59}$$

The maximum shear strain is:

$$\gamma_{\max} = \pm\frac{1}{2}\sqrt{\left(\varepsilon_x - \varepsilon_y\right)^2 + \gamma_{xy}^2} \tag{3.60}$$

Example 3.1.
Consider a plane strain state problem, as shown in Figure 3.7. For the given data, we wish to determine the principal strains and the principal directions.

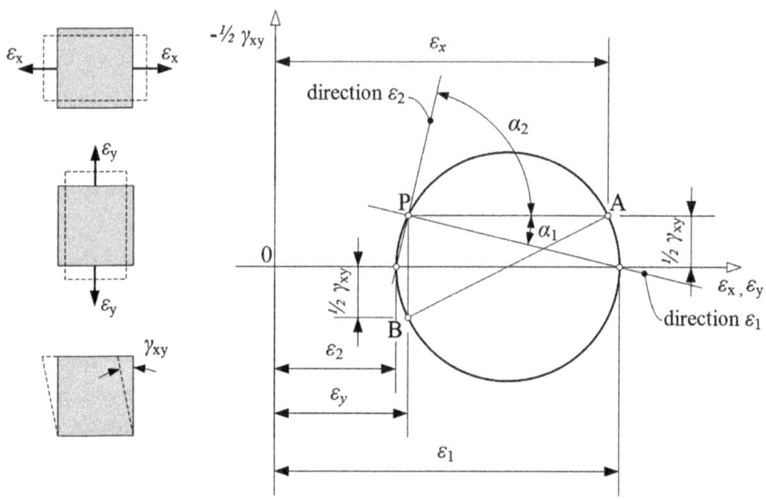

Figure 3.7 Mohr's circle (Example 3.1).

Data:

$$\varepsilon_x = 50\cdot10^{-6},\ \varepsilon_y = 20\cdot10^{-6},\ \gamma_{xy} = -15\cdot10^{-6}.$$

Solution:

$$\varepsilon_{1,2} = \frac{1}{2}\left(\varepsilon_x + \varepsilon_y\right) \pm \frac{1}{2}\sqrt{\left(\varepsilon_x + \varepsilon_y\right)^2 + \gamma_{xy}^2};\ \varepsilon_1 = 51.77\cdot10^{-6};\ \varepsilon_2 = 18.229\cdot10^{-6}$$

Similar to Equation (2.20), here we have:

$$\tan 2\alpha = \frac{\gamma_{xy}}{\varepsilon_x - \varepsilon_y}$$

$$\alpha_1 = \frac{1}{2}\arctan\left(\frac{\gamma_{xy}}{\varepsilon_x - \varepsilon_y}\right) = -13.28^\circ; \tag{3.61}$$

$$\alpha_2 = -13.28^\circ + 90^\circ = 76.72^\circ.$$

Example 3.2.
Consider the rectangular plate in Figure 3.8 in a state of plane strain. Determine the equations through which the displacements shown can be described.

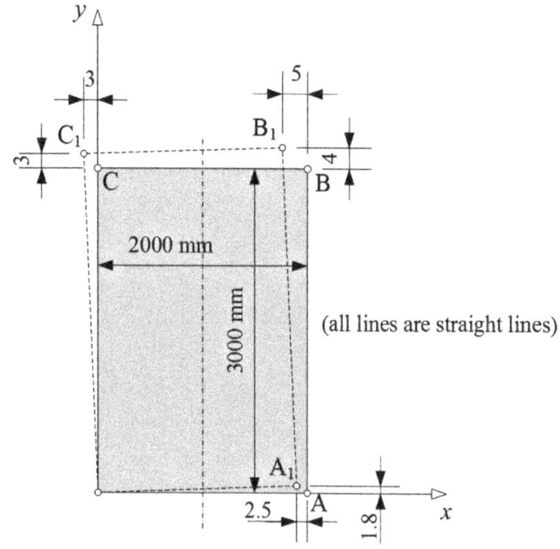

Figure 3.8 Plate in a state of plane strain (Example 3.2).

Data:
The displacements of the plate's vertices are shown in Figure 3.8.

Solution:
In accordance with Figure 3.8, the following data can be written:
A (2 m, 0); B (2, 3); C (0, 3)
The requested equations can be written as:

$$u = C_1x + C_2y + C_3xy \tag{a1}$$
$$v = C_4x + C_5y + C_6xy \tag{a2}$$

If data relating to the coordinates of the points are entered, it follows that:

$$\begin{aligned} -0.0025 &= C_1{\cdot}2 + C_2{\cdot}0 + C_3{\cdot}2{\cdot}0 \\ -0.005 &= C_1{\cdot}2 + C_2{\cdot}3 + C_3{\cdot}2{\cdot}3 \\ -0.003 &= C_1{\cdot}0 + C_2{\cdot}3 + C_3{\cdot}0{\cdot}3 \end{aligned} \tag{b}$$

For simplicity, coordinates are entered in meters.
Solving the above equations, it follows that:

$$C_1 = -0.00125; \quad C_2 = -0.001; \quad C_3 = 0.0000833 \tag{c}$$

Finally, the requested equation for the displacement with respect to the x axis is:

$$u = -0.00125x - 0.001y + 0.0000833xy \tag{d}$$

Using a similar procedure, the equation related to the displacement with respect to the y axis is:

$$v = 0.0009x + 0.001y - 0.000133xy \tag{e}$$

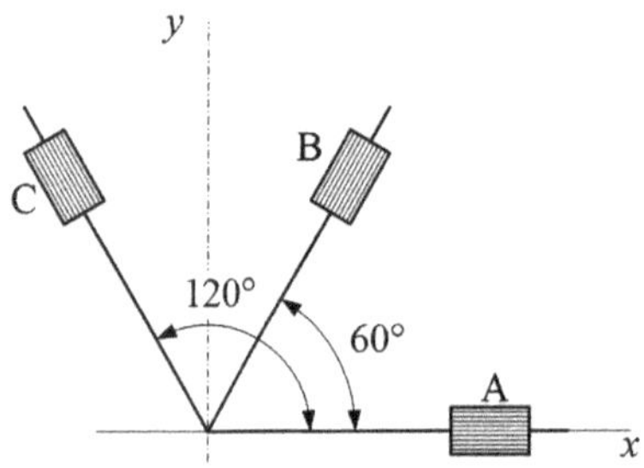

Figure 3.9 A strain rosette (Example 3.3).

Example 3.3.
Consider measured data obtained by a rosette consisting of three strain gages, as shown in Figure 3.9. The rosette was mounted on the engineering element (made of steel). Determine the principal strains and the maximum shear strain at a point.

Data:
The measured strains were:

$$\varepsilon_A = -1000;\ \varepsilon_B = -1500;\ \varepsilon_C = 800$$

Solution:
In accordance with Equation (3.56):

$$\varepsilon(\varphi) = \varepsilon_x \cos^2\varphi + \varepsilon_y \sin^2\varphi + \gamma_{xy} \sin\varphi \cos\varphi$$

For gages A, B and C, this becomes:
Gage A:

$$-1000 = \varepsilon_x \cos^2(0^\circ) + \varepsilon_y \sin^2(0^\circ) + \gamma_{xy} \sin(0^\circ) \cos(0^\circ)$$

Gage B:

$$-1500 = \varepsilon_x \cos^2(60^\circ) + \varepsilon_y \sin^2(60^\circ) + \gamma_{xy} \sin(60^\circ) \cos(60^\circ)$$

Gage C:

$$800 = \varepsilon_x \cos^2(120^\circ) + \varepsilon_y \sin^2(120^\circ) + \gamma_{xy} \sin(120^\circ) \cos(120^\circ).$$

The solution to these three equations follows:

$$\varepsilon_x = -1000;\ \varepsilon_y = -166.7;\ \gamma_{xy} = -2598.$$

Using Equation (3.58), it follows that:

$$\varepsilon_{1,2} = \frac{1}{2}(\varepsilon_x + \varepsilon_y) \pm \frac{1}{2}\sqrt{(\varepsilon_x + \varepsilon_y)^2 + \gamma_{xy}^2}$$

$$\varepsilon_1 = 2007.3;\ \varepsilon_2 = -840.62;\ \gamma_{max} = 1423.97$$

References

1 Brnic, J. and Turkalj, G. (2004) *Strength of Materials I* (in Croatian), Faculty of Engineering, University of Rijeka, Rijeka.
2 Alfirević, I. (1995) *Strength of Materials I* (in Croatian), *ITP* "Tehnička knjiga", D.D., Zagreb.
3 Alfirević, I. (2006) *Linear Structural Analysis*, Pretei d.o.o., Zagreb.
4 Seed, G. M. (2000) *Strength of Materials*, Saxe-Coburg Publications, Edinburgh.
5 Beer, F. P., Johnston, E. R. Jr, DeWolf, J. T. and Mazurek, D. F. (2012) *Mechanics of Materials*, McGraw-Hill, USA.

6 Boresi, A. and Schmidt, R. J. (2003) *Advanced Mechanics of Materials*, John Wiley & Sons, New York.
7 Solecki, R. and Conant, R. J. (2003) *Advanced Mechanics of Materials*, Oxford University Press, New York.
8 Alfirević, I. (1975) *Advanced Strength of Materials* (in Croatian), University of Zagreb, Faculty of Mechanical Engineering and Naval Architecture.
9 Raskovic, D. (1985) *Theory of Elasticity* (in Serbian). Naučna knjiga, Belgrade.
10 Boresi, A. P., Schmidt, R. J. and Sidebottom, O. M. (1993) *Advanced Mechanics of Materials*, John Wiley & Sons, New York.
11 Bowman, K. (2004) *Mechanical Behavior of Materials*, John Wiley & Sons, New York.
12 Boresi, A., Chong, J. D. and Lee, J. D. (2011) *Elasticity in Engineering Mechanics*, John Wiley & Sons, New Jersey.
13 Prelog, E. (1978) *Elasto and Plastomechanics* (in Slovenian), Faculty of Mechanical Engineering, University of Ljubljana, Ljubljana.
14 Brnic, J. (1996) *Elastomechanics and Plastomechanics* (in Croatian), Školska knjiga, Zagreb.
15 Kačanov, L. M. (1969) *Basic Theory of Plasticity* (in Russian), Nauka, Moscow.
16 Štok, B. (1988) *Mechanics of Deformable Bodies* (in Slovenian), part I, Faculty of Mechanical Engineering, University of Ljubljana, Ljubljana, Slovenia.
17 Craig, R. R. (2011) *Mechanics of Materials*, 3rd edition, John Wiley & Sons, New Jersey.
18 Riley, W. F., Sturges, L. D. and Morris, D. H. (2007) *Mechanics of Materials*, John Wiley & Sons, USA.

4

Mechanical Testing of Materials

4.1 Material Properties

In this chapter, we will provide a general, and hopefully informative, outline of certain issues. More detailed information can be found in other chapters of the book, such as Chapters 6, 7 and 9.

Testing materials to determine their mechanical and other properties in order to form a basis for the design of structures in relation to their use began with the development of human society. Nowadays, testing procedures as well as the geometry of specimens for each type of test are defined by appropriate standards, for example, ASTM standards, EU standards or others. Information about material testing and material properties can be found in [1–10].

It is possible to describe a multitude of material properties. However, material properties generally can be divided into two groups: elastic properties and mechanical properties.

Elastic material properties include:

- Modulus of elasticity – the ratio between tensile strength and normal strain (see Equation (3.10));
- Modulus of rigidity/shear modulus – the ratio between shear stress and shear strain (see Equation (3.11));
- Poisson ratio – the ratio between lateral normal strain and longitudinal normal strain (see Equation (3.2));
- Bulk modulus – the ratio between volumetric stress and volumetric strain (see Equation (3.9)).

Mechanical material properties include:

- Strength of the material – the ultimate tensile strength, yield strength (the property of the material which opposes the deformation of an element made of the considered material, that is, which opposes the failure of an engineering element);
- Ductility – the material property which indicates how a material deforms under tensile load;
- Toughness – the ability of a material to absorb energy and be plastically deformed without fracturing;
- Fracture toughness – the property that indicates the ability of a material to withstand crack propagation;

Analysis of Engineering Structures and Material Behavior, First Edition. Josip Brnić.

- Creep – the property which indicates an increase in deformation under constant stress;
- Fatigue – the property that indicates a decrease in the ability to carry repeated stress;
- Hardness – the property that indicates the resistance of a material to deformation/ indentation/penetration by different means: impact, drilling, and so on).

Making the final choice of material that will be used in the manufacture of an engineering element is a very difficult task. During its service life, an engineering component (part of a machine, column, shaft, etc.) must fulfill all of the requirements (safety, functionality, low price, weight, required range of stress and/or strain, etc.) that are placed upon it. It is said that the design of the element must be optimal. There is no doubt that, in order to achieve all of the requirements, the selection of materials must be based on both the knowledge of the designers and on the data about materials behavior which have been obtained through testing.

4.2 Types of Material Testing

Various types of material testing and the expected test results/information are presented in Table 4.1.

4.3 Test Methods Related to Mechanical Properties

In this section, attention will be paid only to the mechanical properties of materials. Test methods related to the mechanical properties are often called mechanical testing methods or, simply, mechanical testing. In determining the mechanical properties of a material, loading of the specimen can be tensile, compressive, flexural, shear or some combination. These material tests (hardness and technological investigations are excluded) may be classified as shown in Table 4.2.

4.4 Testing Machines and Specimens

Testing machines are used to determine some of the material properties or material behavior under certain loads or environmental conditions. Machines can be used to determine, for example, ultimate strength, yield strength, modulus of elasticity, elongations, hardness, toughness, and so on. These investigations can be done at room temperature, elevated temperatures or low temperatures.

4.4.1 Static Tensile Testing Machine and Specimens

Figure 4.1 shows a tensile testing system. The main mechanical properties mentioned above can be determined using this testing system. Also, for example, short-time creep behavior or the relaxation phenomenon of the considered material can be investigated.

Specimens for tensile testing, intended to test the tensile strength, 0.2 offset yield strength, the modulus of elasticity, the uniaxial creep behavior, and so on, are machined according to appropriate standards. A specimen's geometry depends on the purpose of

Table 4.1 Types of testing of materials.

Type of testing	Result/information
Mechanical testing	Mechanical properties (strength of material, hardness, toughness, ...)
Technological testing	Deformability at a certain temperature, weldability, ...
Nondestructive testing	Cracks and other failures
Chemical testing	Material chemical composition, corrosion resistance,...
Physical testing	Thermal and electrical properties, ...
Metallographic testing	Micro and macrostructure, grain size, ...

Table 4.2 Mechanical testing of materials.

Static				Dynamic			
Brief		Time-consuming		Brief		Time-consuming	
At room temp.	At low or elevated temp.	At room temp.	At low or elevated temp.	At room temp.	At low or elevated temp.	At room temp.	At low or elevated temp.

(a) (b)

Figure 4.1 Tensile testing system. (a) Testing machine with furnace; (b) testing machine with chamber. Photographs courtesy of the author.

the testing procedure. The ends of the specimen falling outside the gage length need to be threaded to match the holders of the testing machine.

Figure 4.2 shows a typical specimen used in tensile testing.

In Figure 4.2 (a), the geometry of the specimen is described by: A – a length of constant cross-section; B – a length that fits the grip/holder; C – the end diameter (threaded to

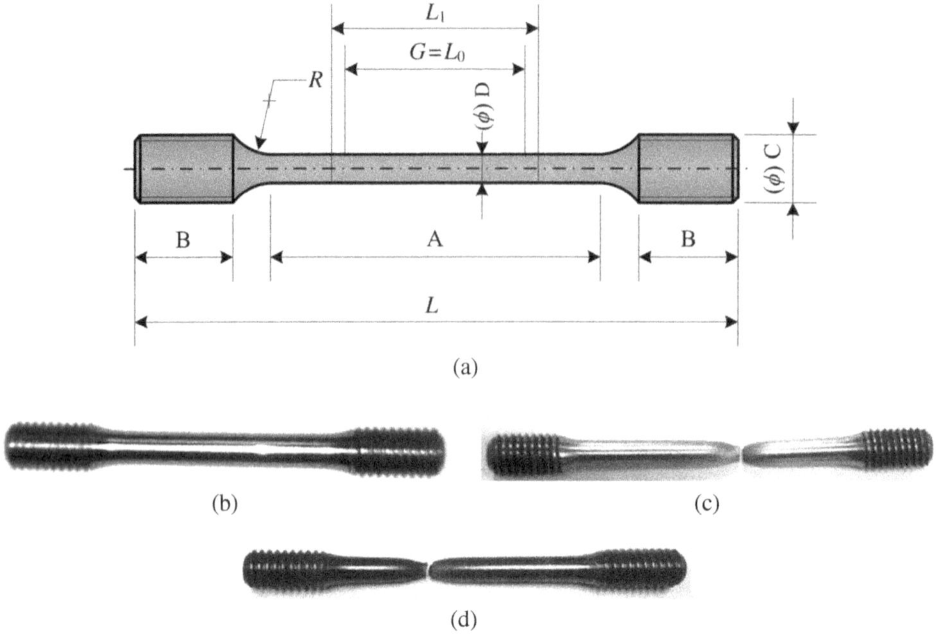

Figure 4.2 Specimen for tensile testing. (a) Specimen geometry; (b) unstrained specimen (room temperature); (c) specimen after fracture, previously subjected to uniaxial testing to determine stress–strain ($\sigma-\varepsilon$) diagram at room temperature; (d) specimen after fracture, previously subjected to uniaxial testing for determination of stress–strain ($\sigma-\varepsilon$) diagram at elevated temperature.Photographs (b), (c) and (d) courtesy of the author.

match the holders); D – the diameter of the specimen; $G = L_0$ – the gage length (measured length); L_1 – the length after elongation, otherwise in the process of testing it is a length that is greater than L_0; R – the radius of the fillet. In this case, using the given symbols, engineering strain is defined as: $\varepsilon = (L_1 - L_0)/L_0$.

4.4.2 Impact Testing Machine and Specimens

Figure 4.3(a) shows a Charpy impact machine while Figure 4.3(b) shows a 2 V-notch specimen, usually used in toughness determination by a Charpy impact machine.

Specimens containing a V-notch are usually used in testing the impact energy for "general" constructional steels, while specimens containing a U-notch are used to determine the impact energy for steels of high quality. In addition, it is possible to distinguish specimens with 3U-notch and 5U-notch (usually used in EN).

4.4.3 Hardness Testing Machine

A hardness testing machine is, unsurprisingly, used to determine the hardness of a material. Figure 4.4 shows a hardness testing machine. Several methods of hardness testing are available. Often, such a machine may be capable of performing several methods of testing the hardness of the material, such as the Brinell (HB), Vickers (HV) or Rockwell methods.

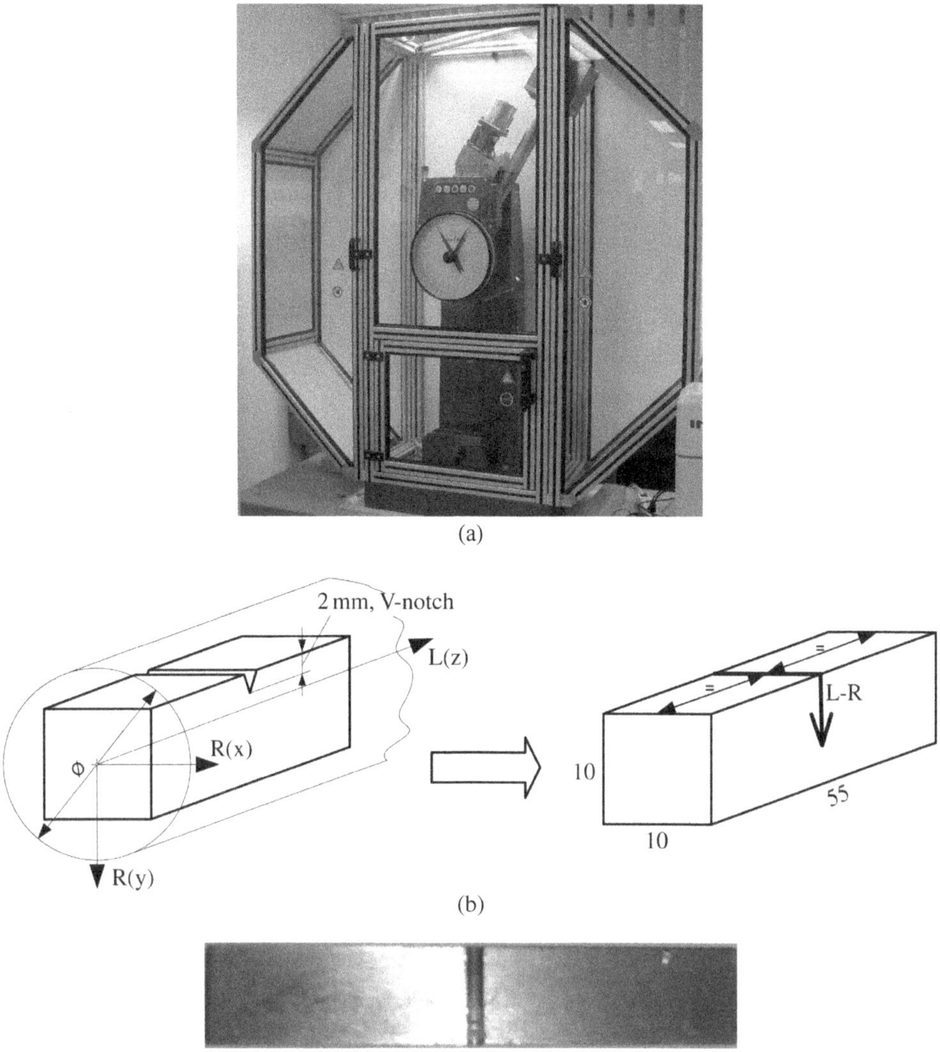

Figure 4.3 Charpy impact testing machine and specimen for toughness testing; L – longitudinal, R – radial. (a) Charpy machine; (b) geometry of a 2 V-notch specimen for toughness testing; (c) manufactured specimen. Photographs (a) and (c) courtesy of the author.

In general, hardness is the resistance of a material to penetration by another, much harder body (indenter). In other words, hardness is the resistance of a material to plastic deformation [4]. The body that is pressed by force into the tested material is a hardened steel ball in the Brinell method, a diamond four-sided pyramid with a top angle of 136° in the Vickers method, and a diamond cone with a top angle of 120° in the Rockwell method, wherein it is not the surface of the indentation that is measured (as when testing the hardness according to the Brinell and Vickers methods), but its depth.

The hardness test may serve as a cheap, nondestructive and simple method for assessment of some material properties like tensile strength, yield strength, and so on [10].

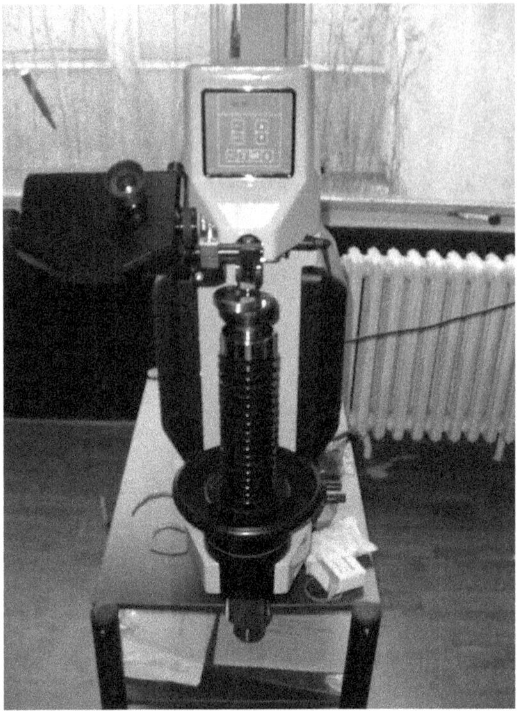

Figure 4.4 Hardness testing machine. Photograph courtesy of the author.

4.4.4 Fatigue Testing Machines

A fatigue testing machine is presented in Figure 4.5.

The machine shown in the figure is intended primarily for tensile testing, although it may be adapted and used to perform bending tests. Fatigue testing machines also exist for torsion testing, and so on.

4.5 Test Results

4.5.1 Static Tensile Test Results

4.5.1.1 Engineering Stress–Strain Diagram

Material properties that are typically obtained by such testing are: ultimate tensile strength, yield strength, modulus of elasticity, strain, and so on. Standards that define the geometry of the specimen and testing procedures, for example, ASTM standards, can be mentioned. The specimen geometry and the testing procedure at room temperature are defined by standard E 8 M-04 while standard E 21-03a defines measures to be used at elevated temperatures. The testing procedure for determination of the modulus of elasticity is presribed by the E111-04 standard.

In engineering practice, most of the materials used are metallic. Further, most of the materials used are subjected to stresses either during manufacturing or during the service life of the structure. Load/stress may cause translation, rotation and deformation.

Figure 4.5 Fatigue testing machine. Photograph courtesy of the author.

Deformation is of significant interest and can be expressed by a change of shape and/or volume. Properties of the material such as strength, stiffness and hardness demonstrate the ability of the material to withstand deformation. The strength of a material is associated with elongation and carrying, stiffness with deformation and hardness with the resistance to indentation. Each material deforms under applied load, and this deformation can be monitored by a stress–strain diagram. A stress–strain diagram/curve describes the relationship between stress and strain of the considered material during the process of deformation. The stress–strain relation for a material is usually obtained experimentally. Many types of experiment which differ in the means of loading the material are standardized. So, we can distinguish tension tests, compression tests, torsion tests, and so on.

After conducting a uniaxial test on a circular cylindrical bar (specimen), or on a flat specimen, a tensile stress–strain diagram can be obtained. The test is usually conducted in such a way that the load is monotonically increased slowly (static loading) from its initial (zero) level to its final value, which leads to fracture. Unstrained and elongated specimens are shown in Figure 4.6.

The stress–strain diagram (Figures 4.7 and 4.8) is considered to determine specific material properties relevant in design. First, the so-called *elastic limit* and *proportional limit* can be distinguished. It is said that a material has been strained within the elastic limit if the strain in the bar returns to zero as the load goes to zero (after unloading). This means that the material has remained perfectly elastic. If the strain in this case is linearly proportional to the point P (boundary), it is said that the material has been strained within the limit of linear elasticity and the stress at the limit of linear elasticity is referred

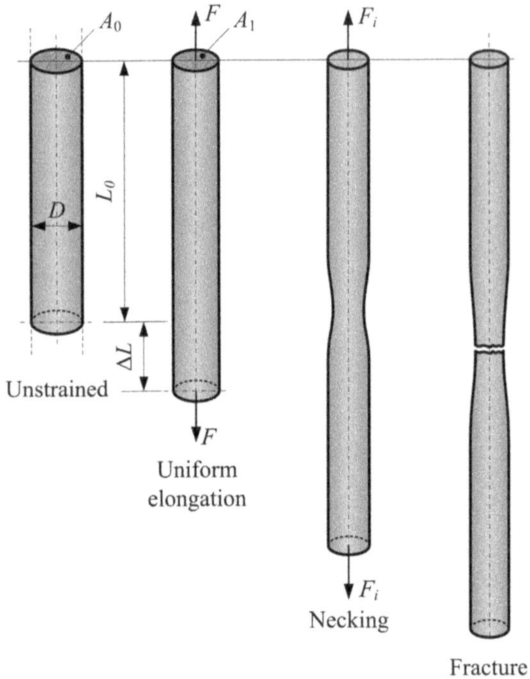

Figure 4.6 Unstrained and elongated specimens.

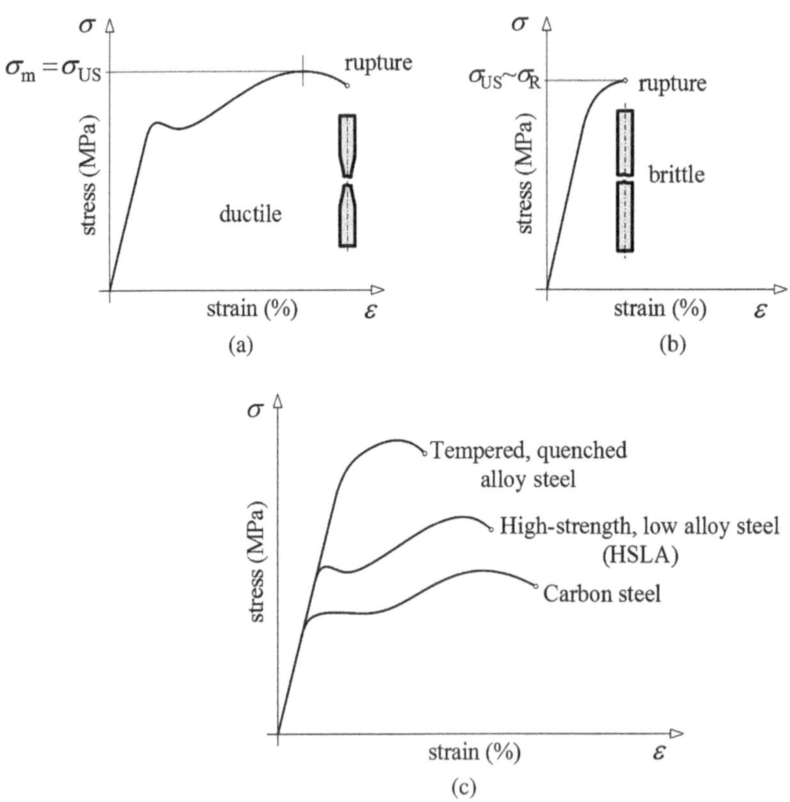

Figure 4.7 Engineering stress–strain diagrams for different materials. a–c) diagrams explained above.

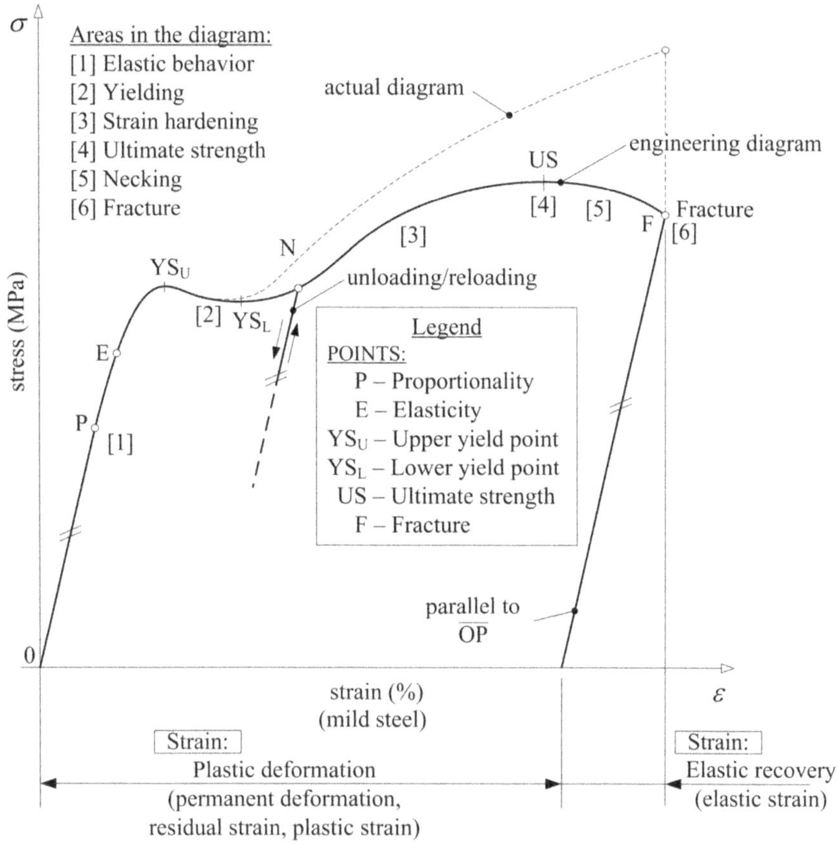

Figure 4.8 Engineering stress–strain diagram for structural steel.

to as the proportional limit. After this point, material behavior under increasing load may also fall within the elasticity area, but if it is no longer linearly proportional, it is only said that the behavior is within the elastic limit (E). Obviously, the elastic limit is larger than the proportional limit, but this has no importance from a practical point of view. The difference between these points – the transition from linear to nonlinear behavior – is very difficult to determine.

Also on the basis of the stress–strain diagram, some properties like ultimate tensile strength – σ_m (*UTS*)/(MPa); 0.2 offset yield strength – $\sigma_{0.2}$ (*YS*)/(MPa); modulus of elasticity – *E*/(GPa); elongation/strain – ε (%), and so on can be obtained.

Engineering stress–strain diagrams (as well as true/real diagrams) of various materials may be different, and different tensile tests conducted on the same material may also yield different results depending upon the temperature, strain rate, and so on. However, it is possible to distinguish two main groups of stress–strain diagrams based on two material groups: ductile and brittle materials. Ductile materials, for example structural steels and alloys of many other metals, are characterized by their ability to yield at room temperature. By contrast, brittle materials, for example cast iron, glass and so on, are characterized by the appearance of fracture without any noticeable elongation. In such cases, ultimate and breaking strength are practically the same. Strain in brittle materials is also much smaller than in ductile materials.

In Figure 4.7, engineering stress–strain diagrams for some materials are presented. A typical stress–strain diagram of structural steel (for example, low-carbon steel) is presented in Figure 4.8. It is clear that the strain after point P is no longer linearly elastic. If, at any point, for example at point N or at point F, in the stress–strain diagram the stress is removed (the material is unloaded), only a portion of the nonlinear (inelastic) strain will be recovered. The path under reloading will be parallel to the elastic modulus line (OP). The strain that remains in the material permanently is called the residual strain/plastic strain/plastic deformation/permanent set. A new loading process (reloading) would take the same line as the first reloading. However, now the proportion limit would be greater than under the original loading and would take the point of the former reloading process. This phenomenon is called *work hardening* (because of the effect of the increasing proportional limit). In general, the rise in the stress–strain curve after the yield point is called strain (or work) hardening; that is, strain hardening is the process whereby the stress that is required for further plastification is increased. On the basis of the stress–strain diagram, so-called *tensile toughness* can be determined and this is a measure of the ability of a material to absorb energy without fracture; it is represented by the area under the entire stress–strain curve.

In some ductile materials, the onset of yield is not characterized by a horizontal portion of the stress–strain curve. For such materials, the yield strength is defined by the offset method, as shown in Figure 4.9.

Properties of a considered material also depend on temperature. This dependence is shown in Figure 4.10.

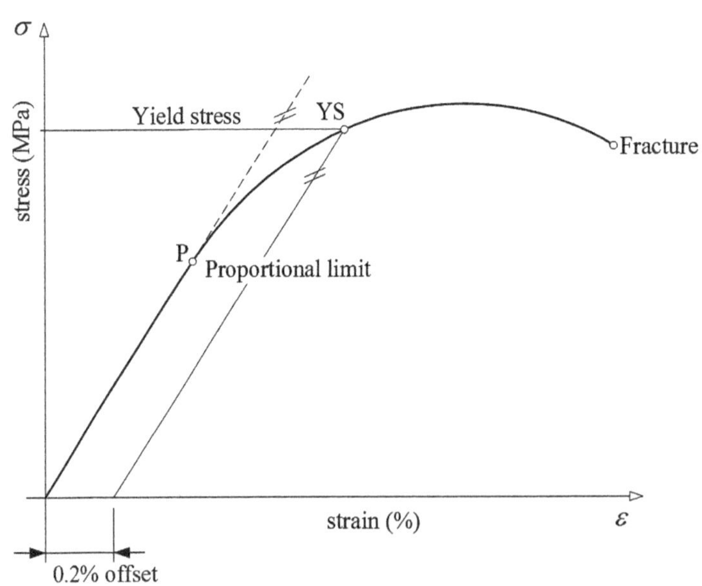

Figure 4.9 Yield strength determinations using the offset method.

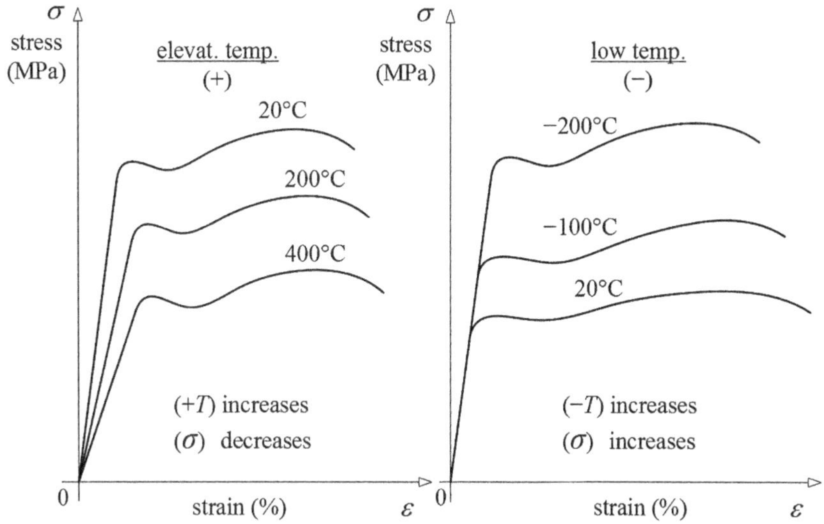

Figure 4.10 Temperature dependence of material properties.

In accordance with all that has been stated previously, during a tensile test (a uniaxial test), a specimen is exposed to loads up to the point of fracture. Measuring the force versus elongation, a stress–strain diagram ($\sigma-\varepsilon$) is plotted in such a way that:

$$\sigma = \frac{F}{A_0}\,(\text{MPa});\quad \varepsilon = \frac{\Delta L}{L_0} = \frac{L_1 - L_0}{L_0}\,(\%) \tag{4.1}$$

where: σ – stress; ε – strain; F – load; A_0 – initial cross-sectional area of the specimen; L_0 – gage length (length between two points of the specimen measured before loading); L_1 – length of the previously mentioned two points of the specimen measured after extension of the specimen, otherwise this is a value L_i that is greater than L_0.

Because the load increases during a tensile test, the cross-sectional area of the specimen decreases. Strain on the specimen also increases. In accordance with this, when a material reaches its yield point, the cross-section of the specimen starts to decrease. At this moment, the actual stress–strain diagram starts to diverge from the engineering stress–strain diagram. In engineering practice, for materials where upper and lower yield points can be recognized, since the upper yield point is transient, the lower yield point should be used as yield strength, that is, as a measure of yield of the considered material. Noticeable necking of the specimen is evident after a material reaches its ultimate tensile strength. Further, related to the temperature dependence of material properties, in general, it needs to be said that, when temperature increases, ultimate tensile strength and 0.2 offset yield strength decrease. The modulus of elasticity also decreases. In cases where the temperature decreases, the ultimate tensile strength and 0.2 offset yield strength increase. In such cases, the modulus of elasticity remains practically constant. The shape of the engineering stress–strain diagram depends on material composition, strain rate, temperature, heat treatment, prior history of plastic deformation, and so on. In engineering practice, the true (actual, realistic) stress–strain relation when plotted as a curve (the true stress–strain curve or true stress–strain diagram) is known as a *flow curve*, since it shows the level of stress required to cause the material to deform in the field of plastic flow, that is, to flow plastically. As mentioned earlier, factors affecting the stress–strain relation may be shown as:

$$\sigma = \sigma(\varepsilon, \dot{\varepsilon}, T, \ldots). \tag{4.2}$$

This means that stress depends on strain, strain rate, temperature, and so on. The most common relation used to describe the material flow is the power-law equation, known as the *Holloman–Ludwig equation* [11]:

$$\sigma = K(\varepsilon)_p^n, \tag{4.3}$$

where K is the strength coefficient (the stress intercept at $\varepsilon_p = 1$) and n is the strain hardening exponent (the line slope) which falls between zero ($n = 0$) for a perfectly plastic solid and $n = 1$ for an elastic solid. For metallic materials, the values are in the range 0.1–0.5.

Some other equations exist that can also describe the stress–strain relationship.

4.5.1.2 Creep Diagram/Curve

Uniaxial creep testing determines a material's creep behavior, usually shown as a creep curve, and this may be short-term creep behavior or long-term creep behavior. Of course, creep behavior and creep resistance may also be defined under bending or torsional loads. Strain versus time behavior during uniaxial creep under constant engineering stress is presented in Chapter 9 related to creep (see, for example, Figure 9.5). Creep testing is defined, for example, by standard E 139-00, while the geometry of the specimen is the same as that already described for uniaxial tensile testing.

Although the level of load encountered in service may be less than the critical one, due to high temperatures, creep may occur. Creep is therefore a subject of interest for structural designers involved in the design of structures that operate in elevated temperatures.

Creep may be defined as inelastic strain in the material that increases with time; that is, it is time-dependent behavior of a material during which deformation continues to increase while the stress (load) is kept constant [12, 13]. In the regions of small strains, the constant load and constant stress creep processes may be treated as the same.

The creep curve is shaped during the time of the creep process and it is a function of the material properties, exposure temperature and the applied load.

The creep process in a metal can be assumed to consist of three stages: primary, secondary and tertiary. During the creep process, the deformation may become too large for a component to perform its function. Although the temperature range at which creep deformation may occur differs in various materials, based on experimental investigations, it has been observed that, in metals, creep is appreciable at temperatures above 0.4 T_m, where T_m is the melting temperature.

4.5.1.3 Relaxation Diagram/Curve

Material relaxation behavior is characterized by a decrease in stress while the strain is kept constant. For a defined material, this phenomenon is presented schematically in Figure 9.14. These two phenomena (creep and relaxation) are considered further in Chapter 9. While our discussion will focus only on short-term creep behavior, for structures or machinery operating under conditions of elevated temperature, it is necessary to select materials carefully with these issues in mind.

4.5.2 Dynamic Test Results

4.5.2.1 Tensile, Flexural and Torsional Test Results

This section will describe some of the typically performed tests. Some details relating to fatigue testing will be mentioned in this section, although, in general, most of the information about this kind of testing features in Chapter 10 relating to fracture mechanics. This is because cracks can arise during fatigue testing and in such cases, crack propagation can be monitored and considered.

S–N *Curve (Wohler Curve)*

A material property called the endurance limit or fatigue limit (σ_f), measured in MPa, can be determined by fatigue testing. The endurance limit is not a constant and varies with the stress ratio, R. Other factors also affect the endurance limit, such as the degree of surface finish, heat treatment, stress concentration and a corrosive environment. Experimental investigation has shown that the endurance limit of a material tested under uniaxial loading is lower than the endurance limit tested under reverse bending, provided that the two loading cases have the same stress ratio, R.

It is well known that the failure of structural elements can occur at stresses below the static tensile strength of the material if the stress applied is time-varying. This phenomenon is called *fatigue*. Facts about the fracture of engineering elements caused by fatigue were already known as far back as 1843 (W. J. M. Rankine). In the period of time between 1856 and 1870, A. Wohler systematically examined the fatigue of materials. His early results were tabulated, but later results were presented graphically in the form of the so-called Wohler curve, or σ–N (S–N) curve (Stress–Number curve). Some of Wohler's curves were obtained using specimens loaded by flexural moments and rotating the constant angular velocity. For one cycle, load is changing according to sinusoidal law. Usually, tests are performed on the basis of tensile loading but using different laws of loads.

Unlike static loading, stress concentration has an important role in fatigue fracture and a strong influence on the endurance limit of materials. Fatigue fracture usually begins at the place of local stress concentration.

Testing of the fatigue of materials is performed on a fatigue testing machine, as shown in Figure 4.5. Some of the types of cycle often used in fatigue testing of materials are presented in Figure 10.2(b). Specimens made of a considered material are usually tested at frequencies in the range of 10–400 Hz. For some metallic materials, the limit on the number of cycles to be applied in the testing procedure is usually in the vicinity of 10^7 cycles. The curve showing the dependence of the number of cycles that a specimen subjected to a given load (stress) can withstand is the aforementioned σ–N (S–N) curve (or Wohler curve). Stress may be applied as tensile stress, tensile–compressive stress, flexural stress or torsional stress. The endurance limit or dynamic strength is the highest applied stress for the considered type of loading at which the specimen can withstand the prescribed number of cycles (usually 10^7) without failure. For example, in determining the fatigue life of a metallic part (specimen) subjected to constant amplitude (σ_a) of cyclic axial load at prescribed stress ratio R, the number of cycles (N_1) is recorded for maximum stress (σ_{max}), as shown in Figure 10.3. In the same figure, the endurance limit/fatigue limit is shown. For fatigue testing, for any stress level, usually 5–10 specimens are used, and for the σ–N (S–N) curve, a minimum of five stress levels need to be applied.

In an S–N curve, on the ordinate is plotted the stress to which the specimen is subjected, on a linear scale (usually), while on the abscissa is plotted the number of cycles which the specimen can withstand before failure, on a logarithmic scale.

4.5.2.2 Toughness Test Results

It is known that structural elements in engineering practice can be stressed statically and dynamically. Some details about static loads on engineering elements can be found in Chapter 6. Special attention needs to be paid to dynamic loads, since the material of the elements subjected to these kinds of load has limited resistance and many failures can arise. For example, fatigue load can affect fatigue failure and the fatigue limit is significantly lower than static strength. Details of fatigue loads as well as the endurance limit/fatigue limit can be found in Chapter 10.

In engineering practice, it is very interesting to consider the toughness of materials. Usually, in the process of material selection, some requirements relate to the toughness of the material, especially when low or high temperatures are involved. As a result, some indication of the toughness is required. Also, it needs to be said that the toughnees of the considered material may serve as a basis for the fracture toughness calculation.

Some relationships between toughness and fracture toughness are established through experiments. Experiments to determine the toughness are simpler than those to determine fracture toughness. Charpy's, Izod's, Brugger's and other methods are used for impact energy determination. Based on the impact energy and taking into consideration the area of the specimen at the notch, the toughness of the material can be calculated. Usually, for the material that will be used, a particular impact energy or toughness is prescribed. For example, for some structural materials, a minimum impact energy of 27 J at a considered temperature is prescribed (the test is performed on a specimen with a 2 V notch and the temperature of the test may be 20 °C, 0 °C, −10 °C).

4.5.2.3 Fracture Toughness Test Results

Fracture toughness, K_{Ic} (MPa$\sqrt{m}$), is a material property that is usually defined as a material's resistance to crack propagation in a tested material that behaves as linearly elastic prior to failure and if the plane strain condition is fulfilled. Generally speaking, the term "fracture" may be used in situations involving brittleness, ductility, creep, fatigue, and so on. However, K_{Ic} (usually defined as the critical value of the stress intensity factor) is denoted as plane strain fracture toughness. In conditions where the portion of the plastic field around the crack tip cannot be neglected, the K_{Ic} property is not an appropriate measure and under these conditions, the J integral or other measures may be used instead. As a result of fracture toughness testing, we may obtain crack resistance curves (*K–R*; *J–R*; *CTOD*, and so on) or a single value for the fracture toughness parameter, such as K, J, etc. In crack resistance curves, the toughness parameter, K, J, etc., is plotted against crack extension. More on fracture toughness and fracture toughness testing is presented in Chapter 10 on fracture mechanics.

References

1 Dowling, N. E. (2013) *Mechanical Behavior of Materials*, Pearson, New York.
2 Philpot, T. A. (2008) *Mechanics of Materials*, John Wiley & Sons, New Jersey.

3 Beer, F. P., Johnston, E. R. Jr, DeWolf, J. T. and Mazurek, D. F. (2012) *Mechanics of Materials*, McGraw-Hill, USA.

4 Smith, W. F. and Hashemi, J. (2010) *Foundations of Material Science and Engineering*, McGraw-Hill, New York.

5 Seed, G. M. (2000) *Strength of Materials*, Saxe-Coburg Publications, Edinburgh.

6 Ashby, M. F. (2011) *Materials Selection in Mechanical Design*, Butterworth Heinemann, New York.

7 ASM Handbook (1997) Vol. **20**, *Materials Selection and Design*, Dieter, G. E. (Volume Chair), ASM International, USA.

8 ASM Handbook (1996) Vol. **19**, *Fatigue and Fracture*, Lampman, S. R. (Technical Ed.), ASM International, USA.

9 Vitez, I. (2006) *Testing of Mechanical Properties of Metallic Materials* (in Croatian), Faculty of Mechanical Engineering Slavonski Brod, University J. J. Stosmayer, Osijek.

10 Pavlina, E. J. and Van Type, C. J. (2008) Correlation of Yield Strength and Tensile Strength with Hardness for Steels. *Journal of Materials Engineering and Performance*, **17**(6), 888–893.

11 Samuel, K. G. and Rodriguez, P. (2005) On power-law type relationships and the Ludwigson explanation for the stress–strain behaviour of AISI 316 stainless steel. *Journal of Materials Science*, **40**, 5727–5731.

12 Riley, W. F., Sturges, L. D. and Morris, D. H. (2007) *Mechanics of Materials*, John Wiley & Sons, USA.

13 Craig, R. R. (2011) *Mechanics of Materials*, 3rd edition, John Wiley & Sons, USA.

5

Material Behavior and Yield Criteria

5.1 Elastic and Inelastic Responses of a Solid

The form of the diagram from a uniaxial tension test depends on the material itself as well as the test conditions. It can be said that for most materials at low enough stress levels, the initial parts of stress–strain curves are quite similar. As the load is increased to a sufficient level, the stress–strain diagram becomes nonlinear.

In general, a material's response may be classified as elastic, plastic, viscoelastic or viscoplastic, as shown in Figure 5.1. However, in engineering practice, fracture can occur at various stress levels and a material's response up to fracture may be almost linear. If the unloading path coincides with the loading path, the material is said to be *elastic*; if this is not the case, the material behavior is said to be *inelastic*. When a material exhibits *plastic* behavior, it is evident that, after the load is released; it does not return to an unstrained state. If, after load removal, the material continues to change with time, the material is exhibiting viscoelastic or viscoplastic behavior. If, over time, the material returns to an unstrained state, we describe this behavior as *viscoelastic*; by contrast, if the material returns only in part, we describe this behavior as *viscoplastic*.

In engineering practice, fracture may occur under different conditions. So, fracture may occur under relatively small inelastic strains, then as a result of crack growth caused by fatigue at a stress level below the yield stress, then at high stress levels that cause microcrack propagation, and so on. If a material is subjected to a particularly dominant stress component, such as uniaxial stress, the uniaxial failure criterion may be appropriate, but in a multiaxial state of stress, a new approach must be applied.

Information about the nonlinear response of a material can be found in [1, 2]. We will now move on to discussing the yield conditions related to isotropic materials.

5.2 Yield Criteria

It is well known that an engineering structure or its components may be subjected to different states of stress, different states of strain, different environmental conditions, and so on. An engineering structure may contain some preexisting failures (for example, some failures in a material), or failures may occur during assembly of engineering components or during the service life of the structure (fatigue, fracture, yielding, creep,

Analysis of Engineering Structures and Material Behavior, First Edition. Josip Brnić.

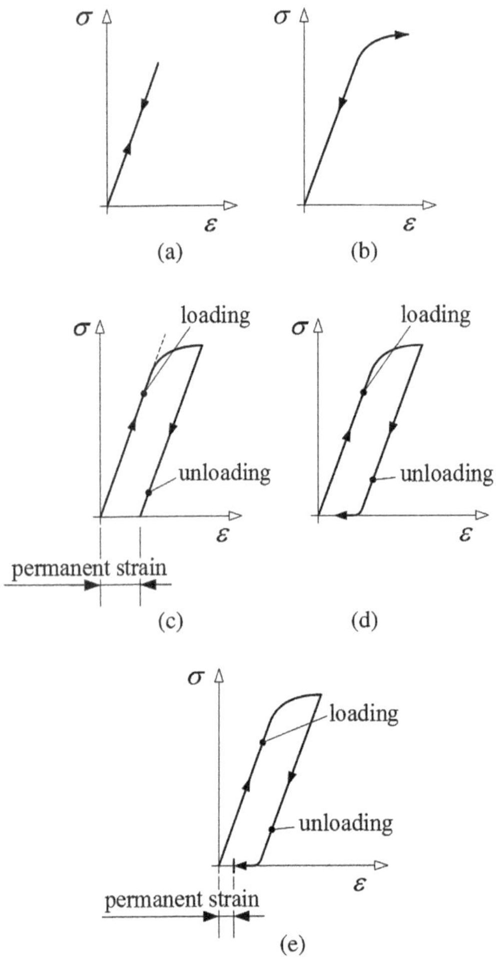

Figure 5.1 Types of material response. (a) Linear elastic; (b) nonlinear elastic; (c) plastic; (d) viscoelastic; (e) viscoplastic.

overload, and so on). It would be very useful to provide a concept of possible control for a structure that would ensure design, manufacture and operating conditions for each considered structure member in such a way that failure occurrence could be avoided. With this in mind, it would be advisible during the design process to choose materials whose properties have previously been determined experimentally and ensure that these properties meet the relevant requirements.

In engineering practice, with regard to the analysis of a structure, it is possible to discuss several issues: the state of stress, the state of strain, strength theories (failure theories, failure hypotheses), yield criteria, the suitability of the structure's maintenance, the regularity of use, and so on. The state of stress may be uniaxial, biaxial or multiaxial. Furthermore, failure hypotheses and yield criteria have a similar form.

In engineering practice, the appearance of yielding in a member subjected to stress is considered a critical state. Failure hypotheses are used as a powerful tool in the design of the member to avoid this critical state. Yield criteria represent statements that define conditions under which yielding will occur. In this section, yield criteria as a standpoint for yielding surface formation will be considered. Later, the effective stress will be taken as an indicator that yielding needs to be avoided (failure hypotheses) or as a factor that yielding will appear and a yield surface will form (yield criterion). The yield criterion may also be treated as a hypothesis concerned with the determination of the limit of elasticity under any combination of stresses, or sometimes it is defined as the criterion that determines the stress level at which plastic deformation begins.

The suitability of the proposed yield criterion must be checked by experimental investigation. The yield criterion can be expressed using the yield function "*f*", on the basis of which the effective (equivalent) stress may be defined. The yield function corresponds to the considered yield criterion. When a hypothesis of failure is under consideration, the setting is such that the effective stress must be less than the yield stress, or, for example, allowable stress, to avoid failure (yielding) in the member. Several specific (critical) factors can be listed that affect the appearance of yielding:

- Principal stress;
- A definite axial strain;
- Maximal shear stress;
- A specific amount of strain energy per unit volume related to a stressed member, etc.

In a member subjected to a uniaxial stress state, it can be assumed that yielding occurs when normal stress becomes equal to yield stress ($\sigma = YS$). When a multiaxial state of stress (or strain) is considered, the situation is complicated. It is too difficult to choose the appropriate yield criterion for yielding prediction [3–14]. The situation is still deteriorating, taking into consideration the various types of materials and load combinations. However, several yield criteria are usually mentioned. The main task is to develop the concept of yield appearance in a multiaxial state of stress; that is, to propose a yield criterion that defines the conditions under which yielding will occur. As such, the appearance of a so-called critical situation (yielding) in a multiaxial stress state is compared to the critical situation (yielding appearance) in a uniaxial stress state. The yield criterion states that yielding (the critical state) occurs (begins) when the critical factor (maximal shear stress, maximal principal stress, and so on) at a point equals the same critical factor at yield under uniaxial tension (compression). A yield criterion is usually written using the aforementioned, so-called yield function, $f(\sigma_{ij}, YS)$, where σ_{ij} describes the state of stress and YS means the yield strength under uniaxial tension. The yield criterion is satisfied when the yield function is equal to zero, i.e. $f(\sigma_{ij}, YS) = 0$. If it is less than zero, $f(\sigma_{ij}, YS) < 0$, the considered stress state is elastic, while if it is greater than zero, the condition is undefined. To develop the described yield criterion, the components of the multiaxial stress state are combined into a single quantity, the so-called *effective* or *equivalent stress*, σ_e, and this is compared with the yield strength under uniaxial tension. In this way, instead of the mentioned multiaxial stress components, using a single quantity (equivalent stress), a general expression for the yield criterion may be written as:

$$f(\sigma_{ij}, YS) = 0 \rightarrow f(\sigma_e, YS) = 0 \tag{5.1a}$$

A general form of the yielding criterion may be given by the following:

$$f(\sigma_{ij}) = C \tag{5.1b}$$

where C is a material parameter that needs to be determined experimentally.

Under this setting, as mentioned above, effective stress can be defined. In the case where failure should be avoided, this must be:

$$\sigma_e \le YS(YS \text{ or } \sigma_{all}) \tag{5.2a}$$

where σ_{all} means allowable stress.

On the other hand, failure is predicted to occur if:

$$\sigma_e \ge YS \tag{5.2b}$$

Here, yield stress is used as failure stress (that is, as the stress at which failure occurs). In engineering practice, two types of materials are usually of interest: ductile and brittle materials. Inelastic deformation in ductile materials (such as metals) is caused by plastic slip along chrystallographic planes, while brittle materials fail by brittle fracture rather than yielding. Experimental investigations indicate that shear plays the dominant role in plastic deformation of a material, which implies that volumetric changes do not affect yielding – in other words, plastic flow is independent of hydrostatic pressure. Based on this, it can be said that deviatoric quantities will define yielding conditions. Consequently, the yield function may be written in the form:

$$f(S_{ij}) = 0 \rightarrow f(J_2, J_3) = 0; \left(J_1 = \sum S_i = 0\right) \tag{5.3}$$

The state of stress at a considered point of a body subjected to a certain load is defined by the principal stresses $\sigma_1, \sigma_2, \sigma_3$ (or by stress components), and is shown by vector $\vec{r} = \vec{p}$, $(\overline{OT})$, as shown in Figure 5.2(a). Vector $\vec{p}$ represents the total stress at the point O. The stress space defined by the principal axes is known as the *Haigh–Westergaard stress space*. In this stress space, we set an octahedral plane whose normal $\vec{n}$ forms the same angles with the principal axes:

$$\begin{gathered} \alpha = \beta = \gamma \rightarrow \cos\alpha = \cos\beta = \cos\gamma; \left(l = m = r = 1/\sqrt{3}\right); \\ \vec{n}\left(1/\sqrt{3}, 1/\sqrt{3}, 1/\sqrt{3}\right) \end{gathered} \tag{a}$$

The intensity of the vector $\vec{r}$ is:

$$r = |\vec{r}| = p = \sqrt{\sigma_1^2 + \sigma_2^2 + \sigma_3^2} \tag{b}$$

The vector of total stress, $\vec{p}$, can be decomposed (resolved) into the vector of normal octahedral stress, $\vec{\sigma}^{oct}$, and shear octahedral stress, $\vec{\tau}^{oct}$. The vector of normal octahedral stress, $\vec{\sigma}^{oct}$, lies on the normal $\vec{n}$ of the octahedral plane. This normal is an axis known as the hydrostatic axis, for which $\sigma_1 = \sigma_2 = \sigma_3$ is valid. Referring back to Chapter 2, these stresses are defined by the following equations: σ^{oct} (2.39); τ^{oct} (2.40a).

In accordance with the experimental fact that plastic flow is independent of hydrostatic pressure, and that shear plays the dominant role in plastic deformation, a geometrical interpretation of the beginning of plastification at the considered point is given by the projection of the vector of total stress, $\vec{p}$, on the deviatoric plane (π plane). This way, the above-mentioned shear octahedral stress, $\vec{\tau}^{oct}$, is obtained. The deviatoric plane is a plane passing through the origin of the coordinate system and is perpendicular to the hydrostatic axis ($\sigma_1 = \sigma_2 = \sigma_3$). Since we are considering isotropic materials (the same properties under tension and compression are expected), the points that define the stress

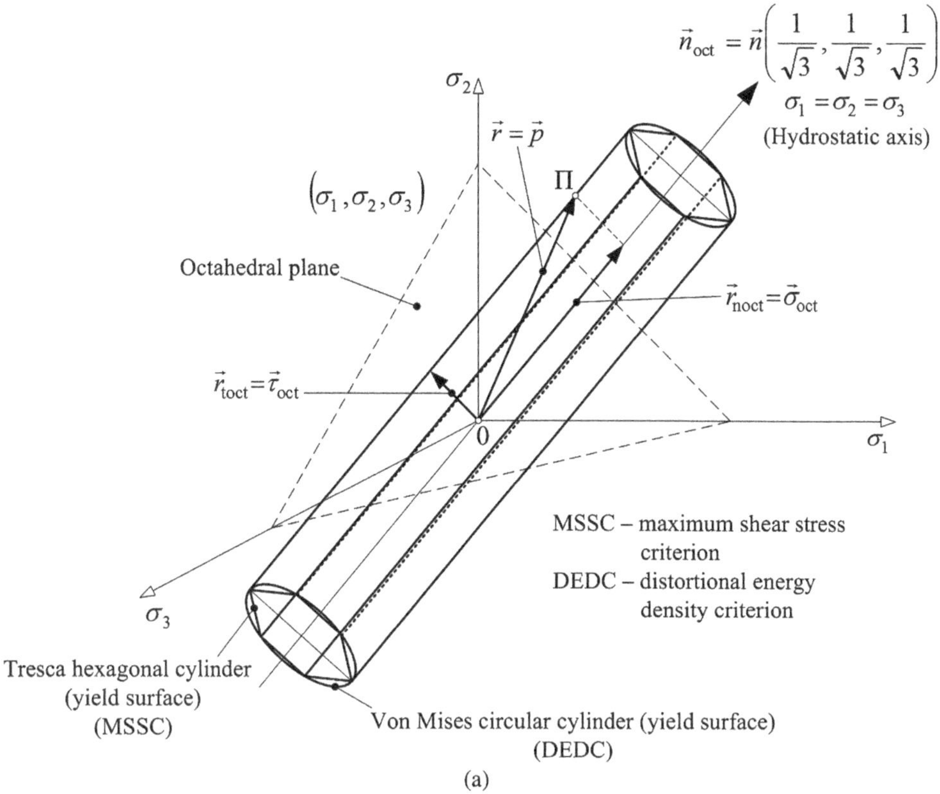

Figure 5.2 Graphical representations of the yield surface: Tresca and von Mises criteria. (a) Yield surface in the principal stress space (multiaxial stress state); (b) yield surface in the case of plane stress (biaxial stress state); (c) deviatoric plane.

at which yielding occurs are placed on the yielding curve formed by $\vec{\tau}^{oct}$ that lies on the deviatoric plane. This yielding curve, placed on the deviatoric plane, must be symmetric in relation to the principal deviatoric axes (S_1, $-S_1$; S_2, $-S_2$; S_3, $-S_3$), and these axes are the projections of the principal axes σ_1, σ_2, σ_3 onto the deviatoric plane, as shown in Figure 5.2(c).

In Equation (5.3), S_i are the principal values of the stress deviator, while J_i are the invariants of the stress deviator (see Chapter 2).

5.2.1 Ductile Materials

5.2.1.1 Maximum Shear Stress Criterion (Tresca Criterion)

The *maximum shear stress criterion* is generally attributed to Tresca. It states that *yielding occurs when the maximum shear stress at a point in a multiaxial stress state equals the maximum shear stress at yield under uniaxial tension (compression).* Observations show that in uniaxial testing of a flat specimen made of mild steel, yielding is accompanied by so-called Lüder's lines that appear on the surface of the tested

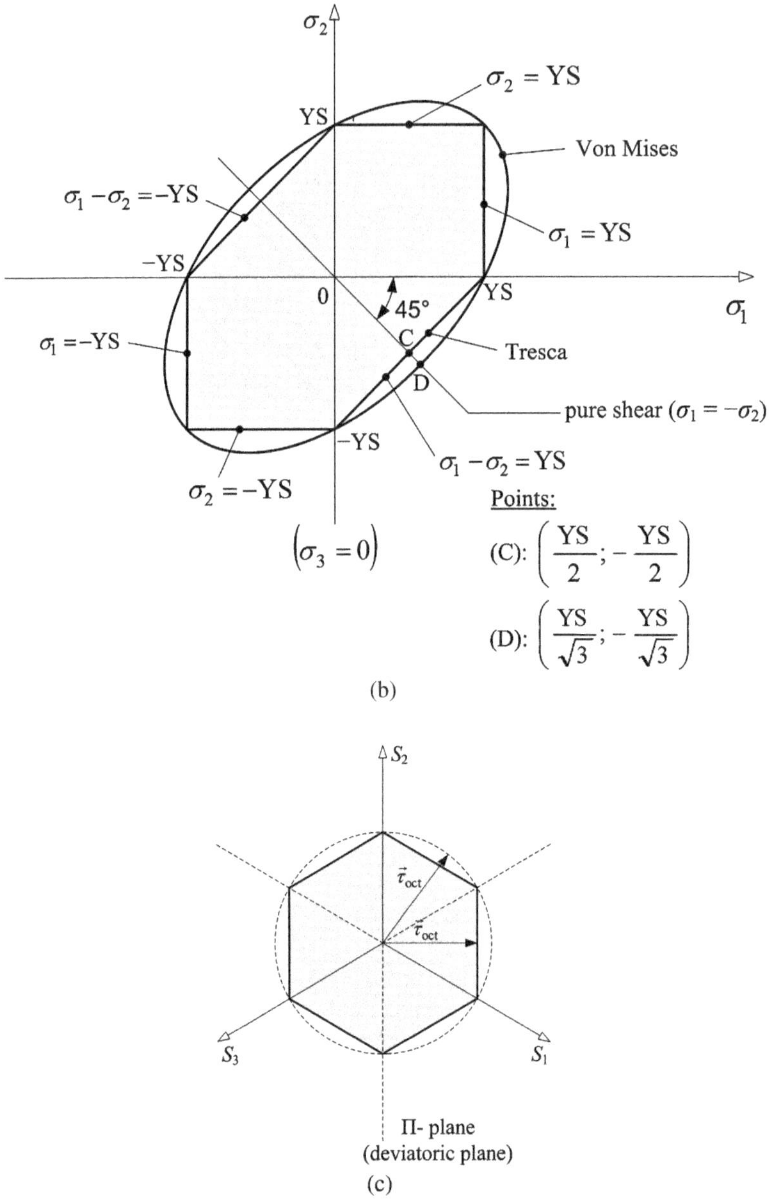

Figure 5.2 *(Continued)*

specimen and are positioned at 45° with respect to the longitudinal axis of the specimen [8]. The occurrence of these lines is caused by slipping that occurs along the planes of randomly ordered material grains. If it is assumed that slip is the failure mechanism associated with yielding of the material, then shear stress on the slip planes is the stress that best characterizes this failure. As is known from the Mohr's cycle that represents the uniaxial tension test (Figure 2.5), the maximum shear stress occurs at 45° with respect to the normal stress direction, just like the Lüder's lines. However, based on observations that plastic deformation is affected by distortion (shape change) and is independent of volumetric change, this leads to the assumption that shear stress affects failure (yielding) and not hydrostatic pressure (normal stress).

In the case of uniaxial tension (Figure 2.5), we have:

$$\sigma_1 = \sigma; \quad \sigma_2 = \sigma_3 = 0; \quad \tau_{max} = \sigma/2, \tag{5.4}$$

Yielding under uniaxial tension begins when $\sigma = YS$ (yielding stress); this means that shear stress associated with yielding (τ_y) under uniaxial tension is:

$$\tau_y = YS/2. \tag{5.5}$$

In this case, an effective stress is:

$$\sigma_e = \tau_{max} \tag{5.6}$$

where the maximum shear stress is:

$$\tau_{max} = \max(|\tau_I|, |\tau_{II}|, |\tau_{III}|) \tag{5.7}$$

The maximum values (the magnitudes of the extreme values) of shear stresses, τ_{max} in a multiaxial stress state are (from Equation (2.31)):

$$\tau_I = \pm\frac{1}{2}(\sigma_2 - \sigma_3), \quad \tau_{II} = \pm\frac{1}{2}(\sigma_3 - \sigma_1), \quad \tau_{III} = \pm\frac{1}{2}(\sigma_1 - \sigma_2), \tag{5.8}$$

We could also use the designations: $\tau_I = \tau_1$; $\tau_{II} = \tau_2$; $\tau_{III} = \tau_3$

In accordance with Equation (5.1a) and Equation (5.5), the yield function related to the maximum shear stress criterion can be written in the form:

$$f = \sigma_e - \frac{YS}{2} \tag{5.9}$$

Finally, comparing Equations (5.6)–(5.9), yielding in a multiaxial stress state can occur for any of the effective stresses:

$$\begin{aligned} &\sigma_2 - \sigma_3 = \pm YS \\ \sigma_e : \rightarrow\ &\sigma_1 - \sigma_3 = \pm YS \\ &\sigma_1 - \sigma_2 = \pm YS \end{aligned} \tag{5.10a}$$

where σ_1, σ_2 and σ_3 are the principal stresses.

In a case where failure needs to be avoided, the failure theorem (or failure hypothesis) known as the maximum shear stress theorem is written in the form:

$$\sigma_e \le YS\,(\text{or} \le \sigma_{all}) \tag{5.10b}$$

In a biaxial stress state ($\sigma_3 = 0$), we have:

$$\sigma_e = \max(|\sigma_2|, |\sigma_1|, |\sigma_1 - \sigma_2|) \le YS\,(\text{or} \le \sigma_{all}). \tag{5.10c}$$

In the case of a biaxial state of stress, the principal stresses are defined by Equation (2.21), while the magnitudes of the extreme values of shear stresses, $\tau_{max/min} = \tau_{I,II}$, are defined by Equation (2.23).

If we consider the case of bending with shear occurring, for example, in a vertical plane, then only normal stress exists along the axial axis, $\sigma_z = \sigma$, and shear stress $\tau_{zy} = \tau$, although this is also a case involving a biaxial state of stress. However, in this case, Equation (2.21) is transformed into the following equation:

$$\sigma_{1,2} = \frac{1}{2}\sigma \pm \frac{1}{2}\sqrt{\sigma^2 + 4\tau^2} \tag{5.11}$$

In accordance with Equation (5.11), in the case of the plane bending with shear, and based on $\sigma_e = |\sigma_1 - \sigma_2|$ given in Equation (5.10), when failure needs to be avoided, the maximum shear stress theorem can be written as:

$$\sigma_e = \sqrt{\sigma^2 + 4\tau^2} \leq YS\left(\text{or} \leq \sigma_{all}\right). \tag{5.12}$$

Yield surfaces related to the maximum shear stress criterion in the principal stress space (also known as the Haigh–Westergaard stress space), are shown in Figure 5.2. It is clear that for points C and D, $\sigma_1 = -\sigma_2 = \sigma$, which corresponds to pure shear, see Figure 5.2(b).

The maximum shear stress criterion (Tresca criterion) may be treated as an appropriate choice for most ductile materials, although experiments performed for a state of pure shear (torsion) on some ductile metals have found that the yield stress is higher (approximately 15% higher) than that predicted by the Tresca criterion.

5.2.1.2 Distortional Energy Density Criterion (von Mises Criterion)

The *distortional energy density criterion* is generally attributed to von Mises. This criterion is also known as the Huber–Mises–Henky criterion (HMH). Sometimes it is also referred to as the maximum octahedral shear stress criterion. It states that *yielding occurs when the distortional strain energy density at a point in a multiaxial stress state equals the distortional strain energy density at yield under uniaxial tension* (*compression*). This kind of energy is associated with a change in the shape of a body.

The strain energy density can be written as:

$$U_0 = \frac{1}{2E}\left[\sigma_1{}^2 + \sigma_2{}^2 + \sigma_3{}^2 - 2\nu\left(\sigma_1\sigma_2 + \sigma_2\sigma_3 + \sigma_3\sigma_1\right)\right], \tag{5.13}$$

and the distortional energy density is:

$$U_{oD} = \frac{1+\nu}{6E}\left[\left(\sigma_1 - \sigma_2\right)^2 + \left(\sigma_2 - \sigma_3\right)^2 + \left(\sigma_3 - \sigma_1\right)^2\right]. \tag{5.14a}$$

In a uniaxial stress state, only $\sigma_1 = \sigma$ exists. In accordance with this, the distortional energy density in a uniaxial stress state can be written in the form:

$$U_{oD,1} = \frac{1+\nu}{3E}\cdot\sigma^2 \rightarrow U_{oD,1} = \frac{1+\nu}{3E}YS^2 \tag{5.14b}$$

since yielding in this case is associated with YS. Based on the equalization of Equation (5.14a) and Equation (5.14b) as a starting point for a criterion for a multiaxial stress state, it follows that:

$$\begin{gathered} U_{oD} = U_{oD,1} \\ \left(\sigma_1 - \sigma_2\right)^2 + \left(\sigma_2 - \sigma_3\right)^2 + \left(\sigma_3 - \sigma_1\right)^2 = 2YS^2, \end{gathered} \tag{5.14c}$$

from which, the effective stress is:

$$\sigma_e = \sqrt{\frac{1}{2}\left(\sigma_1 - \sigma_2\right)^2 + \left(\sigma_2 - \sigma_3\right)^2 + \left(\sigma_3 - \sigma_1\right)^2} \tag{5.15}$$

The yield function for the distorsional energy density criterion (von Mises yield criterion) may be written as:

$$f = \sigma_e^2 - YS^2 \tag{5.16}$$

Equation (5.16) can also be determined in another way. Starting from $U_{oD} = U_{oD,1}$, and dividing both sides by $1 + \nu/E$, we obtain:

$$\frac{1}{6}\left[(\sigma_1 - \sigma_2)^2 + (\sigma_2 - \sigma_3)^2 + (\sigma_3 - \sigma_1)^2\right] = \frac{1}{3}YS^2 \tag{a}$$

Both sides of this equation can be multiplied by -1. In this case, the left-hand side represents the second invariants of the deviatoric stress tensor J_2 from Equation (2.50b). Now, the yield function may be written in the form:

$$f = -J_2 - \frac{1}{3}YS^2 \tag{5.17a}$$

In addition, comparing Equation (2.40a), which represents octahedral shear stress, and Equation (2.50b), which represents the second invariant of the deviatoric stress tensor, it is possible to write:

$$J_2 = -\frac{3}{2}\tau_{oct}^2, \tag{5.17b}$$

and, now, the yield function from Equation (5.16) can be written in the form:

$$f = \tau_{oct} - \frac{\sqrt{2}}{3}YS \tag{5.18}$$

The distorsional energy density (von Mises) criterion is also known as the maximum octahedral shear stress criterion. When $f = 0$, based on Equation (5.18), yielding occurs at a point where octahedral shear stress reaches a value of $0.471\ YS = \sqrt{2}/3$.

In a case involving a biaxial stress state, the effective stress is:

$$\sigma_e = \sqrt{\sigma_1^2 + \sigma_2^2 - \sigma_1\sigma_2} \tag{5.19}$$

When a biaxial stress state is considered under the conditions: $\sigma_1 = -\sigma_2 = \sigma$, the stress state represents pure shear. Based on Equations (5.16) and (5.19), the yield function for the von Mises yield criterion for the above-mentioned conditions of pure shear is:

$$f = \sigma_e^2 - YS^2 = 3\sigma^2 - YS^2. \tag{5.20}$$

Based on this, under pure shear, yielding will occur at:

$$\sigma = \frac{YS}{\sqrt{3}} \rightarrow \tau_y = \frac{YS}{\sqrt{3}} = 0.577\ YS \tag{5.21}$$

For this stress state, $\tau_{max} = |\sigma_1 - \sigma_2|/2 = \sigma$, and at yield $\tau_{max} = \tau_y$.

In a case of plane bending with shear when failure needs to be avoided, the distorsional energy density theorem (von Mises theorem), after Equation (5.11) has been inserted into Equation (5.19), is of the following form:

$$\sigma_e = \sqrt{\sigma^2 + 3\tau^2} \le YS. \tag{5.22}$$

In a case of pure shear ($\sigma_1 = -\sigma_2 = \tau$), based on Equation (5.19), it follows that:

$$\sigma_e = \tau\sqrt{3} \le YS \, (\text{or} \le \sigma_{all}) \tag{5.23}$$

and then

$$\tau_{all} = \frac{\sigma_{all}}{\sqrt{3}} = 0.577 \cdot \sigma_{all} \tag{5.24}$$

A graphical representation of this criterion, that is, the yield surface in the principal stress space, is shown in Figure 5.2.

The distortional energy density criterion (von Mises) is also an appropriate choice for ductile materials, and sometimes is treated as being more accurate than the Tresca criterion in predicting yield when pure shear is considered.

5.2.2 Brittle Materials

Ultimate tensile strength (or fracture strength) may be considered an appropriate limiting stress for brittle materials. Brittle materials tend to fail suddenly, practically without any evidence of yielding unlike ductile materials. Also, tensile strength and compressive strength in brittle materials are often different.

5.2.2.1 Maximum Normal Stress Criterion

This criterion is often called the *maximum principal stress criterion* and is usually attributed to Rankine. It states that *yielding occurs when the maximum principal stress at a point in a multiaxial stress state equals the yield stress under uniaxial tension (compression)*. When a uniaxial, biaxial or multiaxial stress state is being considered, it is clear that only the maximum principal stress is taken into account; that is, this criterion ignores the effect of other principal stresses although they also act at the point. As stated above, brittle materials fail by brittle fracture rather than by yielding, so this criterion, using maximum principal stress, may predict tension fracture.

In accordance with Equation (5.1a), the yield function can be written as:

$$f = \sigma_e - YS, \tag{5.25}$$

where the effective stress is:

$$\sigma_e = \sigma_{max} = \max(|\sigma_1|, |\sigma_2|, |\sigma_3|)$$

Based on $f = 0$, the maximum principal stress yield criterion is defined as:

$$\sigma_1 = \pm YS; \; \sigma_2 = \pm YS; \; \sigma_3 = \pm YS. \tag{5.26}$$

A graphical representation of the yield surface for the maximum principal stress criterion is shown in Figure 5.3.

5.2.2.2 Maximum Normal Strain Criterion

The *maximum normal strain criterion* or maximum principal strain criteiron is often called the Saint Venant criterion. It states that *yielding occurs when the maximum principal strain at a point in a multiaxial stress state equals the maximum principal strain at yield under uniaxial tension.*

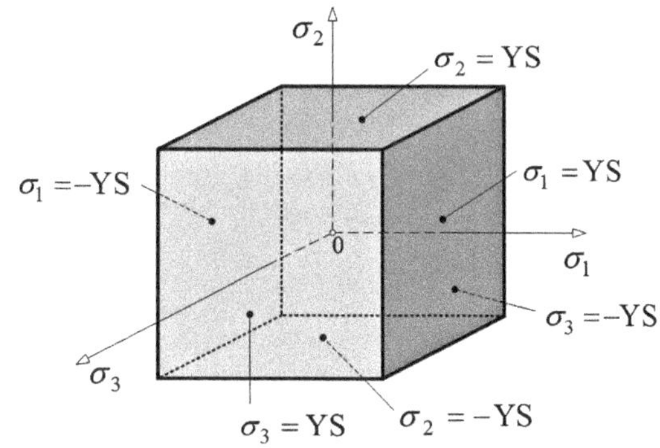

Figure 5.3 Maximum principal/normal stress criterion: yield surface.

In uniaxial tension, stress is defined by σ_1 and yielding begins when $\sigma_1 = YS$. The strain corresponding to this state is: $\varepsilon_1 = YS/E$.

In a multiaxial stress state, the maximum strain (assuming that the principal strains are unordered) is:

$$\varepsilon_{\max} = \max(|\varepsilon_1|, |\varepsilon_2|, |\varepsilon_3|) = YS/E \tag{5.27}$$

$$\begin{aligned} \varepsilon_1 &= \frac{1}{E}(\sigma_1 - \nu(\sigma_2 + \sigma_3)) \\ \varepsilon_2 &= \frac{1}{E}(\sigma_2 - \nu(\sigma_3 + \sigma_1)) \\ \varepsilon_3 &= \frac{1}{E}(\sigma_3 - \nu(\sigma_1 + \sigma_2)) \end{aligned} \tag{5.28}$$

The yield function now may be written as (from Equation (5.1a)):

$$f = \sigma_e - YS$$

$$\begin{aligned} f_1 &= |(\sigma_1 - \nu(\sigma_2 + \sigma_3))| - YS \rightarrow (\sigma_1 - \nu(\sigma_2 + \sigma_3)) = \pm YS \\ f_2 &= |(\sigma_2 - \nu(\sigma_3 + \sigma_1))| - YS \rightarrow (\sigma_2 - \nu(\sigma_3 + \sigma_1)) = \pm YS \\ f_3 &= |(\sigma_3 - \nu(\sigma_1 + \sigma_2))| - YS \rightarrow (\sigma_3 - \nu(\sigma_1 + \sigma_2)) = \pm YS \end{aligned} \tag{5.29}$$

On the basis of Equation (5.29), the effective stress may be defined as:

$$\sigma_e = \max_{i \neq j \neq k} |\sigma_i - \nu(\sigma_j + \sigma_k)| \tag{5.30}$$

For a biaxial stress state, we have:

$$\sigma_e = \max_{i \neq j} |\sigma_i - \nu\sigma_j| \tag{5.31}$$

or, for example, in explicit form:

$$\begin{aligned} \sigma_e &= |\sigma_1 - \nu\sigma_2| \\ \sigma_e &= |\sigma_2 - \nu\sigma_1| \end{aligned} \tag{5.32}$$

If a plane bending with shear is being considered, then, using Equation (2.21), Equation (5.11) and Equation (5.32), the following formula for effective stress can be written:

$$\sigma_e = \frac{1-\nu}{2}\sigma \pm \frac{1+\nu}{2}\sqrt{\sigma^2 + 4\tau^2} \tag{5.33}$$

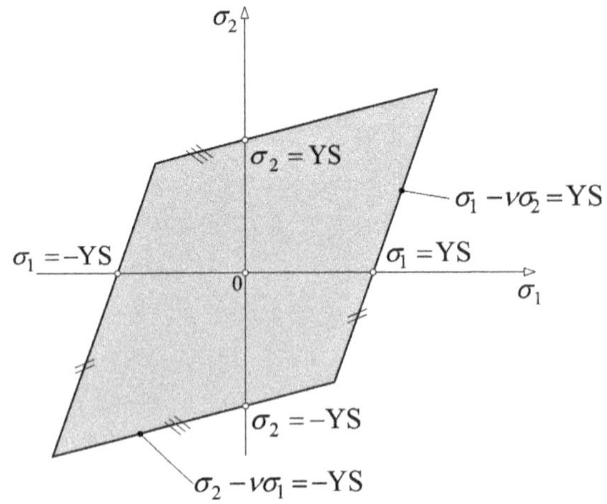

Figure 5.4 Maximum principal strain criterion: yield surface (biaxial stress state).

In cases where the occurrence of yielding needs to be avoided, then, as stated previously, the effective stress needs to be less than *YS* or some allowable stress (σ_{all}). A graphical representation of the yield surface for the maximum principal strain criterion is shown in Figure 5.4.

References

1 Boresi, A. P., Schmidt, R. J. and Sidebottom, O. M. (1993) *Advanced Mechanics of Materials*, John Wiley & Sons, New York.
2 Jirasek, M. and Bažant, Z. P. (2002) *Inelastic Analysis of Structures*, John Wiley & Sons, Chichester.
3 Brnic, J. and Turkalj, G. (2004) *Strength of Materials I* (in Croatian), Faculty of Engineering, University of Rijeka, Rijeka.
4 Alfirević, I. (1995) *Strength of Materials I* (in Croatian), ITP "Tehnička knjiga", D.D., Zagreb.
5 Alfirević, I. (2003) *Introduction to Tensors and Continuum Mechanics* (in Croatian), Golden Marketing, Zagreb.
6 Seed, G. M. (2000) *Strength of Materials*, Saxe-Coburg Publications, Edinburgh.
7 Boresi, A. and Schmidt, R. J. (2003) *Advanced Mechanics of Materials*, John Wiley & Sons, USA.
8 Philpot, T. A. (2008) *Mechanics of Materials*, John Wiley & Sons, New Jersey.
9 Khan, A. S. and Huang, S. (1995) *Continuum Theory of Plasticity*, John Wiley & Sons, New York.
10 Liu, A. F. (2005) *Mechanics and Mechanisms of Fracture*, ASM International, USA.
11 Prelog, E. (1978) *Elasto and Plastomechanics* (in Slovenian), Faculty of Mechanical Engineering, University of Ljubljana, Ljubljana.
12 Bowman, K. (2004) *Mechanical Behavior of Materials*, John Wiley & Sons, New York.
13 Craig, R. R. (2011) *Mechanics of Materials*, 3rd edition, John Wiley & Sons, USA.
14 Riley, W. F., Sturges, L. D. and Morris, D. H. (2007) *Mechanics of Materials*, John Wiley & Sons, USA.

6

Loads Imposed on Engineering Elements

As we know, structural elements subjected to forces, bending moments, torsional moments, pressure and temperature can be deformed. Force, moment of force and couple are known as the basic static elements. When an element is subjected to axial force, it is usually said that it is axially loaded and its deformation may be described by elongation or shortening. Further, if an element is subjected to a bending moment, it is said that this element is subjected to bending/flexure; if an element is subjected to a torsional moment, it is said that this element is twisted. If bending is under consideration, then bending is the type of loading and deflection is its mode of deformation. A particular type of loading (axial loading, bending, torsion, and so on) is described by a particular type (mode) of deformation.

6.1 Axial Loading

During their service life, engineering elements (or structures) can be subjected to different types of load and different environmental conditions. As stated previously, at the time of material selection, it is important for a designer to be familiar with the material properties and material behavior that correspond to both the loads and environmental conditions under consideration.

One of the simplest types of load to which an element can be subjected is axial loading, as shown in Figure 6.1. Assume that the element in Figure 6.1(a), with length L and loaded by F, is a cylindrical element of constant cross-sectional area, A. The load (external force), F, applied to the rod causes an axial deformation of ΔL.

By plotting the magnitude, F, of the load versus the deformation, Δl (or ΔL) in any moment, it is possible to obtain the so-called engineering stress–strain diagram, as shown previously in Figure 4.8. This diagram is the basis by which to obtain many material properties, as was mentioned in Sections 4.4.1 and 4.5.1.1. The ratio:

$$\varepsilon = \varepsilon_z = \frac{\Delta L}{L} = \frac{L_1 - L}{L} (\%) \tag{6.1}$$

is called strain (*engineering strain*), see Figure 6.1, and it is a dimensionless property [1]. Generally speaking, in Equation (6.1), the value L is the length of the considered element

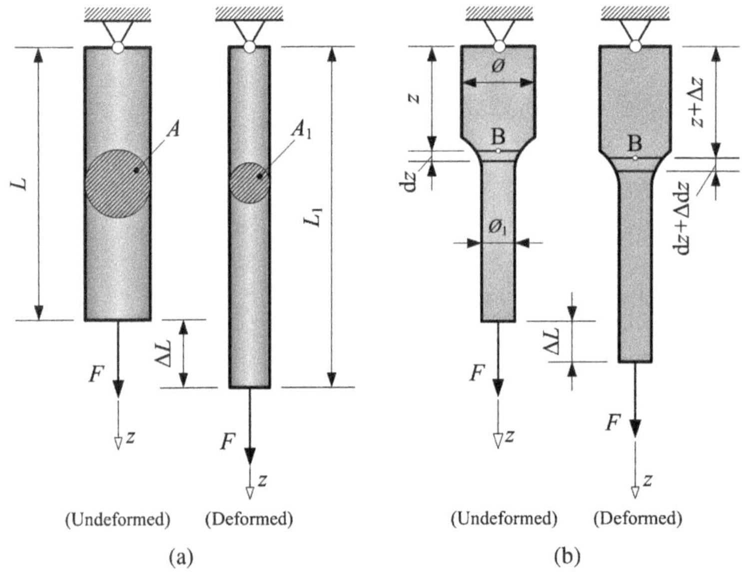

Figure 6.1 Axial loading. (a) Rod with constant cross-section; (b) rod with different cross-sections.

(specimen) in its undeformed (unstrained) state, while L_1 is the length of the element in its deformed (strained) state. Sometimes, instead of labels L and L_1, the labels l and l_1 are used. The extension of the specimen is usually measured by extensometer, and in such cases, only the extension of the portion of a specimen is monitored, as can be seen from Figure 4.2. The length of the portion of a specimen measured for strain is called the *gage length.*

In cases where an element is of variable cross-sectional area, as shown in Figure 6.1(b), the strain of the element dz may be calculated as:

$$\varepsilon_z = \lim_{dz \to 0} \frac{\Delta dz}{dz} \tag{6.2}$$

An axially loaded element must be sized correctly in order to bear the load imposed on it. The task is to determine the dimensions of the cross-section of the loaded element in accordance, for example, with some of the theories of strength. To do this, it is necessary to know the distribution of the stress over the cross-section of the element (beam). Consider the internal forces at the cross-section, as shown in Figure 6.2.

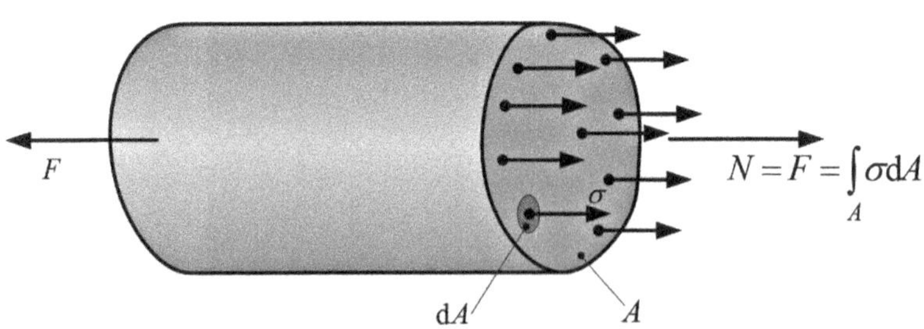

Figure 6.2 Internal forces at the surface of the cross-section.

The equilibrium equation is [1]:

$$\sum F_z = 0 \quad \rightarrow \quad N - F = 0$$
$$\int_A \sigma_z \mathrm{d}A - F = 0 \tag{6.3}$$
$$F = \int_A \sigma_z \mathrm{d}A.$$

6.1.1 Normal Stress

Research shows that the fibers of an axially loaded element are equally extended, and this fact indicates that each fiber is loaded by the same force. This means that the internal forces are equally distributed over the cross-section. Based on this, we can write:

$$\sigma_z = \frac{N}{A} = \pm\frac{F}{A}, \quad N = \pm F, \tag{6.4}$$

where σ_z is the normal stress, N is the resulting axial force acting perpendicularly to the cross-section, and A is the cross-sectional area (+ tension, – compression). Contraction of the cross-sectional area (a reduction in area) and *volumetric deformation* of the axially loaded element are defined as:

$$\psi = \frac{A - A_1}{A} \approx 2\nu\varepsilon_z, \quad \varepsilon_\mathrm{v} = \frac{\Delta V}{V} \approx \varepsilon(1 - 2\nu) \tag{6.5}$$

where the cross-sectional area after the deformation process may be calculated as:

$$A_1 \approx A(1 - 2\nu\varepsilon_z). \tag{6.6}$$

In a case for which we are considering the effect of temperature on the element, we can write:

$$\varepsilon_\mathrm{T} = \frac{\Delta L}{L} = \alpha_\mathrm{T} \cdot \Delta T, \tag{6.7}$$

where α (or sometimes designated as α_T) is the coefficient of thermal expansion (dilatation), (1/°C; 1/K) and ΔT is a change in temperature (increase or decrease).

In accordance with the tensile test, taking into consideration the relationship between stress and strain within the proportional limit, the following equation, known as Hooke's Law, is valid:

$$\sigma = E\varepsilon. \tag{6.8}$$

In Equation (6.8), E is the modulus of elasticity (MPa) and ε is the strain. When, instead of the tensile test, a shear test is considered, Hooke's Law takes the following form:

$$\tau = G\gamma, \tag{6.9}$$

where the so-called shear modulus (the modulus of rigidity) is defined as [2]:

$$G = \frac{E}{2(1 + \nu)} \tag{6.10}$$

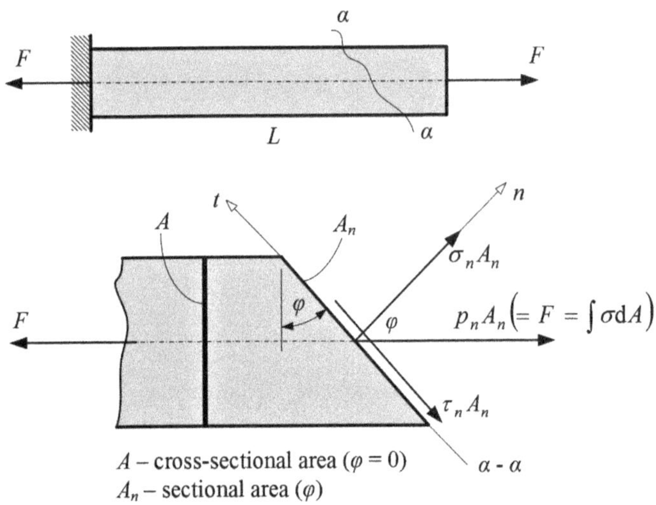

Figure 6.3 Uniaxial stress state.

In accordance with Figure 6.3, it is possible to consider a uniaxial stress state as a result of axial loading.

6.1.2 The Principal Stress

The stress state in the vicinity of the considered point in a tensile-loaded member was also discussed in Section 2.4.1. Let us remember some of the details and consider again the maximum normal stress, since it will be used to determine the size of the cross-sectional area of the considered engineering element.

The equilibrium equation, based on Figure 6.3, is:

$$\sum F_n = 0 \quad \rightarrow \quad \sigma_n \mathrm{d}A_n - \sigma_z \mathrm{d}A_z \cos\varphi = 0$$
$$\sigma_n \mathrm{d}A_n - \sigma_z \mathrm{d}A_n \cos^2\varphi = 0$$

and the change in the normal stress on the inclined section is (from Equation (2.11)):

$$\sigma_n = \sigma_z \cos^2\varphi,$$

or:

$$\sigma_n = \frac{1}{2}\sigma_z(1 + \cos 2\varphi). \tag{6.11}$$

Applying the rule of the extreme to Equation (6.11), as described in Section 2.4.1, the angles of the principal directions, as well as the magnitudes of the principal stresses, can be determined as follows. The first derivative of Equation (6.11) is:

$$\frac{\mathrm{d}\sigma_n}{\mathrm{d}\varphi} = -\sigma_n \sin 2\varphi = 0 \rightarrow \varphi = 0, \frac{\pi}{2}$$

Further, the second derivative of Equation (6.11) is:

$$\frac{d^2\sigma_n}{d\varphi^2} = -2\sigma_z \cos 2\varphi,$$

and for the angles $\varphi = 0$ and $\varphi = \frac{\pi}{2}$, it is:

$$\varphi = 0 \rightarrow \frac{d^2\sigma_n}{d\varphi^2} = -2\sigma_z < 0 \ldots\ldots. \text{ max}$$

$$\varphi = \frac{\pi}{2} \rightarrow \frac{d^2\sigma_n}{d\varphi^2} = 2\sigma_z \ldots > 0 \ldots\ldots. \text{ min}$$

Finally, in accordance with Equation (6.11), the results are:

$$\begin{aligned} &\varphi = 0 \quad \rightarrow \quad \sigma_n = \sigma_z = \sigma_{max} = \sigma_1 \\ &\varphi = \frac{\pi}{2} \quad \rightarrow \quad \sigma_n = 0 = \sigma_{min} = \sigma_2. \end{aligned} \tag{6.12}$$

However, an investigation to find the "principal stresses" was carried out, and it was found that in the case of a bar subjected to a tension load, only one principal stress exists. This means that this case of loading represents the uniaxial stress state ($\sigma_1 \neq 0$, $\sigma_2 = 0$).

The same procedure can be applied in the search for directions of extreme values of the shear stresses and their magnitudes. This is not presented here, since this matter does not belong to the principal stress problem, but it is given in Section 2.4.1. Finally, we have:

$$\begin{aligned} &\varphi = 0 \quad \rightarrow \quad \sigma_n = \sigma_{max} = \sigma_1 \quad \rightarrow \quad \tau_n = 0 \\ &\varphi = \frac{\pi}{2} \quad \rightarrow \quad \sigma_n = \sigma_{min} = \sigma_2 = 0 \quad \rightarrow \quad \tau_n = 0. \\ &\varphi = \mp\frac{\pi}{4} \quad \rightarrow \quad \tau_n = \tau_{max,\ min} \quad \rightarrow \quad \sigma_n = \frac{1}{2}\sigma_z \end{aligned} \tag{6.13}$$

The proposed formula for sizing an axially loaded element is:

$$\sigma_{max} = \left(\frac{F}{A}\right)_{max} \leq \sigma_{all}; \ \Delta L_{max} = \left(\frac{F \cdot L}{AE}\right)_{max} \leq \Delta L_{all}. \tag{6.14}$$

Product AE is known as *axial rigidity*.

Example 6.1.

Consider the behavior of two rods (aluminum and steel) separated from each other by 2 mm and rigidly supported at their ends, as shown in Figure 6.4 (a). It is necessary to determine the temperature increase at which the rods come into contact and to define the elongations and stresses of the rods when the temperature increase is 80 °C.

Data:

Aluminum rod (1), A_1 (cross-sectional area): 2000 mm^2; $E = 70$ GPa; $L_1 = 1000$ mm; Steel rod (2), $A_2 = 1000$ mm^2; $E_2 = 200$ GPa; $L_2 = 700$ mm. The coefficients of thermal dilatations are adopted as: $\alpha_1 = 25.5 \cdot 10^{-6}/°C$ (aluminum), $\alpha_2 = 12.5 \cdot 10^{-6}/°C$ (steel).

Solution:

* Temperature increase at which the rods come into contact (i.e. due to increasing temperature, the rods will be elongated and the gap of 2 mm between them will be closed).

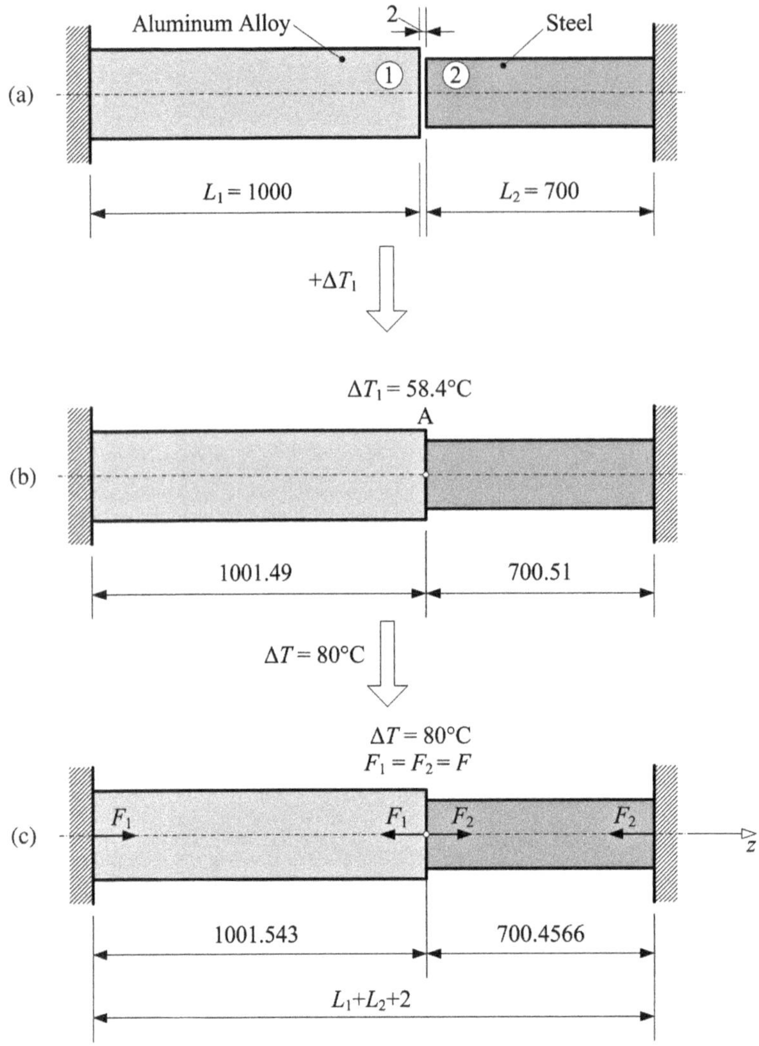

Figure 6.4 Two rods rigidly supported at the ends. (a) The two rigidly supported rods; (b) rods at first contact; (c) rods in contact after a temperature increase of $\Delta T = 80$ °C.

The axial elongations in the two rods due to temperature increase are:

$$\Delta L_{1,T} = \alpha_1 \cdot \Delta T \cdot L_1; \Delta L_{2,T} = \alpha_2 \cdot \Delta T \cdot L_2 \tag{a}$$

$$\Delta L_{1,T} + \Delta L_{2,T} =$$
$$\left(25.5 \cdot 10^{-6}/°\text{C}\right) \cdot \Delta T \cdot 1000 + 12.5 \cdot 10^{-6}/°\text{C}) \cdot \Delta T \cdot 700 = 2 \tag{b}$$
$$\Delta T = 58.4^{\circ}\text{C} = \Delta T_1{}^{*}.$$

*When the temperature increase ($\Delta T = \Delta T_1$) reaches the value of $\Delta T = 58.4$°C, the gap of 2 mm will disappear and the rods will come into contact, as shown in Figure 6.4(b), point A. The elongations of the rods due to a temperature increase of $\Delta T_1 = 58.4$°C, are:

$$\Delta L_{1,\Delta T_1} = 1.49\,\text{mm} \quad \Delta L_{2,\Delta T_1} = 0.51\,\text{mm}$$

As a result of a temperature increase of $\Delta T = 80°\text{C}$, some changes in the rods will arise. These changes relate to rod elongation as well as to the stress that will appear in the rods. First, the rods will elongate according to their thermal properties, and, as determined above, when the gap between them disappears, the rods come into contact. Now, $\Delta T = 80\,°\text{C} > \Delta T = 58.4\,°\text{C}$. Therefore, after contact occurs at $\Delta T = 58.4\,°\text{C}$, due to the further temperature increase ($\Delta T = 80\,°\text{C}$), internal forces in the rods will arise. The equilibrium equation may be considered. In accordance with Figure 6.4 (c), this is:

$$\sum F_z = 0 \to F_1 - F_2 = 0 \to F_1 = F_2 = F \tag{c}$$

The equilibrium that includes the force–temperature relationship now is:

$$\alpha_1 \cdot \Delta T \cdot L_1 + \frac{FL_1}{A_1E_1} + \alpha_2 \cdot \Delta T \cdot L_2 + \frac{FL_2}{A_2E_2} = 2 \tag{d}$$

On the basis of this equation, the force that occurs in each rod ($\Delta T = 80\,°\text{C}$) is:

$$F = -69530.2\,\text{N} = |69.53|\,\text{kN}$$

Check:

The elongation of each rod as a free body due to the temperature increase ($\Delta T = 80\,°\text{C}$) would be:

$$\Delta L_{1,T} = \alpha_1 \cdot \Delta T \cdot L_1 = 2.04\,\text{mm}; \quad \Delta L_{2,T} = \alpha_2 \cdot \Delta T \cdot L_2 = 0.7\,\text{mm}$$

Shortening due to compressive force F is:

$$\Delta L_{1,F} = \frac{FL_1}{A_1E_1} = -0.4966\,\text{mm}; \; \Delta L_{2,F} = \frac{FL_2}{A_2E_2} = -0.24335\,\text{mm}.$$

Finally, the position of the rods in contact at the end of the temperature increase is defined by:

$$\Delta L_1 = \Delta L_{1,T} - \Delta L_{1,F} = 2.04 - 0.4966 = 1.5434\,\text{mm}$$

$$\Delta L_2 = \Delta L_{2,T} - \Delta L_{2,F} = 0.7 - 0.24335 = 0.45665\,\text{mm}$$

As can be seen, the gap of 2 mm between the rods in the starting position (that is, before the increase in temperature) has disappeared, and it is equal to $\Delta L_1 + \Delta L_2$.

Figure 6.4 (c) represents the rods in their deformed state. Stresses in the rods (in fast contact) are:

$$\sigma_1 = \frac{F_1}{A_1} = \frac{F}{A_1} = -34.76\,\text{MPa}; \; \sigma_2 = \frac{F_2}{A_2} = \frac{F}{A_2} = -69.53\,\text{MPa} \tag{e}$$

6.2 Torsion

In engineering practice it is possible to find structural elements that are loaded by torsional moments. A torsional moment (twisting couple, torque) is a moment that seeks to twist a structural member about its longitudinal axis. Here, a prismatic bar of circular cross-section, called a shaft, as a special engineering element often used to transfer torsional moments, may be mentioned. A shaft is used to connect a motor to a pump or similar device. It is necessary to mention that a plane (x, y) through the cross-section of a circular bar (shaft) before twisting remains plane after twisting and radial lines on the considered plane remain straight; that is, circular cross-sections do not warp as they twist – every cross-section remains plane and undistorted. In other cases where the cross-section is noncircular, this assumption is not fulfilled. Torsional members with noncircular cross-sections tend to warp (distort) when subjected to torque. This means that cross-sections do not remain plane. This distortion is known as *warping* [2–6]. The amount of warping may be considered small in a solid bar of noncircular cross-section, but it is noticeable in any thin-walled structural member of noncircular cross-section. Undeformed and deformed (in response to torque) shapes of several bars are shown in Figure 6.5. In general, it is possible to consider the so-called Saint Venant torsion problem ("free torsion", torsion in which warping is not restricted) and torsion problems where warping is restricted.

6.2.1 Elastic Torsion – Shear Stress and Strain Analysis

6.2.1.1 Prismatic Bars: Circular Cross-section

Consider a solid prismatic circular bar [1–3, 5–7] subjected to an external torsional load (torque) with magnitude M_t, as shown in Figure 6.6(a). Depending on the application, there are several names given to an engineering element that is subjected to torsional loading: a shaft, torsion bar, torsion rod or torsion element. If a torque M_t is applied to the free end of the bar, the bar will twist and the free end will rotate for an angle φ called the angle of twist. A longitudinal line (a straight line when the bar is not loaded), like straight line A–B, is twisted into a helix (AB'), as shown in Figure 6.6(b). In a case where loading is within the proportional limit, this helix may be approximated by a straight line. Also, within this proportional limit, the angle of twist is proportional to the magnitude of the torque.

To find the stress distribution over the cross-section of the bar, it is possible to write the equilibrium equation, as shown in Figure 6.6(c), (i = internal):

$$\sum M_z = 0 \quad \rightarrow \quad -M_t + \int_A \tau_\rho \rho \,\mathrm{d}A = 0 \tag{6.15}$$

Because the distribution of the shear stress over the surface is unknown, this equation cannot be solved by the methods of statics. However, by analyzing the deformations that are caused by twisting, some conclusions can be drawn. In accordance with Figure 6.6(d), we have:

$$\mathrm{BB}_1 = \mathrm{d}z \cdot \tan\gamma_R; \quad \mathrm{BB}_1 = R \cdot \mathrm{d}\varphi \tag{6.16a}$$

$$\mathrm{CC}_1 = \mathrm{d}z \cdot \tan\gamma_\rho; \quad \mathrm{CC}_1 = \rho \cdot \mathrm{d}\varphi \tag{6.16b}$$

Assuming that (the angle is considered to be small):

$$\tan\gamma_R \approx \gamma_R, \ \tan\gamma_\rho \approx \gamma_\rho,$$

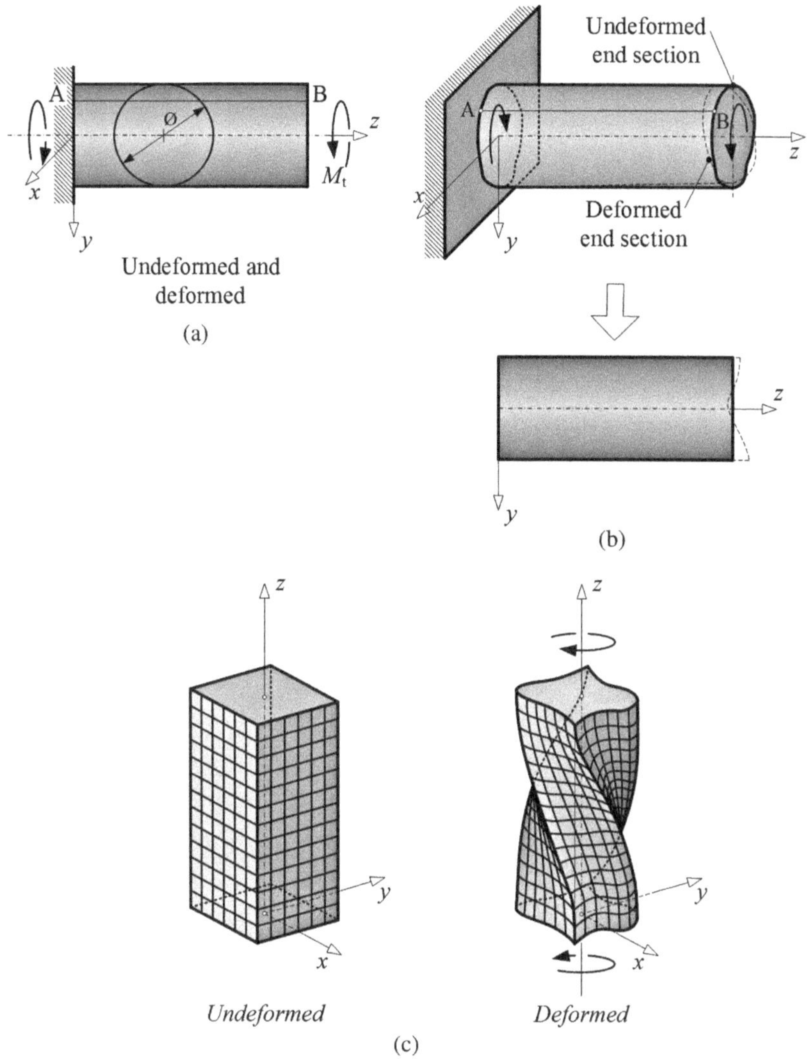

Figure 6.5 Undeformed (unstrained) and deformed shapes of bars subjected to torsional loading (schematic view). (a) Circular cross-section; (b) noncircular cross-section; (c) rectangular cross-section.

it follows that:

$$\gamma_R = R\frac{\mathrm{d}\varphi}{\mathrm{d}z};\ \ \gamma_\rho = \rho\frac{\mathrm{d}\varphi}{\mathrm{d}z} \tag{6.17}$$

The relationship between stress and strain in the case of pure shear, written in the form of Hooke's Law, is (from Equation (6.9)):

$$\tau = G\gamma \tag{a}$$

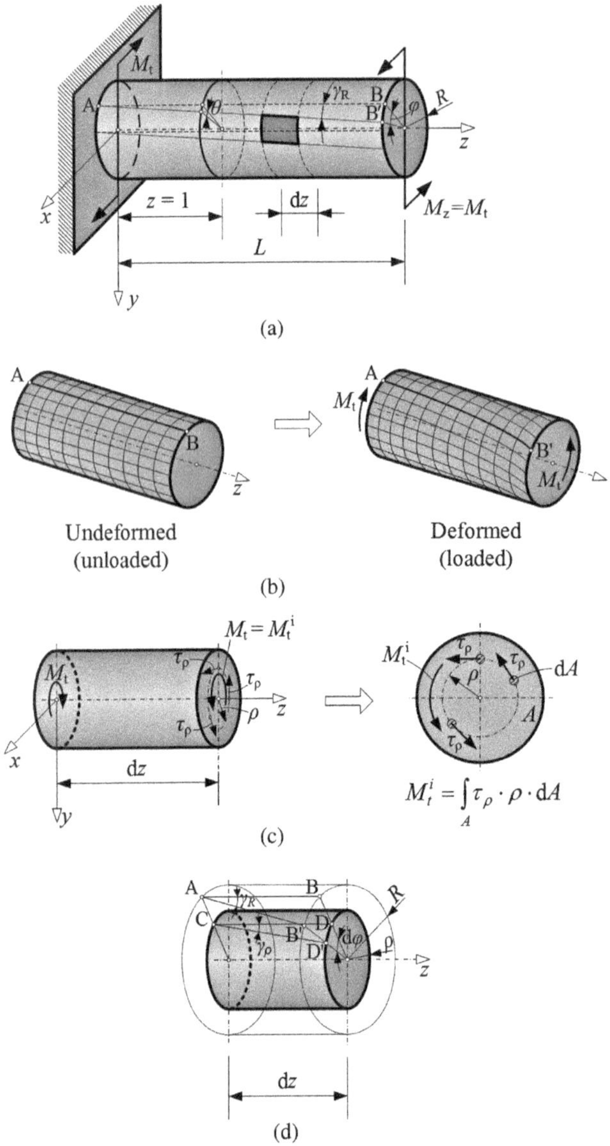

Figure 6.6 Solid prismatic circular bar subjected to torque. (a) Basic data; (b) helix caused by torque; (c) differential part of the bar: equilibrium state; (d) deformation of the differential part.

This assumption is valid if the stress level remains below the proportional limit. On the basis of Equations (6.17) and (6.9), it follows that:

$$\tau_\rho = \tau(\rho) = \tau_R \cdot \frac{\rho}{R}. \tag{6.18}$$

Further, on the basis of Equations (6.15), (6.17) and (6.9 = a), we have:

$$M_t = G\theta I_p, \tag{6.19}$$

where I_p is the *polar moment of inertia*, while θ is the *twisting angle per unit length*.

The twisting angle per unit length is defined as:

$$\theta = \frac{d\varphi}{dz} = \text{const.} \tag{6.20}$$

The polar moment of inertia is defined as:

$$I_p = \int_A \rho^2 dA. \tag{6.21}$$

In a case involving a circular cross-section, the polar moment of inertia is:

$$I_p = \frac{R^4\pi}{2} = \frac{D^4\pi}{32}; \; I_p = \frac{R^4\pi}{2}\left(1-\psi^4\right) = \frac{D^4\pi}{32}\left(1-\psi^4\right); \; \Psi = r/R. \tag{6.22}$$

The twisting angle φ (between two cross-sections of the bar) can be determined as follows:

$$\begin{aligned} \varphi &= \theta \cdot L = \frac{M_t \cdot L}{GI_P} \\ \varphi_{tot} &= \sum \varphi_i \end{aligned} \tag{6.23}$$

Equation (6.23_1) is used to determine the rotation of the cross-section "A" in relation to the cross-section "B". The distance between the considered cross-sections is "L". Within the considered cross-sections, the parameters M_t, G and I_p are considered to be constant.

Equation (6.23_2) can be applied, in general, for consideration of the rotation between the cross-sections "A" and "B" consisting of several segments, where, for each segment within "A" and "B", Equation (6.23_1) is valid.

Product GI_P is known as *torsional rigidity* (or torsional stiffness). On the basis of Equations ($6.9 \equiv a$), (6.17), (6.19) and (6.20), the maximum shear stress appears at $\rho_{max} = R$:

$$\tau = \frac{M_t}{I_p} \cdot \rho \rightarrow \tau_{max} = \frac{M_t}{I_p} \cdot R, \tag{6.24}$$

and can also be expressed as:

$$\tau_{max} = \frac{M_t}{W_p}; \; W_p = \frac{I_p}{R}. \tag{6.25}$$

The elastic torsion formula (Equation (6.24)) is applied only in the case of linear elastic torsion in homogenous and isotropic materials.

If a cross-section is noncircular, instead of the polar moment of inertia (I_p), the so-called *torsional moment of inertia* (I_t) must be used, and this quantity is always lower than the polar moment of inertia. The quantity W_p can be depicted as the *polar moment of resistance*, and analagous to this, W_t can be depicted as the *torsional moment of resistance*.

The Principal Stresses

As we know, Saint Venant's torsion problem (the problem of "free torsion") may be solved using basic observation, similarly to cases involving axial loading; that is, solving is based on the observation of the deformation process. The problem may also be solved by the displacement method using Saint Venant's warping function ψ, which must satisfy the Laplace differential equation (note, label Ψ was used previously to denote the ratio of the radius, r/R).

In addition, the torsion problem may be solved by the method of force using the so-called Prandtl stress function, $\Phi = \Phi\,(x, y)$, also known as the potential stress function. This function must also satisfy some requirements and Poisson's differential equation (known as the differential equation of torsion). To determine the principal stresses, consider the displacements of the point "A" on the cross-section of the beam, as depicted in Figure 6.7. This point is moved (rotated) around the shear center by the torque. Position "A" corresponds to the undeformed (unstrained) state of the section, while position "A_1" corresponds to the deformed state of the section.

Displacements of the point "A" (Figure 6.7) are [1, 8]:

$$-u = x_A - x_{A_1} \rightarrow u = x_{A_1} - x_A = r[\cos(\beta+\varphi) - \cos\beta]$$

$$v = y_{A_1} - y_A = r[\sin(\beta+\varphi) - \sin\beta], \text{ or:}$$

$$u = r(\cos\beta\cos\varphi - \sin\beta\sin\varphi - \cos\beta)$$

$$v = r(\sin\beta\cos\varphi + \cos\beta\sin\varphi - \sin\beta)$$

If $\varphi = \theta\cdot z$ and assuming $\sin\varphi \approx \varphi,\ \ \cos\varphi \approx 1$, it follows that:

$$u = -r\cdot\sin\beta\cdot\varphi = -y\cdot z\cdot\theta = u(y,z)$$
$$v = r\cdot\cos\beta\cdot\varphi = x\cdot z\cdot\theta = v(x,z). \tag{6.26}$$

In general, the distortion of the plane will be described by axial displacements (w_i) of the points on the plane. In accordance with Saint Venant's theory, axial displacement of a considered point is proposed in the form:

$$w = \theta\cdot\Psi(x,y) = w(x,y), \tag{6.27}$$

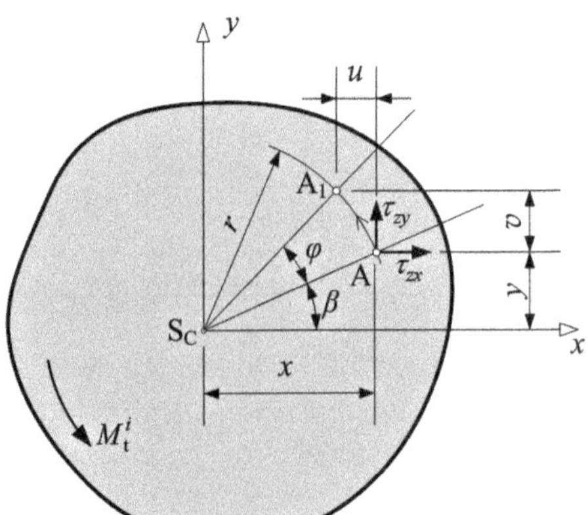

Figure 6.7 Displacements of the considered point of the cross-section.

where ψ is Saint Venant's warping function. Applying Equation (3.16) to Equations (6.26) and (6.27) and using Hooke's Law (Equation (6.9)), it is clear that only shear stresses in the plane of the cross-section exist:

$$\begin{aligned} \tau_{zx} &= G\cdot\gamma_{zx} = G\cdot\theta\left(-y+\frac{\partial\Psi}{\partial x}\right). \\ \tau_{zy} &= G\cdot\gamma_{zy} = G\cdot\theta\left(x+\frac{\partial\Psi}{\partial y}\right), \end{aligned} \tag{6.28}$$

If the equilibrium equation is set with respect to the shear center (point S_c) in such a way that shear stress components are placed at point A in the positive directions of the x and y axes (Figure 6.7), then, using Equation (6.28), the torsional moment (torque) is:

$$\begin{aligned} M_t &= \int\left(\tau_{zy}\cdot x-\tau_{zx}\cdot y\right)\mathrm{d}A = G\theta\iint\left[\left(x+\frac{\partial\psi}{\partial y}\right)x-\left(-y+\frac{\partial\psi}{\partial x}\right)y\right]\mathrm{d}x\mathrm{d}y \\ I_t &= \iint\left(x^2+y^2+x\frac{\partial\psi}{\partial y}-y\frac{\partial\psi}{\partial x}\right)\mathrm{d}x\mathrm{d}y \end{aligned} \tag{6.28a}$$

Since, in the case of free torsion (Saint Venant torsion) there are only shear stresses placed on points of the considered cross-section of the bar, the state of stress may be written by the following tensor:

$$\sigma_{ij} = [\sigma] = \begin{bmatrix} 0 & 0 & \tau_{xz} \\ 0 & 0 & \tau_{yz} \\ \tau_{zx} & \tau_{zy} & 0 \end{bmatrix}. \tag{6.29}$$

The principal stresses can be determined by Equation (2.21), taking into account that the maximum shear stress ($\tau_{max} = \tau_t = \tau$) occurs as a tangent at any point on the boundary of the circle. Based on Equation (2.21), and taking $\tau_{xy} = \tau$, the principal stresses are: $\sigma_{1,2} = \pm\tau$. On the other hand, the principal stresses can also be determined by Equation (2.41):

$$\begin{vmatrix} -\sigma_i & 0 & \tau_{xz} \\ 0 & -\sigma_i & \tau_{yz} \\ \tau_{zx} & \tau_{zy} & -\sigma_i \end{vmatrix} = 0, \tag{6.30}$$

Solving this equation it follows that:

$$\sigma_i\left[\sigma_i^2-\left(\tau_{zx}^2+\tau_{zy}^2\right)\right] = 0;\ \ \sigma_i = \sqrt{\tau_{zx}^2+\tau_{zy}^2};\ \ \sigma_1 = \tau,\ \sigma_2 = -\tau \tag{6.31}$$

and the maximum shear stress is:

$$\tau_{max} = \pm\frac{1}{2}(\sigma_1-\sigma_2) = \pm\sigma_1 \tag{6.32}$$

The stress distribution over the surface of a circular cross-section for some specific cases is presented in Figure 6.8(a)–(f); shear stress in a differential part of a circular element is shown in Figure 6.8(g), while the principal stresses are presented in Figure 6.8(h). Since pure shear is equivalent to tension and compression in two mutually perpendicular directions at the same time, the directions of the principal stresses make an angle of 45° with the longitudinal axis, as depicted in Figure 6.8(h). The same part of the figure shows curves of the principal axes and there are two orthogonal families of curves. The tangent to one family of curves corresponds to the direction of the principal stress, σ_1, while the tangent to the second family corresponds to the direction of principal stress, σ_2, at the considered point.

Criteria for Sizing

The proposed formulas for sizing a shaft subjected to torque using the criterion of allowable stress and the criterion of rigidity are as follows.

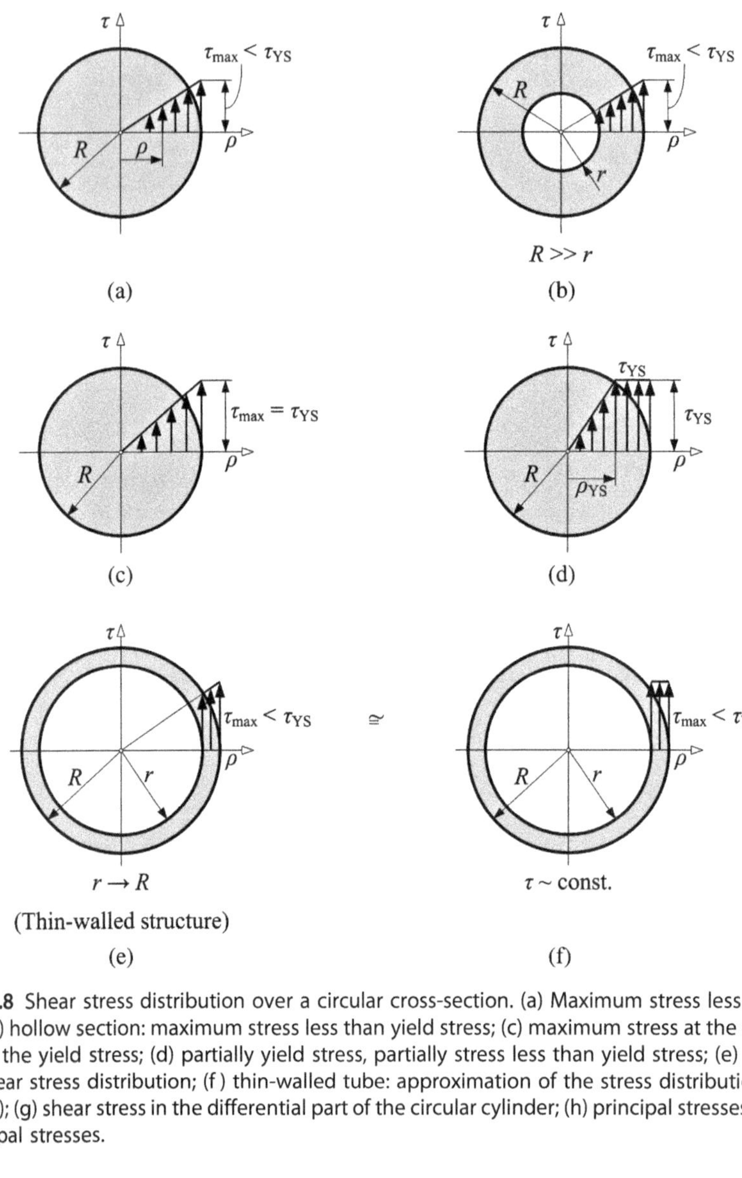

Figure 6.8 Shear stress distribution over a circular cross-section. (a) Maximum stress less than yield stress; (b) hollow section: maximum stress less than yield stress; (c) maximum stress at the outer radius equal to the yield stress; (d) partially yield stress, partially stress less than yield stress; (e) thin-walled tube: linear stress distribution; (f) thin-walled tube: approximation of the stress distribution (as constant); (g) shear stress in the differential part of the circular cylinder; (h) principal stresses and curves of principal stresses.

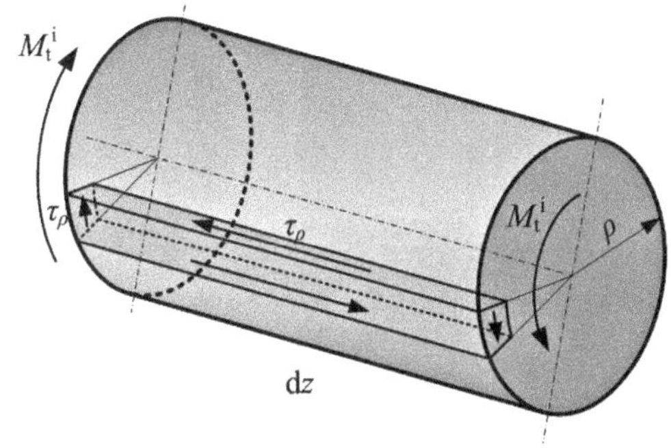

(g)

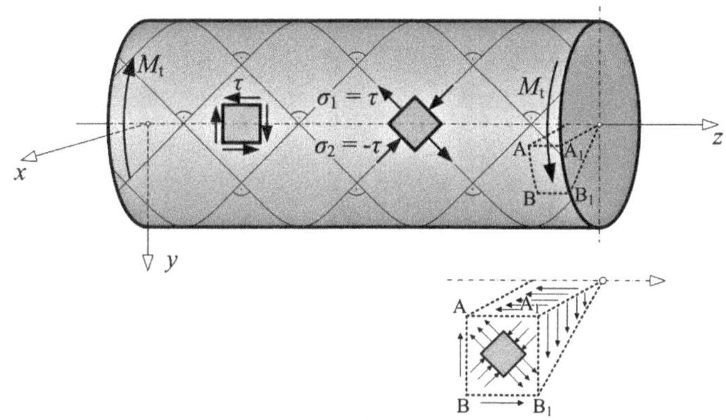

(h)

Figure 6.8 *(Continued)*

Criterion of allowable stress

$$\tau_{\max} = \frac{M_t}{W_p} \leq \tau_{all};\ R \geq \sqrt[3]{\frac{2M_t}{\pi\tau_{all}}}\ ;\ R \geq \sqrt[3]{\frac{2M_t}{\pi\tau_{all}(1-\psi^4)}}; \tag{6.33}$$

Criterion of rigidity

$$\theta = \frac{M_t}{GI_p} \leq \theta_{all};\ R \geq \sqrt[4]{\frac{2M_t}{\pi G\theta_{all}}};\ R \geq \sqrt[4]{\frac{2M_t}{\pi G\theta_{all}(1-\psi^4)}};\ \theta_{all}\ \text{....in}\left(\frac{\text{rad}}{\text{m}} \to \text{m}^{-1}\right). \tag{6.34}$$

When power transmission is considered:

$$R \geq 18250 \cdot \sqrt[3]{\frac{P}{n\tau_{all}}}\ (\text{mm});\ R \geq 18250 \cdot \sqrt[3]{\frac{P}{n\tau_{all}(1-\psi^4)}}\ (\text{mm}), \tag{6.35}$$

In Equation (6.35), the units are: P (kW); n (min^{-1}); τ_{all} (N/m^2).

Example 6.2.
Consider a shaft consisting of two cylindrical parts joined at C, restrained at both ends by rigid walls and subjected to a torsional moment on the joined section C. Determine the maximum shear stress magnitudes in the cylindrical parts and the rotation angle at C, as depicted in Figure 6.9.

Data:
Part I: diameter of 50 mm, material – shear modulus $G = 70000$ MPa, length $L_1 = 1200$ mm; Part II: diameter of 40 mm, material – shear modulus $G = 70000$ MPa, length $L_2 = 1000$ mm. Concentrated torque of 100 Nm.

Solution:
Under consideration is a statically indeterminate system.

The equilibrium equation that relates to torque is:

$$\Sigma M_t = 0 \rightarrow -M_A - M_B + M_t = 0 \tag{a}$$

The compatibility equation is:

$$\varphi_{AB} = \Sigma\varphi_i = 0 \rightarrow \varphi_I + \varphi_{II} = 0 \tag{b}$$

Using Equations (b) and (6.23), it follows that:

$$\frac{M_A \cdot L_I}{GI_{P_I}} + \frac{(M_A - M_t)L_{II}}{GI_{P_{II}}} = 0;\ G_I = G_{II} = G,$$

$$M_A = \frac{M_t \cdot L_{II} \cdot I_{P_I}}{L_I \cdot I_{P_{II}} + L_{II} \cdot I_{P_I}}.$$

The polar moment of inertia (from Equation (6.22)) is:

$$I_p = \frac{\pi D^4}{32};\ I_{P_I} = 613281.25\,\text{mm}^4;\ I_{P_{II}} = 251200\,\text{mm}^4$$

Results:
The twisting moments are:

$$M_A = 67045.7\,\text{Nmm} = 67.046\,\text{Nm};\ M_B = 32.95\,\text{Nm}$$

The stress magnitude is (from Equation (6.24)):

$$\tau_I = \frac{M_A \cdot R_I}{I_{P_I}} = 2.73\,\text{MPa; stress in the second part is}: 2.62\,\text{MPa}.$$

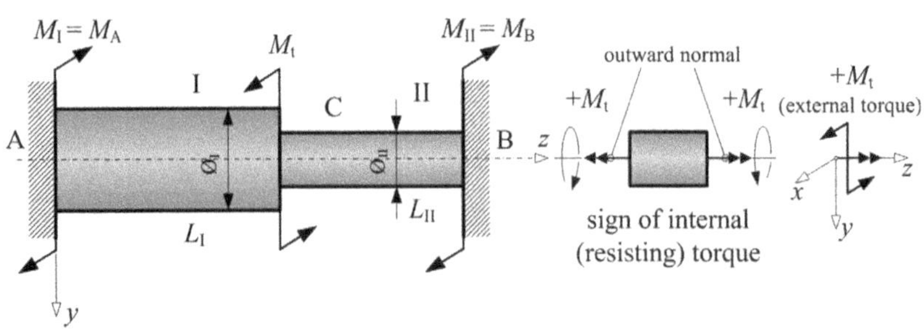

Figure 6.9 A compound shaft.

The twisting angle for segment "I" (position "C") is (from Equation (6.23)):

$$\varphi_{\mathrm{I}}(=\varphi_{\mathrm{AC}}) = \frac{M_{\mathrm{A}} \cdot L_{\mathrm{I}}}{GI_{\mathrm{P_I}}} = 0.00187\ \mathrm{rad}.$$

The calculated angle of twist (rotation) for segment "I", φ_{I}, shows that the cross-section at "C" rotates relative to the cross-section at "A" for this angle.

6.2.1.2 Prismatic Bars: Noncircular Cross-section

Figure 6.10 shows the shear stress distribution over the cross-section of a solid prismatic bar with rectangular cross-section. The maximum stress occurs at the midpoint of the larger side (h) of the cross-section and is zero in the center of the cross-section (O) and at the corners of the cross-section. Instead of the previously used polar moment of inertia (I_{P}), now we need to use the torsional moment of inertia (I_{t}). This quantity is lower than the polar moment of inertia for any cross-section except for a circular cross-section, where $I_{\mathrm{P}} = I_{\mathrm{t}}$.

The magnitudes of the shear stresses, as well as other data, are:

$$\tau_{\max} = \frac{M_{\mathrm{t}}}{\alpha h b^2};\ \tau'_{\max} = \gamma \tau_{\max};\ \theta = \frac{M_{\mathrm{t}}}{\beta G h b^3};\ W_{\mathrm{t}} = \alpha h b^2;\ I_{\mathrm{t}} = \beta h b^3. \tag{6.36}$$

The values of the factors α, β and γ, given in Table 6.1, depend on the ratio (h/b) and can be found in the literature, for example in [1, 2, 5].

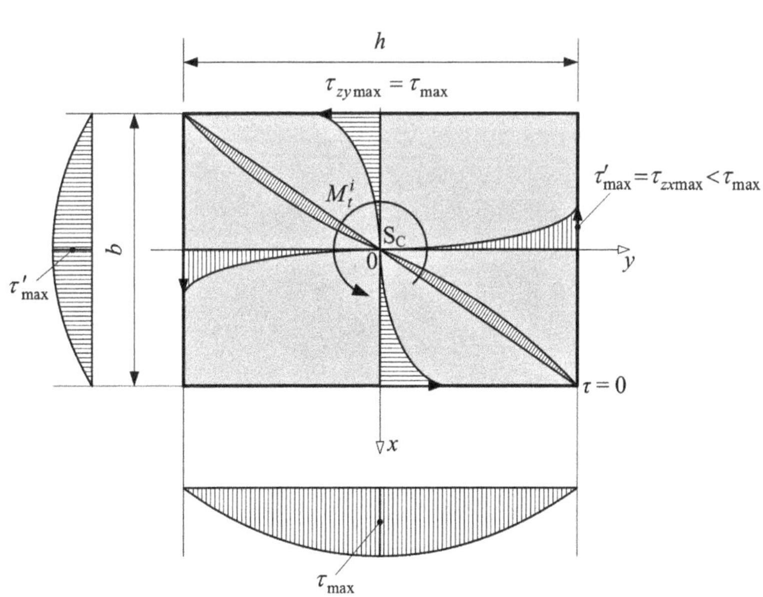

Figure 6.10 Prismatic rectangular bar (solid cross-section): stress distribution.

Table 6.1 Factors α, β and γ as a function of *h/b*.

h/b	1	1.2	1.5	2	2.5	3	4	5	6	8	10	∞
α	0.208	0.219	0.231	0.246	0.258	0.267	0.282	0.291	0.299	0.307	0.312	0.333
β	0.141	0.166	0.196	0.229	0.249	0.263	0.281	0.291	0.299	0.307	0.312	0.333
γ	1	—	0.858	0.796	—	0.753	0.745	—	0.743	0.743	0.743	0.743

6.2.1.3 Thin-walled Structures

Our discussion of the theory of stress and strain on torsionally loaded members has thus far focused primarily on circular cross-sections, although a brief analysis relating to a rectangular cross-section has also been presented. The intention in this section is to cover stress and strain analysis in the field of thin-walled structures, as shown in Figure 6.11.

"Open" Cross-sections

A thin-walled structural member with an "open" cross-section is shown in Figure 6.12.

The differential equation of torsion mentioned in the earlier section on the principal stresses can be obtained using the Prandtl stress function, Φ, which must satisfy the following requirements (see Section 6.2.2) [1, 4]:

$$\tau_{zx} = \frac{\partial \Phi}{\partial y}, \quad \tau_{zy} = -\frac{\partial \Phi}{\partial x} \tag{6.37}$$

If Equations (6.37) are substituted into the Navier equilibrium equations (Equations (2.4)), it is clear that these equations satisfy the Navier equations. The procedure for determining the differential equation of torsion can be found in [1, 8], and here we present the equation in its final form:

$$\frac{\partial^2 \Phi}{\partial x^2} + \frac{\partial^2 \Phi}{\partial y^2} = -2G\theta \ . \tag{6.38}$$

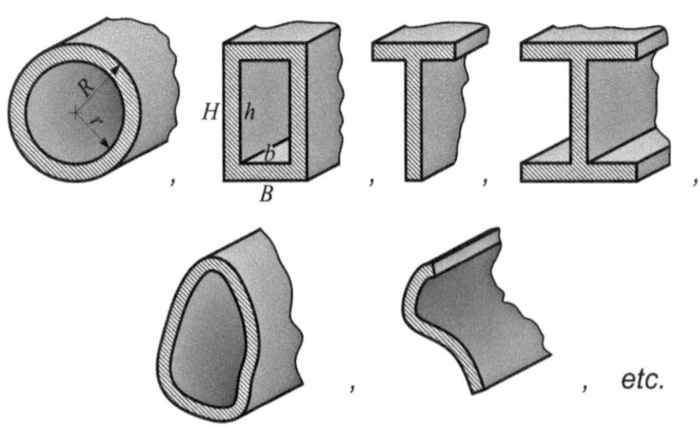

Figure 6.11 Thin-walled structures.

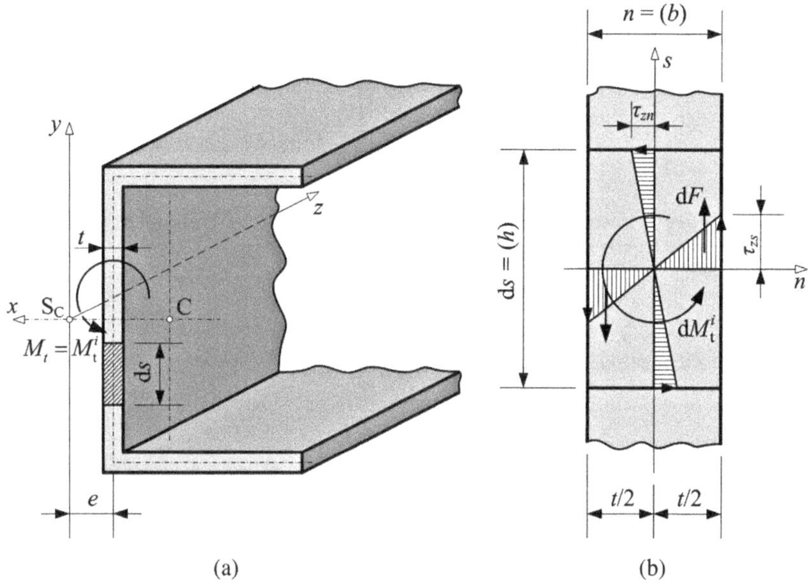

Figure 6.12 A bar with open cross-section ("E" shape – open channel section).

In the case of the thin-walled structure in Figure 6.12(b), the stress components are defined (and designated) as:

$$\frac{\partial \Phi}{\partial s} = \tau_{zn}; \quad \frac{\partial \Phi}{\partial n} = -\tau_{zs} \tag{6.39a}$$

The thickness "t" is treated as being very small compared with the dimension "ds" in the longitudinal direction. As a result, the change of potential function in the "s" direction can be neglected, and thus $\partial\Phi/\partial s = \tau_{zn} = 0$ and $\Phi = \Phi(n)$, $\left(\partial^2\Phi/\partial s^2\right) = 0$. Based on these settings, Equation (6.38) takes the form:

$$\frac{\partial^2 \Phi}{\partial n^2} = \frac{\mathrm{d}^2 \Phi}{\mathrm{d}n^2} = -2G\theta \tag{6.39b}$$

Finally, after first integration:

$$\tau_{zs} = 2G\theta n; \quad \tau_{zs\,(\max)} = G\theta t \tag{6.39c}$$

The torsional moment (torque) can be expressed as a function of the potential function [1, 4]. To obtain this dependence, let us consider the following. Taking Equation (6.28a), and after mathematical calculations, it follows that:

$$M_\mathrm{t} = \int (\tau_{zy}\cdot x - \tau_{zx}\cdot y)\,\mathrm{d}A = -\int_A \left(\frac{\partial \Phi}{\partial x}\cdot x + \frac{\partial \Phi}{\partial y}\cdot y\right)\mathrm{d}x\mathrm{d}y = 2\int_A \Phi\,\mathrm{d}x\mathrm{d}y \tag{6.40a}$$

Further, after a second integration of Equation (6.39b), it follows that:

$$\Phi = G\theta\left(\frac{t^2}{4} - n^2\right) \tag{6.40b}$$

If the quantity in Equation (6.40b) is inserted into Equation (6.40a), it follows that the external torque (based on the mentioned stresses) is:

$$M_t = \frac{G\theta}{3}\int_0^s t^3 \mathrm{d}s\,, \quad t = t(s). \tag{6.40c}$$

In accordance with Figure 6.12(b), the internal torque that arises from stress τ_{zs} is:

$$\left(\mathrm{d}M_t^i\right)_{zs} = 2\mathrm{d}F \cdot \frac{t}{3} = 2\frac{\tau_{zs\,(\max)}}{2} \cdot \frac{t}{2} \cdot \mathrm{d}s \cdot \frac{t}{3} = \tau_{zs\,(\max)} \cdot \frac{t^2}{6} \cdot \mathrm{d}s, \tag{6.41}$$

where "i" means internal. Taking the bar (element) as a whole:

$$\left(M_t^i\right)_{zs} = \frac{G\theta}{6}\int_0^s t^3 \mathrm{d}s. \tag{6.42}$$

Comparing this quantity with that given in Equation (6.40c), it follows that stress τ_{zn} gives another 50% of the torque. Thus, analysis shows that in the equilibrium to external torque, the stress τ_{zs} distributed along the wall thickness (width profile) contributes 50% and the stress τ_{zn} along the height of the profile also contributes 50%.

The torsional moment of inertia is:

$$I_t = \frac{1}{3}\int_0^s t^3 \mathrm{d}s\,, \quad t = t(s), \; I_t = \frac{1}{3}\sum_{i=1}^{i=n} h_i t_i^3 \tag{6.43}$$

If the profile consists of straight parts, then the last of the equations in Equation (6.43) can be used. The maximal shear stress is:

$$\tau_t = \frac{M_t}{I_t} \cdot t \rightarrow \quad \tau_{t\,\max} = \frac{M_t}{I_t} \cdot t_{\max}. \tag{6.44}$$

"Closed" (Hollow) Cross-sections

A thin-walled structural member with hollow cross-section is presented in Figure 6.13.

The shear flow "q" is defined as [5]:

$$q = \tau_1 t_1 = \tau_2 t_2 = \tau_t t = \text{const}. \tag{6.45}$$

The torque, from Figure 6.13, is:

$$M_t = \oint \mathrm{d}F \cdot r \rightarrow M_t = \int_0^S \mathrm{d}F \cdot r$$

Since:

$$\mathrm{d}F = \tau \cdot \mathrm{d}\bar{A}; \;\; \mathrm{d}\bar{A} = t \cdot \mathrm{d}s; \;\; \mathrm{d}A = r \cdot \mathrm{d}s/2,$$

it follows that:

$$M_t = \tau_t t \cdot 2A. \tag{6.46}$$

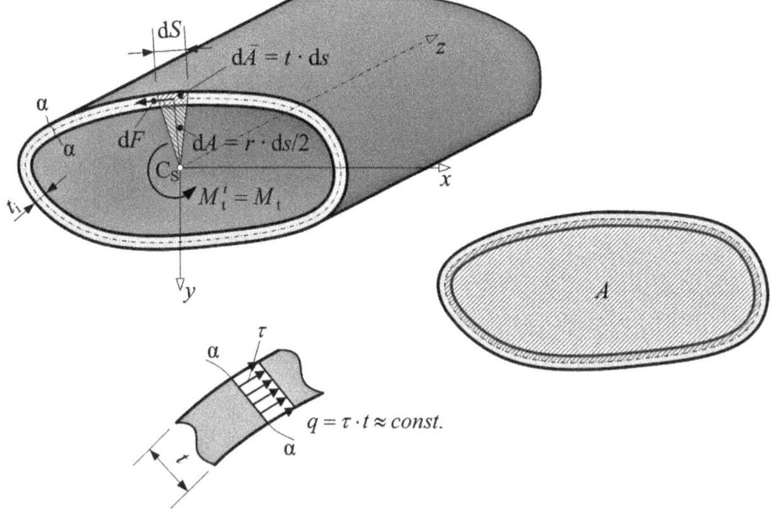

Figure 6.13 Thin-walled hollow structural member.

From Equation (6.46), the stress is:

$$\tau_t = \frac{M_t}{2At} \rightarrow \quad \tau_t \max = \frac{M_t}{2A \cdot t_{min}}, \tag{6.47}$$

where "A" is the area defined by the median line (see Figure 6.13).

Example 6.3.
Consider a multicompartment box beam (multiple-cell member), as shown in Figure 6.14. Determine the stress in the walls of the considered beam.

Data:
The box beam, which consists of three cells, is made of steel and is subjected to torque M_t; the shear modulus is G; the thickness of each wall is "t"; the dimensions of each compartment are: $b \times h$.

Solution/Results:
Stress in a hollow box is defined as (from Equation (6.47)):

$$\tau = \frac{M_t}{2At} \rightarrow M_t = 2A\tau t = 2Aq; \quad q = \tau t \tag{a}$$

$$M_t = \sum M_{t_i} = 2\sum A_i q_i \tag{a1}$$

On the basis of the equivalence of the external and internal works, we can obtain the twisting angle per unit length:

$$\frac{1}{2}M_t \cdot d\varphi = \frac{1}{2}\int \sigma \cdot \varepsilon \cdot dV = dz\int \frac{\tau^2}{2G} t \cdot ds \tag{b}$$

$$\theta = \int \frac{M_t}{4A^2 G}\frac{ds}{t} \tag{c}$$

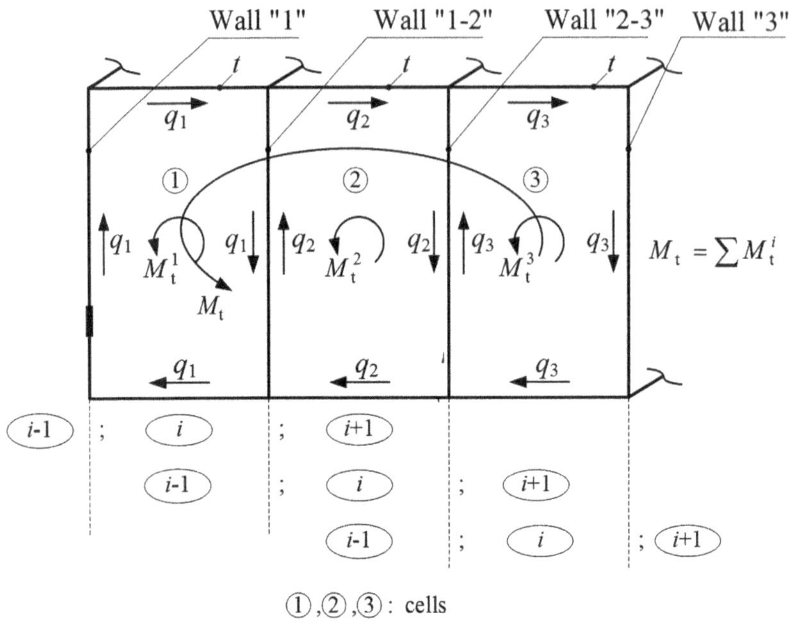

Figure 6.14 Multicompartment box beam.

Comparing Equation (a) and Equation (c), it follows that:

$$2A\theta = \int \frac{q}{G}\frac{ds}{t} \tag{d}$$

$$2A\theta - \frac{q}{G}\int \frac{ds}{t} = 0. \tag{e}$$

Applying this relation to the first box, we have:

$$2A_1\theta = \frac{q_1}{G}\oint_{S_1} \frac{ds}{t} - \frac{q_2}{G}\int_1^2 \frac{ds}{t}, \tag{f}$$

or, applying it to the box "i":

$$2A_i\theta = \frac{q_i}{G}\oint_{S_i} \frac{ds}{t} - \frac{q_{i-1}}{G}\int_{i-1}^{i} \frac{ds}{t} - \frac{q_{i+1}}{G}\int_{i}^{i+1} \frac{ds}{t}. \tag{g}$$

Taking:

$$X_i = \frac{q_i}{\theta}, \tag{h}$$

it follows that:

$$2A_iG = X_i\oint_{S_i} \frac{ds}{t} - X_{i-1}\int_{i-1}^{i} \frac{ds}{t} - X_{i+1}\int_{i}^{i+1} \frac{ds}{t}. \tag{i}$$

Equations (a_1) and (i) yield:

$$M_t = 2\theta \sum A_i X_i, \tag{j}$$

and then the torsional rigidity is:

$$GI_t = 2\sum A_i X_i. \tag{k}$$

Now, for each compartment of the box beam, (i) becomes:

$$2A_1 G = X_1 \cdot 2\frac{b+h}{t} - X_2\frac{h}{t}, \tag{l}$$

$$2A_2 G = X_2 \cdot 2\frac{b+h}{t} - X_1\frac{h}{t} - X_3\frac{h}{t}, \tag{m}$$

$$2A_3 G = X_3 \cdot 2\frac{b+h}{t} - X_2\frac{h}{t} \tag{n}$$

Since:

$$A_1 = A_2 = A_3 = b \cdot h,$$

from Equations (l), (m) and (n), it follows that:

$$X_1 = X_3; \; X_2 = f(X_1). \tag{o}$$

After the values for X_1, X_2 and X_3 have been determined, we can use the formula:

$$q_i = \frac{M_t \cdot X_i}{2\sum A_i \cdot X_i}, \tag{p}$$

to determine the flow q_1, q_2, q_3. Finally, the stresses may be calculated as follows:

$$\text{wall "1" and wall "3"}: \tau_1 = \tau_3 = \frac{q_1}{t} = \frac{q_3}{t}, \tag{r}$$

while the stresses in the central walls of the multicompartment box beam, i.e. in wall "(1–2)" and in wall "(2–3)", are:

$$\tau_{(1-2)} = \frac{q_1 - q_2}{t}; \; \tau_{(2-3)} = \frac{q_2 - q_3}{t}. \tag{s}$$

6.2.2 Warping (Distortion) of a Cross-section

As stated previously (in the section on the principal stresses within Section 6.2.1), the state of strain at a point in an element subjected to torque is defined by shear strains, see Equation (6.28):

$$\begin{aligned} \gamma_{zx} &= \theta\left(-y + \frac{\partial \Psi}{\partial x}\right) \\ \gamma_{zy} &= \theta\left(x + \frac{\partial \Psi}{\partial y}\right), \end{aligned} \tag{6.48}$$

These equations are obtained by substitution of Equations (6.26) and (6.27) into Equations (3.16). If, now, the equation for γ_{zx} is differentiated with respect to y, the equation for γ_{zy} is differentiated with respect to x, but using displacements u, v and w, and the last result is subtracted from the first, it follows that:

$$\frac{\partial \gamma_{zx}}{\partial y} - \frac{\partial \gamma_{zy}}{\partial x} = -2\theta \tag{6.49}$$

Also, since only shear stress components are nonzero, the equilibrium equationss (Equations (2.4)) applied to the problem of torsion become:

$$\frac{\partial \tau_{zx}}{\partial x} + \frac{\partial \tau_{zy}}{\partial y} = 0 \tag{6.50}$$

$$\frac{\partial \tau_{zx}}{\partial z} = 0; \quad \frac{\partial \tau_{zy}}{\partial z} = 0 \tag{6.51}$$

It is evident that shear stress components are independent of z and must satisfy Equation (6.50). This is a necessary and sufficient condition for introducing the so-called Prandtl stress function (potential function) from Equation (6.37) as follows:

$$\tau_{zx} = \frac{\partial \Phi}{\partial y}, \quad \tau_{zy} = -\frac{\partial \Phi}{\partial x}$$

Substituting these equations into Equation (6.49) yields the torsion differential equation (Equation (6.38)):

$$\frac{\partial^2 \Phi}{\partial x^2} + \frac{\partial^2 \Phi}{\partial y^2} = -2G\theta$$

Function Φ is defined as the product of a constant (C) and a function (F) that describe the boundary of the cross-section for a given torsional member. If a prismatic circular bar subjected to torque is being considered, and the Prandtl stress function is introduced/assessed in the following form [1]:

$$\Phi = \mathrm{C}\left(\frac{x^2}{R^2} + \frac{y^2}{R^2} - 1\right), \tag{6.52}$$

then, applying Equation (6.38) it follows that:

$$\mathrm{C}\left(\frac{2}{R^2} + \frac{2}{R^2}\right) = -2G\theta \;\Rightarrow\; \mathrm{C} = -\frac{G\theta R^2}{2}. \tag{6.53}$$

Now, the stress function takes the form:

$$\Phi = -\frac{1}{2}G\theta R^2\left(\frac{x^2}{R^2} + \frac{y^2}{R^2} - 1\right). \tag{6.54}$$

In accordance with Equations (6.37), we have:

$$\tau_{zx} = \frac{\partial \Phi}{\partial y} = -G\theta y, \quad \tau_{zy} = -\frac{\partial \Phi}{\partial x} = G\theta x. \tag{6.55}$$

Also, using Hooke's Law, we have:

$$\gamma_{zx} = \frac{\tau_{zx}}{G} = -\theta y, \quad \gamma_{zy} = \frac{\tau_{zy}}{G} = \theta x, \tag{6.56}$$

and applying Equations (3.16):

$$\gamma_{zx} = \frac{\partial u}{\partial z} + \frac{\partial w}{\partial x}, \quad \gamma_{zy} = \frac{\partial v}{\partial z} + \frac{\partial w}{\partial y},$$

it follows that:

$$\begin{aligned} \frac{\partial w}{\partial x} &= \gamma_{zx} + \theta y = -\theta y + \theta y = 0, \\ \frac{\partial w}{\partial y} &= \gamma_{zy} - \theta x = \theta x - \theta x = 0 \end{aligned} \tag{6.57}$$

which results in:

$$\frac{\partial w}{\partial x} = \frac{\partial w}{\partial y} = 0. \tag{6.58}$$

This result indicates that axial displacement has a constant value. Further, only a quantity of $w = 0$ may be a real result/solution because any other result means that a free body subjected to torque will move translationally. However, a prismatic circular bar does not warp when subjected to torque. In a similar manner, the distortion, for example of an elliptical or other cross-sectional shape, can be determined.

6.2.3 Inelastic Torsion and Residual Stress

The formula proposed in Equation (6.24), using which the stress in a member subjected to torque may be calculated, is no longer valid at the moment when the linear stress distribution is no longer valid. The same is true of Equation (6.19). In Figure 6.15 non-linear and linear shear stress-strain diagrams are shown. This is shown directly on the circular cross-section. Center C is also shear center. Consider a circular cross-section

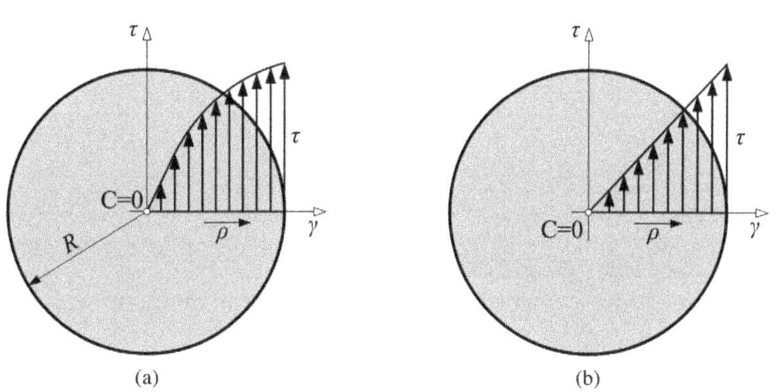

Figure 6.15 Nonlinear and linear shear stress-strain diagrams. (a) nonlinear; (b) linear.

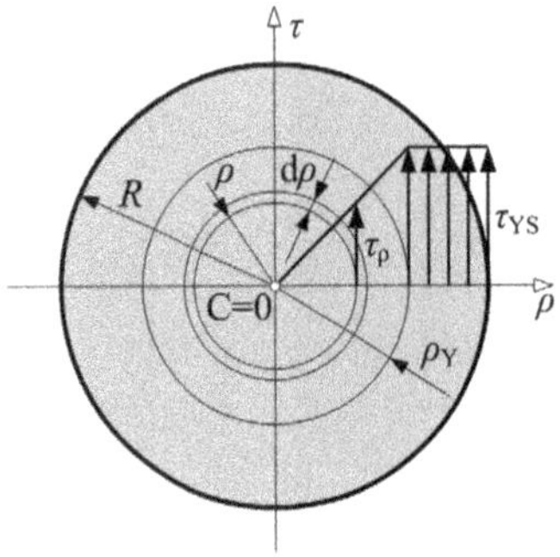

Figure 6.16 Cross-section in elastic–plastic torsion.

subjected to partially elastic, partially plastic stresses, as shown in Figure 6.16. (Figures 6.15 and 6.16 show the shear stress).

The torque that defines an elastoplastic state of the cross-section is:

$$M_t^{elpl} = M_t^{el} + M_t^{pl}. \tag{6.59}$$

The torque moment is defined as (from Equation (6.15)):

$$M_t = \int \tau \cdot dA \cdot \rho = \int \tau \cdot 2\pi\rho d\rho \cdot \rho \tag{6.60}$$

In Equation (6.60), $\tau = \tau\,(\rho)$. In accordance with Figure 6.16(a), we have:

$$\frac{\tau_\rho}{\tau_{YS}} = \frac{\rho}{\rho_Y}, \tag{6.61}$$

where ρ_Y is a radius that corresponds to the boundary of the elastic state while τ_{YS} corresponds to the yield strength. The torque defined in Equation (6.59) is now:

$$M_t^{elpl} = \int_0^{\rho_Y} 2\pi \frac{\tau_{YS}}{\rho_Y}\rho^2 \cdot \rho \cdot d\rho + \int_{\rho_Y}^{R} 2\pi \cdot \tau_{YS} \cdot \rho^2 \cdot d\rho, \tag{6.62}$$

and after solving this integral, it follows that:

$$\begin{aligned} M^{elpl} &= 2\pi \cdot \tau_{YS}\left(\int_0^{\rho_Y} \frac{\rho^3}{\rho_Y} d\rho + \int_{\rho_Y}^{R} \rho^2 d\rho\right) \\ &= 2\pi \cdot \tau_{YS}\left(\frac{\rho_Y^3}{4} + \frac{R^3 - \rho_Y^3}{3}\right) = \frac{2\pi \cdot \tau_{YS} \cdot R^3}{3}\left(1 - \frac{1}{4k^3}\right) \end{aligned} \tag{6.63}$$

Equation (6.63) can also be written in another form.

Shafts are common engineering elements and can be subjected to different loads. In Equation (6.63),

$$\frac{1}{k} = \frac{\rho_Y}{R} = \frac{\gamma_Y}{\gamma_{max}} \tag{6.64}$$

For a fully elastic state (the entire cross-section is elastic), the torque moment is:

$$M_t^{el} = \tau_{YS} \cdot W_p = \tau_{YS} \cdot \frac{R^3 \pi}{2}. \tag{6.65}$$

This is the maximum elastic torque, and this value is also achieved by setting $k = 1$ in Equation (6.63). In this case, τ_{YS} occurs only at the radius which is seen from the Figure 6.16, that is, the yield at R ($R = \rho_Y$).

It is known that, (see Equations (6.21) and (6.25)):

$$W_p = \frac{I_p}{R}$$

For a fully plastic cross-section ($\rho_Y = 0$), the torque moment follows from Equation (6.62):

$$M_t^{pl} = \frac{2}{3} R^3 \pi \cdot \tau_{YS} \tag{6.66}$$

This is the maximum plastic torque, known as the fully plastic moment.

If the maximum elastic moment (from Equation (6.65)) is designated the fully elastic moment, that is, $M_{t_{max}}^{el} = M_t^{FE}$ and the maximum plastic moment (from Equation (6.66)) as the fully plastic moment, that is, $M_{t_{max}}^{pl} = M_t^{FP}$, it is clear that their ratio is:

$$\frac{M_t^{FP}}{M_t^{FE}} = \frac{4}{3}. \tag{6.67}$$

Torque that corresponds to the fully elastic state is torque that corresponds to the state just at the beginning of yielding at radius R. We can see that torque that corresponds to the state when a part of the cross-section is elastic and a part of the cross-section is plastic is given by Equation (6.63). For this case, the torque caused by stress in the elastic region and the torque caused by stress in the plastic region may be written in the form:

$$M_t^{el} = \tau_{YS} \cdot \frac{\rho_Y^3 \cdot \pi}{2}; \quad M_t^{pl} = \frac{2}{3} \pi \cdot \tau_{YS} \left(R^3 - \rho_Y^3 \right). \tag{6.68}$$

6.2.3.1 Residual Stress

Consider a member subjected to torque. After its unloading from an elastic–plastic state, residual shear stress remains in the considered element. Residual stress is usually determined under the assumption that unloading will proceed elastically.

Example 6.4.

Consider a prismatic circular shaft subjected to torque M_t. It is necessary to determine: (a) Whether or not the shaft has yielded; (b) if yielding has occurred, what is the radius, r_Y, to the elastic–plastic boundary; (c) the magnitude of the torque, M_t; (d) the residual stress after elastic unloading.

Data:

Diameter $D = 50$ mm; length $L = 2000$ mm; modulus of rigidity $G = 75$ GPa; yield shear stress, $\tau_{YS} = 150$ MPa. The angle of twist between the ends is: $\varphi = 0.22$ rad.

Solution/Results (see Figure 6.17):

a) The maximum shear strain of the element subjected to torque may be determined using Equations (6.9), (6.19), (6.23) and (6.24):

$$\gamma_\rho = \frac{\tau_\rho}{G} = \frac{M_t \cdot \rho}{GI_P} = \frac{G\theta I_P \cdot \rho}{GI_P} = \theta \cdot \rho = \frac{\varphi}{L}\rho$$

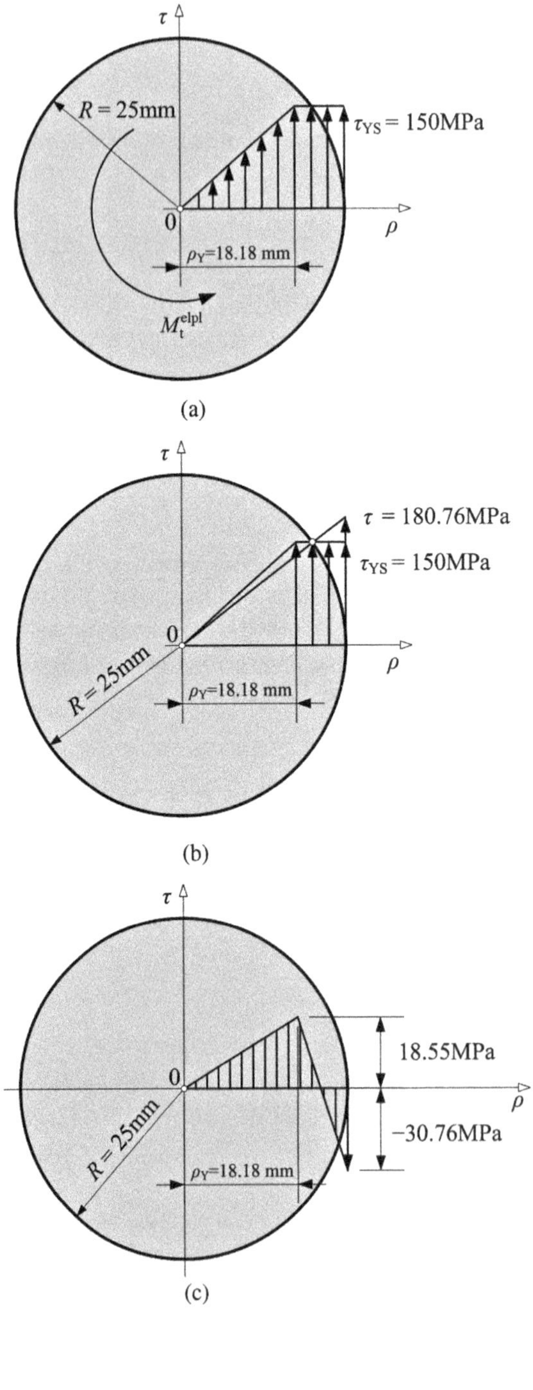

Figure 6.17 Graphical representation of the solution to Example 6.4. (a) Shear stress distribution under elastic–plastic torsion and the elastic–plastic boundary; (b) shear stress distribution due to torque, M^{elpl}; (c) residual shear stress distribution.

In this case, $\rho = R$, $\gamma_\rho = \gamma_R = \gamma_{max}$

$$\gamma_{max} = \frac{R}{L} \cdot \varphi = 0.00275. \tag{a}$$

The elastic strain corresponds to the yield stress:

$$\gamma_Y = \frac{\tau_{YS}}{G} = 0.002 \tag{b}$$

According to Equation (6.64), it follows that:

$$k = \frac{\gamma_{max}}{\gamma_Y} = 1.375,\ k > 1\text{...shaft has yielded.}$$

b) Radius r_Y: elastic–plastic boundary, from Equation (6.64):

$$\rho_Y = \frac{R}{k} = 18.18\,\text{mm}$$

c) The torque magnitude is (from Equation(6.63)):

$$M^{elpl} = \frac{2\pi \cdot \tau_{YS} \cdot R^3}{3}\left(1 - \frac{1}{4k^3}\right) = 4.4344\,\text{kNm}$$

Under the assumption that the previously calculated torque has caused linear shear stress distribution, then the shear stress at the boundary of the shaft (outer edge) is (from Equation (6.65)):

$$\tau = \frac{2M_t}{\pi R^3} = 180.76\,\text{MPa}$$

d) The residual stress is calculated as follows: from the elastic–plastic stress distribution is substracted the linear stress distribution (Figure 6.17(b)). Thus, we have:

$$150\,\text{MPa} - 180.76\,\text{MPa} = -30.76\,\text{MPa}$$

Because of the linear stress distribution, stress at the point $\rho_Y = 18.18$ mm is calculated as:

$$180.76\,\text{MPa} : 25\,\text{mm} = x(\text{MPa}) : 18.18\,\text{mm};\quad x = 131.45\,\text{MPa}.$$

So, at the radius ρ_Y, the stress magnitude is:

$$150\,\text{MPa} - 131.45\,\text{MPa} = 18.55\,\text{MPa}.$$

Example 6.5.
Consider a shaft that is subjected to torque which is increased until the cross-section of the bar has fully yielded. We need to determine the residual stress distribution if the shaft is elastically unloaded.

Data:
Known data: diameter D ($D = 2R$); material: steel (elastic–perfectly plastic); modulus of rigidity G; yield shear stress τ_{YS} .

Solution/Results (see Figure 6.18)

As the torque increases ($M_t \uparrow$), shear stress also increases ($\tau_t \uparrow$). First, shear stress reaches the value of yield stress (τ_{YS}) at the boundary ($\rho = \rho_Y = R$) of the circle (the cross-section of the shaft is a circle). After that, yielding spreads inward into the cross-section.

When the cross-section of the shaft has fully yielded ($\rho_Y = 0$), the shaft is elastically unloaded. In accordance with the presented torsion theory and considerations related to fully elastic and fully plastic as well as elastic–plastic states, it is clear that the following relationship is established (from Equation (6.67)).

When yielding occurs, then $\rho_Y = R$, and the torque with yielding stress at the boundary is (from Equation (6.68_1)):

$$M_t^{el} = \tau_{YS}\frac{R^3}{2}\pi, \left(=M_t^{FE}\right) \tag{a}$$

while torque in the case of a fully yielded cross-section (fully plastic region) is (from Equation (6.68_2)):

$$M_t^{pl} = \frac{2}{3}\pi\tau_{YS}R^3 \left(=M_t^{FP}\right) \tag{b}$$

Taking $M_t^{pl} = M_t^{FP}$ as the maximum plastic torque (fully plastic), then the ratio between the fully plastic moment and the fully elastic moment is:

$$\frac{M_t^{FP}}{M_t^{FE}} = \frac{\frac{2}{3}\pi\tau_{YS}R^3}{\tau_{YS}\frac{R^3}{2}\pi} = \frac{4}{3} \tag{c}$$

On a similar basis to what we did in Example 6.4, that is, under the assumption that the previously calculated torque (fully plastic) has caused a linear shear stress distribution, then the shear stress at the boundary of the shaft (outer edge) is:

$$\tau_{YS}^{FP} = 1.33\tau_{YS}^{FE} \tag{d}$$

The residual stress is calculated as:

$$\tau_{YS}^{FE} - \tau_{YS}^{FP} = -1/3\tau_{YS} \tag{e}$$

That is, from the fully elastic stress distribution we subtract the linear stress distribution (fully plastic). A graphical representation of the solution of this problem is presented in Figure 6.18.

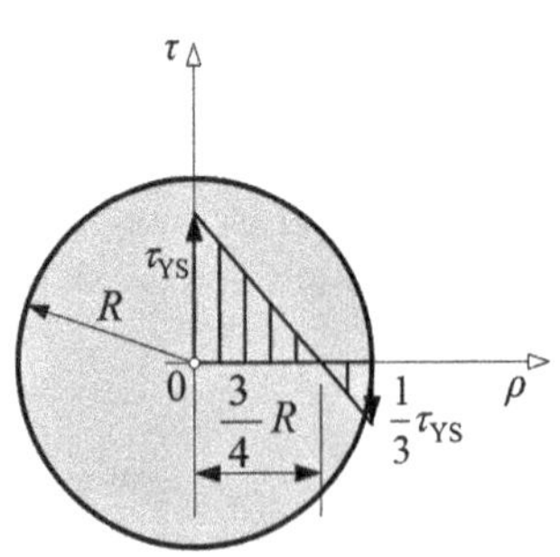

Figure 6.18 Graphical representation of the solution to Example 6.5.

6.3 Bending

Bending can be viewed from the standpoint of the load, deformation and stress. Bending (flexure) is the type of load acting perpendicular to the longitudinal axis of the beam and it is very common in engineering practice.

Bending as a load may be caused by a couple (a concentrated bending moment), M_f (Nm), by a concentrated force (shear force), F (N), or by a distributed load, $q\left(\frac{\mathrm{N}}{\mathrm{m}}\right)$. A distributed load is usually applied in the form of a uniformly distributed load, a linearly varying distributed load or a trapezoidal distributed load.

A beam can be a standalone element or may be a part of another engineering structure such as a frame or similar. The previously mentioned concentrated force can act perpendicular to the longitudinal axis of the beam or in an inclined direction. As such, a concentrated force can be divided into its transverse component (transverse force) that acts perpendicularly and an axial force.

A beam is a very common engineering element supporting loads at different positions along it and providing resistance to bending. A beam may be prismatic or nonprismatic. The cross-section of the beam can be of a solid, thin-walled or thick-walled type and can also take different shapes, such as a circle, square or any other. The cross-section of a beam can be constant or variable. Describing the beam as a load-bearing element, it is possible to say that a beam is predominantly a straight element whose length is much greater than its cross-sectional dimensions. Curved beams can also serve as engineering elements.

In general, a beam can support loads acting transversely to its longitudinal axis, such as shear force and bending moments as well as torque. Bending moments and shear force acting perpendicularly (transversely) to the longitudinal axis of the beam cause bending/flexure of the beam and its shear deformation, and, as a consequence, normal and shear stresses arise. The longitudinal straight axis of a beam subjected to bending deforms into a curved line called an elastic curve (deflection curve). In other words, put simply, loads on a beam cause it to flex (bend).

Depending on the position of the load applied to the beam, it is possible to distinguish symmetrical (plane) bending and nonsymmetrical bending [1–4, 9–10]. In a case of symmetrical bending, the loads and deformation (deflection) of the beam lie in the same plane. This is not the case in nonsymmetrical bending. In both mentioned groups, we may distinguish pure bending and nonuniform bending (also known as bending with shear).

In engineering practice, but also in the literature, we find that certain loads are associated with particular engineering elements. This is usually based on the ability of the element to carry a certain load. As such, axial loads are usually associated with engineering elements known as axial elements/rods/truss elements/bars, while torque (torsional moments) is usually associated with shafts/bar elements of different cross-sections, and bending/flexure is associated with beam elements. Beam elements can carry loads of any kind [3, 8–9].

6.3.1 Beam Supports, Types of Beams, Types of Loads

Figure 6.19 shows different types of support; Figure 6.20 shows different types of beam and Figure 6.21 different types of load. Point G shown in Figure 6.20 (a composed beam) designates the so-called Gerber joint.

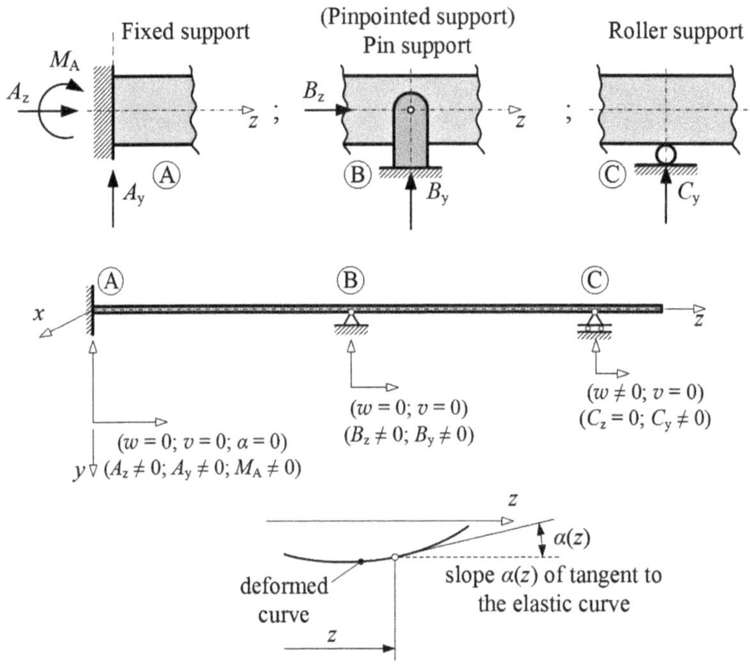

Figure 6.19 Types of support.

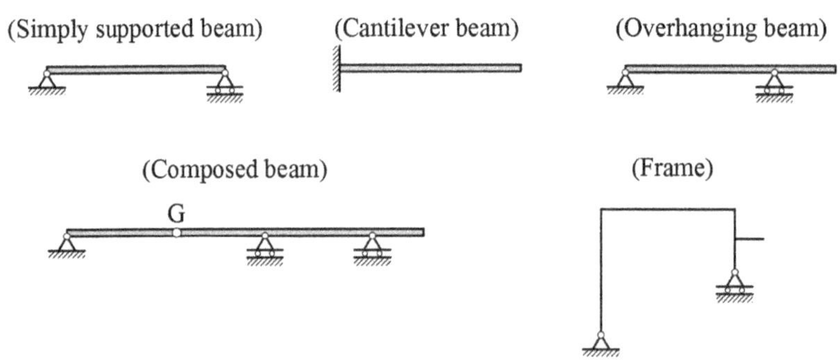

Figure 6.20 Types of beam.

Classification of beams is usually made according to the types of their supports and loads, but beams can also be classified in accordance with static indeterminacy. The body (engineering element, beam, and so on) is considered statically indeterminate if it has more reactions (unknowns) than available (usable) equilibrium equations to ensure their determination. The excess reactions are called redundants and the number of redundants is referred to as the degree of indeterminacy. If all of the beam reactions can be determined by usable equilibrium equations, the beam is statically determinate.

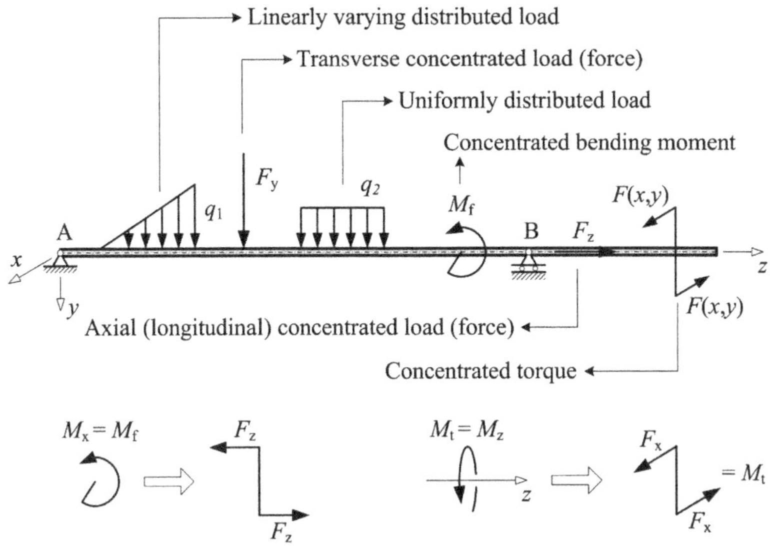

Figure 6.21 Types of load.

6.3.2 Internal Forces – Bending Moments (M_f), Shear Force (Q), Distributed Load (q)

Figure 6.22 shows internal forces and their proposed signs.

In Figure 6.23, the equilibrium state of a differential part (segment), dz, of the beam is considered. Although this differential segment is cut from the part of the beam which is loaded by a trapezoidal distributed load, due to the very small length, it is considered to be uniformly loaded. The following dependences between internal forces are established (see Figure 6.23(b)).

Based on the equilibrium equations, we have:

$$\Sigma F_y = 0$$

$$-Q + Q + \mathrm{d}Q + q \cdot \mathrm{d}z = 0 \rightarrow \frac{\mathrm{d}Q}{\mathrm{d}z} = -q \tag{6.69a}$$

$$\Sigma M_{f,A} = 0$$

$$M_f + Q \cdot \mathrm{d}z - q \cdot \mathrm{d}z \cdot \frac{\mathrm{d}z}{2} - M_f - \mathrm{d}M_f = 0 \rightarrow \frac{\mathrm{d}M_f}{\mathrm{d}z} = Q \tag{6.69b}$$

$$\left[\text{A product}\left(q \cdot \mathrm{d}z \cdot \frac{\mathrm{d}z}{2}\right) \text{ is neglected}\right].$$

Also, on the basis of Equations (6.69), it follows that:

$$\frac{\mathrm{d}^2 M_f}{\mathrm{d}z^2} = -q \tag{6.69c}$$

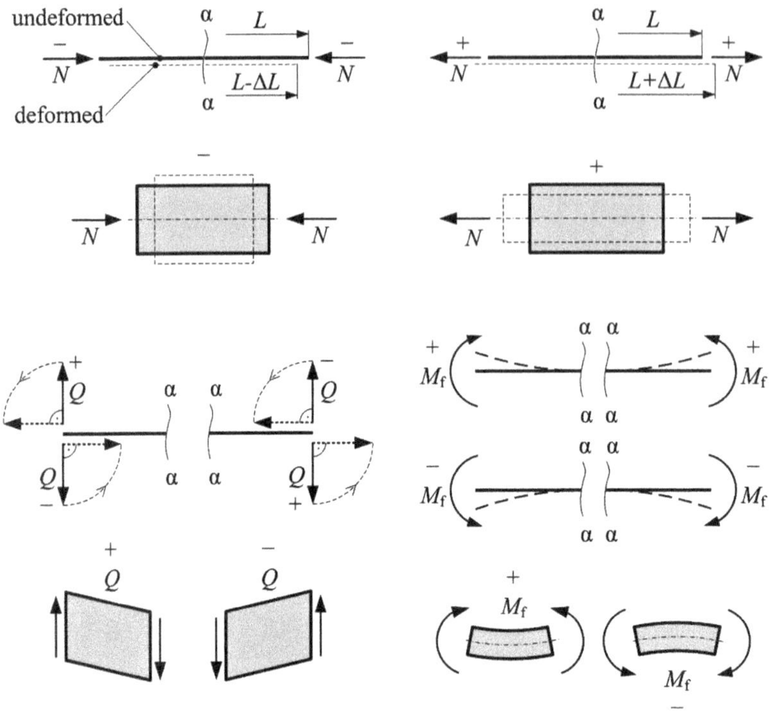

Figure 6.22 Internal forces.

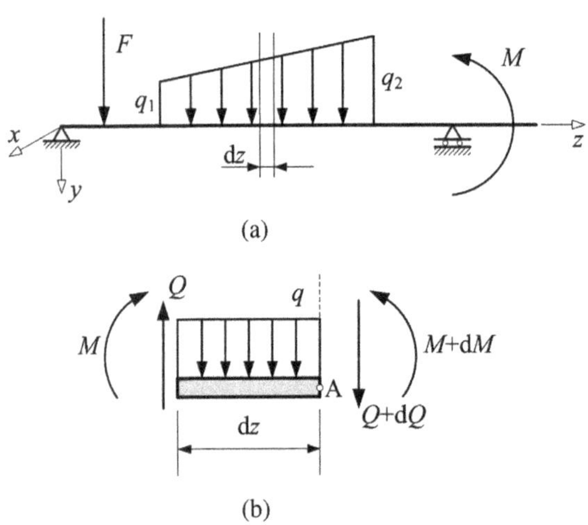

Figure 6.23 Consideration of the equilibrium state (a) A beam subjected to different types of loads; (b) internal forces acting on the differential segment of a beam.

6.3.3 Principal Moments of Inertia of an Area (I_1, I_2) and Extreme Values of Product of Inertia (I_{xy}) of an Area

For every cross-sectional area along the longitudinal axis of the beam there are two orthogonal axes for which the product of inertia of the cross-sectional area (the centrifugal moment of inertia, the deviation moment of inertia) is equal to zero; that is, $I_{xy} = 0$ [1–3, 8,

9, 11], and these axes are known as the principal axes of inertia of the cross-sectional area [2] (see Figure 6.24). The principal axes can be considered at the center of gravity (centroid) or at an arbitrary point of an area, for example point N (see Figure 6.24(b)). Designation of the principal axes of inertia as well as the principal moments of inertia, as applied here, can be seen in Figure 6.24. The principal axes of inertia together with the centroidal longitudinal axis make two centroidal principal planes of inertia. Further, the bending axis is a longitudinal axis (line) on the beam which represents the beam deformation (deflection) and is known as the elastic curve. A beam subjected to bending (flexure) will deflect about the neutral axis of the cross-section of the beam and this axis may be treated as an axis for which the cross-section of the beam provides the lowest resistance to deformation. The so-called section modulus, W (Equation (6.87)), as a measure of a section's resistance to deflection, will be calculated in relation to this axis. A lower value of the section modulus corresponds to higher bending stress given the same loading conditions and characteristics of the cross-section of the beam.

Second moments of area (the moments of inertia, axial moments of inertia of an area, area moments of inertia), I_x, I_y, and *the product of inertia of an area* (the centrifugal moment of inertia, the deviation moment of inertia), I_{xy}, are defined as follows [1–3, 8–9, 11]:

$$dI_x = y^2 dA \;\rightarrow\; I_x = \int_A y^2 dA \tag{6.70a}$$

$$dI_y = x^2 dA \;\rightarrow\; I_y = \int_A x^2 dA \tag{6.70b}$$

$$dI_{xy} = xy dA \;\rightarrow\; I_{xy} = \int_A xy dA, \tag{6.70c}$$

In Equations (6.70), A is the considered (cross-sectional) area of the beam. The first moments of area, S_x, S_y, are mentioned in Section 6.3.4.1 (Pure bending).

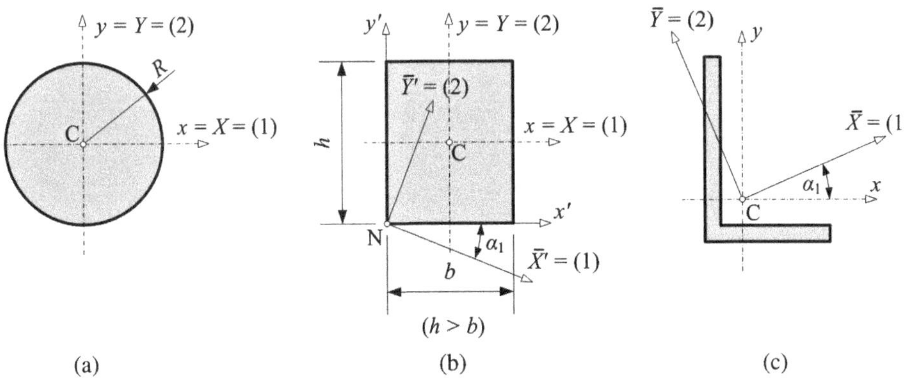

Figure 6.24 Principal axes of a cross-section. (a) Principal axes at the centroid (center of gravity) in the case of a two-dimensional (biaxial) symmetric section; (b) principal axes at arbitrary point "N" in the case of a biaxial symmetric section; (c) principal axes at the centroid in the case of an asymmetrical section.

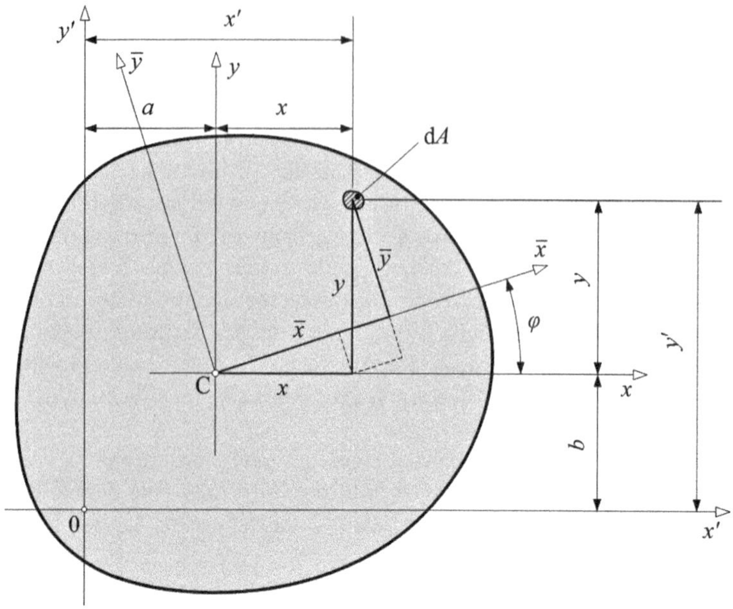

Figure 6.25 Coordinate systems: $(x-y)$; $(x'-y')$; $(\bar{x}-\bar{y})$; translation and rotation.

Changes in the mentioned moments of inertia depending on the possible translation or rotation of the coordinate system (Figure 6.25) can be presented in the form shown in Section 6.3.3.1 below. Here, x and y are the coordinate axes placed in the centroid (C) of the area (sometimes called centroidal axes), x' and y' are axes parallel to the centroidal axes, while $\bar{x}$ and $\bar{y}$ are rotated axes relative to the x and y axes.

6.3.3.1 Axes Parallel to the Centroidal Axes

On the basis of the known moments of inertia (I_x, I_y) of an area (the axial moments of inertia of an area) as well as the known centrifugal moment of inertia (I_{xy}) of an area for the centroidal axes, using the parallel axis theorem, the mentioned moments of inertia of the considered area for the axes parallel to the centroidal axes can be calculated [1] (see Figure 6.25).

$$x' = x + a;\; y' = y + b \tag{6.71a}$$

$$I'_x = \int_A (y')^2 \mathrm{d}A\,,\quad I'_y = \int_A (x')^2 \mathrm{d}A\,,\quad I'_{xy} = \int_A x'y'\,\mathrm{d}A \tag{6.71b}$$

$$I'_x = \int_A y^2 \mathrm{d}A + b^2 \int_A \mathrm{d}A = I_x + b^2 A \tag{6.72a}$$

$$I'_y = \int_A x^2 \mathrm{d}A + a^2 \int_A \mathrm{d}A = I_y + a^2 A \tag{6.72b}$$

$$I'_{xy} = \int_A xy \mathrm{d}A + ab \int_A \mathrm{d}A = I_{xy} + abA \tag{6.72c}$$

6.3.3.2 Rotation of the Coordinate Axes at the Observed Point (Rotated Axes)

On the basis of Figure 6.25, we have:

$$\begin{aligned} \bar{x} &= x\cos\varphi + y\sin\varphi \\ \bar{y} &= y\cos\varphi - x\sin\varphi. \end{aligned} \tag{6.73}$$

$$\bar{I}_x = \int_A \bar{y}^2 \mathrm{d}A\,, \quad \bar{I}_y = \int_A \bar{x}^2 \mathrm{d}A\,, \quad \bar{I}_{xy} = \int_A \bar{x}\bar{y}\mathrm{d}A \tag{6.74}$$

$$\begin{aligned} \bar{I}_x &= I_x\cos^2\varphi + I_y\sin^2\varphi - 2I_{xy}\sin\varphi\cos\varphi \\ \bar{I}_y &= I_x\sin^2\varphi + I_y\cos^2\varphi + 2I_{xy}\sin\varphi\cos\varphi \\ \bar{I}_{xy} &= \left(I_x - I_y\right)\sin\varphi\cos\varphi + I_{xy}\left(\cos^2\varphi - \sin^2\varphi\right). \end{aligned} \tag{6.75}$$

Introducing substitutions (Equation (2.19)) gives:

$$\cos^2\varphi = \frac{1}{2}(1 + \cos 2\varphi)\,; \quad \sin^2\varphi = \frac{1}{2}(1 - \cos 2\varphi); \quad \sin\varphi\cos\varphi = \frac{1}{2}\sin 2\varphi,$$

It follows that:

$$\begin{aligned} \bar{I}_x &= \frac{1}{2}\left(I_x + I_y\right) + \frac{1}{2}\left(I_x - I_y\right)\cos 2\varphi - I_{xy}\sin 2\varphi \\ \bar{I}_y &= \frac{1}{2}\left(I_x + I_y\right) - \frac{1}{2}\left(I_x - I_y\right)\cos 2\varphi + I_{xy}\sin 2\varphi \\ \bar{I}_{xy} &= \frac{1}{2}\left(I_x - I_y\right)\sin 2\varphi + I_{xy}\cos 2\varphi. \end{aligned} \tag{6.76}$$

Similarly to the determination of the directions of principal stresses (principal directions) in Equation (2.20), here, for the determination of the directions of principal axial moments of inertia, we have:

$$\tan 2\alpha = -\frac{2I_{xy}}{I_x - I_y} \rightarrow \alpha = \alpha_1; \quad \alpha_2 = \alpha_1 + \frac{\pi}{2} \tag{6.77}$$

Equation (6.77) may be applied in a case using a cross-sectional centroid (point "C"), as in Figure 6.24(a), (b) and (c), but also in the case of an arbitrary point "N", as shown in Figure 6.24(b). Finally, in general, we have:

$$I_{1,2} = \frac{1}{2}\left(I_x + I_y\right) \pm \frac{1}{2}\sqrt{\left(I_x - I_y\right)^2 + 4I_{xy}^2} \tag{6.78}$$

In the process of searching for the principal axes of inertia (and the principal moments of inertia), it is possible to distinguish, in general, two different cases: (a) the determination of the principal axes of inertia for the center of gravity (centroid), C and (b) for a specific point that is not the centroid. In addition, case (a) can be further divided into two possibilities: (1) the central axes (two mutually perpendicular axes) passing through the centroid (center of gravity) and one of these is the axis of symmetry. In this case, these central axes are at the same time the principal centroidal axes of inertia, that is, $x = X = (1)$ and $y = Y = (2)$, taking $I_x > I_y$, and the moments of inertia for these axes are the principal centroidal moments (axial moments) of inertia; (2) the central axes passing through the centroid but neither is an axis of symmetry. In this case, the principal axes of inertia are also called the principal centroidal axes of inertia, but now: $\bar{x} = \bar{X} = (1)$ and $\bar{y} = \bar{Y} = (2)$, that is, the principal centroidal axes are rotated. The principal centroidal moments of inertia also relate to these principal centroidal axes of inertia. When, in case (b), we are considering any point that is not the centroid, then: $\bar{x}' = \bar{X}' = 1$ and $\bar{y}' = \bar{Y}' = 2$. In this case, it is necessary to determine the axial moments of inertia for this point in advance using the parallel axis theorem and then to find rotated axes that correspond to the principal axes. The same is true when the product of inertia (the centrifugal moment of inertia) is being considered. In any case, for the same point of an area (for example, for the centroid), the following is valid:

$$I_x + I_y = \bar{I}_x + \bar{I}_y = I_1 + I_2 \tag{6.78a}$$

Since the determination of the stress level as well as the deformation level of the engineering element is of importance in the design procedure, it is necessary to determine the minimum resistance of the cross-section to bending. Minimal resistance of the cross-section is defined by one of the principal centroidal axes.

Similarly to Equation (2.23):

$$\tan 2\beta = \frac{I_x - I_y}{2I_{xy}}, \ \beta_{\mathrm{I,II}} = \alpha_1 \pm \frac{\pi}{2} \tag{6.79}$$

Here, we may determine the extreme values of the deviation moment of inertia:

$$\left(I_{xy}\right)_{\min}^{\max} = \pm\frac{1}{2}\sqrt{\left(I_x - I_y\right)^2 + 4I_{xy}^2} = \pm\frac{1}{2}(I_1 - I_2) \tag{6.79a}$$

6.3.4 Symmetrical Bending

The analysis presented here refers to an engineering element which has at least one plane of symmetry and where the load which causes bending acts in that plane. The two mutually perpendicular planes passing through the center of gravity (centroid) of the cross-section of the considered element, where at least one of them is the plane of symmetry, are the principal centroidal axes of inertia of the cross-section of the element. The planes of the element containing the principal centroidal axes of inertia and its longitudinal axis are the principal centroidal planes of inertia of the element. The engineering element (member) under consideration loaded in the plane of symmetry would bend (flex) in that plane of symmetry, or, put another way, the load and the deflection of the member occur in this plane. First we will consider so-called pure bending and after that nonuniform bending. As can be seen from Figure 6.26, and in accordance with the explanations above, the longitudinal axis is the z axis, and the principal centroidal planes are the $y-z$ and $x-z$ planes.

6.3.4.1 Pure Bending

The beam shown in Figure 6.26(a) is subjected to bending. Part of it, between points C and D, is subjected to pure bending.

As can be seen, between points C and D there are no shear forces, that is, the bending moment is of constant value and it can be said that this part is subjected to pure bending. Since the intention is to obtain a formula for designing the cross-section of a beam subjected to pure bending, we analyze the deformation of a differential prismatic element (dz), cut from the considered part CD and subjected to equal and opposite couples M acting at its ends in a plane of symmetry (Figure 6.26(b)). Due to couples $M = M_f$, as indicated in Figure 6.26(b), between the upper and lower surfaces of the beam there exists a neutral surface where stresses and strains are zero.

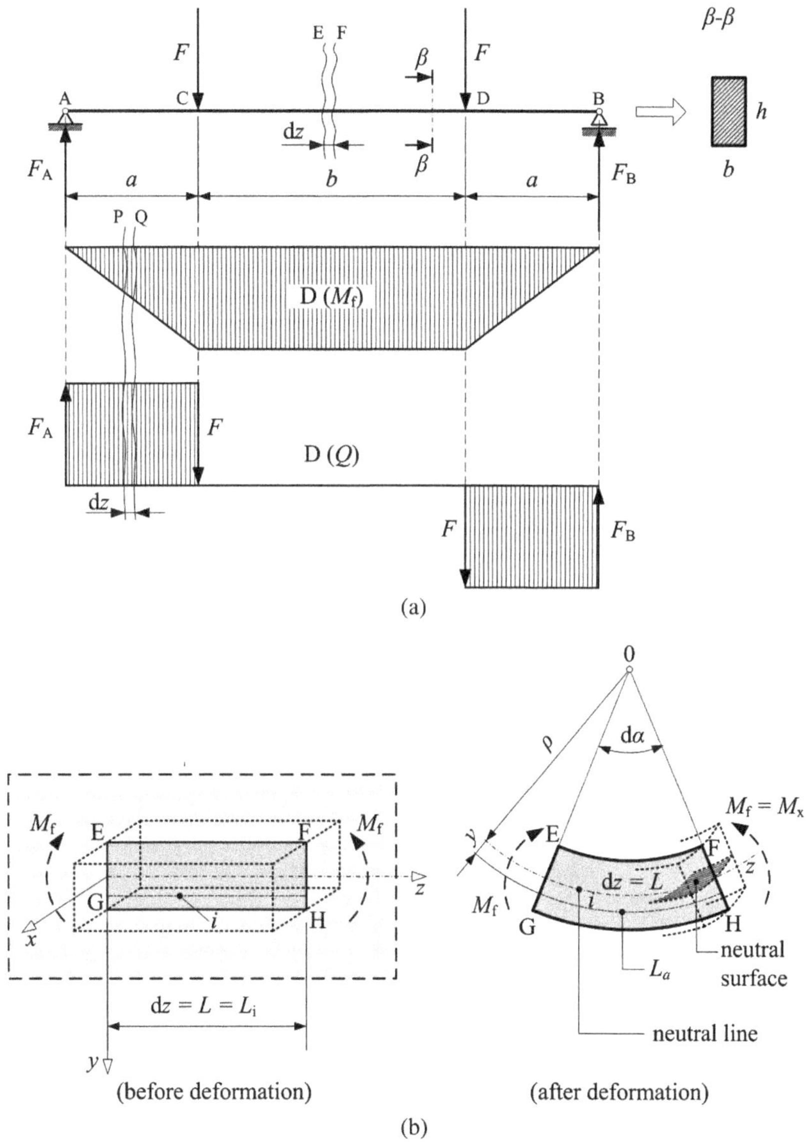

Figure 6.26 A beam subjected to bending. (a) Diagrams of internal forces (bending moments, shear forces) of a beam loaded in the vertical principal centroidal plane of inertia; (b) a segment of the beam loaded by pure bending; (c) distribution of normal stress in the cross-section – shown by the dashed line (highlighted only in the vertical principal centroidal axis of inertia).

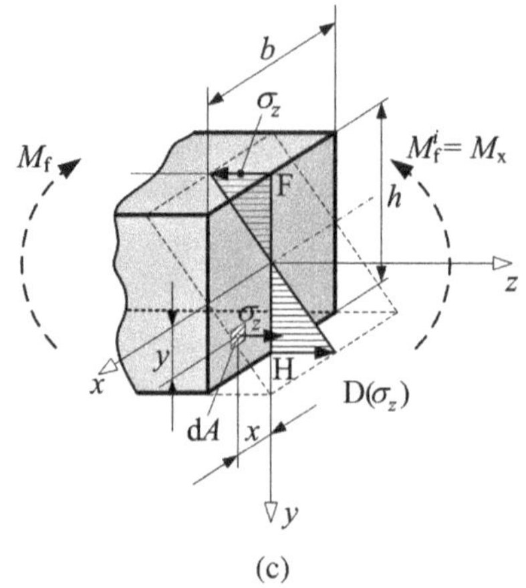

Figure 6.26 *(Continued)*

In accordance with Figure 6.26(b), consider the changes in the lengths of fibers due to deformation in the vertical principal centroidal plane of inertia. The length of the fiber on the neutral line (the longitudinal axis) before and after deformation is kept constant and let's denote it L $(dz = L)$. The length of the fiber "i" before deformation is of the same length as the fiber on the neutral line, let's denote this $L_i = L$. After deformation, the length of the fiber "i" is L_a. The strain on fiber "i" after the deformation process has finished is:

$$\varepsilon = \varepsilon_z = \frac{\Delta L}{L} = \frac{L_a - L}{L} = \frac{(\rho + y)\cdot d\alpha - \rho \cdot d\alpha}{\rho \cdot d\alpha} = \frac{y}{\rho}. \tag{6.80}$$

Applying Hooke's Law, we have:

$$\sigma_z = E \cdot \varepsilon_z; \;\; \sigma_z = E \cdot \frac{y}{\rho} \tag{6.81}$$

As can be seen, stress in fibers varies linearly with the distance y from the neutral surface (Figure 6.26(c)). Using equilibrium equations, relating to the differential part of the beam (Figure 6.26(c)), the following conclusions may be drawn (equations: $\sum F_x = 0; \sum F_y = 0; \sum M_z = 0$, are automatically fulfilled):

a) Equilibrium of forces in the z direction

$$\sum F_z = 0 \rightarrow \int_A \sigma_z dA = 0 \rightarrow \frac{E}{\rho}\int_A y dA = 0. \tag{6.82}$$

Since $E/\rho \neq 0$, we must have $\int y dA = 0$. The value $\int y dA = S_x$ is known as *the first moment of area* and it is equal to zero for the centroidal axis. This means that axis x is the centroidal axis and it is the neutral axis of bending (that is, the beam is deflected about the x axis) of the cross-section. Similarly, when bending about the y axis is considered, $\int x dA = S_y$ is the first moment of area related to bending about the y axis. The section of the vertical plane $(x - y)$ and the neutral surface give the neutral axis of the cross-section, and in this case it is the x axis. As such, it can also be said that stress in fibers varies linearly with the distance y from the neutral axis (x).

b) Equilibrium of bending moments about the x axis (flexure moment M_f in this case is M_x)

$$\sum M_x = 0 \rightarrow \int_A \sigma_z y \mathrm{d}A = M_x \rightarrow \frac{E}{\rho}\int_A y^2 \mathrm{d}A = M_x, \tag{6.83}$$

It follows that:

$$\frac{1}{\rho} = \frac{M_x}{EI_x}, \tag{6.84}$$

where EI_x is known as the *flexural rigidity* of a beam.

c) Equilibrium of bending moments about the y axis

$$\sum M_y = 0 \rightarrow \int_A \sigma_z x \mathrm{d}A = 0 \rightarrow \frac{E}{\rho}\int_A xy \mathrm{d}A = 0, \tag{6.85}$$

That is, axes x and y are the principal centroidal axes of inertia. Comparing Equations (6.81) and (6.84), we have:

$$\sigma_z = \frac{M_x}{I_x} \cdot y \rightarrow \sigma_z \max = \frac{M_x}{I_x} \cdot y_{\max} = \frac{M_x}{W_x}, \tag{6.86}$$

where

$$W_x = \frac{I_x}{y_{\max}} \tag{6.87}$$

is known as the *section modulus* [2].

According to Equations (6.70) and (6.87), for the cross-sections in Figure 6.27, the axial moments of inertia and the section modulus are:

a) $\rightarrow I_x = I_y = \dfrac{\pi D^4}{64}; \quad W_x = W_y = \dfrac{\pi D^3}{32}$

b) $\rightarrow I_x = I_y = \dfrac{\pi}{64}\left(D^4 - d^4\right); \quad W_x = W_y = \dfrac{\pi}{32} \cdot \dfrac{\left(D^4 - d^4\right)}{D}$ (6.88)

or: $I_x = I_y \approx \pi R_{sr}^3 t; \; W_x = W_y \approx \pi R_m^2 t; \; R_m = \frac{(D+d)}{2}$

c) $\rightarrow \begin{array}{l} I_x = h^3 b/12 \quad W_x = h^2 b/6 \\ I_y = b^3 h/12; \quad W_y = b^2 h/6 \end{array}$

d) $\rightarrow I_x = \dfrac{h^3 b - h_1^3 b_1^3}{12}, \; W_x = \dfrac{h^3 b - h_1^3 b_1}{6h}$

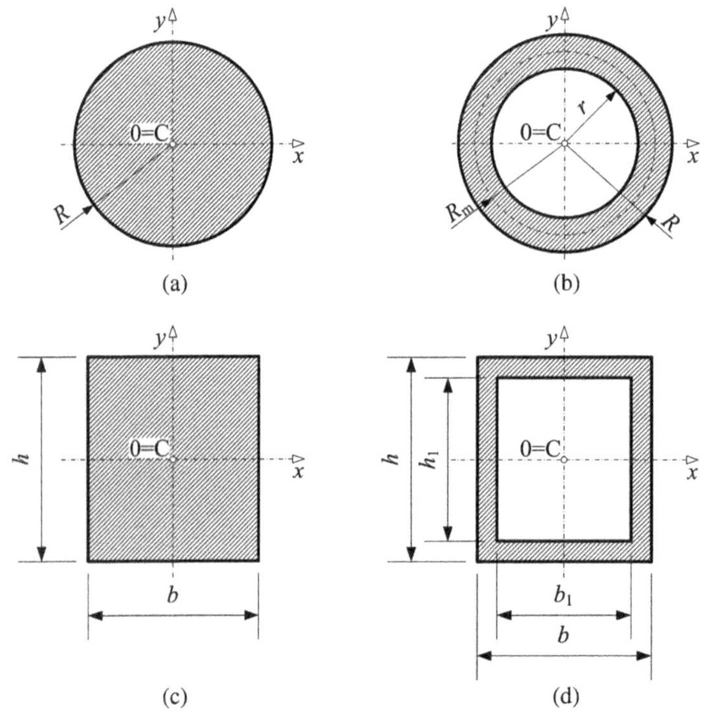

Figure 6.27 Different types of beam cross-section. (a) Circular cross-section; (b) circular thin-walled cross-section; (c) rectangular cross-section; (d) thin-walled rectangular cross-section.

Example 6.6.

Consider the beam shown in Figure 6.28(a), for which it is necessary to determine the bending moment M_{YS} when yield first occurs and the bending (flexure) moment M_f when the flanges have become fully plastic. After the plastification of the flanges, determine the residual stresses after unloading under the assumption that unloading will proceed elastically.

Data:

The material of the beam is assumed to be elastic–perfectly plastic with $\sigma_{YS} = 150\,\text{MPa}$. The geometry of the beam's cross-section is shown in Figure 6.28(a).

Solution:

The axial (centroidal) moment of inertia is (from Equation (6.88c)):

$$I_x = \frac{100 \cdot 170^3}{12} - \frac{(100-6) \cdot 150^3}{12} = 14504166.7\,\text{mm}^4$$

The bending moment that corresponds to the occurrence of yield in the outer fibers (Figure 6.28(c)), is (from Equation (6.86)):

$$M_{YS} = \frac{\sigma_{YS} \,.\, I_x}{\frac{H}{2}} = \frac{150 \cdot 14504166.7}{\frac{170}{2}} = 25595588.3\text{ Nmm}$$

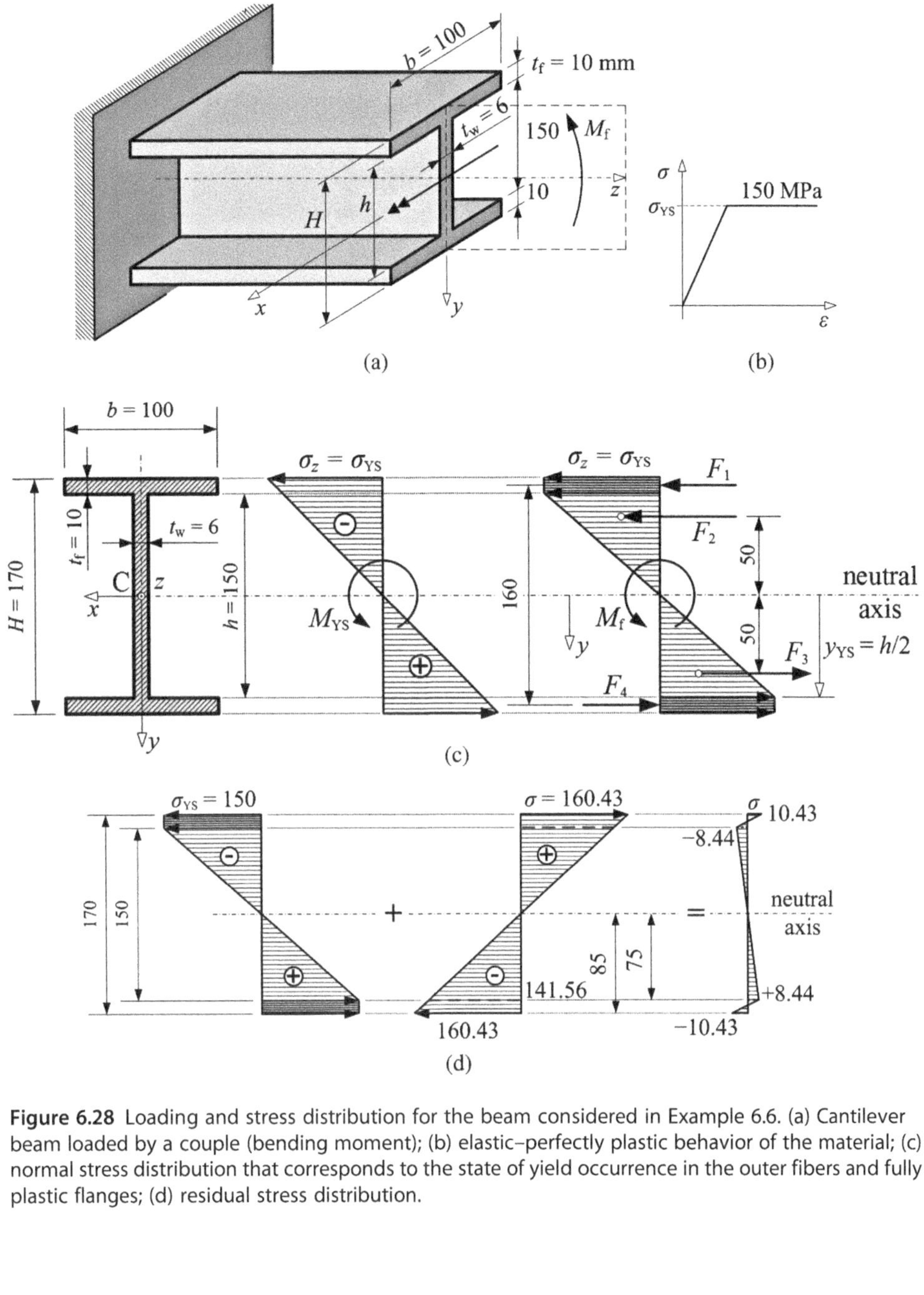

Figure 6.28 Loading and stress distribution for the beam considered in Example 6.6. (a) Cantilever beam loaded by a couple (bending moment); (b) elastic–perfectly plastic behavior of the material; (c) normal stress distribution that corresponds to the state of yield occurrence in the outer fibers and fully plastic flanges; (d) residual stress distribution.

The bending moment that corresponds to the fully plastic flanges is:

$$M_f = F_1 \cdot 160 + F_2 \cdot 100$$

$$= \sigma_{YS} \cdot b \cdot t_f \cdot 160 + \frac{1}{2} \cdot \frac{h}{2} \cdot \sigma_{YS} \cdot t_w \cdot 100 = 27375000\,\text{Nmm}$$

The result can also be obtained as follows:

$$M_{\mathrm{f}} = 2b \int_{y=h/2}^{y=H/2} \sigma_{YS} \cdot y \cdot \mathrm{d}y + 2t_{\mathrm{w}} \int_{y=0}^{y=h/2} \left(\frac{\sigma_{YS}}{y_{YS}} \cdot y \right) \cdot y \cdot \mathrm{d}y$$

$$= 2 \cdot 100 \cdot 150 \cdot \frac{1}{2} \left[85^2 - 75^2 \right] + 2 \cdot 6 \cdot \frac{150}{75} \cdot \frac{75^3}{3} = 27375000\,\mathrm{Nmm}$$

Designations are as follows: t_{f}– thickness of the flange; t_{w} – thickness of the web.

If unloading proceeds elastically, then the bending moment M_{f} causes the following stresses:

$$\bar{\sigma} = \frac{M_{\mathrm{f}} \cdot \frac{H}{2}}{I_x} = 160.43\,\mathrm{MPa}$$

The sign of the stress is shown in Figure 6.28. Residual stresses are shown in Figure 6.28(d).

6.3.4.2 Nonuniform Bending

Consider again the beam in Figure 6.26(a). The part AC (also DB) of this beam is loaded by bending moments but at the same time it is loaded by transverse forces (see the diagrams of internal forces). When a transverse (shear) force acts on a beam, it causes normal and shearing stresses in the beam. Assume that this loading causes elastic behavior of the beam. The level of the normal stress is defined by Equations (6.86), but what about the shearing stress distribution?

Vertical Shearing Stresses

In considerations relating to the beam, the coordinate system is placed, for example, as shown in Figure 6.19. The shearing stress distribution along the web of the beam (that is, along the y axis), known as the vertical shearing stress, will be considered first. Consider the sections $PQ = \mathrm{d}z$ within the part AC of the beam, shown in Figure 6.26. Data relating to the explanation of the vertical shear stress distribution can be seen in Figure 6.29.

The resultant of shearing stresses (see Figure 6.29) is:

$$Q_y = \int_A \tau_{zy} \mathrm{d}A \tag{6.89}$$

Further,

$$\mathrm{d}T = \int_{A_z} \tau_{yz} \mathrm{d}A_z \tag{6.90}$$

Assuming that the shear stress is uniformly distributed along b, and since $\mathrm{d}z$ is very small, we can write:

$$\mathrm{d}T = \tau_{yz} \cdot A_z = \tau_{yz} \cdot b \cdot \mathrm{d}z; \;\; b = b(y) \rightarrow b_y \tag{6.91}$$

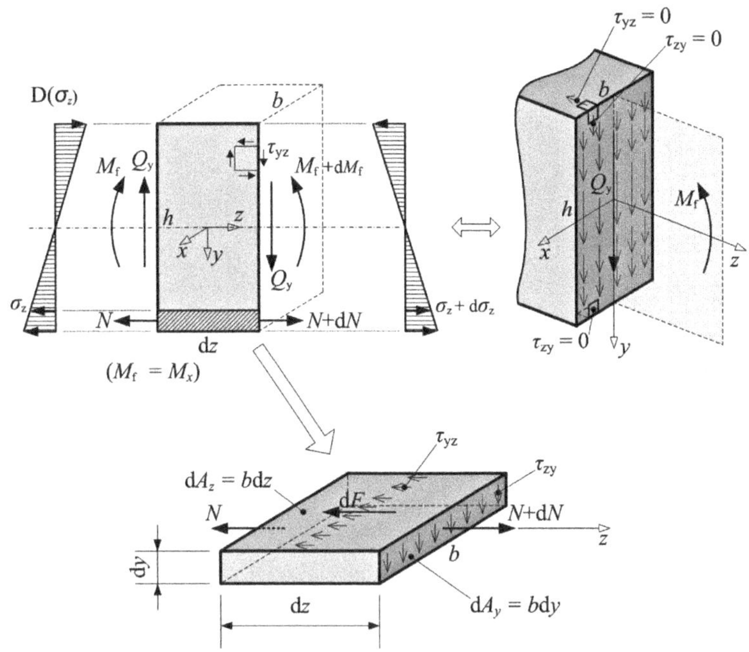

Figure 6.29 Vertical shearing stresses – explanation.

The equilibrium equation of forces in the z direction (see Figure 6.29) is (and further, b is written):

$$\sum F_z = 0 \quad \rightarrow \quad N + \mathrm{d}N - N - \mathrm{d}T = 0 \rightarrow \mathrm{d}N = \mathrm{d}T \tag{6.92}$$

and then it follows that:

$$\int_{A_y} \frac{\mathrm{d}M_x}{I_x} \cdot y \cdot \mathrm{d}A_y = \tau_{yz} \cdot b \cdot \mathrm{d}z.$$

Taking (from Equation (6.69b)):

$$\frac{\mathrm{d}M_x}{\mathrm{d}z} = Q_y$$

finally, we have:

$$\tau_{yz} = \tau_{zy} = \frac{Q_y\, S_x}{I_x\, b_y}, \tag{6.93}$$

where the width of the cross-section at the point under consideration, at the coordinate y, is: $b = b_y$. Also, in Equation (6.93), S_x represents the first (static) moment of the area with respect to the x axis [1–3, 8–9]:

$$\int_{Ay} y\mathrm{d}A_y = S_x. \tag{6.94}$$

The shear stress distribution for some cross-sectional shapes is presented in Figure 6.30.

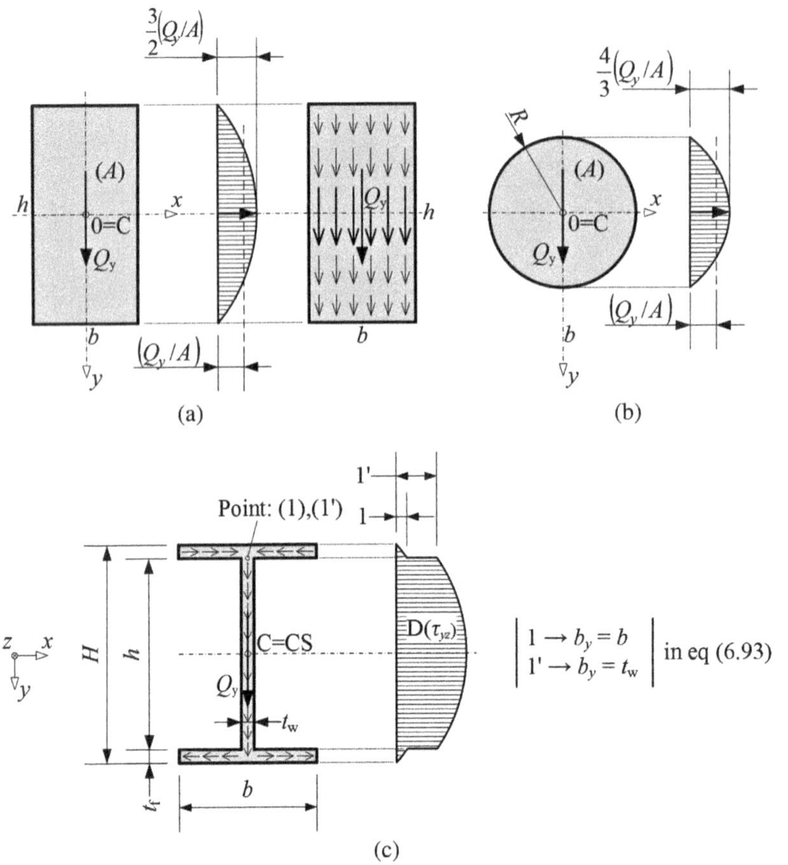

Figure 6.30 Shear stress (τ_{yz}) distribution for different cross-sections. (a) Rectangular cross-section; (b) circular cross-section; (c) I profile.

Horizontal Shearing Stresses

A vertical shear force (Q_y) causes shear stresses not only in the vertical direction (y axis, that is, along the web) but also in the horizontal direction (x axis, that is, along the flange), as shown in Figure 6.31.

If the equilibrium equation of forces in the z direction (part of the flange: $dz \cdot c \cdot t_f$) is written in a similar way to that presented in Equation (6.92), when the distribution of shear forces τ_{yz} along the web of the profile was considered, it follows that:

$$\sum F_z = 0 \;\rightarrow\; N + dN - N - dT = 0 \rightarrow dN = dT \tag{6.95}$$

and further:

$$dN = \int_{A_x} d\sigma_z \cdot dA_x\,,\quad A_x = c \cdot t_f,\quad c = \frac{b}{2} - x,$$

$$dT = \int_{A_z} \tau_{xz}\, dA_z = \tau_{xz} \cdot A_z,\quad A_z = t_f \cdot dz.$$

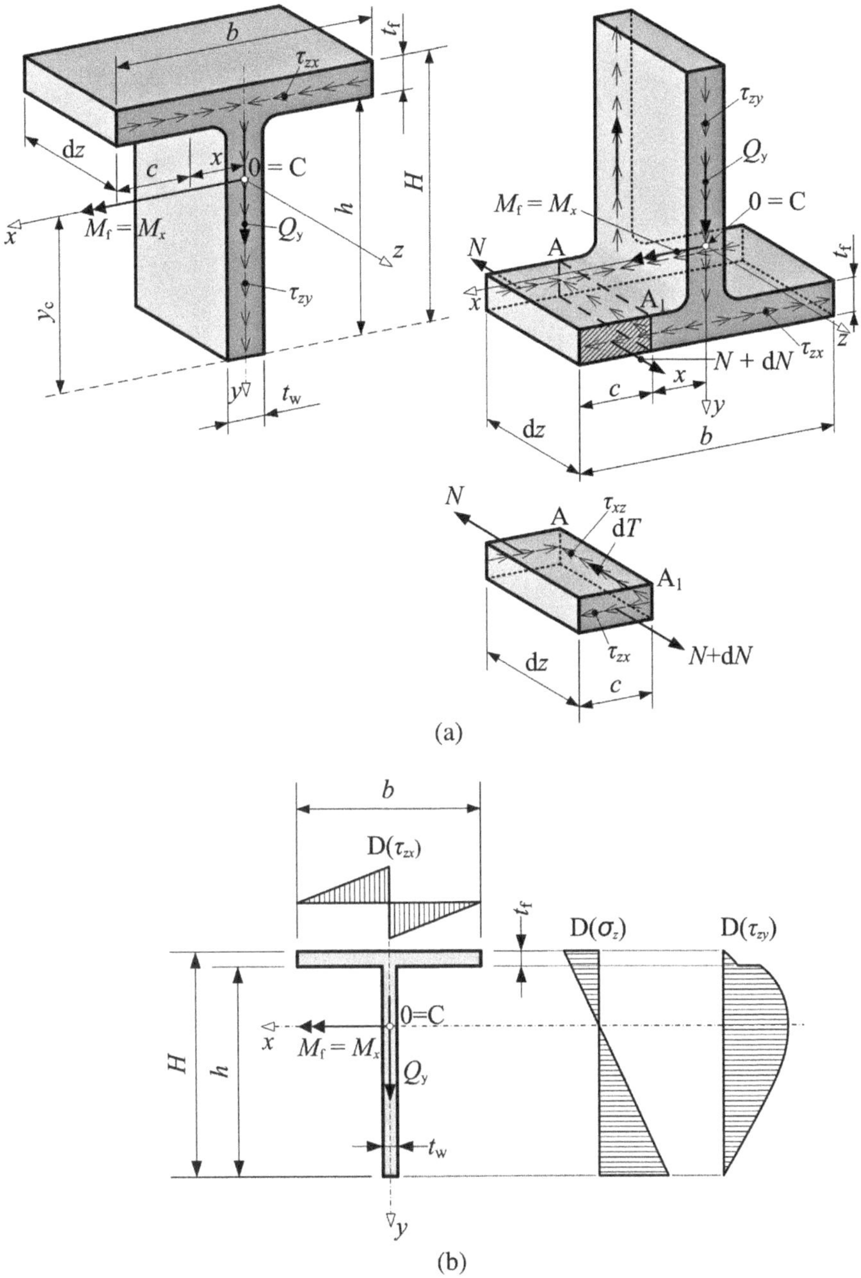

Figure 6.31 Distribution of shear stresses in the horizontal direction (along the flange of the profile) caused by vertical shear force. (a) Beam loaded by vertical shear force and bending moment: T profile; (b) stress distribution: normal stress distribution caused by the couple, shear stress distribution along the web (*y* direction) caused by vertical shear force, horizontal shear stress distribution along the flange (*x* direction) caused by vertical shear force.

After some mathematical calculation, it follows that:

$$\frac{Q_y}{I_x t_f}\int_{A_x} y\,\mathrm{d}A_x = \tau_{xz}.$$

Since $\int_{A_x} y\,\mathrm{d}A_x$ is the first moment of the area $A_x\ (=c\cdot t_f)$ related to the neutral axis, and it is defined as:

$$S_x = A_x\cdot y_{C(A_x)},$$

it follows that this quantity is:

$$S_x = c\cdot t_f\left(h-y_c+\frac{t_f}{2}\right)$$

Finally, the result is:

$$\tau_{xz}=\tau_{zx}=\frac{Q_y\cdot S_x}{I_x\cdot t_f}. \tag{6.96}$$

In the equation above, the quantity $y_{C(A_x)}$ denotes the center of gravity of the area A_x, while y_c is the coordinate y of the centroid. In addition, Equations (6.93) and (6.96) are similar and determine the distribution of the shear stress τ_{yz} along the web of the profile, and the distribution of the shear stress τ_{zx} along the flange of the profile. The former stress, τ_{yz}, is in the direction of the shear force, while stress τ_{xz} corresponds to the fluid flow in the channel.

Neglecting $(t_f/2)$ in comparison with h in the equation related to S_x, finally gives:

$$\tau_{xz}=\tau_{zx}=\frac{Q_y\cdot c(h-y_C)}{I_x}. \tag{6.97}$$

It is clear that the shear stress distribution along the flange (x direction) is linear, as shown in Figure 6.31.

6.3.5 Nonsymmetrical Bending

As stated previously, symmetrical bending of a member occurs when the load acts in a principal centroidal plane of inertia which is, at the same time, the plane of symmetry, and bending of the member will also occur in that plane, see Figure 6.32(a). However, in engineering practice, there are situations when bending moments and/or forces do not act in a principal centroidal plane of inertia that is also the plane of symmetry. This may be due to the load acting in a plane that is not the plane of symmetry or because the considered element does not possess a plane of symmetry (see Figure 6.32(b)).

Consider an arbitrary cross-section of a beam which does not possess a plane of symmetry (Figure 6.32(c)). Assume that the bending moments, M_x and M_y, act as demonstrated; point C is the centroid of the cross-section.

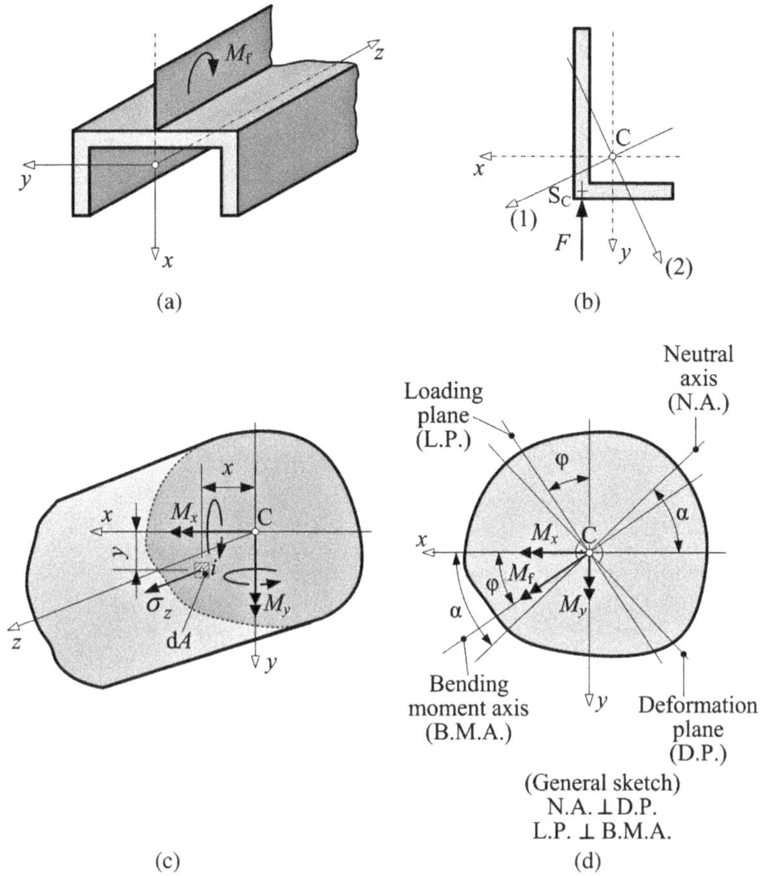

Figure 6.32 Symmetrical and nonsymmetrical bending. (a) Open channel ([profile, open, thin-walled cross-section) loaded by a bending moment acting in the plane of symmetry – symmetrical bending; (b) an unequal L shape (unequal leg-angle shape) subjected to force – nonsymmetrical bending; (c) arbitrary cross-section loaded by bending moments – nonsymmetrical bending; (d) loading plane, deformation plane, neutral axis.

The following equilibrium equations must be satisfied:

$$\int_A \sigma_z \cdot \mathrm{d}A = 0 \tag{6.98}$$

$$\int_A \sigma_z \cdot \mathrm{d}A \cdot y = M_x \tag{6.99}$$

$$\int_A \sigma_z \cdot \mathrm{d}A \cdot x = -M_y \tag{6.100}$$

Based on Equations (6.98)–(6.100), the external moments must be equal to the moments caused by internal forces. Thus, for example, the external moment M_y, which is positive

(Figure 6.32(c)), must be equal to the moment of internal forces, as indicated in Equation (6.100). However, the moment of internal forces ($\sigma_z dA \cdot x$) about the y axis is negative (that is, $-\sigma_z dA \cdot x$), and this results in Equation (6.100).

The bending moments shown (M_x and M_y) cause beam curvature in the (y–z) and (x–z) planes. In beam-bending analysis, according to the Euler–Bernoulli beam theory [1, 2], shear deformation is excluded. Therefore, in accordance with this theory, it is assumed that plane cross-sections perpendicular to the beam's longitudinal axis remain plane and perpendicular to the longitudinal axis after deformation. So, the transverse displacement is the only variable. The normal strain, ε_z, in the longitudinal direction for any fiber "i" in a beam of arbitrary cross-section, (for an analogy, see, for example, Equation (6.80) and Figure 6.26), can be written as:

$$\varepsilon_z = -\frac{x}{\rho_x} + \frac{y}{\rho_y}. \tag{6.101}$$

If we consider the bending to be elastic, the stress distribution over the cross-section is defined by:

$$\sigma_z = E \cdot \varepsilon_z. \tag{6.102}$$

Since the bending is considered elastic, Hooke's Law can be applied.

If Equations (6.101) and (6.102) are inserted into Equation (6.98), we obtain:

$$\begin{aligned} &\int_A \sigma_z \cdot dA = 0 = \int_A E\left(-\frac{x}{\rho_x} + \frac{y}{\rho_y}\right) dA = 0 \rightarrow \\ &\rightarrow -\frac{1}{\rho_x}\int_A x \cdot dA + \frac{1}{\rho_y}\int_A y \cdot dA = 0 \end{aligned} \tag{a}$$

As we can see, Equation (a) can be satisfied only if the neutral axis, NA (Figure 6.32(d)) passes through the centroid of the cross-section. The slope of the neutral axis (NA) as a straight line will be determined later. Entering Equations (6.101) and (6.102) into Equations (6.99) and (6.100) gives:

$$\frac{I_x}{\rho_y} - \frac{I_{xy}}{\rho_x} = \frac{M_x}{E} \tag{b}$$

$$\frac{I_{xy}}{\rho_y} - \frac{I_y}{\rho_x} = -\frac{M_y}{E}. \tag{c}$$

Taking Equations (b) and (c), and solving each of them for $1/\rho_y$, it follows that:

$$\frac{1}{\rho_y} = \frac{M_x}{EI_x} + \frac{I_{xy}}{\rho_x I_x} \tag{d}$$

$$\frac{1}{\rho_y} = -\frac{M_y}{EI_{xy}} + \frac{I_y}{\rho_x I_{xy}}. \tag{e}$$

On the basis of Equations (d) and (e), it follows that:

$$\frac{1}{\rho_x} = \frac{M_y I_x + M_x I_{xy}}{E\left(I_x I_y - I_{xy}^2\right)} \tag{f}$$

Also, taking Equations (b) and (c) and solving each of them for $1/\rho_x$, it follows that:

$$\frac{1}{\rho_x} = -\frac{M_x}{EI_{xy}} + \frac{I_x}{\rho_x I_{xy}} \tag{g}$$

$$\frac{1}{\rho_x} = \frac{M_y}{EI_y} + \frac{I_{xy}}{\rho_y I_y} \tag{h}$$

Considering Equations (g) and (h) gives us:

$$\frac{1}{\rho_y} = \frac{M_y I_{xy} + M_x I_y}{E\left(I_x I_y - I_{xy}^2\right)} \tag{i}$$

Now, expressions (f) and (i), as curvature expressions, may be substituted into Equation (6.102), including Equation (6.101), to represent a relationship for normal stresses occurring in a beam of arbitrary cross-section subjected to bending moments M_x and M_y [8]:

$$\sigma_z = \frac{\left(M_y I_{xy} + M_x I_y\right)y}{\left(I_x I_y - I_{xy}^2\right)} - \frac{\left(M_y I_x + M_x I_{xy}\right)x}{\left(I_x I_y - I_{xy}^2\right)} \tag{6.103}$$

The orientation of the neutral axis can be obtained on the basis that stresses on the neutral surface (that is, stresses in fibers passing through the neutral axis) are zero. Setting Equation (6.103) equal to zero, yields:

$$-\left(M_y I_x + M_x I_{xy}\right)x + \left(M_y I_{xy} + M_x I_y\right)y = 0 \tag{j}$$

and then (see Figure 6.32(c)), the orientation of the neutral axis is defined by:

$$\tan\alpha = \frac{y}{x} = \frac{M_y I_x + M_x I_{xy}}{M_y I_{xy} + M_x I_y}. \tag{6.104}$$

This equation is applicable for positive prescribed values of M_x, M_y, as indicated in Figure 6.32(c). If any value has a different sign than previously mentioned, it must be entered. In a case when the beam cross-section has at least one axis of symmetry, the considered axes passing through the center of gravity, C (Figure 6.32(c)) are the principal centroidal axes, and in this case, $I_{xy} \equiv I_{12} = 0$. Now, according to Equation (6.103), we have:

$$\sigma_z = \frac{M_x}{I_x}y - \frac{M_y}{I_y}x. \tag{6.105}$$

In addition, for this case, Equation (6.104) becomes:

$$\tan\alpha = \frac{M_y I_x}{M_x I_y}; \left(\tan\alpha = \tan\varphi\frac{I_x}{I_y}\right) \tag{6.106}$$

All of the equations are derived in accordance with Figure 6.32(c) and Figure 6.32(d).

If a load (bending moment or force) is acting in a different direction, then its sign must be entered. As can be seen, the plus sign of φ is taken from the $+x$ axis (since M_y and M_x are positive, and now the plus sign of α is in the same direction).

A similar solution may be obtained if the coordinate system is as indicated in Figure 6.33(b), that is, if the principal centroidal axes are considered. In this situation, Equation (6.105), for example, may be obtained on the basis of the influence of the load on the fibers of the first (I) quadrant (Figure 6.33(b)). If fibers subjected to the load are elongated, then the load has a plus sign, and vice versa. Of course, the principal centroidal axes must be determined previously. In general, if the principal centroidal axes are used, the equation for determining stress is written as follows:

$$\sigma_z = \pm\sigma_z^{M_x} \pm \sigma_z^{M_y} = \pm\frac{M_x}{I_x}y \pm \frac{M_y}{I_y}x \tag{6.107}$$

Stress at the considered point of the profile (or section) is defined in accordance with the influence of the moment on the fibers of quadrant I (+ for extension; – for stretching) and taking into consideration the sign of the coordinates (x, y) of the considered point.

Example 6.7.
Consider the beam cross-section presented in Figure 6.33(a). This is an open-channel shape (open, thin-walled profile; [profile). Determine the neutral axis and the highest stress of the cross-section.

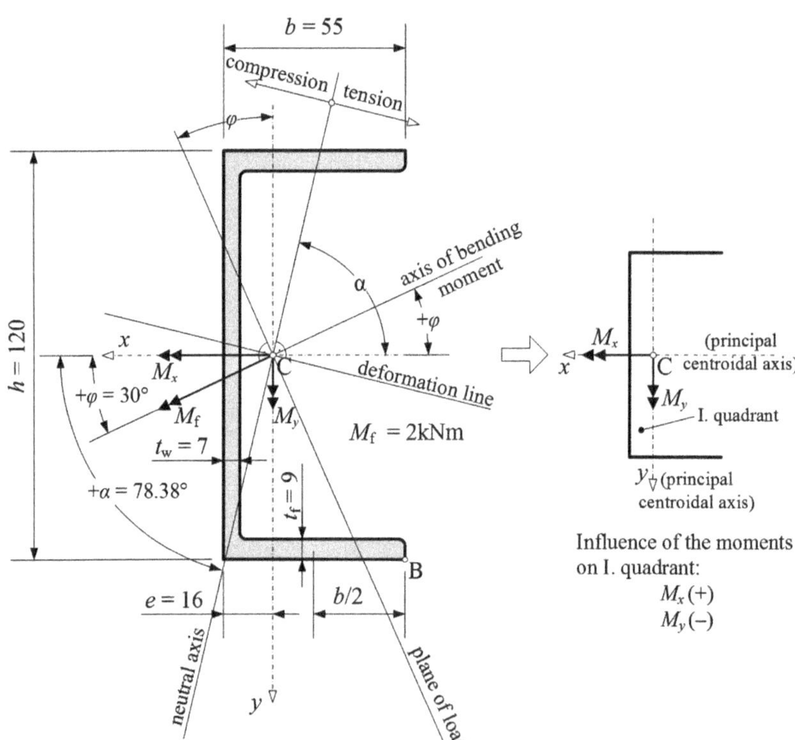

Figure 6.33 Nonsymmetrical bending of the open, uniaxial, symmetric profile (open-channel shape) considered in Example 6.7. (a) Geometry of the channel, the resultant bending moment, neutral axis; (b) the influence of the components of the moment on the fibers of quadrant I.

Data:

For the given shape of the cross-section of the beam, the following data are given: resultant bending moment $M_f = M = 2$ kNm; $\varphi = 30°$; axial moments of inertia: $I_x = 364$ cm^4, $I_y = 43.2$ cm^4; height of the profile $h = 120$ mm; width of the profile $b = 55$ mm; thickness of the web $t_w = 7$ mm; $e = 16$ mm; thickness of the flange $t_f = 9$ mm at the center of the flange. (This profile may be found in the standards).

Solution:

A graphical interpretation of the solution is presented in Figure 6.33(a).

Since the considered profile (shape) has an axis of symmetry, the orientation (slope of direction) of the neutral axis can be calculated using Equation (6.106):

$$\tan\alpha = \frac{M_y I_x}{M_x I_y} = \frac{M \cdot \sin 30° \cdot I_x}{M \cdot \cos 30° \cdot I_y} = 4.86 \rightarrow \alpha = 78.38°$$

As can be seen, the maximum stress will occur at point B. As stated, the shape has an axis of symmetry and the normal stress (caused by bending moments) can be determined using Equation (6.105):

$$\sigma_{z(B)} = \frac{M_x}{I_x} y - \frac{M_y}{I_y} x$$

$$= \frac{2\text{kNm} \cdot \cos 30°}{364 \cdot 10^{-8}\text{m}^4} 0.06\text{m} - \frac{2\text{kNm} \cdot \sin 30°}{43.2 \cdot 10^{-8}\text{m}^4}(-0.039\text{m}) = 118.8\text{MPa}$$

Example 6.8.

Consider the beam cross-section shown in Figure 6.34(a). This is an unequal "**L**" shape (unequal-leg angle shape). Determine the neutral axis and the stress at the point "D".

Data:

For the given shape of the cross-section of the beam, the following data are given: bending moment $M_f = M = 2$ kNm; $\varphi = 180°$ (It is clear from Figure 6.34(a) that the bending moment is actually: -2 kNm); axial moments of inertia: $I_x = 226$ cm^4, $I_y = 80.8$ cm^4; height of the profile $h = 120$ mm; width $b = 80$ mm; thickness of the web $t_w = 8$ mm; $e_x = 18.8$ mm, $e_y = 38.3$ mm; thickness of the flange $t_f = 8$ mm. (This profile may be found in the standards). For this geometry, $I_{xy} = -80.6$ cm^4 is calculated for a given coordinate system.

Solution:

The direction of the neutral axis may be calculated according to Equation (6.104):

$$\tan\alpha = \frac{y}{x} = \frac{M_y I_x + M_x I_{xy}}{M_y I_{xy} + M_x I_y} = \frac{M_x I_{xy}}{M_x I_y}$$

$$= -\frac{80.6\text{cm}^4}{80.8\text{cm}^4} = -0.9975 \rightarrow \alpha = -44.93°$$

Stress at point "D" can be calculated in accordance with Equation (6.103):

$$\sigma_{z(D)} = \frac{\left(M_y I_{xy} + M_x I_y\right) y}{\left(I_x I_y - I_{xy}^2\right)} - \frac{\left(M_y I_x + M_x I_{xy}\right) x}{\left(I_x I_y - I_{xy}^2\right)}$$

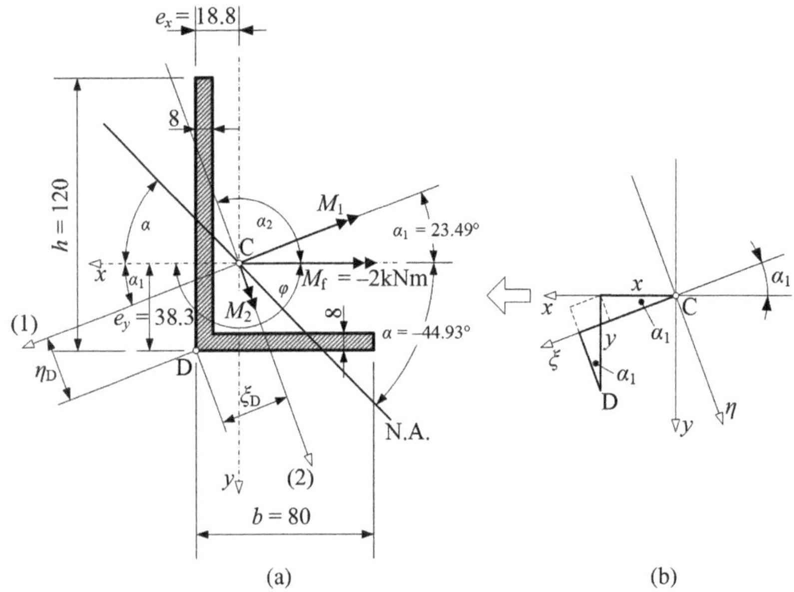

Figure 6.34 Nonsymmetrical bending of the open ("**L**") asymmetrical profile considered in Example 6.8. (a) An unequal "**L**" shape (geometry in mm), bending moment, principal centroidal axes of inertia; (b) the two coordinate systems.

Since

$$M_y = 0 \rightarrow \sigma_{z(D)} = \frac{M_x I_y \,.\, y - M_x I_{xy} \,.\, x}{I_x I_y - I_{xy}^2}$$

$$= \frac{-2 \cdot 10^6 \cdot 8.08 \cdot 10^5 \cdot 38.3 - \left(-2 \cdot 10^6 \cdot \left(-8.06 \cdot 10^5\right) \cdot 18.8\right)}{2.26 \cdot 10^6 \cdot 8.08 \cdot 10^5 - \left(-8.06 \cdot 10^5\right)^2}$$

$$\cong -78.3\,\text{MPa}$$

(Entered data: M (Nmm); I_x, I_y(mm^4); x, y(mm)

The result is also obtainable if the problem is considered with respect to the centroidal principal axes. To do this, it is necessary to determine the directions of the principal centroidal axes, the principal bending moments and the coordinates of the point "D" in this new coordinate system. The directions of the principal centroidal axes can be calculated according to Equation (6.77) for $I_x > I_y$ as:

$$\tan 2\alpha = -\frac{2I_{xy}}{I_x - I_y} = -\frac{2(-80.6\,\text{cm}^4)}{226\,\text{cm}^4 - 80.8\,\text{cm}^4} \rightarrow 2\alpha = 47.99°;$$

$$\alpha_1 = 23.99°$$

The directions of the principal centroidal axes of inertia designated (1) and (2) are shown in Figure 6.34. The designation "α" used in this case differs from the previously used symbol "α" in the determination of the neutral axis direction. The principal centroidal moments of inertia which correspond to these directions can be determined by Equation (6.78):

$$I_{1,2} = \frac{1}{2}\left(I_x + I_y\right) \pm \frac{1}{2}\sqrt{\left(I_x - I_y\right)^2 + 4I_{xy}^2} \rightarrow$$

$$I_{1,2} = \frac{1}{2}(226 + 80.8) \pm \frac{1}{2}\sqrt{(226 - 80.8)^2 + 4(80.6)^2}$$

$$I_1 = 261.87\,\text{cm}^4;\ \ I_2 = 44.9\,\text{cm}^4$$

Stress at the point "D" can now be calculated by the principal bending moments M_1 and M_2 which act on the principal centroidal axes of inertia. The level of stress is defined by Equation (6.107):

$$\sigma_z = \pm\frac{M_1}{I_1}\eta \pm \frac{M_2}{I_2}\xi$$

and for the point D, we have:

$$\sigma_{z(D)} = -\frac{M_1}{I_1}\eta_D - \frac{M_2}{I_2}\xi_D$$

The minus sign means that the bending moments shorten the fibers of the first quadrant. The principal bending moments (absolute values) are:

$$M_1 = M\cos\alpha_1 = 1.827\,\text{kNm};\ \ M_2 = M\cos\alpha_1 = 0.813\,\text{kNm}$$

The coordinates of the point "D" (with respect to the centroid C) in the $(\xi - \eta)$ coordinate system (Figure 6.34(b)) are:

$$\xi_D = x_D\cos\alpha_1 + y_D sin\alpha_1$$

$$= 18.8\ \cos 23.99^\circ + 38.3\ \sin 23.99^\circ = 32.75\,\text{mm}$$

$$\eta_D = y_D\cos\alpha_1 - x_D\sin\alpha_1$$

$$= 38.3\ \cos 23.99^\circ - 18.8\ \sin 23.99^\circ = 27.35\,\text{mm}$$

The stress level is:

$$\sigma_{z(D)} = -\frac{M_1}{I_1}\eta_D - \frac{M_2}{I_2}\xi_D$$

$$= -\frac{1827000\,\text{Nmm}}{2618700\,\text{mm}^4}\cdot 27.35 - \frac{813000\,\text{Nmm}}{449990\,\text{mm}^4}\cdot 32.75 \cong -78.3\,\text{MPa}$$

As can be seen, results obtained using different approaches match.

6.3.6 Loading of Thin-walled Engineering Elements; Shear Center

Section 6.3.4 looked at symmetrical bending, while Section 6.3.5 considered nonsymmetrical bending. However, in each of these sections, only a bending response was considered. Let's consider again Figure 6.33(a). If a load (bending moment or force) is applied in

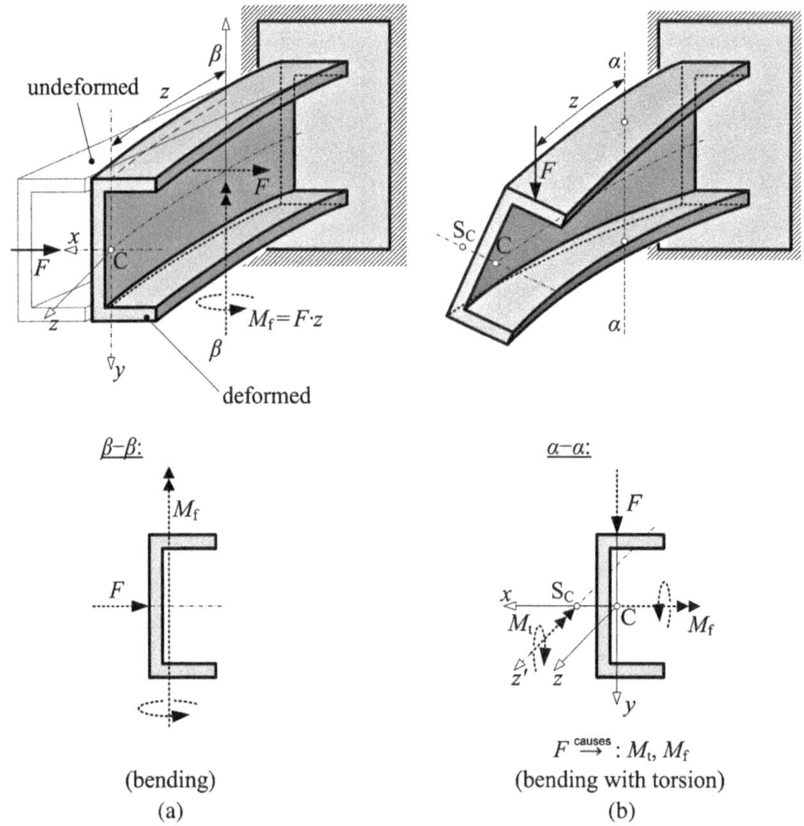

Figure 6.35 A profile possessing one plane of symmetry. (a) [profile loaded by a force acting in the principal centroidal plane of inertia that is a plane of symmetry; (b) [profile loaded by a force acting in the principal centroidal plane of inertia that is not a plane of symmetry – bending with torsion.

a plane which is, at the same time, a plane of symmetry (that is, this plane of symmetry is also the principal centroidal plane of inertia) of this profile, then only bending will arise in the plane of loading. In this scenario, the weight of the beam is neglected. We usually say that the bending (deflexion or deformation) occurs in the plane of the load. In a case when a force is applied as shown in Figure 6.33(a), in any considered section of the beam, the bending moment and shear force will act and this will result in normal and shear stresses. If a load is applied in a plane that is not the plane of symmetry (although it is the principal centroidal plane of inertia), as in Figure 6.35(b), the member will be observed to bend and twist under the applied load.

6.3.6.1 Shear Center

Consider the open-channel shape shown in Figure 6.33(a), which possesses one plane of symmetry (x–z). Let a force "F" act in a plane (y–z) that is not a plane of symmetry although it is the principal centroidal plane of inertia (see Figure 6.35(b)). It is clear that bending is accompanied by torsion [3].

One can try to discover whether it is possible to find a position where the force "F" must act in such a way that a member will be subjected to bending only (Figure 6.36). If the considered element bends without twisting, then the shear force must act as shown in Figure 6.36(b) (see also Figure 6.36(c)). This point is known as the *shear center* (S_C, or sometimes designated C_S). The shear stress distribution is shown in Figure 6.36(d). For a complete picture of stress, the distribution should still include a display of the distribution of normal stresses along the web of the profile. These normal stresses along the web

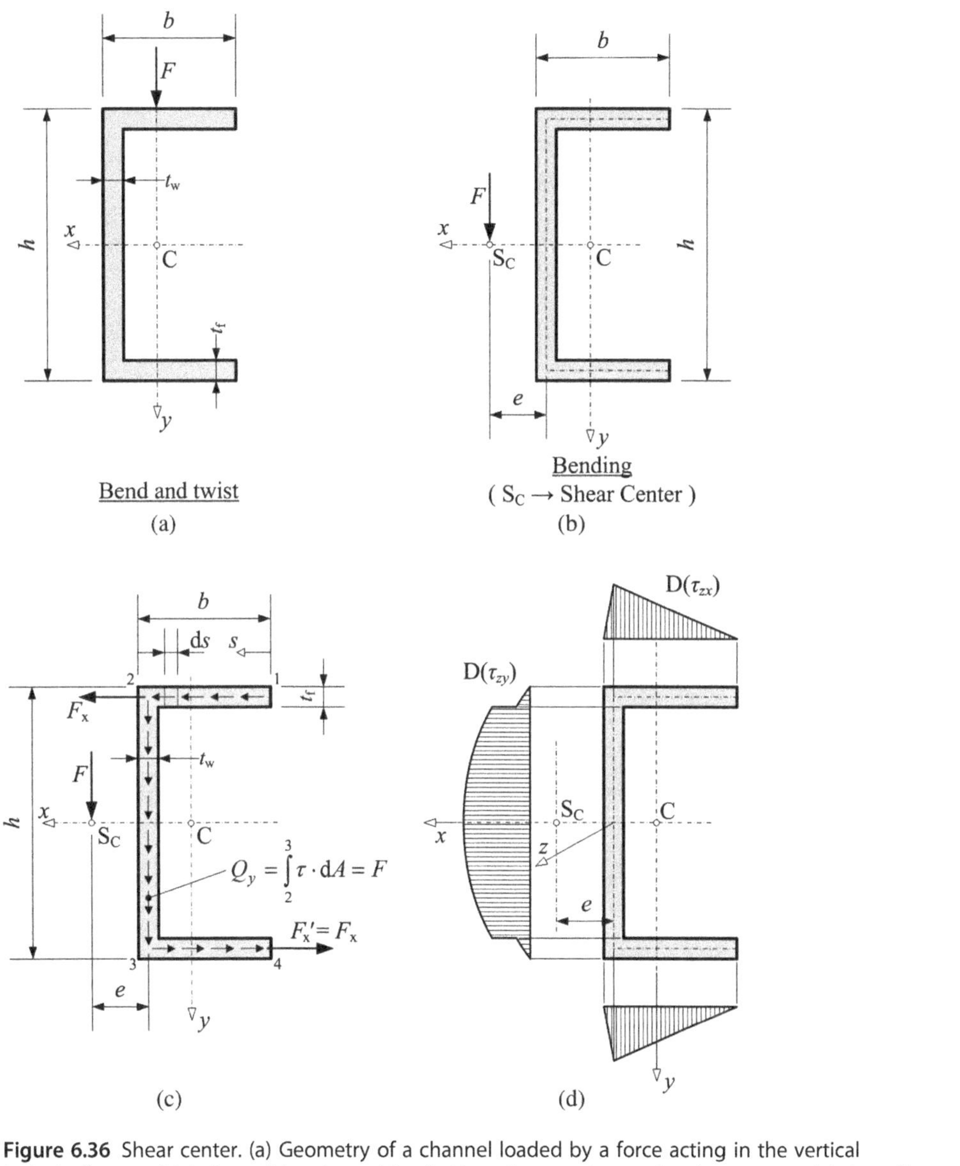

Figure 6.36 Shear center. (a) Geometry of a channel loaded by a force acting in the vertical (principal) centroidal plane; (b) a channel loaded by a force acting at the shear center; (c) shear flow in a channel; (d) shear stress distribution along the web and along the flange of the profile.

are caused by the bending moment. The distribution of normal stresses is linear along the web of the profile with zero value at the center of gravity (centroid) and with the peak occurring in the fibers that are farthest from the centroid.

In accordance with Figure 6.36(c), we have:

$$F = \int_2^3 \tau \cdot \mathrm{d}A = \int_2^3 \tau \cdot t \cdot \mathrm{d}s = \int_2^3 q \cdot \mathrm{d}s = Q_y \tag{6.108}$$

$$F_x = \int_1^2 \tau \cdot \mathrm{d}A = \int_1^2 \tau \cdot t \cdot \mathrm{d}s = \int_1^2 q \cdot \mathrm{d}s = \left| -F'_x \right| = \int_3^4 q \cdot \mathrm{d}s \tag{6.109}$$

The thickness, t, in the above equations refers to the thickness of the web (t_w), or the flange (t_f).

Based on the equilibrium equation of moments of all the forces for the shear center (S_C), Figure 6.36(c) yields:

$$\begin{aligned} &\sum M_{S_C}^{F_i} = 0 \rightarrow Q_y \cdot e = F_x \cdot h \rightarrow \\ &\rightarrow e = \frac{F_x}{Q_y} \cdot h; \left(Q_y \cong F\right), \end{aligned} \tag{6.110}$$

where, for the considered profile ([), force F_x is:

$$F_x = \frac{Q_y \cdot b^2 \cdot h \cdot t_f}{4I_x} \tag{6.111}$$

Now, the position of the shear center is:

$$e = \frac{b^2 h^2 t_f}{4I_x}. \tag{6.112}$$

The axial moment of inertia of the considered profile is:

$$\begin{aligned} &I_x = I_W + 2I_f \rightarrow \\ &I_x = \frac{t_w h^3}{12} + 2\left(\frac{b\, t_f^3}{12} + b t_f \left(\frac{h}{2}\right)^2\right), \end{aligned} \tag{a}$$

where the following designations are used: w = web, f = flange.

Finally, for this channel ([), the shear center, neglecting (t_f^3), can be determined by:

$$e = \frac{3b^2 t_f}{h t_w + 6 b t_f}. \tag{6.113}$$

6.3.7 Beam Deflections

In the design of beams, it is important to know the distribution of stresses in their cross-sections but also their deflections. The deflection of a beam is associated with the function for which it is intended. A deflection depends on the flexural stiffness (flexural rigidity), see Equation (6.84), applied loads and the types of supports (Figure 6.37) [1, 3–4]. The axial rigidity depends on the material used and the dimensions of the cross-section.

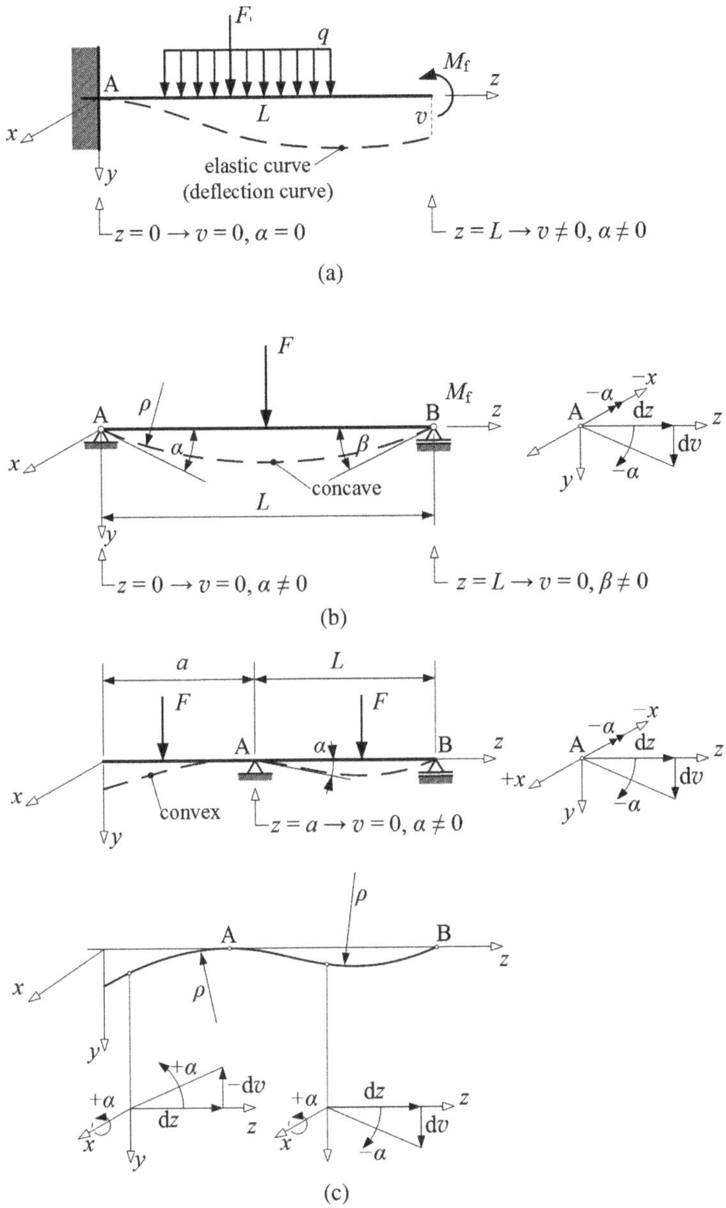

Figure 6.37 Boundary conditions. (a) Cantilever beam; (b) simply supported beam; (c) beam with overhang.

Figure 6.37 shows some types of possible boundary conditions. As we know, curvature is defined by:

$$\kappa = \frac{1}{\rho} = \pm \frac{d^2 v / dz^2}{\left[1 + (dv/dz)^2\right]^{3/2}} \tag{6.114a}$$

Since only small deflections are considered (that is, the slope (dv/dz) may be treated as small), we may write:

$$\frac{1}{\rho} \cong \pm \frac{d^2 v}{dz^2} \tag{6.114b}$$

In accordance with Figure 6.37, it follows that:

$$dz = \rho d\alpha \rightarrow \frac{1}{r} = \frac{d\alpha}{dz} = \frac{d}{dz}\left(-\frac{dv}{dz}\right) \rightarrow \frac{1}{\rho} = -\frac{d^2 v}{dz^2} \tag{6.115}$$

Since a small deformation of the beam is being considered (and also a small deflection), the arc length, ds, of the deformed longitudinal axis of the beam can be substituted by dz. The sign (–) on the right-hand side of Equation (6.115) is defined in accordance with the coordinate system used in Figure 6.37.

It is clear that a positive increase in the z direction and a positive increase in the y direction cause a negative rotation around the x axis (that is, a negative angle, α).

Comparing Equation (6.84) and Equations (6.115), it follows that:

$$\frac{1}{\rho} = \frac{M_x}{EI_x} = -\frac{d^2 v}{dz^2}$$
$$EI_x v'' = -M_x; \; (M_x = M_f) \tag{6.116}$$

This equation is known as the moment–curvature relationship and assumes small deflections.

In this equation, the bending moment is considered to be positive if the elastic curve of the beam created by it is of concave form in the plane $(y–z)$ (see Figure 6.37 or Figure 6.43 (c)). Equation (6.116) is also known as the differential equation of the elastic curve. As we know, the following relationships between bending moment, shear force and distributed load can be written (from Equation (6.69)):

$$M_x = -EI_x v'' \tag{6.117a}$$

$$Q_y = \frac{dM_x}{dz} = -EI_x v''' \left(= -EI_x \frac{d^3 v}{dz^3}\right) \tag{6.117b}$$

$$q = -\frac{dQ}{dz} = EI_x v'''' \left(= EI_x \frac{d^4 v}{dz^4}\right). \tag{6.117c}$$

Example 6.9.

Consider the cantilever beam shown in Figure 6.38, subjected to the force F and bending moment M at the free end. Determine the deflection and slope of the beam at B.

Data:

Known length of the beam, L, modulus, E and axial moment of inertia, I_x.

Solution:

In accordance with the coordinate system established in this case (Figure 6.38), the differential equation (Equation (6.116)) is now:

$$EI_x v'' = -M_x \tag{a}$$

Figure 6.38 Cantilever beam, slope and deflection at point B (Example 6.9). (a) Cantilever beam with prescribed loads; (b) deflection due to force; (c) deflection due to bending moment.

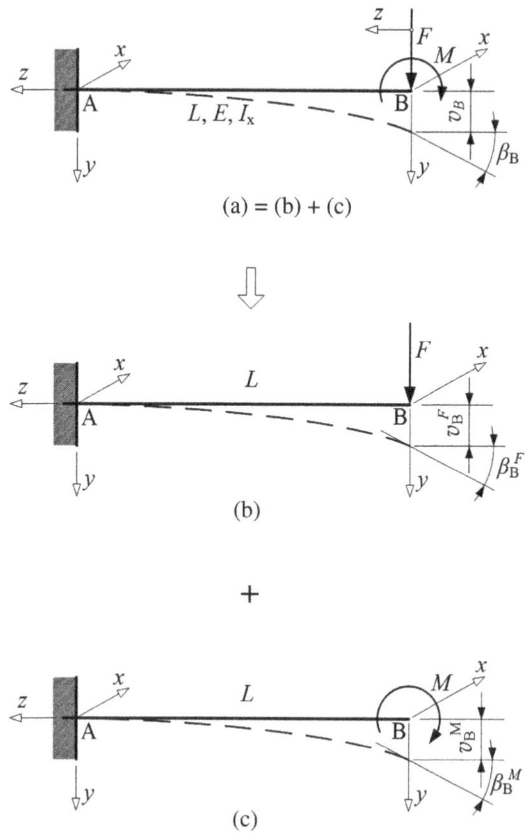

Since:

$$M_x = -F \cdot z - M \rightarrow EI_x v'' = F \cdot z + M$$

Coordinate "z" is taken from the point B on the left side; that is, in this case, the origin of the coordinate system is placed at B. A first integration of the value defined by Equation (a) gives:

$$EI_x v' = \frac{F \cdot z^2}{2} + M \cdot z + C_1 \tag{b}$$

The boundary conditions are:

1) $z = L \rightarrow v' = 0 \rightarrow C_1$

$$C_1 = -\frac{FL^2}{2} - ML$$

A second integration of the value in Equation (b) gives:

$$EI_x v = \frac{F \cdot z^3}{6} + \frac{M \cdot z^2}{2} + C_1 \cdot z + C_2 \tag{c}$$

2) $$z = L \rightarrow v = 0 \rightarrow C_2$$

$$C_2 = \frac{FL^3}{3} + \frac{ML^2}{2}$$

Entering constants C_1 and C_2 into Equations (b) and (c) gives the slope and deflection at B $(z = 0)$ as:

$$\beta_B = -v' = \frac{1}{EI_x}\left(\frac{FL^2}{2} + ML\right);\ v = \frac{1}{EI_x}\left(\frac{FL^3}{3} + \frac{ML^2}{2}\right) \tag{d}$$

6.3.8 Bending of Curved Elements

All of our previous analyses of stress due to bending have been restricted to straight engineering elements. Even when initial curvature of the engineering element exists but is very small, the stress distribution obtained using the formulas for a straight member is a good approximation. Under a small curvature of an engineering element, it is understood that its radius of curvature is large compared to the depth of its cross-section. When the initial curvature of the element is no longer small compared to the depth of its cross-section, then a new method of stress analysis has to be applied. Our analysis will be focused on initially curved beams possessing a plane of symmetry on which equal and positive opposite couples are applied. Couple M is positive when it tends to increase the curvature of the considered member. A curved element of uniform cross-section which is symmetric to the y axis in its undeformed (unstrained) state is shown in Figure 6.39(a). The deformed state of this curved element subjected to mutually opposing couples of the same intensity, M, acting in the plane of symmetry of the element is shown in Figure 6.39(c).

In this consideration, $M = M_\varphi$ is the bending moment. The stress state is considered to be uniaxial. In addition, similar to the pure bending of a symmetric element (the member possesses a symmetric cross-section), here we can also assume that any perpendicular cross-section that is planar before deformation remains planar after deformation, and the radius of curvature after deformation will be transformed to a new radius. An assumed cross-sectional shape is shown in Figure 6.39(b). It is known from experiments that opposing couples with intensity M cause extension of the fibers on the surfaces which are farther than the neutral surface relative to the center of curvature, and shortening of the fibers on the surfaces nearer than the neutral surface [3]. Obviously, the length of the fibers on the neutral surface remains unchanged (constant).

The position of the neutral surface needs to be determined. Let $\bar{G}\bar{H}$ be the neutral line of the element shown in the deformed position in Figure 6.39(c). It is obtained by the intersection of the vertical plane (y–z) with the assumed neutral surface. Figure 6.39 (a) shows the mapped position of the neutral surface in the undeformed position, just to display its imaginary position. A consideration includes the following settings:

1) Deformation is considered to be small.
2) $\Delta\varphi > 0 \text{ for } M > 0$.

In accordance with the mentioned elongation and shortening of the fibers, it follows that: $\bar{A}\bar{B} < AB$, and $\bar{E}\bar{F} > EF$. It is necessary to determine the position of the neutral surface in the cross-section of the element (beam). The fact that the length of the neutral line remains constant, that is, $GH = \bar{G}\bar{H}$, can be expressed in the form [3]:

$$\rho \cdot \varphi = \bar{\rho} \cdot \bar{\varphi} \tag{6.118}$$

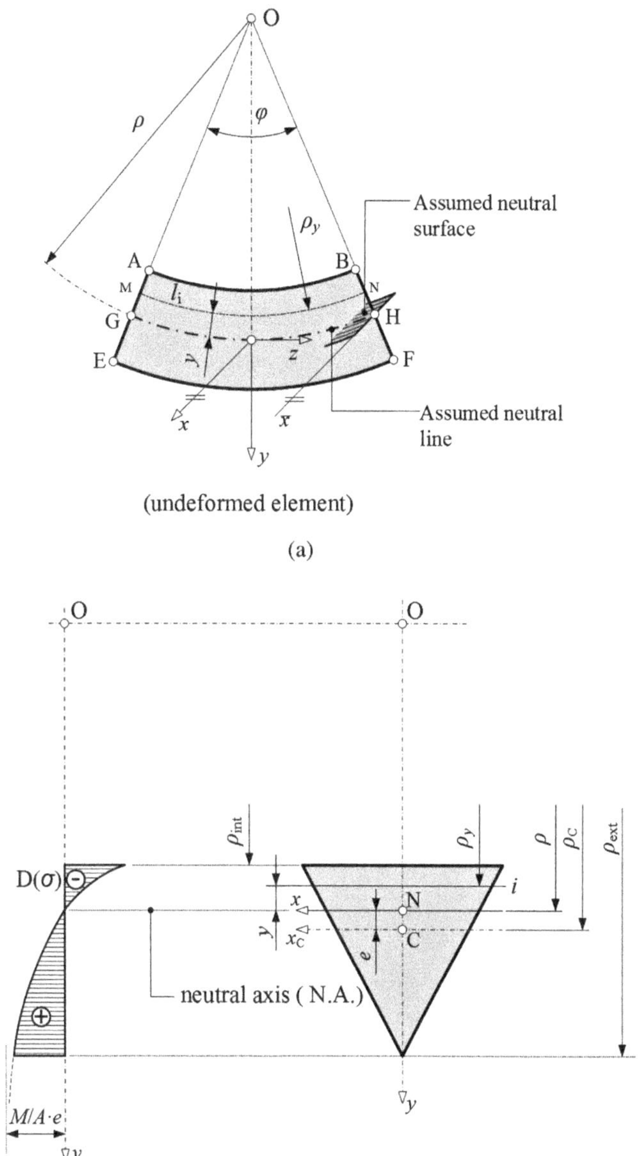

Figure 6.39 Curved engineering element. (a) A segment of the undeformed curved element; (b) the cross-section of the curved element; (c) a segment of the deformed curved element; (d) different cross-sectional shapes of the curved element – neutral axis.

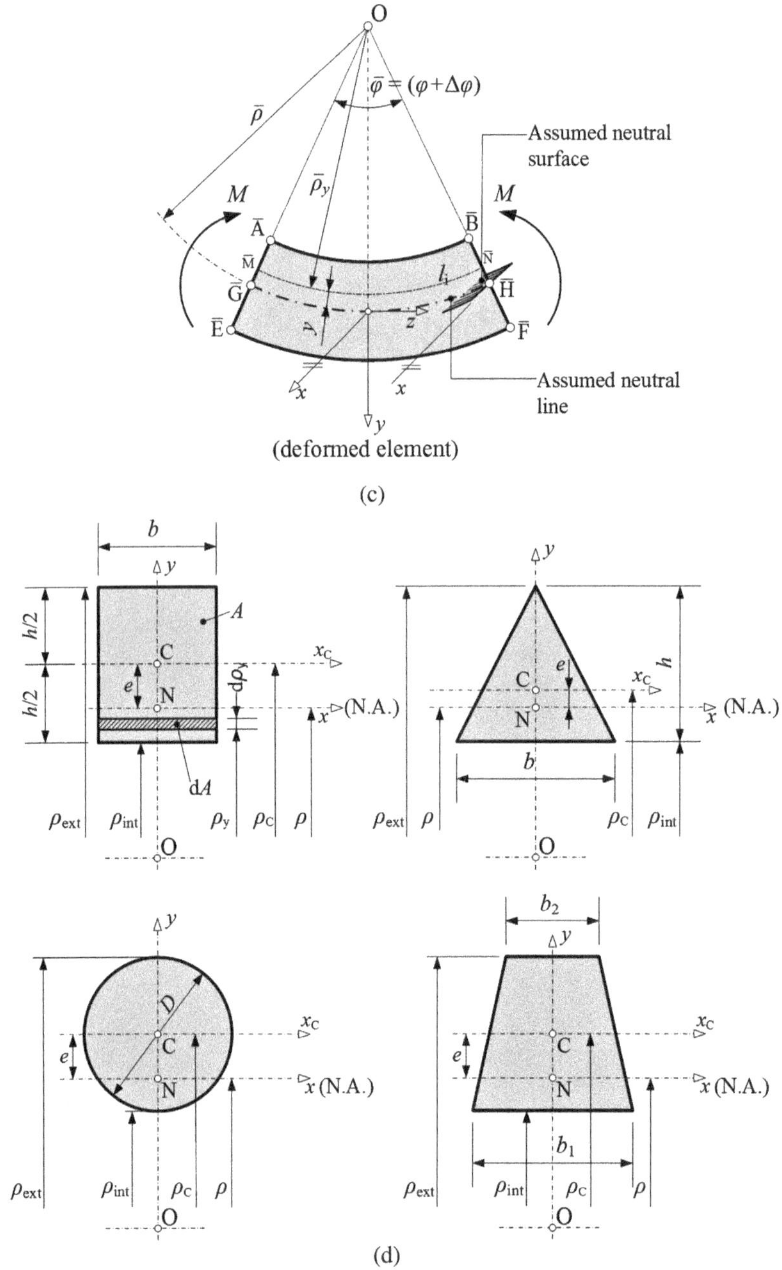

Figure 6.39 *(Continued)*

Consider the deformation of the "ith" fiber (MN, $\bar{M}\bar{N}$) whose radii are: ρ_y and $\bar{\rho}_y$. This deformation can be obtained as the quotient of the length of the considered fiber after and before the couples have been applied. This gives:

$$\varepsilon_z = \frac{\bar{\rho}_y \cdot \bar{\varphi} - \rho_y \cdot \varphi}{\rho_y \cdot \varphi}. \tag{6.119}$$

Since:

$$\bar{\rho}_y = \bar{\rho} - y; \;\; \rho_y = \rho - y; \;\; \bar{\varphi} - \varphi = \Delta\varphi, \tag{a}$$

and using Equation (6.118), it follows that:

$$\varepsilon_z = -y\frac{\Delta\varphi}{\rho_y \cdot \varphi} = -\frac{\Delta\varphi}{\varphi}\frac{y}{\rho - y} \tag{6.120a}$$

It is clear that the normal strain, ε_z, does not change linearly along the vertical axis, y, which is perpendicular to the neutral axis (NA) (that is, with the distance y from the neutral surface).

Entering the second of the Equations (a) into Equation (6.120a), it follows that:

$$\varepsilon_z = -\frac{\Delta\varphi}{\varphi}\frac{\rho - \rho_y}{\rho_y} \tag{6.120b}$$

Using Hooke's Law, we have:

$$\sigma_z = E \cdot \varepsilon_z = -E\frac{\Delta\varphi}{\varphi}\frac{\rho - \rho_y}{\rho_y}, \tag{6.121}$$

And now it is clear that the stress distribution along the y axis (that is, with the distance y from the neutral surface) is hyperbolic. Furthermore, by applying the first of the equilibrium equations (Equation (6.82)), it follows that:

1) $$\sum F_z = 0 \;\rightarrow\; \int \sigma_z \mathrm{d}A = 0. \tag{6.122}$$

By entering the value from Equation (6.121) into Equation (6.122), the following expression can be written:

$$\begin{gathered}
\int -E\frac{\Delta\varphi}{\varphi}\frac{\rho - \rho_y}{\rho_y}\mathrm{d}A = 0 \\
E\frac{\Delta\varphi}{\varphi} \neq 0 \;\rightarrow\; -\int \frac{\rho - \rho_y}{\rho_y}\mathrm{d}A = 0 \\
\rho\int\frac{\mathrm{d}A}{\rho_y} - \int \mathrm{d}A = 0 \;\rightarrow \\
\rho = \frac{A}{\int \frac{\mathrm{d}A}{\rho_y}}
\end{gathered} \tag{6.123}$$

Equation (6.123) defines the position of radius "ρ" of the neutral axis with respect to the center of curvature (O). Let us consider the rectangular cross-section shown in Figure 6.39(d). The first (static) moment of the area (A) of this rectangular cross-section with respect to the x axis passing through the center of curvature (O) is:

$$S_{x(\mathrm{C})} = A \cdot \rho_{\mathrm{C}} = \int \rho_y \cdot \mathrm{d}A \rightarrow \rho_{\mathrm{C}} = \frac{\int \rho_y \cdot \mathrm{d}A}{A} \tag{6.124}$$

where ρ_C is the distance between the centroid (C) of the cross-section and the center of curvature (O).

Further, for a rectangular cross-section, we have:

$$A = b \cdot h \tag{b}$$

$$\mathrm{d}A = b \cdot \mathrm{d}\rho_y \tag{c}$$

In accordance with Equation (6.123), we have:

$$\rho = \frac{b \cdot h}{\int_{\rho_{int}}^{\rho_{ext}} \frac{b \cdot \mathrm{d}\rho_y}{\rho_y}} = \frac{b \cdot h}{b \int_{\rho_{int}}^{\rho_{ext}} \frac{\mathrm{d}\rho_y}{\rho_y}} = \frac{h}{\ln \rho_{ext} - \ln \rho_{int}} \tag{d}$$

$$\rho = \frac{h}{\ln \frac{\rho_{ext}}{\rho_{int}}} \tag{e}$$

In Equation (e), ρ_{ext} and ρ_{int} represent the external and internal radius of the cross-section, respectively.

In accordance with Figure 6.39(c), it is clear that, for $M > 0 \rightarrow \Delta\varphi > 0$. Referring to Equations (6.123) and (6.124), it follows that $\rho_C - \rho > 0$ (this can be verified at any cross-section). This means that the neutral axis of a considered cross-section is located between the centroid (C) of the section and the center of curvature (O) of the element. For some typically used cross-sectional shapes, data relating to the position of the neutral axis are given in Figure 6.39(d), while the mathematical expressions are as follows:

Rectangle

$$\rho = \frac{h}{\ln \frac{\rho_{ext}}{\rho_{int}}} \tag{6.125}$$

Triangle

$$\rho = \frac{1}{2} \cdot \frac{b \cdot h}{\frac{b}{h} \cdot \rho_{ext} \cdot \ln \frac{\rho_{ext}}{\rho_{int}} - b} \tag{6.126}$$

Circle

$$\rho = \frac{D^2}{4\left(2\rho_C - \sqrt{4\rho_C^2 - D^2}\right)} \tag{6.127}$$

Trapezoid

$$\rho = \frac{1}{2} \cdot \frac{(b_1 + b_2) \cdot h}{\left(b_2 + \frac{b_1 - b_2}{h} \cdot \rho_{ext}\right) \cdot \ln \frac{\rho_{ext}}{\rho_{int}} - (b_1 - b_2)} \tag{6.128}$$

By applying the second of the equations of equilibrium (Equation (6.83)), it follows that:

2) $$\sum M_x = 0 \rightarrow \int \sigma_z \cdot (-y) \mathrm{d}A = M. \tag{6.129}$$

If the data for the stress, σ_z, (Equation (6.121)) and for $y\left(\rightarrow y = \rho - \rho_y\right)$, obtained from Equation (a), are entered into Equation (6.129), it follows that:

$$E \frac{\Delta\varphi}{\varphi} \int \frac{\left(\rho - \rho_y\right)^2}{\rho_y} \mathrm{d}A = M \tag{6.130}$$

After some mathematical calculations, and taking into consideration Equations (6.123) and (6.124), we obtain:

$$E\frac{\Delta\varphi}{\varphi}=\frac{M}{A(\rho_C-\rho)}=\frac{M}{A\cdot e}. \tag{6.131}$$

The product "$A\cdot e$" is the first moment of the area of the cross-section of a member with respect to the neutral axis passing through the point N, that is, for the neutral axsis ($A\cdot e=S_{x(\mathrm{NA})}$). Finally, the stress ($\sigma_z=\sigma_\varphi$) is defined as (from Equation (6.121)):

$$\sigma_z=E\cdot\varepsilon_z=-\frac{M\cdot y}{A\cdot e(\rho-y)}, \tag{6.132a}$$

or:

$$\sigma_z=\frac{M\cdot\left(\rho_y-\rho\right)}{A\cdot e\cdot\rho_y}. \tag{6.132b}$$

$$\sigma_z=\frac{M}{A\cdot e}\left(1-\frac{\rho}{\rho_y}\right) \tag{6.132c}$$

Based on Equation (6.132c), it is clear that when:

$$\rho_y\rightarrow 0;\quad \sigma_z=\sigma_\varphi\rightarrow-\infty$$

$$\rho_y\rightarrow\infty;\quad \sigma_z=\sigma_\varphi\rightarrow\frac{M}{A\cdot e}=\frac{M}{S_x}$$

and, in accordance with this explanation, the normal stress distribution is presented in Figure 6.39(b) for the case where a curved beam is subjected to a positive bending moment. As stated, a bending moment is considered to be positive when it seeks to increase the curvature of the beam.

In Equation (6.132b), ρ_y is the distance between the considered point in the cross-section and the center of curvature (O). The maximal and minimal values of the radius, ρ_y are: $\rho_{y,\,\max}=\rho_{\mathrm{ext}}$ and $\rho_{y,\,\min}=\rho_{\mathrm{int}}$, respectively. The stress distribution is presented in Figure 6.39(b). The normal stress, σ_z, that is perpendicular to the cross-section of the member sometimes may be designated σ_φ. This is done by pointing out that it was a stress in an element with an initial curvature.

Finally, by applying the third of the equations of equilibrium (the moment about the y axis), analysis shows that axis "y" is the axis of symmetry, which was our assumption at the beginning. Thus:

$$\sum M_y=0\rightarrow\int\sigma_z\cdot x\cdot dA=0$$

Example 6.10.

Consider a curved beam loaded as shown in Figure 6.40. Determine the extreme values of stress.

Data:

Data defining the geometry of the beam as well as the load can be found in Figure 6.40.

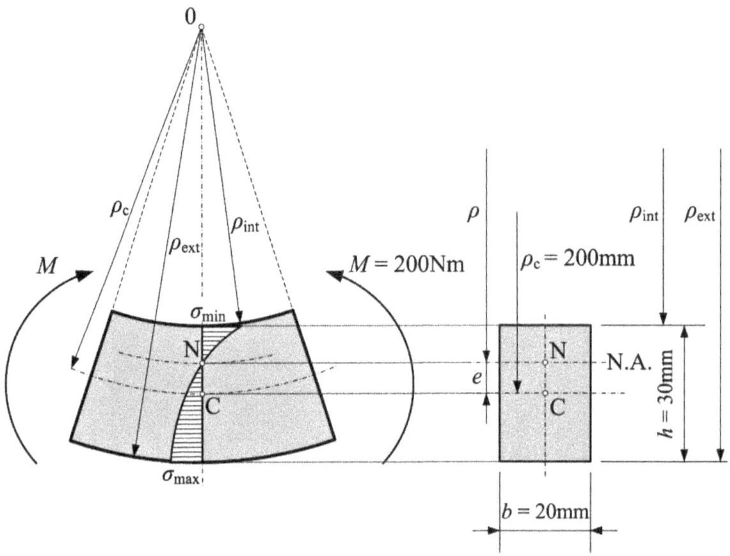

Figure 6.40 A curved beam subjected to a bending moment – cross-sectional geometry, load and stress distribution (Example 6.10).

Solution:

The distance of the neutral surface (neutral line) from the center of curvature (O) is (from Equation (6.125)):

$$\rho = \frac{A}{\int \frac{dA}{\rho_y}} = \frac{h}{\ln \frac{\rho_{ext}}{\rho_{int}}} = 199.62\,\text{mm}$$

The area of the cross-section is:

$$A = b \cdot h = 600\,\text{mm}^2,$$

and the distance between the centroid and the neutral axis ($\overline{NC}$, Figure 6.40) is:

$$e = \rho_C - \rho = 200 - 199.62 = 0.38\,\text{mm}.$$

The stresses are:

$$\sigma_{max} = \frac{M(\rho_{ext} - \rho)}{A \cdot e \cdot \rho_{ext}} = \frac{200000\,\text{Nmm}(215 - 199.62)}{600 \cdot 0.38 \cdot 215} = 62.75\,\text{MPa}$$

$$\sigma_{min} = \frac{M(\rho_{int} - \rho)}{A \cdot e \cdot \rho_{int}} = \frac{200000\,\text{Nmm}(185 - 199.62)}{600 \cdot 0.38 \cdot 185} = -69.32\,\text{MPa}$$

$$|\sigma_{max}| < |\sigma_{min}|$$

Example 6.11.

Consider a crane hook (Figure 6.41(a)) made of steel that has a yield stress of $YS = 520$ MPa. Assume that the hook has been designed with a factor of safety of $SF = 2$ against the initiation of yielding. Determine the maximum load F that can be carried by the hook.

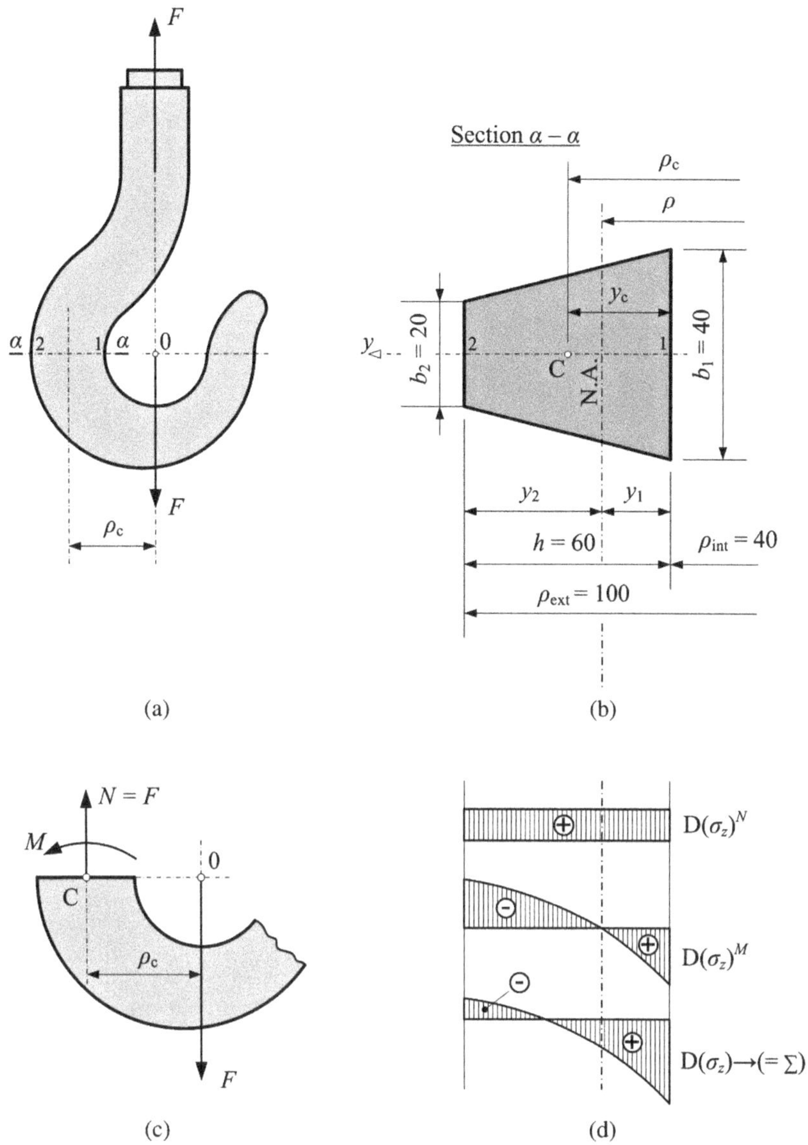

Figure 6.41 The crane hook considered in Example 6.11. (a) The hook shape and load; (b) the geometry of the cross-section; (c) a segment of the hook; (d) the stress distribution.

Data:

The cross-sectional shape of the hook is shown in Figure 6.41(b).

Solution:

The distance between the neutral axis and the center of curvature (O) is (from Equation (6.128)):

$$\rho = \frac{1}{2} \cdot \frac{(b_1 + b_2) \cdot h}{\left(b_2 + \frac{b_1 - b_2}{h} \cdot \rho_{\text{ext}}\right) \cdot \ln \frac{\rho_{\text{ext}}}{\rho_{\text{int}}} - (b_1 - b_2)}$$

$$= \frac{1}{2} \frac{(40 + 20)60}{\left(20 + \frac{40 - 20}{60} \cdot 100\right) \cdot \ln \frac{100}{40} - (40 - 20)} = 62.38\,\text{mm}$$

The position of the centroid (from the b_1 basis) is:

$$y_C = \frac{h}{3} \frac{b_1 + 2b_2}{b_1 + b_2} = 26.67\,\text{mm},$$

and the position of the centroid relative to the center of curvature is:

$$\rho_C = \rho_{\text{int}} + y_C = 66.67\,\text{mm}$$

The area of the cross-section is:

$$A = h \cdot \frac{b_1 + b_2}{2} = 1800\,\text{mm}^2$$

The distance between the centroid and the neutral axis is:

$$e = \rho_C - \rho = 66.67 - 62.38 = 4.29\,\text{mm}$$

Stress at the points "1" and "2" of the section "$\alpha - \alpha$" is:

$$\sigma = \sigma^N + \sigma^M;\ \sigma^M \rightarrow \text{Equation}\,(6.132\text{b})$$

Since the direction of the moment is opposite to that shown in Figure 6.39(c), it follows that:

$$\sigma = \frac{F}{A} - \frac{M \cdot \left(\rho_y - \rho\right)}{A \cdot e \cdot \rho_y}$$

Because ρ_y is ρ_{int} or ρ_{ext}, it is evident that the maximum stress occurs at the point "1":

$$\sigma_{(1)} = \frac{F}{A} - \frac{M \cdot \left(\rho_y - \rho\right)}{A \cdot e \cdot \rho_y}$$

The allowable force, F, taking into account the factor of safety, is:

$$\sigma_{(1)} = \frac{F}{A} - \frac{F \cdot \rho_C \left(\rho_y - \rho\right)}{A \cdot e \cdot \rho_y} \leq \frac{YS}{2}$$

$$F = \frac{\frac{YS}{2} \cdot A \cdot e \cdot \rho_y}{e \cdot \rho_y - \rho_C \left(\rho_y - \rho\right)} \cong 48.3\,\text{kN}$$

$$\frac{F}{A} \cong 26.8\,\text{MPa}, \sigma_{(1)} \cong 260\,\text{MPa}$$

The magnitudes of the stress can also be determined as:

$$\sigma_z = \frac{F}{A} - \frac{M\left(\rho_y - \rho\right)}{A \cdot e \cdot \rho_y} = \frac{F}{A} - \frac{M}{S_x}\frac{y_i}{\rho_{yi}}$$

where:

$$M = F \cdot \rho_C; \; S_x = A \cdot e; \; y_i = (y_1, y_2) = \rho_{yi} - \rho; \; \rho_{yi} = \rho_{\text{int}}, \rho_{\text{ext}}$$

For example, normal stress, $\sigma_{z(1)}$, is:

$$\begin{aligned}\sigma_{z(1)} &= \frac{F}{A} - \frac{M}{S_x}\frac{y_1}{\rho_{\text{int}}} = \frac{F}{A} - \frac{F \cdot \rho_C}{A \cdot e} \cdot \frac{y_1}{\rho_{\text{int}}} \\ &= \frac{48300}{1800} - \frac{48300 \cdot 66.67}{1800 \cdot 4.29} \cdot \frac{40 - 62.38}{40} \\ &\cong 260\,\text{MPa}\end{aligned}$$

$$\begin{aligned}\sigma_{z(2)} &= \frac{F}{A} - \frac{M}{S_x}\frac{y_2}{\rho_{\text{ext}}} = \frac{F}{A} - \frac{F \cdot \rho_C}{A \cdot e} \cdot \frac{y_2}{\rho_{\text{ext}}} \\ &= \frac{48300}{1800} - \frac{48300 \cdot 66.67}{1800 \cdot 4.29} \cdot \frac{100 - 62.38}{100} \\ &\cong -16.7\,\text{MPa}\end{aligned}$$

The stress distribution is shown in Figure 6.41(d).

6.4 Stability of Columns

Previous analyses of engineering elements were based, in general, on designs that met the allowable stress and/or allowable deformation values but neglected the possibility of changes in their configurations [1–3]. Consider a straight column (Figure 6.42(a)) subjected to compressive force passing through the centroids of the column's cross-sections. The column is made of linearly elastic material that follows Hooke's Law. If the level of the stress, $\sigma = F/A$ (F is force, A is the cross-sectional area) is less than the allowable stress and the deformation, ΔL, is also within the allowable range, it may be said that the column has been designed properly. Usually it can be said that under sufficiently small compressive load, the column remains in equilibrium in its straight position. However, a possible change in its configuration has not been considered in the previous discussion. It is of interest to investigate what happens in the case where the above-mentioned constraints related to the stress and deformation are fulfilled but under applied load, the column is no longer straight – that is, the column ceases to be stable.

To derive the level of compressive force that causes instability in the column, that is, the level of the so-called critical force, consider the column in Figure 6.42(b). In this procedure, the critical load, or so-called critical Euler's force, will be derived and it relates to global buckling. As can be seen from Figure 6.42(b), the column is subjected to compressive force F, which is assumed to be less than the critical force F_{cr}, usually known as Euler's critical force, and to a very small lateral (disturbance) force, ΔF. The column will

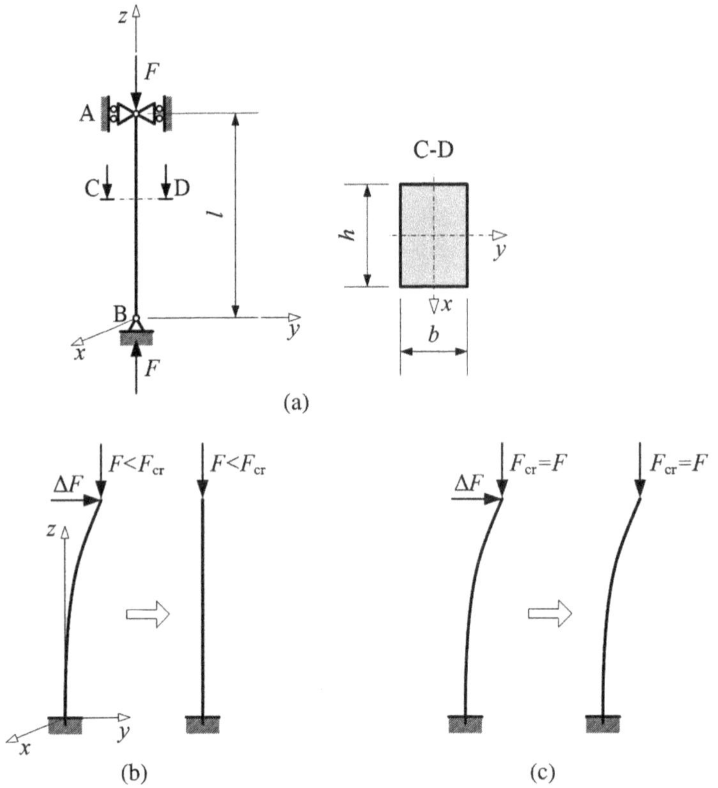

Figure 6.42 A column subjected to compressive force. (a) A simply supported column subjected to compressive force; (b) clamped-free column subjected to lateral disturbance force and compressive force that is less than the critical force; (c) Clamped-free column subjected to lateral disturbance force and compressive force that is equal to the critical force.

deflect laterally. When the lateral force ΔF is removed, the column will return to its straight position. The column is said to be in a stable equilibrium state. If the compressive force gradually increases to the critical level, and the column is subjected to this force and to the lateral (disturbance) force, the column will bend laterally, but after the lateral force is removed, the elastic restorative forces are only large enough to maintain equilibrium in the buckled (lateral) position; that is, the column will not return to its straight position (Figure 6.42(c)). The force F that keeps the column in a buckled state is called the *critical force* (or Euler's critical force) and it is designated F_{cr}. This type of buckling relates to the so-called Euler's (global) buckling. Any further lateral deflection may lead to the fracture of the column. A rod, bar or beam may serve as a column or strut.

6.4.1 Critical Buckling Force in the Elastic Range

6.4.1.1 Pin-ended Columns

Consider the column with pinned ends (also known as a pinned–pinned column or a simply supported column) shown in Figure 6.43(a). The column is in a deformed equilibrium state due to critical compressive force.

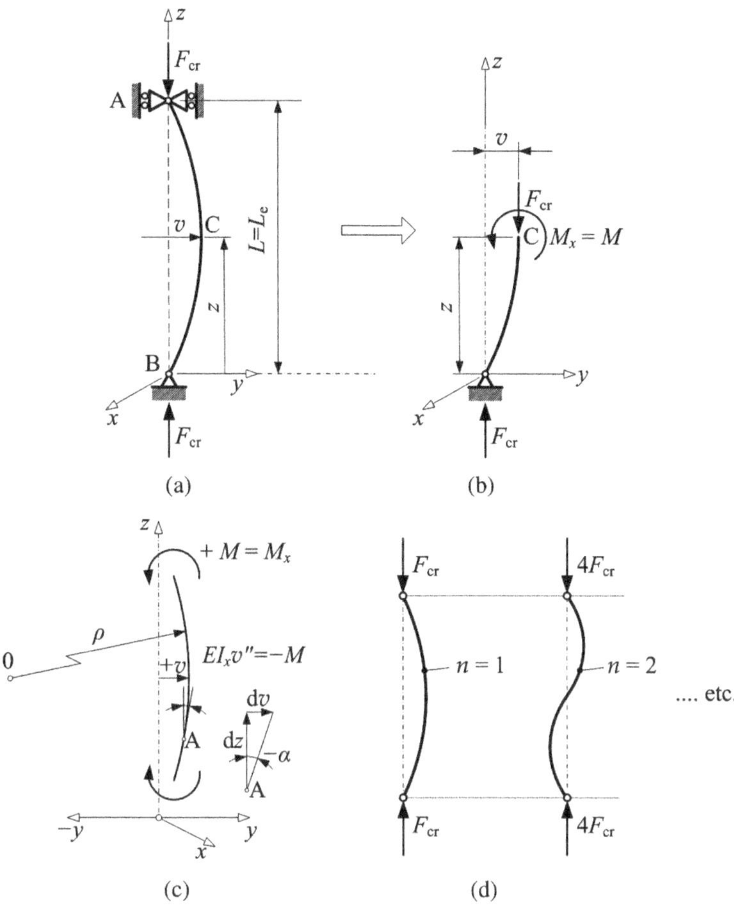

Figure 6.43 Buckling of a pin-ended column. (a) The column in its deformed equilibrium state (buckled form); (b) the forces acting on a segment of the column; (c) the sign of the bending moment and rotation; (d) critical forces in accordance with the modes (forms) of deformation.

The buckled form of a column is usually considered a form that is caused by a bending moment. As such, the differential equation of the elastic curve of the column in its buckled form is defined by Equation (6.116). When a *pin-ended column* is being considered, then, in its buckled form, the momentum equilibrium equation (the equilibrium of moments about point C) can be written in accordance with Figure 6.43(b):

$$\sum M_C = -F_{cr} \cdot v + M_x = 0 \rightarrow M_x = F_{cr} \cdot v \tag{6.133}$$

The same result is obtained by taking a point "B" as the momentum point.

Using differential Equation (6.116) that describes the moment–curvature relationship, and assuming a small deflection, we have:

$$EI_x v'' = -F_{cr} \cdot v. \tag{6.134}$$

The minus sign on the right-hand side of Equation (6.134) (or the minus sign in Equation (6.116)) is prescribed in accordance with Figure 6.37(b) or 6.37(c) and also with Equation (6.115).

Introducing:

$$k^2 = \frac{F_{cr}}{EI_x}, \tag{6.135}$$

into Equation (6.134), it follows that:

$$v'' + k^2 \cdot v = 0 \tag{6.136}$$

This equation is a linear homogeneous differential equation of second order with constant coefficients. The general solution of this equation can be written in the form:

$$v = \mathrm{A} \cdot \sin kz + \mathrm{B} \cdot \cos kz \tag{6.137}$$

In Equation (6.137), A and B are constants which need to be determined in accordance with the boundary conditions:

$$z = 0 \rightarrow v = v(0) = 0; \quad z = L \rightarrow v = v(L) = 0 \tag{a}$$

Based on the first boundary condition, it follows that B = 0. Further, based on the second boundary condition, deflection "v" can be equal to zero if A = 0 or sinkz = 0. A case where A = 0 is a trivial solution, since it implies that deflection "v" is zero and the column remains straight. A possible solution is:

$$\sin kz = 0, \tag{6.138}$$

which implies that:

$$kL = n\pi, \; n = 0, 1, 2, \ldots. \tag{6.139}$$

On the basis of Equations (6.139) and (6.135), the critical force is:

$$F_{cr} = \frac{n^2 \pi^2 EI_x}{L^2} \tag{6.140}$$

In Equation (6.140), I_x is the minimum axial moment of inertia (the second moment of area), which corresponds to Figure 6.43(a), while L is the so-called effective buckling length. Both the axial moment of inertia, I_x, and the effective buckling length, $L_e = L$, correspond to Figure 6.43. This implies that when considering an element (column) with a rectangular cross-section, its longer side (designated by h) coincides with the "x" coordinate axis. Further, the shape of the buckled form for "$n = 1$" (half a sinusoid) is shown in Figure 6.43(d). In the same figure, the second buckling mode is also shown and this corresponds to $n = 2$. However, in engineering practice, the first buckling mode is of importance. Higher orders of buckled forms may be the subject of laboratory investigations. With this in mind, in a general form, the critical force for $n = 1$ can be written as:

$$F_{cr} = \pi^2 \frac{EI_{min}}{L_e^2} \tag{6.141}$$

Since buckling occurs about an axis for which the axial moment of inertia has minimal value (in this case, it is the x axis), in Equation (6.141) the minimal moment of inertia is designated I_{min}. Only some irregularity in the material, such as a geometric irregularity or something similar, can cause buckling about an axis of the cross-section which is not the axis of the minimal moment of inertia. In Equation (6.141), L_e is the so-called effective (or free) buckling length of the column, which is the length/distance between two successive points of the column of zero internal bending moment [2]. These two points of zero internal bending moment are known as *inflection points*. Usually, instead of L_e, the value of kL can be written. In a case involving pinned ends (a pinned-pinned column), factor "k" is equal to 1. The elastic curve of the buckled column, also for $k = 1$ and for the minimal value of the axial moment of inertia, is defined by:

$$v = \mathrm{A} \cdot \sin\frac{\pi z}{l}, \tag{6.142}$$

and is shown in Figure 6.43(d). However, Euler's buckling formula (Equation (6.141)) was derived for an "ideal" column – that is, this consideration assumed that the force acted at the centroid of the cross-section and that the column was perfectly straight. In engineering practice, this assumption is rarely fulfilled. This result, however, may give quite a good prediction for a long or slender column.

6.4.1.2 Columns with Other End Conditions

Figure 6.44 shows columns with other end conditions.

Consider, for example, the column with pinned-fixed ends in Figure 6.44(a). The momentum equilibrium equation (the equilibrium of moments about point C) with respect to the point C (section z), as shown in Figure 6.44(a), is:

$$\begin{gathered} \sum M_{\mathrm{C}} = -M_x + F_{\mathrm{cr}} \cdot v + Q(L-z) = 0 \rightarrow \\ M_x = F_{\mathrm{cr}} \cdot v + Q(L-z) \end{gathered} \tag{6.143}$$

In accordance with Equation (6.116), the differential equation of the elastic curve is:

$$EI_x v'' = -M_x.$$

Based on Equation (6.143), the differential equation of the elastic curve (the moment–curvature relationship) for a pinned-fixed column in its buckled form is:

$$EI_x v'' = -F_{\mathrm{cr}} \cdot v - Q(L-z) \tag{6.144}$$

Finally, using Equation (6.135), the differential equation of the elastic curve for the buckled column presented in Figure 6.44(a) is:

$$v'' + k^2 \cdot v = -k^2 \frac{Q}{F_{\mathrm{cr}}}(L-z) \tag{6.145}$$

This equation is a linear, nonhomogeneous differential equation of second order with constant coefficient. Since the left-hand members of Equations (6.137) and (6.145) are identical, it may be concluded that the general solution of Equation (6.145) can be obtained by adding a particular solution of Equation (6.145) to the earlier obtained solution for Equation (6.137).

The solution of Equation (6.145) may be assumed to take the form:

$$v(z) = \mathrm{A} \cdot \sin kz + \mathrm{B} \cdot \cos kz - \frac{Q}{F_{\mathrm{cr}}}(L-z) \tag{6.146}$$

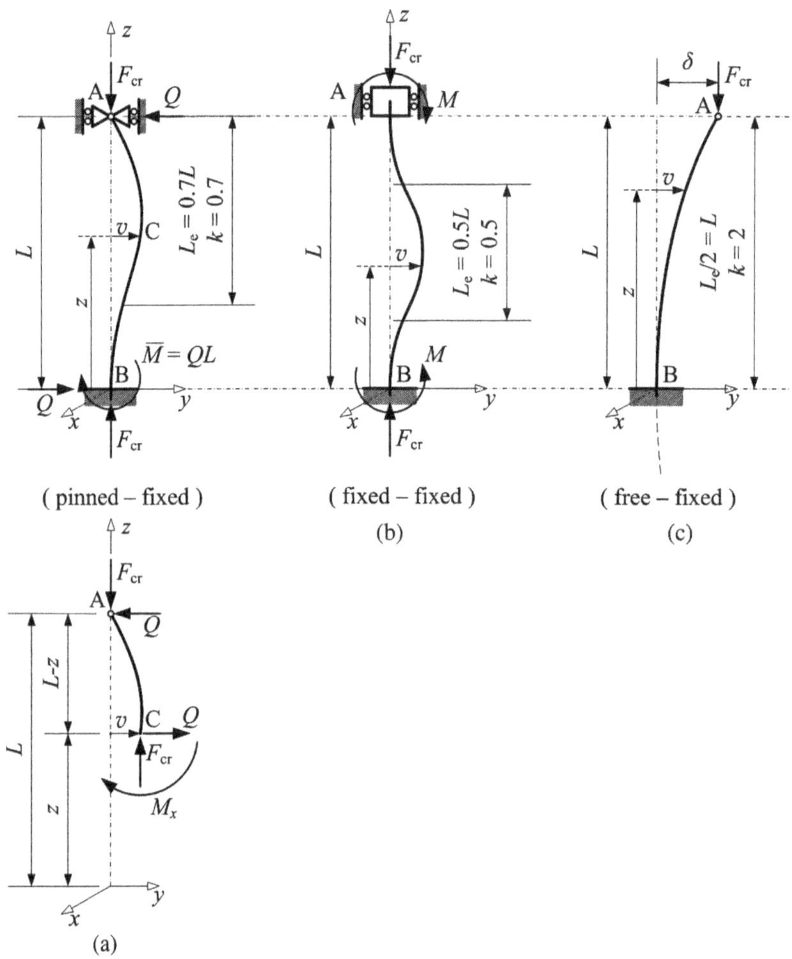

Figure 6.44 Buckling of the column – the effect of end conditions. (a) Column: pinned–fixed; (b) column: fixed–fixed; (c) column: free–fixed.

The boundary conditions are:

$$z=0,\quad v=v(0)=0,\quad v'=v'(0)=0. \qquad\qquad z=L,\quad v=v(L)=0. \tag{6.147}$$

On the basis of the boundary conditions, the constants A and B in Equation (6.146) can be determined. Based on the first boundary condition, we have:

$$z=0,\ v=v(0)=0$$

$$v(0)=0\ \rightarrow \mathrm{A}\cdot 0+\mathrm{B}\cdot 1-\frac{QL}{F_{\mathrm{cr}}}=0\rightarrow \mathrm{B}=\frac{QL}{F_{\mathrm{cr}}} \tag{6.147a}$$

Since the first derivative of Equation (6.146) is:

$$v'=\frac{\mathrm{d}v}{\mathrm{d}z}=\mathrm{A}\cdot k\cdot \cos kz-\mathrm{B}\cdot k\cdot \sin kz+\frac{Q}{F_{\mathrm{cr}}}, \tag{6.147b}$$

it follows that:

$$v'(0) = 0 \rightarrow \mathrm{A} \cdot k \cdot 1 - \mathrm{B} \cdot k \cdot 0 + \frac{Q}{F_{cr}} = 0 \rightarrow \mathrm{A} = -\frac{Q}{kF_{cr}} \tag{6.147c}$$

Further, on the basis of the second boundary condition, we have:

$$v(L) = \mathrm{A} \cdot \sin kL + \mathrm{B} \cdot \cos kL = 0 \tag{6.147d}$$

If the values for constants A and B from Equations (6.147c) and (6.147a) are inserted into Equation (6.147d), it follows that:

$$v(L) = -\frac{Q}{kF_{cr}} \sin kL + \frac{QL}{F_{cr}} \cos kL = 0 \rightarrow \tan kL = kL \tag{6.148}$$

Since Equation (6.148) is a transcendental equation, an explicit solution cannot be found. From the mutual intersection of the functions tan kL and kL, it follows that:

$$\begin{aligned} &kL = 4.4934 \\ &k^2L^2 = 4.4934^2 \cong \frac{\pi^2}{0.7^2} \\ &F_{cr} = \frac{\pi^2 E I_x}{(0.7L)^2}; \quad L_e = 0.7L; \quad I_x = I_{min} \end{aligned} \tag{6.149}$$

To gain further insight into buckling of columns, some analysis can also be found in [12–13].

6.4.2 Critical Buckling Stress in the Elastic Range

In engineering practice, several categories of column can be identified: short, intermediate and long columns, for example. Short columns can be loaded up to the yield stress of the material and these columns will not buckle. In the second category, the applied load can cause compression stress that exceeds the proportional limit and these columns may fail by yielding and by buckling. Long (slender) columns under applied load can buckle elastically. These columns may be treated as ineffective since the buckling stress can be much lower than the material's proportional limit [2]. The critical stress can be derived as the quotient of the critical force (F_{cr}) and the cross-sectional area (A) of the column [1]:

$$\sigma_{cr} = \frac{F_{cr}}{A} \tag{6.150}$$

In accordance with Equation (6.141), we have:

$$\sigma_{cr} = \frac{\pi^2 E I_{min}}{L_e^2 A} = \pi^2 E \left(\frac{i_{min}}{L_e}\right)^2 = \pi^2 \frac{E}{\lambda^2} \tag{6.151}$$

where

$$\lambda = \frac{L_e}{i_{min}} \tag{6.152}$$

is the slenderness of the column, while i_{min} is the minimum radius of inertia. In the coordinate system (σ, λ), Equation (6.151) is a hyperbolic curve. At very high levels of slenderness, this curve tends to zero, which is in accordance with experimental investigations – that is, very slender columns buckle under very small stresses. In cases involving low levels of slenderness, Equation (6.141) is not adequate for determining the critical force. In low-slenderness scenarios, experiments do not coincide with the results obtained by using Equations (6.141) and (6.151). In the procedure for determining the differential equation of the elastic curve, the modulus of elasticity is assumed to be constant. This means that the solution represented by Equation (6.151) is valid when $\sigma_{cr} < \sigma_P$, where σ_P is the stress at the proportional limit. So, the determination of Euler's critical force according to Equation (6.141) may also be applied only in the case when $\sigma_{cr} < \sigma_P$. As such, the slenderness at which the critical buckling force can be determined using the previous procedure is $\lambda = \lambda_P$, and then $\sigma_{cr} = \sigma_P$. On the basis of this explanation, the level of slenderness that corresponds to the stress at the material's proportional limit (σ_P) is:

$$\lambda_P = \pi\sqrt{\frac{E}{\sigma_P}} \tag{6.153}$$

Since, in engineering practice, the stress at the proportional limit is quite difficult to determine, some other approximation in the critical stress determination can be used.

6.4.3 Buckling – Plastic Range

When the applied stress is higher than the stress at the proportional limit ($\sigma_{cr} > \sigma_P$), the stress–strain relationship is no longer linear, and this implies that the modulus of elasticity (E) is no longer constant. In this case, other solutions can be used. For example, the critical stress can be determined by the formula that defines the so-called Tetmayer's straight:

$$\sigma_{cr} = \frac{F_{cr}}{A} = a - b \cdot \lambda. \tag{6.154}$$

In Equation (6.154), a and b are experimentally determined coefficients depending on the material used and can be found in the literature [1, 5, 14]. Also, a recommendation for a solid circular column subjected to compressive load (stress) may be found in [15]. As stated there, to avoid buckling, a reasonable compromise is $L/D = 3$ for ductile materials, and $L/D = (1.5–2)$ for brittle materials, if the end effects are small. Here, D is the diameter of the column. Interesting information related to the stability of structures, including thin-walled structures, experiments and so on, may be found in [16–20].

Example 6.12.
Consider the pin-ended column shown in Figure 6.45(a), which is made of steel. Buckling of the column is restrained in point D, as shown. Determine the maximum compressive load F (for both cross-sections shown) the column can support if a factor of safety (against buckling) of 3 is required. Assume that buckling is possible only about the weak axis (x). Consideration of buckling about the strong axis, y, is not required.

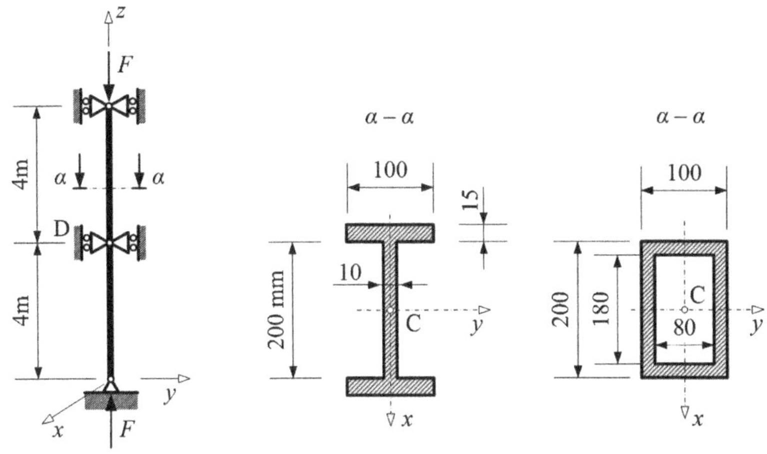

Figure 6.45 Stability of the pin-ended column with different cross-sections considered in Example 6.12. (a) Column geometry and load; (b) I cross-sectional shape; (c) rectangular, thin-walled cross-section (box profile).

Data:
Dimensions are shown in Figure 6.45. The modulus of elasticity, $E = 200$ GPa.

Solution:
The following section properties can be calculated:

- For an "I" profile (Figure 6.45(b)):

$$I_x = 2\frac{100^3 \cdot 15}{12} + \frac{10^3 \cdot 200}{12} = 2516666.7\,\text{mm}^4$$

Using Equation (6.141), the critical force may be calculated as:

$$F_{\text{cr}} = \frac{\pi^2 \cdot 200000 \cdot 2516666.7}{4000^2} = 103388.9\,\text{N}$$

$$F_{\max} \le F_{\text{cr}}/3\,(\le 34463\,\text{N})$$

- For a box profile (Figure 6.45(c)):

$$I_x = \frac{200 \cdot 100^3}{12} - \frac{80^3 \cdot 180}{12} = 8986666.7\,\text{mm}^4$$

$$F_{\text{cr}} = \frac{\pi^2 \cdot 200000 \cdot 8986666.7}{4000^2} = 1107561.7\,\text{N}$$

$$F_{\max} \le F_{\text{cr}}/3\,(\le 369187\,\text{N})$$

6.4.3.1 Local Buckling of the Column

Consideration in the previous sections has focused on the determination of Euler's critical force, and this can be defined as a global buckling problem. However, if a column has a cross-section that is formed of several thin-walled parts, like a [-beam, an L-beam, and so on, then this column may fail by local buckling of the web or flange. This local buckling, in contrast to Euler's (global) buckling has different names, for example, wrinkling, crimping, and so on. So, a column may fail by wrinkling on one side before it buckles as an Euler column (see Figure 6.46).

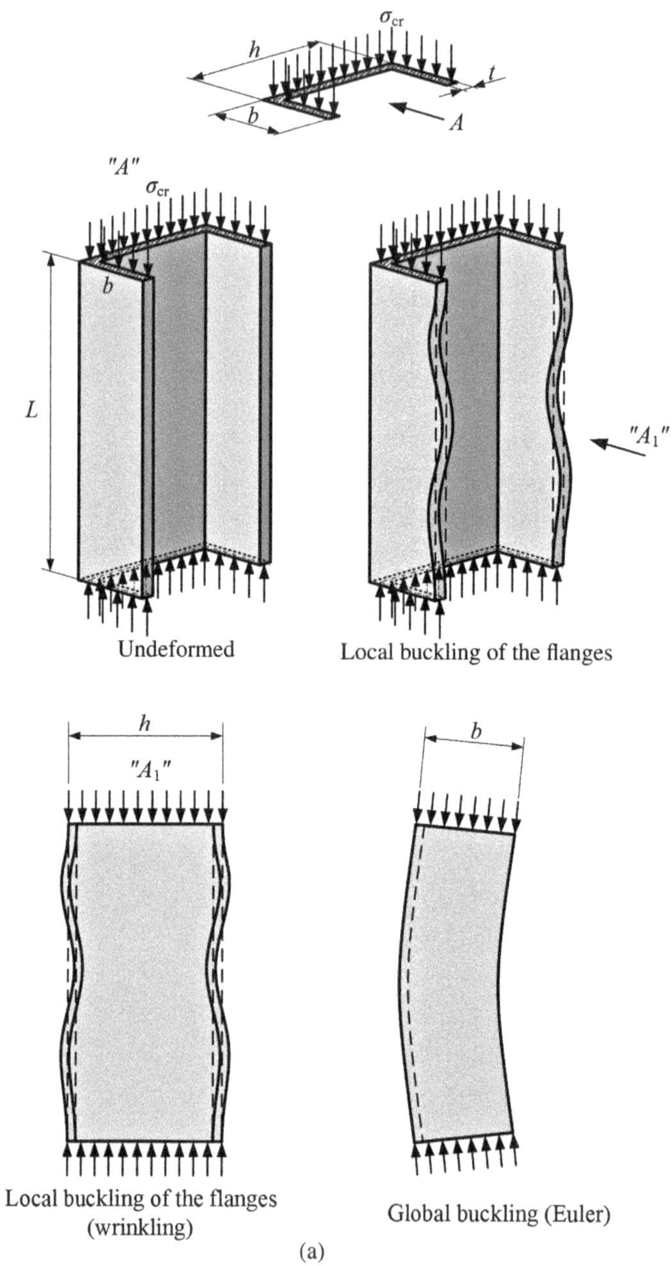

Figure 6.46 Global and local buckling of a thin-walled structure. (a) Channel-shaped column – global and local buckling; (b) L-shaped column – global and local buckling.

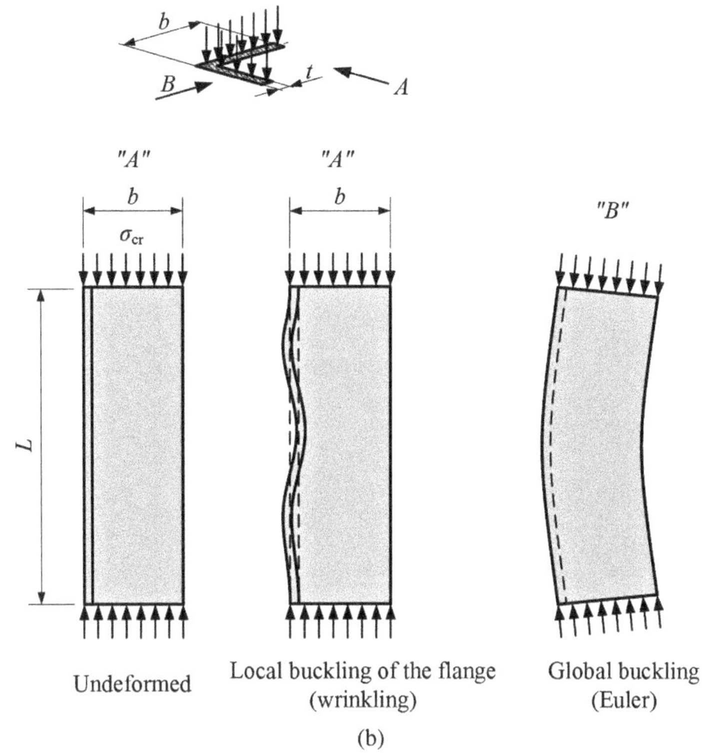

Figure 6.46 *(Continued)*

6.5 Eccentric Axial Loads

An engineering element subjected to an axial load passing through the centroids of the cross-sections of the element is exposed to normal stress. The axial load may be compressive or tensile. If the element is subjected to an eccentric axial load, that is, to a load whose line of action does not pass through the centroids of the cross-sections, then the considered element is exposed to the normal stresses caused by the axial load in addition to the normal stress caused by a bending moment, as shown in Figure 6.47 [1–3].

6.5.1 Eccentric Axial Load Acting in a Plane of Symmetry

Consider first the case when an eccentric axial load acts in a plane of symmetry, as shown in Figure 6.47(a) and Figure 6.47(b). In this case, the level of stress may be calculated as follows:

$$\sigma_z = \pm\frac{F}{A} \pm \frac{|M_x|}{I_x} y \leq \sigma_{\text{all}}; \quad M_x = F \cdot e \tag{6.155}$$

The sign of the normal stress is defined in accordance with Equation (6.155). First, the sign of the normal stress caused by force F is defined with respect to its tensile (+) or

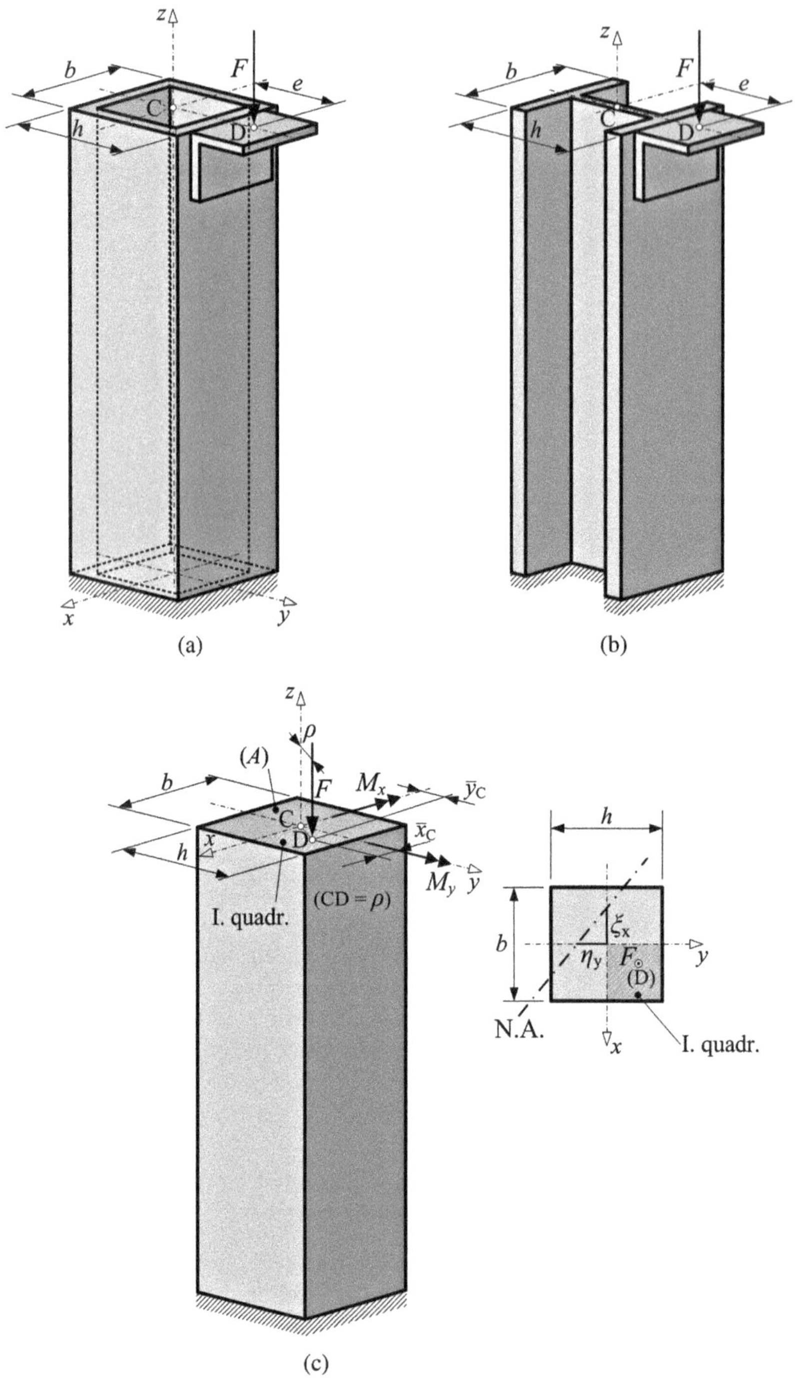

Figure 6.47 Eccentric axial compressive load. (a) Box girder; (b) I girder; (c) rectangular girder of solid section.

compressive (−) action. Next, the sign of the normal stress caused by the bending moment is defined with respect to its action on the fibers of the first quadrant (that is, for extension (+), and for compression (−)). Finally, the sign of the coordinate y must also be taken into consideration.

6.5.2 General Case of an Eccentric Axial Load

A general case of an eccentric axial load is shown in Figure 6.47(c). The following loads relating to the centroid occur:

$$\text{Axial force: compressive force}(-F)\text{, coordinates of the force application point are }(-\bar{x}_C;\ \bar{y}_C)$$

The signs of the bending moments with respect to the direction of their vectors are:

$$M_x = -F\cdot\bar{y}_C;\ M_y = F\cdot\bar{x}_C$$

The signs of the bending moments with respect to their action on the fiber of the first quadrant are:

$$\text{Bending moments: } M_x = -F\cdot\bar{y}_C \text{ (shortens the fibers of the I quadrant);}$$
$$M_y = -F\cdot\bar{x}_C \text{ (shortens the fibers of the I quadrant)}$$

The resulting level (magnitude) of the stress is:

$$\sigma_z = \pm\sigma_z^N \pm \sigma_z^{M_x} \pm \sigma_z^{M_y} \tag{6.156}$$

The internal axial force is: $N = F$.

Equation (6.156) can also be written in the following form:

$$\sigma_z = \pm\frac{F}{A} \pm \frac{|M_x|}{I_x}y \pm \frac{|M_y|}{I_y}x \tag{6.157}$$

As explained above, the sign (+ or −) in Equation (6.157) is determined depending on the action of the load on fibers of the I quadrant of the coordinate system as well as on the sign of coordinates of the considered point on the cross-section. Once the influence of the bending moment on the fibers of the I quadrant has been defined, the moment is taken in its absolute value as indicated in Equation (6.157). It is known that:

$$I_x = i_x^2\cdot A,\ \ I_y = i_y^2\cdot A, \tag{6.158}$$

where i_x and i_y are radii of gyration [1, 2, 7].

Based on Figure 6.47 it is clear that both of the bending moments (M_x and M_y) shorten the fibers of the first quadrant (negative sign). Force F also shortens the fibers of the first quadrant. In accordance with this, Equation (6.157) can be written in the form:

$$\sigma_z = -\frac{F}{A} - \frac{F\cdot\bar{y}_C}{I_x}\cdot y - \frac{F\cdot\bar{x}_C}{I_y}\cdot x \tag{6.159}$$

As can be seen, the bending moments are inserted in accordance with their influences on the fibers of the first quadrant. Now, Equation (6.159) can be written as:

$$\sigma_z = -\frac{F}{A} - \frac{F \cdot \bar{y}_C}{i_x^2 \cdot A} \cdot y - \frac{F \cdot \bar{x}_C}{i_y^2 \cdot A} \cdot x \tag{6.160}$$

Based on Equation (6.160), the stress at any point of the cross-section can be derived. This equation can be written in the form:

$$\sigma_z = -\frac{F}{A}\left(1 + \frac{\bar{x}_C}{i_y^2} \cdot x + \frac{\bar{y}_C}{i_x^2} \cdot y\right) \tag{6.161}$$

Since the stress in each fiber which passes through the neutral axis is equal to zero, and F/A is not zero, then, from Equation (6.160), it follows that:

$$\sigma_z = 0, \text{ and } F/A \neq 0 \rightarrow 1 + \frac{\bar{x}_C}{i_y^2} \cdot x + \frac{\bar{y}_C}{i_x^2} \cdot y = 0. \tag{6.162}$$

Equation (6.162) is the equation of the neutral axis (NA). It is clear that it does not pass through the centroid. Slices of the neutral axis on the coordinate axes can be obtained as follows. For Figure 6.47(c):

$$\begin{aligned} y = 0 \rightarrow x = \xi_x = -\frac{i_y^2}{\bar{x}_C} \\ x = 0 \rightarrow y = \eta_y = -\frac{i_x^2}{\bar{y}_C} \end{aligned} \tag{6.163}$$

It is clear (from Figure 6.47(c)) that the neutral axis is positioned diametrically opposite to the point of application of the compressive force. However, the stress level must be less than the allowable stress,

$$\sigma_{z_{\max}} \leq \sigma_{\text{all}}.$$

Example 6.13.
Consider the column in Figure 6.48(a), which is made of steel. Determine the level of the stress in the point E.

Data:
The dimensions are shown in Figure 6.48(a). Force $F = 50\,\text{kN}$.

Solution:
The bending moments are:

$$M_x = F \cdot y_C = F \cdot 55 = 2750000\,\text{Nmm}$$

$$M_y = -F \cdot x_C = -F \cdot 20 = -1000000\,\text{Nmm}$$

The equation representing the stress, taking into account that both of the bending moments extend the fibers of the first (I) quadrant, while the force shortens the fibers of the first quadrant, in this case, is:

$$\sigma_z = -\frac{F}{A} + \frac{M_x}{I_x} y + \frac{|M_y|}{I_y} x \tag{a}$$

After mathematical calculations, the following data can be obtained.

The area of the cross-section is:

$$A = 120 \cdot 10 + 2 \cdot 58 \cdot 10 = 2360\,\text{mm}^2$$

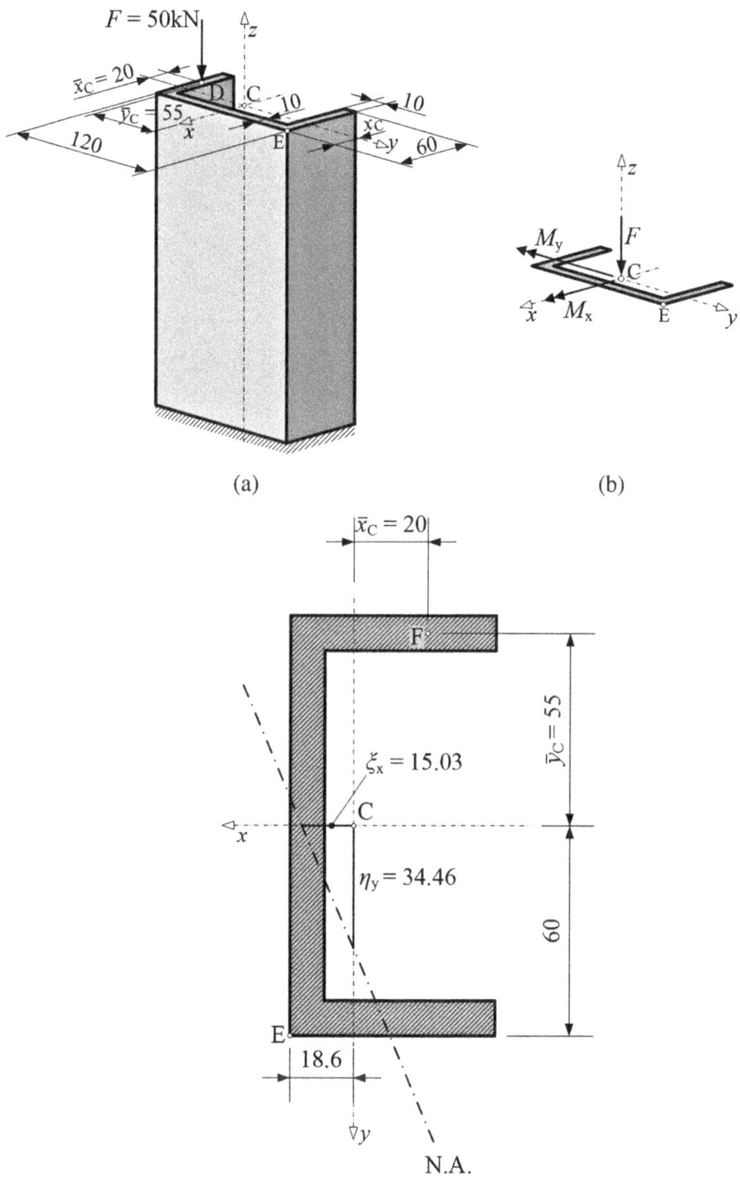

Figure 6.48 Channel loaded by an eccentric axial force. (a) Geometry and load; (b) force and moment components; (c) position of the neutral axis.

The coordinates of the centroid are:

$$x_C = \frac{2 \cdot 60 \cdot 10 \cdot 30 + 100 \cdot 10 \cdot 5}{2 \cdot 60 \cdot 10 + 100 \cdot 10} = 18.6\,\text{mm};$$

$$y_C = 60\,\text{mm}$$

Both coordinates (x_C, y_c) are measured from the point E in Figure 6.48.

The moments of inertia are:

$$I_x = \frac{2 \cdot 10^3 \cdot 60}{12} + 2 \cdot 60 \cdot 10 \cdot 55^2 + \frac{100^3 \cdot 10}{12} = 4473333.3\,\text{mm}^4$$

$$I_y = \frac{2 \cdot 60^3 \cdot 10}{12} + 2 \cdot 60 \cdot 10\,(30 - 18.6)^2 + \frac{10^3 \cdot 100}{12} + 100 \cdot 10\,(18.6 - 5)^2$$

$$= 709245.3\,\text{mm}^4$$

The coordinates of the point E are:

$$x = x_E = x_C = 18.6\,\text{mm},\; y = y_E = 60\text{ mm}$$

In accordance with Equation (a), since both coordinates of the point "E" are positive, the stress at point "*E*" is:

$$\sigma_z = -\frac{50000}{2360} + \frac{2750000}{4473333.3}60 + \frac{1000000}{709245.3}18.6$$

$$\sigma_z = 41.92\,\text{MPa}$$

The neutral axis is defined by Equation (6.162):

$$\xi_x = -\frac{i_y^2}{\bar{x}_C} = -\frac{{}^{I_y}/_{A}}{-20} = -\frac{{}^{709245.3}/_{2360}}{-20} = 15.03\,\text{mm}$$

$$\eta_y = -\frac{i_x^2}{\bar{y}_C} = -\frac{{}^{I_x}/_{A}}{-55} = -\frac{{}^{4473333.3}/_{2360}}{-55} = 34.46\,\text{mm}$$

References

1 Brnic, J. and Turkalj, G. (2004) *Strength of Materials I* (in Croatian), Faculty of Engineering, University of Rijeka, Rijeka.

2 Philpot, T. A. (2008) *Mechanics of Materials*, John Wiley & Sons, New Jersey.

3 Beer, F. P., Johnston, E. R. Jr, DeWolf, J. T. and Mazurek, D. F. (2012) *Mechanics of Materials*, McGraw-Hill, USA.

4 Boresi, A. and Schmidt, R. J. (2003) *Advanced Mechanics of Materials*, John Wiley & Sons, USA.

5 Alfirević, I. (1995) *Strength of Materials I* (in Croatian), ITP "Tehnička knjiga", D.D. Zagreb.

6 Alfirević, I. (2006) *Linear Structural Analysis*, Pretei d.o.o., Zagreb.

7 Solecki, R. and Conant, R. J. (2003) *Advanced Mechanics of Materials*, Oxford University Press, New York.

8 Boresi, A. P., Schmidt, R. J. and Sidebottom, O. M. (1993) *Advanced Mechanics of Materials*, John Wiley & Sons, New York.
9 Seed, G. M. (2000) *Strength of Materials*, Saxe-Coburg Publications, Edinburgh.
10 Sun, C. T. (2006) *Mechanics of Aircraft Structures*, John Wiley & Sons, New York.
11 Alfirević, I., Saucha, J., Tonković, Z. and Kodvanj, J. (2010) *Introduction to Mechanics II* (in Croatian), Golden Marketing,Tehnička knjiga, Zagreb.
12 Craig, R. R. (2011) *Mechanics of Materials*, 3rd edition, John Wiley & Sons, USA.
13 Riley, W. F., Sturges, L. D. and Morris, D. H. (2007) *Mechanics of Materials*, John Wiley & Sons, USA.
14 Rašković, D. (1971) *Resistance of Materials* (in Serbian), Naučna knjiga, Belgrade.
15 Dowling, N. E. (2013) *Mechanical Behavior of Materials*, Pearson, New York.
16 Galambos, T. V. (1988) *Guide to Stability Design – Criteria for Metal Structures*, 4th edition, John Wiley & Sons, New York.
17 Kumar, A. (1985) *Stability Theory of Structures*, McGraw-Hill, New Delhi.
18 Bažant, Z. and Cedolin, L. (1991) *Stability of Structures*, Oxford University Press, New York.
19 Timoshenko, S. P. and Gere, J. M. (1988) *Theory of Elastic Stability*, McGraw-Hill, London.
20 Singer, J., Arbocz, J. and Weller, T. (1998) *Buckling Experiments: Experimental Methods in Buckling of Thin-Walled Structures*, John Wiley & Sons, New York.

7

Relationships Between Stress and Strain

7.1 Fundamental Considerations

When a solid body is subjected to a load, it experiences changes in its volume and/or shape. The analysis of the stress in solid bodies (deformable bodies) is of interest in the design process. On the basis of previous analyses, it is evident that the state of stress may be defined by the stress tensor and the state of strain may be defined by the strain tensor. In the solving of problems in the field of elastostatics, there are six unknowns in the stress tensor (Equation (2.6)), six unknowns in the strain tensor (Equation (3.19)) and three unknowns in the displacement vector (Equation (3.20)). The above implies that the tensor components are symmetric with respect to the main diagonal of the tensor. There are fifteen unknowns in total. This means that one needs to have fifteen equations if the unknowns need to be derived. However, the theory of stress and the theory of strain are derived "in general". This means that the theories have been derived independent of the material properties; that is, independent of the material from which the considered body is made. As such, the theory of stress is based on the equilibrium of forces while the theory of deformation is based on the compatibility of geometry. It is to be expected that the stress occurring in a body is associated with the deformation and the material properties. The question that arises is what the relationship between stress and strain is. It is very difficult to find universal equations that can be used to summarise the relationship between stress and strain for all of types of solid bodies, and especially if all of types of continuum are considered. If solid bodies are being considered, tests indicate that stress depends on strain, strain rate, temperature, time and other factors, and this fact can be expressed in the form:

$$\sigma_{ij} = \sigma_{ij}\left(\varepsilon_{ij}, \dot{\varepsilon}_{ij}, T, t, p_i\right). \tag{7.1}$$

In Equation (7.1), ε_{ij} represents strain; $\dot{\varepsilon}_{ij}$ represents the strain rate; T represents temperature; t represents time; p_i represents the material parameters. Equations of this type are known as constitutive equations. In many cases, for example in metallic materials, but also many others, Equation (7.1) may be written as:

$$\sigma_{ij} = \sigma_{ij}\left(\varepsilon_{ij}\right). \tag{7.2}$$

If the problem involves the temperature, then the temperature distribution needs to be known or it needs to be solved for using the equation of heat transfer. However, compatibility conditions need to be fulfilled.

Analysis of Engineering Structures and Material Behavior, First Edition. Josip Brnić.

Solid bodies, with regard to their responses, are usually divided into the following categories:

- Elastic bodies
- Plastic bodies
- Viscoelastic bodies
- Viscoplastic bodies.

Depending on whether the material properties of the considered body at an arbitrary point depend on the direction or not, bodies can be further divided into:

- Isotropic bodies
- Orthotropic bodies
- Anisotropic bodies.

In Section 5.1 some material responses were shown. Sometimes, relationships represented by Equations (7.1) and (7.2) can be quite complex, and diagrams that represent real material responses can sometimes be shown in an idealized form. If we restrict our considerations to linear elastic bodies, which implies that, after unloading, the body returns to its original form and size (that is, the relationship between stress and strain is unique, or the work of internal forces in a closed cycle is equal to zero), the so-called generalized Hooke's Law is usually written in the form (tensor notation) [1–3]:

$$\sigma_{ij} = C_{ijkl} \cdot \varepsilon_{kl} \tag{7.3}$$

In order to define mutual dependence of the components of stress (σ_{ij}) and the components of strain (ε_{kl}), it is necessary to determine the structure of the matrix of elasticity (C_{ijkl}); that is, to define the elastic constants of elasticity. If the body is exposed to a temperature increase in addition to a mechanical load, we can assume that the deformation of the body is prevented. Let deformation be small and linearly dependent on the temperature. Then, Equation (7.3), in a general case (not, for example, only for isothermal conditions), can be written:

$$\sigma_{ij} = C_{ijkl} \cdot \varepsilon_{kl} - \beta_{ij} \Delta T \tag{7.3a}$$

Here, β_{ij} are coefficients that depend on the thermal properties of the material and ΔT represents an increase (or decrease) in temperature. The reverse form of Hooke's Law presented in Equation (7.3), with respect to strain, is:

$$\varepsilon_{ij} = S_{ijkl} \cdot \sigma_{kl} \tag{7.4}$$

If the body is subjected to a high/low temperature, this may be written:

$$\varepsilon_{ij} = S_{ijkl} \cdot \sigma_{kl} + \alpha_{ij} \Delta T \tag{7.4a}$$

C_{ijkl} is a fourth-order tensor and it is known as the *elasticity tensor* (or the elastic matrix or stiffness matrix), while S_{ijkl} is also a fourth-order tensor and is known as the *compliance tensor*. Coefficients α_{ij} are the coefficients of thermal expansion of the material. The elasticity tensor, C_{ijkl}, consists of elastic constants (or elastic coefficients). The stress tensor has nine components (six different components), the strain tensor has nine components (six different components) and the displacement vector has three components. Finally, to solve for the previously mentioned fifteen unknowns, the following equations are allowable: three equilibrium equations (Equation (2.4)), six equations relating to strain components and displacement components (known as geometric equations; Equation (3.16)) and six equations for Hooke's Law (Equation (7.3)) [1–5]. In this sense, it is said that the problem can be solved.

7.2 Anisotropic Materials

Materials that are used in engineering practice cannot be considered perfectly isotropic. When there are different values of material properties for different directions through the same point of the body, the material is considered to be *anisotropic*. In other words, it can be said that a material involving a number of independent elastic coefficients in its stress–strain relationships is said to be anisotropic. As such, the assumption that the considered material exhibits isotropic behavior is no longer a reasonable approximation of the material's behavior.

Consider now a general anisotropic linear elastic material. The material properties at a point vary with direction (or depend on the orientation of reference axes). Given that nine components of stress (the stress tensor σ_{ij}) are connected with nine components of strain (the strain tensor ε_{ij}), this would imply the existence of eighty-one components in the elasticity tensor, C_{ijkl}. However, only six different components exist in the stress tensor and only six different components exist in the strain tensor, which gives thirty-six different components in the elasticity tensor, and these stress–strain relationships depict the behavior of anisotropic materials. The thirty-six coefficients are called elastic coefficients and materials whose stress–strain relationships are described by them are said to be *anisotropic materials*.

In accordance with Hooke's Law, each component of the stress tensor depends linearly on each component of the strain tensor. When Hooke's Law is written in the x, y, z coordinate system, Equation (7.3) may be written as:

$$\begin{aligned}&\sigma_{xx} = C_{1111}\cdot\varepsilon_{xx} + C_{1112}\cdot\varepsilon_{yy} + \cdots C_{1116}\gamma_{xz}\\&\ldots.\end{aligned} \tag{7.5a}$$

When Hooke's Law is presented in the principal axes system (that is, in a 1, 2, 3, system where shear stresses are equal to zero, i.e. $\sigma_{12} = \sigma_{21} = \sigma_{13} = \sigma_{31} = \sigma_{23} = \sigma_{32} = 0$), it follows that:

$$\begin{aligned}&\sigma_{11} = (\sigma_1) = C_{1111}\cdot\varepsilon_{11} + C_{1122}\cdot\varepsilon_{22} + C_{1133}\cdot\varepsilon_{33},\\&\sigma_{22} = (\sigma_2) = \ldots,\\&\sigma_{33} = (\sigma_3) = \ldots.\end{aligned} \tag{7.5b}$$

On the basis of Equation (7.5b), it is clear that there are nine coefficients. Thus, instead of 81 coefficients in the x, y, z coordinate system, now, in the principal axes system, there are nine coefficients. In order to define the stress–strain relationships, it is necessary to determine the coefficients (constants). The property of symmetry of elastic constants in stress–strain relationships can be determined using the first law of thermodynamics:

$$\delta W + \delta Q = \delta U + \delta K \tag{7.6}$$

Equation (7.6) expresses the following: the increase in internal and kinetic energies is equal to the work performed on a mechanical system by external forces plus the heat that flows into the system from the outside. As such, in Equation (7.6), δW is the work of the external forces performed in the body's deformation, δQ is the heat that flows into the body (for example, the heat absorbed by the body from the environment), δU is the increase in potential energy of deformation (strain) energy and δK is the increase in kinetic energy. The first law of thermodynamics in a case of static equilibrium and adiabatic conditions may be written in the form:

$$\delta W = \delta U. \tag{7.6a}$$

This means that the variation in the work of external forces is equal to the variation in the work of internal forces (the work of the internal forces is also termed the potential energy of deformation, or the strain energy). The variation in the external forces can be divided into variations belonging to surface forces and variations belonging to volume forces.

Let's consider the arbitrary point N on an incremental surface area dA. The surface force components (F_x, F_y, F_z) acting at the considered point N of the surface produce the stress vector (stress components), the work of which on the displacement variations (δu, δv, δw) is:

$$\delta W_A = \int_A \sigma_{Nx} \delta u \, dA + \int_A \sigma_{Ny} \delta v \, dA + \int_A \sigma_{Nz} \delta w \, dA \tag{7.7a}$$

N is the considered point, x, y and z are coordinate axes and u, v and w are displacements on the coordinate axes. The work of the body (volume) forces is:

$$\delta W_V = \int_V \left(f_x \cdot \delta u + f_y \cdot \delta v + f_z \cdot \delta w \right) dV \tag{7.7b}$$

The variations in the strains are:

$$\delta \varepsilon_x = \frac{\partial (\delta u)}{\partial x}; \quad \delta \varepsilon_y = \frac{\partial (\delta v)}{\partial y}, \text{ etc.} \tag{7.7c}$$

If the surface integral (Equation (7.7a)) is converted into the volume integral using the divergency theorem, then taking this transformation and Equation (7.7b), (7.7c) and (2.4), it follows that the work done by external forces is:

$$\delta W = \int_V (\sigma_x \delta \varepsilon_x + \ldots\ldots\ldots\ldots + 2\tau_{zx} \delta \gamma_{zx}) dV \tag{7.7d}$$

The variation in strain energy may be written in terms of strain energy density:

$$\delta U = \int \delta U_0 dV. \tag{7.8}$$

By substituting Equations (7.7d) and (7.8) into Equation (7.6a), the variation in the strain energy density in terms of stress components and the variation in the strain components may be written as:

$$\delta U_0 = \sigma_x \delta \varepsilon_x + \sigma_y \delta \varepsilon_y + \cdots \tag{7.9}$$

Since:

$$U_0 = U_0\left(\varepsilon_x, \varepsilon_y, \ldots\ldots.x, y, z, T\right), \tag{7.9a}$$

it follows that:

$$\delta U_0 = \frac{\partial U_0}{\partial \varepsilon_x}\delta\varepsilon_x + \frac{\partial U_0}{\partial \varepsilon_y}\delta\varepsilon_y + \ldots. \tag{7.9b}$$

and

$$\sigma_x = \frac{\partial U_0}{\partial \varepsilon_x}, \sigma_y = \frac{\partial U_0}{\partial \varepsilon_y}, \cdots \tag{7.10}$$

By differentiating Equation (7.5), taking into account Equation (7.10), it follows that:

$$\frac{\partial^2 U_0}{\partial \varepsilon_x \partial \varepsilon_y} = C_{1112} = C_{2221} \tag{7.11}$$

$$\ldots$$

In accordance with Equation (7.11) it is evident that the elastic coefficients are symmetric; that is, $C_{ij} = C_{ji}$. On the basis of this consideration it is evident that the general anisotropic linear elastic material has only 21 elastic coefficients (six on the diagonal plus (36 – 6)/2), instead of the 81 coefficients mentioned earlier, or the 36 based on symmetry of the stress and strain tensors. Using generalized vectors, Equations (7.5) can be written as:

$$\sigma_i = C_{ij}\varepsilon_j \tag{7.12}$$

where:

$$\sigma_1 = \sigma_x, \sigma_2 = \sigma_y, \ldots\ldots..\sigma_6 = \tau_{xz}$$

$$\varepsilon_1 = \varepsilon_x, \varepsilon_2 = \varepsilon_y, \ldots\ldots..\varepsilon_6 = \varepsilon_{xz}$$

$$\left(\varepsilon_{ij} = (1/2)\gamma_{ij}\right) \tag{7.13}$$

The matrix form of Equation (7.12) is shown by Equation (7.19) and is usually known as the contracted form. Here: $\sigma_{11} = \sigma_1$, $\sigma_{xx} = \sigma_x$, ….., $\varepsilon_{11} = \varepsilon_1$, $\varepsilon_{xx} = \varepsilon_x$, ……, $C_{1111} = C_{11}$….

7.3 Isotropic Materials

Consider an elastic engineering element made of an isotropic material. An isotropic material is characterized by the fact that its elastic and mechanical properties do not depend on the direction (straight line) across a considered point. Since an infinite number of straight lines can pass through the point, this corresponds to an infinite number of coordinate systems as well as to an infinite number of planes through the same point. Due to the isotropy, the elastic constants (elastic coefficients) are invariants, that is, they do not change when the direction is changed. This implies that isotropic materials possess an infinite number of planes of material symmetry through a considered point. As stated, the elastic and mechanical properties of isotropic materials do not depend on the direction (straight line) across a considered point; this means that the elastic coefficients are invariant under arbitrary rotation of the coordinate system [2, 6–7]. This implies that the coefficients C_{ij} in Equation (7.12) in the x_i coordinate system (x, y, z) remain unchanged (invariants due to isotropy) in the new, rotated x_φ system $(\bar{x}, \bar{y}, \bar{z})$. So, in the new system, we can write:

$$\bar{\sigma} = \sigma_\varphi = C_{\varphi\alpha}\varepsilon_\alpha, \tag{7.14}$$

and, due to the isotropy:

$$C_{\varphi\alpha} = C_{ij}. \tag{7.15}$$

To find the structure of the matrix C_{ij}, consider its behavior during the process of transformation (in this case it is rotation) of the coordinate system. Transformation of the tensor matrix (stress tensor) is prescribed by Equation (2.55):

$$\left[\sigma_\varphi\right] = [a][\sigma][a]^{\mathrm{T}}, \tag{7.16}$$

where $[a]$ is the transformation matrix, while $\left[\sigma_\varphi\right] = [\bar{\sigma}]$ is the stress tensor in the new, rotated $(\bar{x}, \bar{y})$ coordinate system and $[\sigma]$ is the stress tensor in the old, original (x, y) coordinate system. Assume that the x–y system (or the plane x–y, or, for short, the x_i system) rotates for angle φ around the "z" axis which remains unchanged $(z = \bar{z})$; that is, angle φ is the angle between the x_i and the x_α system. This means that the considered systems are:

$$x_i \rightarrow x, y, \rightarrow \sigma: \text{original (old) system}$$

$$x_\alpha \rightarrow \bar{x}, \bar{y} \rightarrow \bar{\sigma}: \text{rotated system for } \varphi \text{ with respect to } x_i$$

If angle $\varphi = 180^\circ$, Equation (7.16) now becomes [2, 7]:

$$\left[\bar{\sigma}_{ij}\right] = \left[\sigma_\varphi\right] = \begin{bmatrix} \bar{\sigma}_x & \bar{\tau}_{xy} & \bar{\tau}_{xz} \\ \bar{\tau}_{yx} & \bar{\sigma}_y & \bar{\tau}_{yz} \\ \bar{\tau}_{zx} & \bar{\tau}_{zy} & \bar{\sigma}_z \end{bmatrix}$$

$$= \begin{bmatrix} \cos\varphi & \sin\varphi & 0 \\ -\sin\varphi & \cos\varphi & 0 \\ 0 & 0 & 1 \end{bmatrix} \begin{bmatrix} \sigma_x & \tau_{xy} & \tau_{xz} \\ \tau_{yx} & \sigma_y & \tau_{yz} \\ \tau_{zx} & \tau_{zy} & \sigma_z \end{bmatrix} \begin{bmatrix} \cos\varphi & -\sin\varphi & 0 \\ \sin\varphi & \cos\varphi & 0 \\ 0 & 0 & 1 \end{bmatrix} \tag{7.17}$$

$$= \begin{bmatrix} \sigma_x & \tau_{xy} & -\tau_{xz} \\ \tau_{yx} & \sigma_y & -\tau_{yz} \\ -\tau_{zx} & -\tau_{zy} & \sigma_z \end{bmatrix}.$$

The same procedure is valid for the strain tensor. If generalized vectors are used, then Equation (7.17), respecting Equation (7.13), can be written in the form:

$$\left[\bar{\sigma}_{ij}\right] = \left[\sigma_\varphi\right] = \begin{bmatrix} \bar{\sigma}_1 & \bar{\sigma}_4 & \bar{\sigma}_6 \\ \bar{\sigma}_4 & \bar{\sigma}_2 & \bar{\sigma}_5 \\ \bar{\sigma}_6 & \bar{\sigma}_5 & \bar{\sigma}_3 \end{bmatrix} = \begin{bmatrix} \sigma_1 & \sigma_4 & -\sigma_6 \\ \sigma_4 & \sigma_2 & -\sigma_5 \\ -\sigma_6 & -\sigma_5 & \sigma_3 \end{bmatrix} \tag{7.18}$$

Further, using generalized vectors, Equation (7.12) can be given in the form:

$$\begin{bmatrix} \sigma_1 \\ \cdot \\ \cdot \\ \cdot \\ \cdot \\ \sigma_6 \end{bmatrix} = \begin{bmatrix} C_{11} & \cdot & \cdot & \cdot & \cdot & \cdot \\ \cdot & \cdot & \cdot & \cdot & \cdot & \cdot \\ \cdot & \cdot & \cdot & \cdot & \cdot & \cdot \\ \cdot & \cdot & \cdot & \cdot & \cdot & \cdot \\ \cdot & \cdot & \cdot & \cdot & \cdot & \cdot \\ \cdot & \cdot & \cdot & \cdot & \cdot & C_{66} \end{bmatrix} \begin{bmatrix} \varepsilon_1 \\ \cdot \\ \cdot \\ \cdot \\ \cdot \\ \varepsilon_6 \end{bmatrix}, \tag{7.19}$$

and Equation (7.14) takes a similar form:

$$\begin{bmatrix} \bar{\sigma}_1 \\ \cdot \\ \cdot \\ \cdot \\ \cdot \\ \bar{\sigma}_6 \end{bmatrix} = \begin{bmatrix} C_{11} & \cdot & \cdot & \cdot & \cdot & \cdot \\ \cdot & \cdot & \cdot & \cdot & \cdot & \cdot \\ \cdot & \cdot & \cdot & \cdot & \cdot & \cdot \\ \cdot & \cdot & \cdot & \cdot & \cdot & \cdot \\ \cdot & \cdot & \cdot & \cdot & \cdot & \cdot \\ \cdot & \cdot & \cdot & \cdot & \cdot & C_{66} \end{bmatrix} \begin{bmatrix} \bar{\varepsilon}_1 \\ \cdot \\ \cdot \\ \cdot \\ \cdot \\ \bar{\varepsilon}_6 \end{bmatrix}. \tag{7.20}$$

Based on Equations (7.19) and (7.20), stress–strain relationships can be written for the original (x_i) and the rotated (x_α) coordinate systems. If the obtained equations are compared, that is, σ_1 with $\bar{\sigma}_1$, σ_2 with $\bar{\sigma}_2$, etc., then, taking into consideration Equation (7.18), it follows that:

$$\begin{aligned} \sigma_1 &= C_{11}\varepsilon_1 + C_{12}\varepsilon_2 + C_{13}\varepsilon_3 + C_{14}\varepsilon_4 + C_{15}\varepsilon_5 + C_{16}\varepsilon_6 \\ \bar{\sigma}_1 &= C_{11}\bar{\varepsilon}_1 + C_{12}\bar{\varepsilon}_2 + C_{13}\bar{\varepsilon}_3 + C_{14}\bar{\varepsilon}_4 + C_{15}\bar{\varepsilon}_5 + C_{16}\bar{\varepsilon}_6. \end{aligned} \tag{a}$$

Since Equation (7.18) is also valid for the strain tensor, it follows that:

$$\bar{\varepsilon}_1 = \varepsilon_1, \bar{\varepsilon}_2 = \varepsilon_2, \bar{\varepsilon}_3 = \varepsilon_3, \bar{\varepsilon}_4 = \varepsilon_4, \bar{\varepsilon}_5 = -\varepsilon_5, \bar{\varepsilon}_6 = -\varepsilon_6. \tag{b}$$

If the quantities in Equation (b) are entered into Equations (a), we find that $C_{15} = C_{16}$. By further comparisons – σ_2 with $\bar{\sigma}_2$, etc. – finally, for the case when the coordinate system x_i (that is, $x-y$) is rotated by 180°, it follows that:

$$\begin{aligned} C_{15} &= C_{16} = C_{25} = C_{26} = C_{35} = C_{36} = C_{45} = C_{46} = C_{51} = C_{52} \\ &= C_{53} = C_{54} = C_{61} = C_{62} = C_{63} = C_{64} = 0. \end{aligned}$$

As such, based on the above-mentioned consideration that the x–y system was rotated (transformed) by $\varphi = 180°$ about the z axis, the matrix of elastic constants, C_{ij}, takes the following form:

$$C_{ij} = \begin{bmatrix} C_{11} & C_{12} & C_{13} & C_{14} & 0 & 0 \\ C_{21} & C_{22} & C_{23} & C_{24} & 0 & 0 \\ C_{31} & C_{32} & C_{33} & C_{34} & 0 & 0 \\ C_{41} & C_{42} & C_{43} & C_{44} & 0 & 0 \\ 0 & 0 & 0 & 0 & C_{55} & C_{56} \\ 0 & 0 & 0 & 0 & C_{65} & C_{66} \end{bmatrix}. \tag{7.21}$$

In addition, after analogous transformations of the coordinate systems $(y–z)$ and $(z–x)$, that is, rotations of planes $(y–z)$ and $(z–x)$ also by $\varphi = 180°$, the matrix of the elastic constants becomes:

$$C_{ij} = \begin{bmatrix} C_{11} & C_{12} & C_{13} & 0 & 0 & 0 \\ C_{21} & C_{22} & C_{23} & 0 & 0 & 0 \\ C_{31} & C_{32} & C_{33} & 0 & 0 & 0 \\ 0 & 0 & 0 & C_{44} & 0 & 0 \\ 0 & 0 & 0 & 0 & C_{55} & 0 \\ 0 & 0 & 0 & 0 & 0 & C_{66} \end{bmatrix}. \tag{7.22}$$

On the basis of Equations (7.22) and (7.3), it can be stated that:

- The normal stress components do not depend on the shear strain components and the shear stress components do not depend on the normal strain components (that is, $\sigma = f(\varepsilon)$, $\tau = f(\gamma)$).
- Each component of the normal stress depends on all of the normal strain components but any one of the shear stress components depends only on a particular shear strain component.

If the analog transformations are now implemented for the angle $\varphi = 90^{\circ}$ in the $(x–y)$ plane, and after that in the $(y–z)$ and $(z–x)$ planes using the same procedures, finally we have:

$$C_{ij} = \begin{bmatrix} \bar{\bar{C}} & \bar{C} & \bar{C} & 0 & 0 & 0 \\ \bar{C} & \bar{\bar{C}} & \bar{C} & 0 & 0 & 0 \\ \bar{C} & \bar{C} & \bar{\bar{C}} & 0 & 0 & 0 \\ 0 & 0 & 0 & C & 0 & 0 \\ 0 & 0 & 0 & 0 & C & 0 \\ 0 & 0 & 0 & 0 & 0 & C \end{bmatrix} \tag{7.23}$$

This is the matrix of elastic constants.

On the basis of Equation (7.23), it is possible to write equations of the type shown in Equation (7.5). The constants of elasticity have the same values (isotropic values) for any rotated angle, φ. After analysis of equations of the type in Equation (7.5a) that are written by quantities from Equation (7.23) for the original and the rotated angle, it can be shown that:

$$\bar{\bar{C}} - \bar{C} = C. \tag{7.24}$$

It is clear that only two independent elastic constants exist. Lame has introduced the new constants in such way that:

$$\bar{\bar{C}} = \lambda + 2\mu; \quad \bar{C} = \lambda; \quad C = 2\mu \tag{7.25}$$

and these constants (λ, μ) are known as *Lame constants*. Using Equations (7.25), (7.23) and (7.21), we can write:

$$\begin{aligned} \sigma_x &= (2\mu+\lambda)\varepsilon_x+\lambda\left(\varepsilon_y+\varepsilon_z\right), \quad \tau_{xy}=\mu\gamma_{xy} \\ \sigma_y &= (2\mu+\lambda)\varepsilon_y+\lambda\left(\varepsilon_x+\varepsilon_z\right), \quad \tau_{yz}=\mu\gamma_{yz} \\ \sigma_z &= (2\mu+\lambda)\varepsilon_z+\lambda\left(\varepsilon_x+\varepsilon_y\right), \quad \tau_{zx}=\mu\gamma_{zx} \end{aligned} \tag{a}$$

and these equations represent Hooke's Law. Since volumetric strain, ε_v, is defined by Equation (3.5b):

$$\varepsilon_v=\varepsilon_x+\varepsilon_y+\varepsilon_z,$$

Hooke's Law can be written in the explicit form:

$$\begin{aligned} \sigma_x &= 2\mu\varepsilon_x+\lambda\varepsilon_v, \quad \tau_{xy}=\mu\gamma_{xy} \\ \sigma_y &= 2\mu\varepsilon_y+\lambda\varepsilon_v, \quad \tau_{yz}=\mu\gamma_{yz} \\ \sigma_z &= 2\mu\varepsilon_z+\lambda\varepsilon_v, \quad \tau_{zx}=\mu\gamma_{zx}. \end{aligned} \tag{7.26a}$$

One of the common written versions of the generalized Hooke's Law for isotropic materials, written in tensor notation is:

$$\sigma_{ij}=2\mu\varepsilon_{ij}+\lambda\varepsilon_{kk}\delta_{ij}, \tag{7.26b}$$

$$\varepsilon_{ij}=\frac{1}{2\mu}\sigma_{ij}-\frac{\lambda}{2\mu}\varepsilon_{kk}\delta_{ij} \tag{7.26c}$$

where δ_{ij} is the Kronecker tensor. However, there are other versions.

7.3.1 Determination of Hooke's Law – Method of Superposition

In the previous section, the generalized Hooke's Law was determined (Equation (7.26)), although its general form was mentioned earlier in Equation (7.2). As was found in Equation (7.22), normal stress depends on normal strain only and shear stress depends on shear strain only. Based on these statements, Hooke's Law will now be determined by the principle of superposition. Recall that, in Section 6.1, Equation (6.8) was introduced to consider the relationship between stress and strain in a uniaxial state of stress. This equation represents Hooke's Law for a uniaxial state of stress. If this law is extended to two-dimensional and three-dimensional states of stress, as shown in Figure 7.1, then Hooke's Law may be written as shown in the three parts of Table 7.1 [1, 2, 5].

In the derivation of Hooke's Law, it is understood that deformations are in the elastic domain and the principle of superposition may be applied.

If Hooke's Law is presented with respect to stress, then it follows (from Equation (6.8)) and on the basis of Table 7.1a, that:

$$\sigma_z=E\cdot\varepsilon_z \tag{7.27}$$

In a uniaxial state of stress for a linear elastic material, the proportionality factor E (the modulus of elasticity) relates stress and strain. This means that, in the case of a unaxial state of stress, only one material property is required to relate stress and strain.

Consider now a two-dimensional (plane) state of stress (Figure 7.1(b)). A result of the mutual dependency between stress and strain is presented in Table 7.1b.

As shown in Table 7.1b, Hooke's Law is defined by $\varepsilon_x, \varepsilon_y, \varepsilon_z, \gamma_{xy}$. If the equations in Table 7.1b are solved with respect to stress, then it follows that:

$$\sigma_x = \frac{E}{1-\nu^2}\left(\varepsilon_x + \nu\varepsilon_y\right); \ \sigma_y = \frac{E}{1-\nu^2}\left(\varepsilon_y + \nu\varepsilon_x\right); \ \tau_{xy} = G\gamma_{xy} \tag{7.28}$$

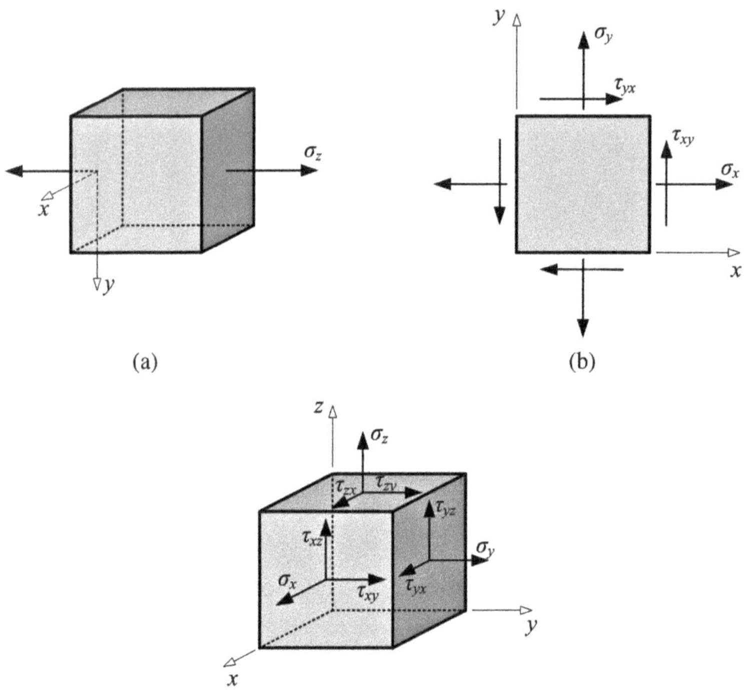

Figure 7.1 Body subjected to stress. (a) Uniaxial (one-dimensional) state of stress; (b) two-dimensional (plane, biaxial) state of stress; (c) three-dimensional state of stress.

Table 7.1a Hooke's Law determination for a uniaxial (one-dimensional) state of stress.

strain \ stress	σ_x	σ_y	σ_z	Σ $\varepsilon = \varepsilon(\sigma)$
ε_x	—	—	$-\nu\frac{\sigma_z}{E}$	$-\nu\frac{\sigma_z}{E}$
ε_y	—	—	$-\nu\frac{\sigma_z}{E}$	$-\nu\frac{\sigma_z}{E}$
ε_z	—	—	$\frac{\sigma_z}{E}$	$\frac{\sigma_z}{E}$

Table 7.1b Hooke's Law determination for a two-dimensional state of stress.

stress / strain	σ_x	σ_y	τ_{xy}	Σ	
				$\varepsilon = \varepsilon(\sigma)$	$\gamma = \gamma(\tau)$
ε_x	$\frac{\sigma_x}{E}$	$-\nu\frac{\sigma_y}{E}$	—	$\frac{1}{E}\left[\sigma_x - \nu\sigma_y\right]$	—
ε_y	$-\nu\frac{\sigma_x}{E}$	$\frac{\sigma_y}{E}$	—	$\frac{1}{E}\left[\sigma_y - \nu\sigma_x\right]$	—
ε_z	$-\nu\frac{\sigma_x}{E}$	$-\nu\frac{\sigma_y}{E}$	—	$-\frac{\nu}{E}\left(\sigma_x + \sigma_y\right)$	—
γ_{xy}	—	—	$\frac{\tau_{xy}}{G}$	—	$\frac{\tau_{xy}}{G}$

Table 7.1c Hooke's Law determination for a three-dimensional state of stress.

stress / strain	σ_x	σ_y	σ_z	Σ $\varepsilon = \varepsilon(\sigma)$
ε_x	$\frac{\sigma_x}{E}$	$-\nu\frac{\sigma_y}{E}$	$-\nu\frac{\sigma_z}{E}$	$\frac{1}{E}\left[\sigma_x - \nu\left(\sigma_y + \sigma_z\right)\right]$
ε_y	$-\nu\frac{\sigma_x}{E}$	$\frac{\sigma_y}{E}$	$-\nu\frac{\sigma_z}{E}$	$\frac{1}{E}\left[\sigma_y - \nu\left(\sigma_x + \sigma_z\right)\right]$
ε_z	$-\nu\frac{\sigma_x}{E}$	$-\nu\frac{\sigma_y}{E}$	$\frac{\sigma_z}{E}$	$\frac{1}{E}\left[\sigma_z - \nu\left(\sigma_x + \sigma_y\right)\right]$

When a three-dimensional state of stress is considered, the result of mutual dependency between stress and strain is given in Table 7.1c. Taking also shear strains, we have:

$$\gamma_{xy} = \frac{\tau_{xy}}{G}; \; \gamma_{yz} = \frac{\tau_{yz}}{G}; \; \gamma_{zx} = \frac{\tau_{zx}}{G} \tag{7.29a}$$

Table 7.1c and Equation(7.29a) represent Hooke's Law for a three-dimensional state of stress that is presented in the form of strain components which are functions of stress components. Solutions with respect to stress follows (from Equation (2.30)):

$$\left.\begin{aligned} \sigma_x &= \frac{E}{(1+\nu)(1-2\nu)}\left[(1-\nu)\varepsilon_x + \nu(\varepsilon_x + \varepsilon_z)\right], \quad \tau_{xy} = G\gamma_{xy} \\ \sigma_y &= \frac{E}{(1+\nu)(1-2\nu)}\left[(1-\nu)\varepsilon_y + \nu(\varepsilon_x + \varepsilon_z)\right], \quad \tau_{yz} = G\gamma_{yz} \\ \sigma_z &= \frac{E}{(1+\nu)(1-2\nu)}\left[(1-\nu)\varepsilon_z + \nu\left(\varepsilon_x + \varepsilon_y\right)\right], \quad \tau_{zx} = G\gamma_{zx} \end{aligned}\right\} \tag{7.29b}$$

Equations (7.27) to (7.29) and Table 7.1 define Hooke's Law as derived by the principle of superposition. Comparing Equations (7.26) and Equation (7.29b), it is clear that $\mu = G$.

Instead of the matrix of elastic constants (C_{ij}) written in Equations (7.23) and (7.3), a compliance matrix (S_{ij}), mentioned in Equation (7.4) can be written. In accordance with the equations for the generalized Hooke's Law (Table 7.1c and Equation (7.29a)), this compliance matrix is [8]:

$$[S_{ij}] = \begin{bmatrix} \frac{1}{E} & -\frac{\nu}{E} & -\frac{\nu}{E} & 0 & 0 & 0 \\ -\frac{\nu}{E} & \frac{1}{E} & -\frac{\nu}{E} & 0 & 0 & 0 \\ -\frac{\nu}{E} & -\frac{\nu}{E} & \frac{1}{E} & 0 & 0 & 0 \\ 0 & 0 & 0 & \frac{1}{G} & 0 & 0 \\ 0 & 0 & 0 & 0 & \frac{1}{G} & 0 \\ 0 & 0 & 0 & 0 & 0 & \frac{1}{G} \end{bmatrix}. \tag{7.30}$$

7.3.2 Engineering Constants of Elasticity

The elastic constants (or elastic coefficients) that form the elasticity tensor (elastic matrix or stiffness matrix) are not values that are directly measured from laboratory tests on a material. On the other hand, engineering constants like Young's modulus/modulus of elasticity (E), shear modulus/modulus of rigidity (G), Poisson's ratio (ν) and the bulk modulus (K) can be measured from laboratory tests. The modulus of elasticity and Poisson's ratio can be determined by a tension test while the shear modulus may be determined by a shear test and the bulk modulus by a hydrostatic compression test. Since constants C_{ijkl} and S_{ijkl} do not have a particular physical meaning, in engineering literature, the above-mentioned constants, E, G, ν and K, are called the engineering constants of elasticity and they are usually used in engineering practice [9]. As already stated, in a case involving isotropic materials, two constants of elasticity are introduced: λ and μ (the Lame constants). If Equations (7.26a):

$$\begin{aligned} \sigma_x &= 2\mu\varepsilon_x + \lambda\varepsilon_v, \quad \tau_{xy} = \mu\gamma_{xy} \\ \sigma_y &= 2\mu\varepsilon_y + \lambda\varepsilon_v, \quad \tau_{yz} = \mu\gamma_{yz}\ , \\ \sigma_z &= 2\mu\varepsilon_z + \lambda\varepsilon_v, \quad \tau_{zx} = \mu\gamma_{zx} \end{aligned} \tag{a}$$

are applied to a uniaxial state of stress ($\sigma_x = \sigma_y = \tau_{xy} = \tau_{y_z} = \tau_{zx} = 0$), it follows that:

$$\begin{aligned} 0 &= 2\mu\varepsilon_x + \lambda\varepsilon_v, \\ 0 &= 2\mu\varepsilon_y + \lambda\varepsilon_v, \\ \sigma_z &= 2\mu\varepsilon_z + \lambda\varepsilon_v, \end{aligned} \tag{b}$$

Summing equations (b) yields:

$$\sigma_z = (3\lambda + 2\mu)\varepsilon_v. \tag{c}$$

On the basis of Equation (c), the volumetric strain is:

$$\varepsilon_V = \frac{\sigma_z}{3\lambda + 2\mu}. \tag{d}$$

When the quantity from Equation (d) is entered into the last of Equations (b), the modulus of elasticity is:

$$E = \frac{2\mu + 3\lambda}{\mu + \lambda}\mu, \tag{7.31}$$

and, the second of Equations (b) yields Poisson's ratio:

$$\nu = \frac{\lambda}{2(\mu + \lambda)}. \tag{7.32}$$

The relationships between the Lame constants and the modulus of elasticity may also be given in the form:

$$G = \mu = \frac{E}{2(1+\nu)}, \quad \lambda = \frac{\nu E}{(1+\nu)(1-2\nu)}. \tag{7.33}$$

The first value (μ) in Equations (7.33) is the well-known shear modulus, G (from Equation (6.10)). Taking Equations (7.31) and (7.32), the bulk modulus (K) from Equation (3.9) can also be written as:

$$K = \frac{E}{3(1-2\nu)} = \lambda + \frac{2}{3}\mu \tag{7.34}$$

The relationship between the shear modulus and the modulus of elasticity can also be determined on the basis of simple considerations. Imagine an element whose axes are the principal strain axes. In this case, shear strains are equal to zero:

$$\gamma_{xy} = \gamma_{yz} = \gamma_{zx} = 0$$

At the same time, the shear stresses (Equation (7.29a)) are also equal to zero. For an isotropic material, the principal stress axes and principal strain axes coincide. Further, based on this fact, it can be shown that the shear modulus is not an independent constant. Let's consider a two-dimensional problem. From Table 7.1b, we have:

$$\varepsilon_x - \varepsilon_y = \frac{1}{E}(1+\nu)(\sigma_x - \sigma_y) \tag{e}$$

The directions of the principal stresses are defined as (from Equation (2.20)):

$$\tan 2\alpha(\sigma) = \frac{2\tau_{xy}}{\sigma_x - \sigma_y}$$

The directions of the principal strains are defined as (from Equation (3.61)):

$$\tan 2\alpha(\varepsilon) = \frac{\gamma_{xy}}{\varepsilon_x - \varepsilon_y} \tag{f}$$

Table 7.2 The mutual dependence of engineering constants in an anisotropic linear elastic solid [4, 10].

	E, G	***E, K***	***E, ν***	***G, K***	***G, λ***	***G, ν***	***K, ν***
E	—	—	—	$\frac{9GK}{3K+G}$	$\frac{G(3\lambda+2G)}{G+\lambda}$	$2G(1+\nu)$	$3K(1\text{-}2\nu)$
G	—	$\frac{3EK}{9K-E}$	$\frac{E}{2(1+\nu)}$	—	—	—	$\frac{3K(1-2\nu)}{2(1+\nu)}$
K	$\frac{EG}{9G-3E}$	—	$\frac{E}{3(1-2\nu)}$	—	$\frac{2}{3}G+\lambda$	$\frac{2G(1+\nu)}{3(1-2\nu)}$	—
ν	$\frac{E}{2G}-1$	$\frac{3K-E}{6K}$	—	$\frac{3K-2G}{6K+2G}$	$\frac{\lambda}{2(G+\lambda)}$	—	—
λ	$\frac{(E-2G)G}{3G-E}$	$\frac{3K(3K-E)}{9K-E}$	$\frac{E\nu}{(1+\nu)(1-2\nu)}$	$K-\frac{2}{3}G$	—	$\frac{2G\nu}{1-\nu}$	$\frac{3K\nu}{1+\nu}$

Since the principal axes coincide, the quantities in Equations (e) and (f) are equal. On the basis of equalization of these relations, it follows that:

$$\tau_{xy} = \frac{E}{2(1+\nu)}\gamma_{xy} = G\gamma_{xy}, \tag{g}$$

and hence it follows (from Equations (3.11), (6.10) and (7.33)) that:

$$G = \frac{E}{2(1+\nu)} \tag{h}$$

A review of the mutual dependence of constants is given in Table 7.2.

7.4 Orthotropic Materials

In the literature one can find many details relating to structural analysis. Most of the details relate to constructions which are made from isotropic materials. However, there are materials that are sometimes more acceptable in the design of a particular type of structure, primarily due to their mechanical properties and performance. We are talking here about composite materials, which are usually defined as materials consisting of two or more phases. The fibers that make up the dispersed phase of these materials are stiffer and are known as reinforcements, while the continuous phase that is less stiff is called the matrix (see Figure 7.2). The properties of a composite material are controlled by the fibers.

The main role of the matrix is to protect the fibers and provide local stress transfer between fibers. A fiber-reinforced composite consists of fiber-reinforced layers [11]. The long, unidirectional (or woven) continuous fibers surrounded (and protected) by a matrix give the structure the label lamina or layer (or ply).

Carbon fibers are most commonly used in composites. Their strength and stiffness depend on the production process. These types of composites are usually known as advanced composites. Glass fibers belong to the group of composites in the low to medium performance range. Most fibers (carbon, boron, and so on) behave linearly to failure. Fiber-reinforced composites may be classified into broad categories in

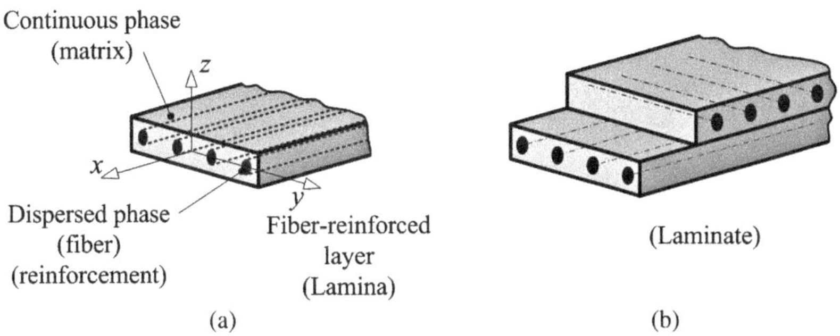

Figure 7.2 Composite material. (a) Fiber-reinforced layer (lamina); (b) laminate.

accordance with the matrix used: metal-matrix composites, polymer-matrix composites, ceramic-matrix composites, carbon-matrix composites. Two or more unidirectional laminae, which can be of various thicknesses, when stacked together at various orientations make a laminate.

The performance and properties of composite materials in many cases are superior to those that have their constituents acting separately. A low density in addition to high strength, a suitable level of stiffness and a long fatigue life mean that composite materials are very desirable in many fields of engineering (aerospace, automotive, marine and other applications). However, composites also have some limitations when compared to monolithic materials. The differences between composites and monolithic materials can be viewed in macromechanics, manufacturing technology, mechanical characterization, and so on.

If metals, ceramics and polymers are counted as conventional materials, then a composite material is treated as a combination of two or more materials. At the laminar level, analysis is usually made by taking average values of material properties and it is considered to be homogenous, although anisotropic [12]. Materials such as laminated plastics, various composite materials, reinforced concrete and so on are classified as *orthotropic materials* and they possess three orthogonal planes of material symmetry, in contrast to isotropic materials which posses an infinite number of planes of material symmetry. In short, materials with three mutually perpendicular planes of elastic symmetry are known as orthotropic materials (or orthogonally anisotropic materials). The three mutually orthogonal axes, x, y, z, that define these planes are called the *orthotropic axes* [13].

Let the fibers lie in the x–y plane and let this plane be the plane of symmetry. In this case, the x axis is usually referred to as the longitudinal axis of the lamina and coincides with the longitudinal axis of the fibers. As for other types of materials, it is necessary to establish the relationships between stress and strain, that is, it is desirable to define the coefficients of elasticity (the elastic constants, the matrix of elasticity), C_{ij}, in a manner similar to that used for isotropic materials. We know that the elasticity matrix is a diagonally symmetric matrix, based on Equation (7.11). However, the elastic constants (coefficients) are invariants at a point under rotation of $\varphi = 180°$ about the orthotropic axis. In this case, for a point on the $(x-y)$ plane, this can be viewed as a reflection. This means that the mirror image of the point placed in the $(-y)$ plane with respect to the plane $(x-y)$ where $z=0$ (and this plane is a plane of symmetry) can be written as: $\bar{x}\to x,\ \bar{y}\to y,\ \bar{z}\to -z$. Two other reflections are known: $\bar{x}\to x,\ \bar{y}\to -y,\ \bar{z}\to z$; $\bar{x}\to -x,\ \bar{y}\to y,\ \bar{z}\to z$. Using the direction cosines for the mentioned transformations (as given, for example, by

Equation (2.38) relating to normal stresses), finally, as a result for the stress–strain relationships for orthotropic materials possessing three mutually perpendicular planes of elastic symmetry, we can write:

$$C_{ij} = \begin{bmatrix} C_{11} & C_{12} & C_{13} & 0 & 0 & 0 \\ C_{12} & C_{22} & C_{23} & 0 & 0 & 0 \\ C_{13} & C_{23} & C_{33} & 0 & 0 & 0 \\ 0 & 0 & 0 & C_{44} & 0 & 0 \\ 0 & 0 & 0 & 0 & C_{55} & 0 \\ 0 & 0 & 0 & 0 & 0 & C_{66} \end{bmatrix} \tag{7.35}$$

However, the stress–strain relationships for a material with three mutually perpendicular planes of elastic symmetry, as shown in Equation (7.35), are the same as those that can be given for two planes of symmetry [11]. As can be seen from Equation (7.35), there are nine elastic coefficients relative to the orthotropic axes (x, y, z). In this case, for orthotropic materials there are the following Young's moduli:

$$E_x \neq E_y \neq E_z, \tag{a}$$

orthotropic shear moduli:

$$G_{xy}, G_{yz}, G_{zx} \tag{b}$$

and generalized Poisson ratios:

$$\nu_{xy}, \nu_{xz}, \nu_{yx}, \nu_{yz}, \nu_{zx}, \nu_{zy}. \tag{c}$$

In Equation (c), the first subscript indicates the direction in which the considered element is stressed, while the second refers to the direction of transverse strain. As with an isotropic material, an extension in the z direction causes contractions in both the x and y axes:

$$\varepsilon_x = -\nu_{zx}\varepsilon_z; \quad \varepsilon_y = -\nu_{zy}\varepsilon_z. \tag{d}$$

The stress–strain relationships using the constants shown in Equation (7.35) can be written as:

$$\begin{aligned} \sigma_x &= C_{11}\varepsilon_x + C_{12}\varepsilon_y + C_{13}\varepsilon_z \\ &\ldots\ldots\ldots \\ \tau_{zx} &= C_{66}\gamma_{zx} \end{aligned} \tag{7.36}$$

If Equations (a)–(d) are applied, the stress–strain relationships, shown by Equation (7.3) or Equation (7.36), but now with respect to strain components, can be written in the form:

$$\begin{aligned} \varepsilon_x &= \frac{\sigma_x}{E_x} - \frac{\nu_{yx}}{E_y}\sigma_y - \frac{\nu_{zx}}{E_z}\sigma_z \\ \varepsilon_y &= \frac{\sigma_y}{E_y} - \frac{\nu_{xy}}{E_x}\sigma_x - \frac{\nu_{zy}}{E_z}\sigma_z \\ &\vdots \\ \gamma_{zx} &= \frac{\tau_{zx}}{G_{zx}} \end{aligned} \tag{7.37}$$

As has been stated, for orthotropic materials, the elasticity matrix, C_{ij}, in Equation (7.35), which relates stress with strain, and the matrix S_{ij}, which relates strain with stress, are symmetric matrices, as is to be expected.

If Equations (7.37) are considered, it is clear that double equality arises because of the mentioned symmetry. The mentioned equations, for example, yield:

$$S_{12} = -\frac{\nu_{yx}}{E_y} = -\frac{\nu_{xy}}{E_x} \tag{7.38}$$

and similar is true for S_{13} and for S_{23}. In contrast to isotropic materials, where stress and strain are defined by the same quantities E and ν, here it is not so. In addition, consider a plane stress state of an orthotropic element with (x, y, z) orthotropic axes. Let the element be subjected to tensile stresses in the (z, x) plane and in the direction of the z and x axes as well as to in-plane shear stresses (Figure 7.3). In accordance with Equations (7.37), for a plane stress state we have:

$$\begin{aligned}
\varepsilon_x &= \frac{\sigma_x}{E_x} - \frac{\nu_{zx}}{E_z}\sigma_z \\
\varepsilon_y &= -\frac{\nu_{xy}}{E_x}\sigma_x - \frac{\nu_{zy}}{E_z}\sigma_z \\
\varepsilon_z &= \frac{\sigma_z}{E_z} - \frac{\nu_{xz}}{E_x}\sigma_x \\
\gamma_{xz} &= \frac{\tau_{xz}}{G_{xz}}
\end{aligned} \tag{7.38a}$$

Solving Equations (7.38a) with respect to stress, it follows that:

$$\begin{aligned}
\sigma_x &= \frac{E_x}{1-\nu_{xz}\nu_{zx}}(\varepsilon_x + \nu_{zx}\varepsilon_z) \\
\sigma_z &= \frac{E_z}{1-\nu_{xz}\nu_{zx}}(\varepsilon_z + \nu_{xz}\varepsilon_x) \\
\tau_{xz} &= G_{xz}\gamma_{xz}
\end{aligned} \tag{7.38b}$$

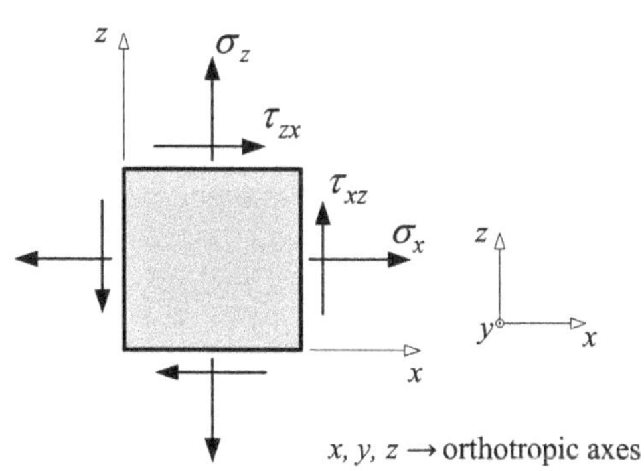

Figure 7.3 An orthotropic element in a plane stress state.

7.5 Linear Stress–Strain–Temperature Relations for Isotropic Materials

In engineering practice there are different types of structural analysis. Structural analysis can be focused on the level of stress, strain, vibration, temperature, creep behavior, crack propagation, fatigue, service life assessment, and so on. In many cases, structural analysis refers to a static load or a dynamic load but often does not include temperature effects. However, stresses in a body can be caused not only by external loads but also by temperature increases (decreases). Temperature effects are of importance for any technological process, manufacturing process or any service life assessment for a structure. These effects are of importance when determining a material's responses, since a material subjected to high temperatures can change its properties.

Temperature effects can be considered in structures where one can expect fatigue problems, fatigue crack growth problems, creep problems, and so on. Fracture toughness and notch impact energy also depend on temperature effects. The influence of temperature on mechanical properties is clear from the engineering stress–strain diagrams presented in Chapter 4. Creep in metallic materials is considered in Chapter 9. In a structure subjected to cycle loading, in the presence of an elevated temperature, a smaller number of cycles to failure can be expected [8]. In addition, at temperatures above about half of the absolute melting temperature, nonlinear deformation may be expected due to time-dependent phenomena (creep). In such a case, fatigue–creep interaction is also present. Higher temperatures give rise to a faster fatigue crack growth rate. As for fracture toughness, in general, it increases with an increase in temperature. The same is true for impact energy. Materials like metals, polymers, and so on with low impact energy are prone to brittle behavior. In some materials, significant changes in impact energy can be observed with temperature changes. So-called temperature-transition behavior is of importance when considering the use of materials. It is not recommended to use a material at a temperature where it exhibits low impact energy. It is usually said that the DBTT (ductile to brittle transition temperature) marks the transition from brittle to ductile behavior of a material (Figure 10.28). This temperature is especially notable in metals with a BCC (body cubic centered) crystal structure.

It is very difficult to cover all the influences of temperature on material behavior and to connect them to an appropriate stress–strain–temperature relationship. In this short section, the impact of temperature change on a body will be considered from two angles: when the body is subjected to a temperature increase and is free to deform, and when free deformation of the body is prevented.

Let an unconstrained engineering element made from an isotropic elastic material be subjected to a small uniform temperature increase (that is, to a temperature change) ΔT, and let the body be allowed to expand freely. According to experimental observation, in this case, the considered element undergoes equal extensions along its initial directions. In other words, the change in temperature will cause a change in volume but not a change in shape. This means that only normal strains will be associated with this temperature change.

If the change in temperature is:

$$\Delta T = T - T_0, \tag{7.39}$$

using the principle of superposition, the total strain can be written as:

$$\varepsilon_{ij} = \varepsilon_{ij}^{\sigma} + \varepsilon_{ij}^{T}. \tag{7.40}$$

In Equation (7.39), T is a new (for example, higher) temperature while T_0 is a reference (basic) temperature. The first term on the right, $\varepsilon_{ij}^{\sigma}$, indicates the mechanical strains due to stresses, and the second item, ε_{ij}^{T}, indicates the strains caused by the increase in temperature. Strains caused by an increase in temperature, based on Equation (7.4a) and taking $\sigma_{kl} = 0$, are: $\varepsilon_{ij} = \alpha_{ij}\Delta T$. On this basis, for an isotropic material we have:

$$\varepsilon_{ij}^{T} = \alpha \cdot \Delta T \cdot \delta_{ij}. \tag{7.41}$$

Consider the generalized Hooke's Law (Equation (7.26b)) relating to the stress caused by the load and presented in the form:

$$\sigma_{ij} = 2\mu\varepsilon_{ij} + \lambda\varepsilon_{kk}\delta_{ij}. \tag{a}$$

Now let this law be written in the form with respect to strain (Equation (7.26c)):

$$\varepsilon_{ij} = \frac{1}{2\mu}\sigma_{ij} - \frac{\lambda}{2\mu}\varepsilon_{kk}\delta_{ij}. \tag{b}$$

If this quantity is multiplied by δ_{ij}, then, for ε_{kk}, we obtain:

$$\varepsilon_{kk} = \frac{1}{2\mu + 3\lambda}\sigma_{kk}. \tag{7.42}$$

Note that $\delta_{ij} \cdot \delta_{ij} = \delta_{ii} = 3$, $\varepsilon_{ij} \cdot \delta_{ij} = \varepsilon_{ii} = \varepsilon_{kk}$, $\sigma_{ij} \cdot \delta_{ij} = \sigma_{ii} = \sigma_{kk}$

Our previously conducted considerations were based on strains due to mechanical loads. Consider now Hooke's Law which also includes strains due to thermal loads. If Equation (b) is added to the quantity in Equation (7.41), then Hooke's Law takes the form:

$$\varepsilon_{ij} = \frac{1}{2\mu}\left(\sigma_{ij} - \frac{\lambda}{2\mu + 3\lambda}\sigma_{kk}\delta_{ij}\right) + \alpha \cdot \Delta T \cdot \delta_{ij}. \tag{7.43a}$$

Since (from Equations (7.32) and (7.33)):

$$\frac{\lambda}{2\mu + 3\lambda} = \frac{\nu}{1+\nu} \rightarrow \frac{\lambda}{2\mu + 3\lambda} \cdot \frac{1}{2\mu} = \frac{\nu}{1+\nu} \cdot \frac{1}{2\mu} = \frac{\nu}{E}, \tag{c}$$

Equation (7.43a) can also be written in the form:

$$\varepsilon_{ij} = \frac{1}{2\mu}\sigma_{ij} - \delta_{ij}\left(\frac{\nu}{E}\sigma_{kk} - \alpha \cdot \Delta T\right). \tag{7.43b}$$

If Equation (7.43b) is multiplied by δ_{ij}, taking $\mu = G$, $G = E/2(1+\nu)$, the following result is obtained:

$$\varepsilon_{kk} = \frac{1-2\nu}{E}\sigma_{kk} + 3\alpha \cdot \Delta T \tag{7.43c}$$

or:

$$\sigma_{kk} = \frac{E}{1-2\nu}\varepsilon_{kk} - \frac{3E \cdot \alpha \cdot \Delta T}{1-2\nu} \tag{7.43d}$$

Equations (7.43) represent a general form of Hooke's Law which takes into consideration the impact of temperature on the deformation of isotropic materials. Coefficient α (1/°C) is the linear coefficient of thermal expansion (dilatation) of the material. In accordance with the above consideration, it is clear that a thermal stress of size $E\alpha\Delta T/1-2\nu$, as given in Equation (7.43d), must also be included in Equation (a) (that is, the earlier mentioned Equation (7.26b)) when this equation is used as a basis for the determination of stresses that include thermal load. Thus, Equation (7.26b), in this case, takes the form:

$$\sigma_{ij} = 2\mu\varepsilon_{ij} + \lambda\varepsilon_{kk}\delta_{ij} - \alpha\cdot\Delta T(2\mu + 3\lambda)\delta_{ij} \tag{7.43e}$$

since:

$$2\mu + 3\lambda = \frac{E}{1-2\nu} \tag{d}$$

In fact, Equations (7.43a), (7.43b) and (7.43e) are the basic stress–strain relationships in thermoelasticity for isotropic materials and are known as the *Duhamel–Neumann equations.* Depending on the field in which analysis needs to be performed (the elastic field, the plastic field, and so on), it is sometimes convenient/desirable to write Hooke's Law in the form in which spherical stress components are associated with spherical strain components or in the form in which deviatoric stress components are associated with deviatoric strain components. These forms of Hooke's Law can be presented using Lame constants or using engineering constants:

Spherical form (7.44a)

$$\sigma_{kk} = 3K\varepsilon_{kk} \quad \text{(using engineering constants)}$$

$$\sigma_{kk} = (3\lambda + 2\mu)\varepsilon_{kk} \quad \text{(using Lame constants)}$$

Deviatoric form (7.44b)

$$S_{ij} = 2Ge_{ij} \quad \text{(using engineering constants)}$$

$$S_{ij} = 2\mu e_{ij} \quad \text{(using Lame constants)}$$

References

1 Prelog, E. (1972) *Elastomechanics and Plastomechanics* (in Slovenian), Faculty of Mechanical Engineering, University of Ljubljana.
2 Brnic, J. (1996) *Elastomechanics and Plastomechanics* (in Croatian), Školska knjiga Zagreb, Zagreb.
3 Brnic, J. (2013) *Analysis of Deformable Bodies by Finite Element Method* (in Croatian), Fintrade & Tours, d.o.o. Rijeka and Faculty of Engineering, University of Rijeka, Rijeka.
4 Alfirević, I. (2006) *Linear Structural Analysis*, Pretei d.o.o., Zagreb.
5 Philpot, T. A. (2008) *Mechanics of Materials*, John Wiley & Sons, New Jersey.
6 Boresi, A. and Schmidt, R. J. (2003) *Advanced Mechanics of Materials*, John Wiley & Sons, USA.
7 Štok, B. (1988) *Mechanics of Deformable Bodies II* (in Slovenian), Faculty of Mechanical Engineering, Ljubljana.
8 Dowling, N. E. (2013) *Mechanical Behavior of Materials*, Pearson, New York.

9 Alfirević, I. (2003) *Introduction to Tensors and Continuum Mechanics* (in Croatian), Golden Marketing, Zagreb.
10 Findley, W. N., Lai, J. S. and Onaran, K. (1989) *Creep and Relaxation of Nonlinear Viscoelastic Materials*, Dover Publications, Inc., New York.
11 Solecki, R. and Conant, R. J. (2003) *Advanced Mechanics of Materials*, Oxford University Press, New York.
12 Daniel, I. M. and Ishai, O. (2006) *Engineering Mechanics of Composite Materials*, Oxford University Press, New York.
13 Boresi, A. P., Schmidt, R. J. and Sidebottom, O. M. (1993) *Advanced Mechanics of Materials*, John Wiley & Sons, New York.

8

Rheological Models

8.1 Introduction

In engineering practice, structures and technological processes are two major features. When it comes to structures, it is possible to talk about their design, relevant manufacturing processes, the material from which they are made, maintenance, testing, their response to stress/deformation, their purpose and so on. We have already looked at the responses of structures subjected to stress, temperature or specific environmental conditions (see, for example, Section 5.1). In addition, it is necessary to consider how the response of the structure can be modeled. This is important since the behavior of a structure which will be subjected to certain conditions can be predicted in advance. When it comes to technological processes, we must differentiatebetween processes for manufacturing structures and processes for obtaining other products, such as processes for the production of certain fuels, food products and so on.

Whether the focus is directed toward the response of a structure to a load or to any other subject, it is possible to say that there are real responses/real processes and modeled (or simulated/predicted) responses and processes. The latter, that is, responses that can be modeled, can be of great importance in the design of structures. Clearly, in essence, we are talking about predicting the behavior of materials in the structure that will be subjected to certain possible loads. Modeling the behavior of the structure reduces the price of the structure, since it removes the need to consider experimental investigation of the structural behavior. It is no longer necessary to manufacture a model, the testing of which would determine the responses of interest.

Modeling of structural responses, that is, the responses of a material from which a structure is made, can sometimes be done by using analytical or numerical models but is usually addressed by using so-called *rheological models*, which establish the relationship between stress and strain. From a deformation point of view, elastic and inelastic behavior of engineering materials may be discussed (as introduced in Section 5.1). However, a material's response is characterized by a stress–strain diagram.The term "inelastic" is used to denote a material response that is nonlinear. An elastic response is associated, for example, with elongation (extension) or with shortening but not with fracture, although an elastic response can be linearly or nonlinearly elastic. If an inelastic response is time independent, it is said that the material is displaying plastic deformation. Viscoelastic or viscoplastic deformation modes are associated with time-dependent deformation. Creep, as a special type of material behavior that is a

Analysis of Engineering Structures and Material Behavior, First Edition. Josip Brnić.

time-dependent process, may have three stages: transient, steady-state and accelerating creep, usually ending with fracture. This behavior can be shown in a strain–time diagram.

Stress–strain relationships are usually called stress–strain behavior, but are also known as constitutive equations and they are needed in structural design. Classical design, based on the strength of a material often uses linear stress–strain relationships. The design of an engineering element or part of a machine in which time-dependent behavior may occur (for example, creep) involves complex relationships between stress and strain. Some mechanical devices (or elements) such as springs, viscous dashpots and frictional sliders may be used to describe different types of mechanical behavior (deformation). These elements are usually known as rheological models. These elements (models) can sometimes be used individually and sometimes associated with assembled models. Briefly, an appropriate definition of rheology can be mentioned [1]: "Rheology is the study of the response of certain materials to the stresses imposed on them".

Constitutive relationships (stress–strain relationships) exist for time-dependent and time-independent problems. It is known that, in engineering practice, the mutual relationship between stress and strain in an elastic or plastic regime is defined by the theory of elasticity or by the theory of plasticity. Deformation problems are usually considered at room temperature and for short periods of observation. This means that the effect of time is neglected. However, in some cases, observations have shown that a deformation does not reach a final value immediately after the application of constant stress. Similar changes in stress apply for a body subjected to constant strain. If the behavior of the material is treated as time-dependent, it is monitored for a time; as a result of this monitoring, a curve of the behavior is obtained, then, using some models, the curve itself can be modeled. In other words, a constitutive equation expressed in the form of the curve can be modeled by a constitutive model called a rheological model. As such, a material's properties, such as its elastic, plastic and viscoelastic properties, are usually called rheological properties. Constitutive equations (stress–strain equations) that can be determined by rheological models may also be called rheological equations.

8.2 Time-independent Behavior Modeling

Further, when the behavior of a model under stress is considered, stress will be considered constant.

8.2.1 Elastic Deformation Modeling

The elastic deformation of an engineering element, that is, an elastic response of a material in a considered element, may be represented by the elastic behavior of a simple linear spring.

8.2.1.1 Hooke's Element (H Model)

In this case, a linear spring represents the model, known as Hooke's element (model). Deformation is always proportional to the load/force. Hooke's model and its response are shown in Figure 8.1. Elastic deformation of the engineering element is associated with the elongation of the bonds between its atoms.

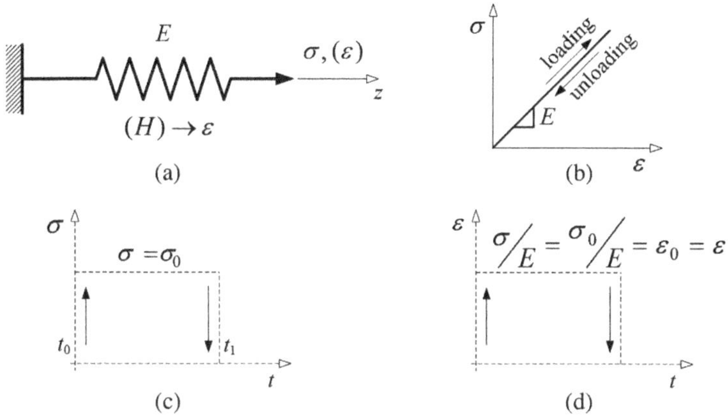

Figure 8.1 Hooke's model and its response. (a) H model; (b) stress–strain diagram; (c) stress–time diagram; (d) strain–time diagram.

There are many constitutive equations (stress–strain equations, rheological equations) that can describe the behavior of a material which obeys Hooke's Law, as detailed in Chapter 7. Consider Equation (7.43e), which represents Hooke's Law when the temperature impact is also included:

$$\sigma_{ij} = 2\mu\varepsilon_{ij} + \lambda\varepsilon_{kk}\delta_{ij} - \alpha\cdot\Delta T(2\mu + 3\lambda)\delta_{ij} \tag{a}$$

If the values defined by Equations (2.8a) and (3.47d), and the value:

$$2\mu + 3\lambda = \frac{E}{1-2\nu} \tag{b}$$

are inserted into Equation (7.43e), it follows that the rheological equation of Hooke's Law in deviatoric form (that is, written using deviatoric stress and deviatoric strain) is:

$$S_{ij} = 2\mu e_{ij};\quad G = \mu \tag{8.1}$$

As can be seen, this equation is independent of the impact of temperature. This implies that a complete image of Hooke's element behavior can be gained when the two equations (8.1) and (7.43d) are written in combination:

$$\begin{aligned} S_{ij} &= 2\mu e_{ij} \\ \sigma_{kk} &= \frac{E}{1-2\nu}\varepsilon_{kk} - \frac{3E}{1-2\nu}\alpha\Delta T \end{aligned} \tag{8.1a}$$

Finally, the general relationship between stress and strain for Hooke's element is given by Equations (8.1a), or by Equation (7.43e):

$$\sigma_{ij} = 2\mu\varepsilon_{ij} + \lambda\varepsilon_{kk}\delta_{ij} - \alpha\cdot\Delta T(2\mu + 3\lambda)\delta_{ij} \tag{c}$$

However, the stress–strain relationship representing the behavior of Hooke's element in a linear elastic field is:

$$\sigma = E\varepsilon. \tag{8.2}$$

8.2.2 Deformation Modeling after the Elastic Limit

The material ceases to behave elastically when the stress rises above the elastic limit. If an element is subjected to stress above the elastic limit, the strain that does not disappear after unloading of the element is called permanent strain (or plastic strain). Plastic deformation in metallic materials occurs by one or more processes [2, 3]: slip, twining, diffusional creep or grain boundary sliding. Slip (sliding) is the most common mechanism of plastic deformation and involves one plane of atoms moving over another.

8.2.2.1 Saint Venant Element (SV Model)

A model that is commonly used to describe ideal plastic behavior of a material is the so-called Saint Venant Model (Figure 8.2). Commonly, the rheological equation related to this model is written:

$$S_{ij} = 2G\dot{e}_{ij} \tag{8.3}$$

Figure 8.3 shows rheological models and their responses in accordance to different strain inputs. In each of the models considered, the plastic phase of the material's response can be seen.

8.2.2.2 Saint Venant Element-Spring/(SV–Spring)

A model of a Saint Venant element and a spring in series configuration is presented in Figure 8.3(a) and this serves to describe a linear elastic–perfectly plastic response (Figure 8.3(b)).

8.2.2.3 Saint Venant Element | Spring – Spring/(SV | Spring – Spring)

A model where a Saint Venant element is connected in parallel with one spring, and after that this combination is connected in series with a second spring is shown in Figure 8.3(c). This model is used to describe a linear elastic–linear hardening response (Figure 8.3(d)).

One-dimensional models that are appropriate for describing the behavior of real materials are presented in Figure 8.3(a) and Figure 8.3(c). The first (Figure 8.3(a)) is a combination of a linear spring and a frictional slider in series, while the second (Figure 8.3(c)) is a combination of a linear spring connected in parallel with a frictional slider and then assembled in series with a further spring. Figures 8.3(b) and 8.3(d) show idealized stress–strain curves for the models shown in Figures 8.3(a) and 8.3(c). In both cases (Figures 8.3(a) and 8.3(c)), the material initially deforms in accordance with the elastic modulus, E_1. Further, when the element is subjected to stress that corresponds

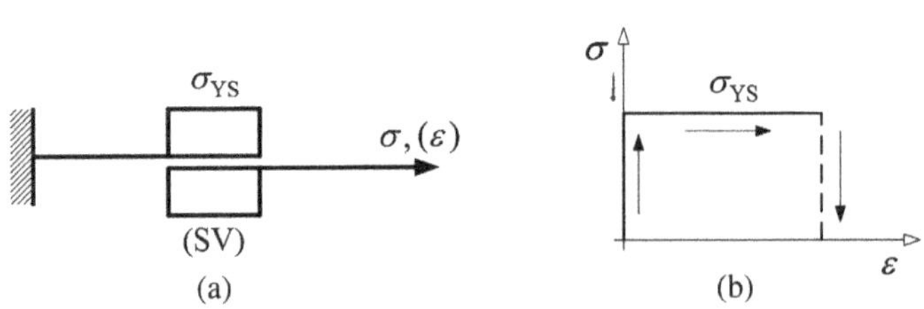

Figure 8.2 Saint Venant model (rigid–perfectly plastic behavior). (a) Model; (b) stress–strain diagram.

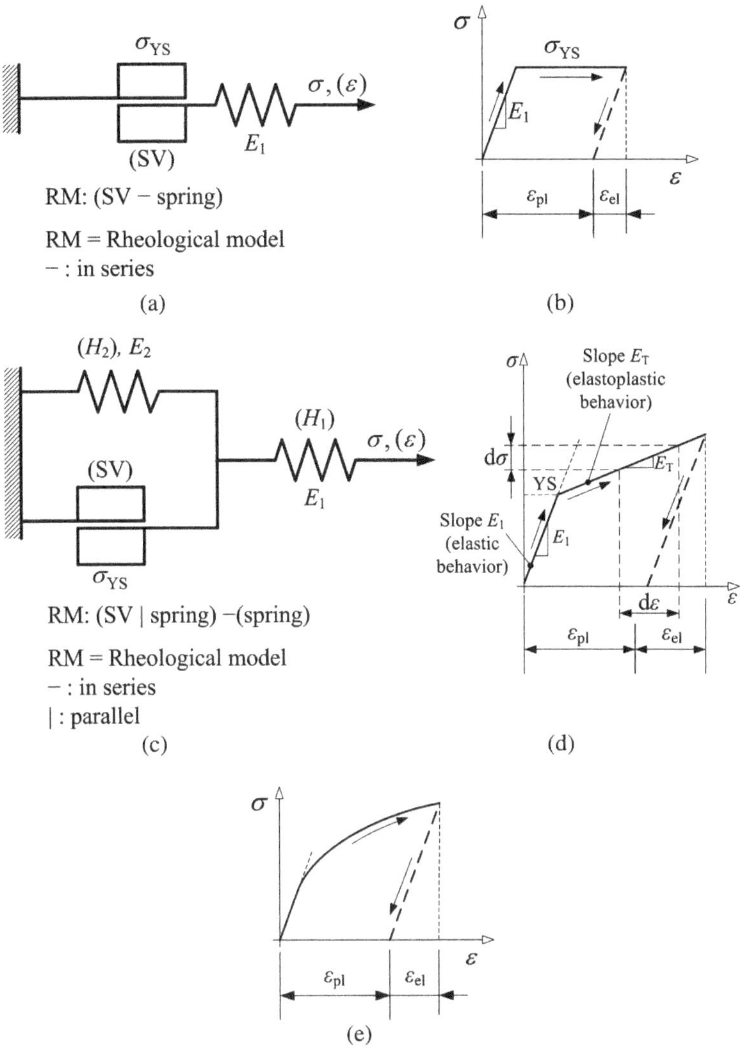

Figure 8.3 Material responses that include plastic behavior. (a) Model: elastic–perfectly plastic behaviour; (b) stress–strain diagram for linear elastic–perfectly plastic behaviour; (c) model: linear elastic–linear hardening behaviour; (d) stress–strain diagram for linear elastic–linear hardening behaviour; (e) stress–strain diagram for nonlinear hardening behavior.

to the YS stress level (the uniaxial yield stress), as shown in Figure 8.2(a) and Figures 8.3 (a) and (c), the frictional slider becomes active. When the frictional slider becomes active ($\sigma \geq \sigma_{YS}$), that is, the YS stress level has been reached; this stress level remains in the cases shown in Figure 8.2 and Figure 8.3(a). In Figure 8.3(c), after the frictional slider has become active ($\sigma \geq \sigma_{YS}$), the overload (stress higher than the YS stress level) will be transferred to the second spring (E_2), that is, due to an increase in the load, the material is assumed to display linear elastic–linear hardening, characterized by the so-called tangential modulus, E_T, as shown in Figure 8.3(d). If the stress level is reduced to below the YS stress level ($\sigma < \sigma_{YS}$), the spring (E_1) will take up the stress. The strain (ε), that is, the overall strain in the model, shown in Figure 8.3(c), is the sum of the strain (ε_1) in the first spring (E_1) and the strain (ε_2) in the parallel item (spring E_2 plus slider) and will be, as stated when $\sigma \geq \sigma_{YS}$:

$$\varepsilon = \frac{\sigma}{E_1} + \frac{\sigma - \sigma_{YS}}{E_2}, \quad \sigma \geq \sigma_{YS} \tag{8.4a}$$

In the case when $\sigma \leq \sigma_{YS}$,

$$\varepsilon = \varepsilon_1 = \frac{\sigma}{E_1}. \tag{8.4b}$$

Plastic deformation can also occur when the response is nonlinear strain hardening (see Figure 8.3(e)). However, the strain that occurs after unloading in all of these models, as shown in Figure 8.3, consists of elastic and plastic parts.

8.3 Time-dependent Behavior Modeling

So far, our considerations have not included the time-dependent deformation process. Such a review is justified by the fact that some materials exhibit linear, nearly linear or even nonlinear behavior at room temperature and at appropriate stress levels but do not exhibit time-dependent behavior. On the other hand, metallic materials, for example, exposed to a high enough stress level at high temperatures exhibit so-called viscoelastic behavior, that is, these materials demonstrate both elastic and viscous features [4, 5]. Several phenomena can be listed with respect to such a viscoelastic material's behavior: creep under constant stress, relaxation under constant strain, delayed recovery, permanent set, and so on.

In engineering practice, especially for metallic materials that are used in the design of engines and which are exposed to high temperatures, a special phenomenon can occur – creep. Creep is usually defined as a phenomenon of slow, continuous deformation of a material under constant stress if the material is exposed to a high enough temperature [4]. Creep may also occur in other materials under different conditions. Creep as a phenomenon will be discussed in the next chapter.

Some of the stress–strain relationships which can be used to describe certain types of viscoelastic material behavior are developed to fit experimentally obtained curves. Rheological models (constitutive models) that are conventionally used to describe a viscoelastic material's behavior are considered in the next part of this chapter. In Figure 5.1(d), we saw the stress–strain diagram representing viscoelastic material behavior. It is clear that, after a period of time, the strain will disappear. Now, some of the rheological models we can use to model the time-dependent behavior of the materials will be considered. Rheological models and the obtained rheological (constitutive) equations are also sometimes called viscoelastic constitutive equations. Material constants such as E_i and η_i may be determined from experiments.

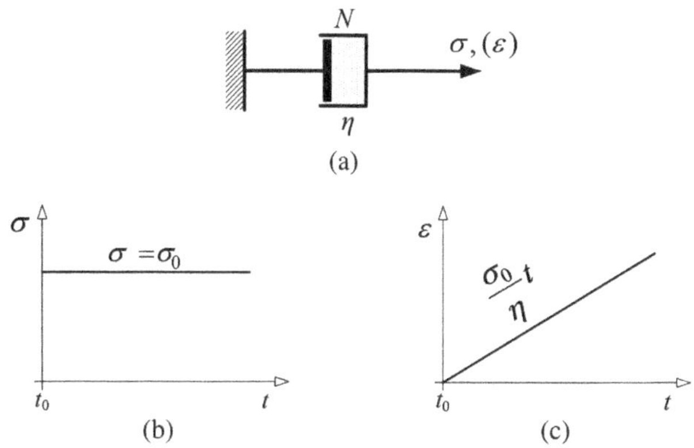

Figure 8.4 A Newton element (model): linear viscous dashpot. (a) Model; (b) stress–time diagram (constant stress imposed in the model); (c) strain–time diagram.

8.3.1 Newton Element (N Model): Linear Viscous Dashpot Element

A Newton element is shown in Figure 8.4.

The following equation is valid for a linear viscous dashpot element:

$$\sigma = \eta \frac{d\varepsilon}{dt} = \eta \dot{\varepsilon}, \tag{8.5}$$

where η is a coefficient of viscosity (Newtonian viscosity) and $\dot{\varepsilon}$ is the strain rate. The stress in a viscous element depends not only upon strain but also on the strain rate. As can be seen in Figure 8.4(c), the element continuously deforms when it is subjected to constant stress (σ). However, it is impossible to subject the element to constant strain instantaneously, since that would involve the occurrence of infinite stress, which is impossible in engineering practice [4].

8.3.2 Maxwell Model (M = H − N)

Here, M stands for Maxwell's model, H stands for Hooke's model and N stands for Newton's model, while the label (−) between them indicates a serial connection. A linear spring element (H) in series with a linear viscous dashpot element (N) constitutes Maxwell's model (M), and this is shown in Figure 8.5.

A model is subjected to constant force/stress (σ), as shown in Figure 8.5. Since both members (spring and dashpot) of the model are in series, the same level of stress will appear in each element ($\sigma_1 = \sigma_2 = \sigma$) and, according to Equations (8.2) and (8.5), the stress level is:

$$\sigma = E\varepsilon_1 \tag{8.6}$$

$$\sigma = \eta \dot{\varepsilon}_2 \tag{8.7}$$

but the overall strain will be the sum of the strain in the spring and that in the dashpot ($\varepsilon_1 \neq \varepsilon_2$). The response of the model with respect to deformation is:

$$\varepsilon = \varepsilon_1 + \varepsilon_2. \tag{8.8}$$

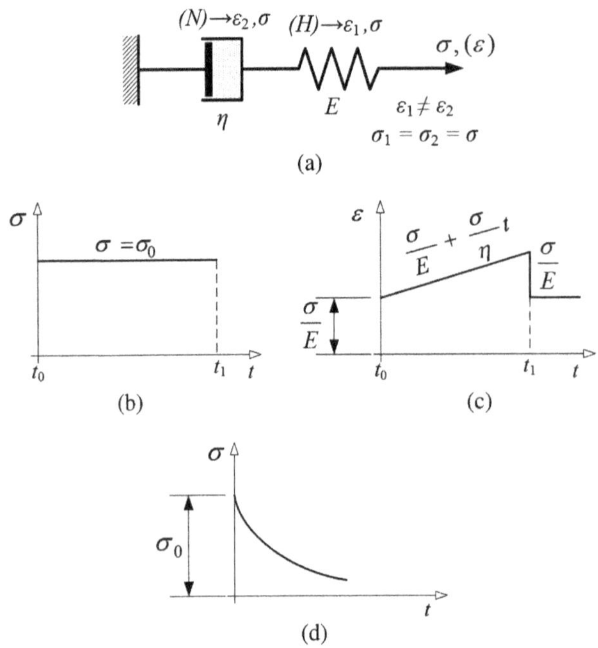

Figure 8.5 A Maxwell model (spring and linear viscous dashpot in series). (a) Model; (b) stress–time diagram; (c) strain–time diagram; (d) stress relaxation curve.

Based on Equations (8.6), (8.7) and (8.8), the stress–strain response of the model can be written as:

$$\varepsilon = \frac{\sigma}{E} + \frac{\sigma}{\eta}t. \tag{8.9}$$

Equation (8.9) represents behavior in the Maxwell model. This response is shown in Figure 8.5(c). It is clear that, after the instantaneous strain (elastic or elastic–plastic) that occurs in the elastic element (spring), further strain in the model increases linearly with time. Reloading of the model at time t_1 indicates that the stress (σ/E) in the spring will disappear while the stress in the dashpot remains.

Instead of the consideration when the model is subjected to constant stress (σ), a similar consideration can be made when the model is subjected to constant strain or to a constant strain rate. For example, if Equation (8.8), including Equations (8.6) and (8.7), is written in the form:

$$\dot{\varepsilon} = \frac{\dot{\sigma}}{E} + \frac{\sigma}{\eta} \tag{8.10}$$

and if our consideration is focused on the Maxwell model subjected to constant strain $\varepsilon = \varepsilon_0 =$ const., that is, to the process where $\varepsilon =$ const., $\dot{\varepsilon} = 0$, then the differential Equation (8.10) becomes:

$$\dot{\sigma} + \frac{E}{\eta}\sigma = 0 \tag{8.11}$$

The solution of this differential equation is:

$$\sigma = C \cdot e^{-Et/\eta} \tag{8.12}$$

The boundary condition now is:

$$t = 0 \rightarrow \sigma = \sigma_0. \tag{a}$$

At the beginning of the process ($t = 0$) at which point the strain is constant, it follows that $= \sigma_0$, which represents the known (initial) level of stress.

On the basis of Equation (8.12), it follows that:

$$C = \sigma_0 ,$$

and finally, the solution of differential Equation (8.11) is:

$$\sigma = \sigma_0 \cdot e^{-Et/\eta} \tag{8.13}$$

Equation (8.13) describes a process in which the model is subjected to constant strain ($\varepsilon = \varepsilon_0 =$ const.) with initial stress, σ_0; that is, the stress level at the beginning of the process is the mentioned constant strain. This process is presented in Figure 8.5(d) and is known as *stress relaxation*. The relaxation process is characterized by a decrease in stress with time while strain is kept constant. So, Maxwell's model can simulate stress relaxation in a real material, as shown in Figure 8.5(d), that is, it can simulate the function: $\sigma = \sigma(\sigma_0)$, where σ_0 is an initial stress, but it cannot simulate the creep process adequately. The Maxwell model simulates deformation behavior, $\varepsilon = \varepsilon(t)$, as shown in Figure 8.5(c), while the curve of the real, uniaxial creep process, $\varepsilon = \varepsilon(t)$, is of the form shown later in Figure 9.5. The discrepancy in creep behavior description between the Maxwell model and the behavior of real materials occurs because in most real materials, the strain rate decreases with time and the strain tends to a finite value, while in the Maxwell model, the strain rate is constant (from Equation (8.10), if $\sigma = \text{const.}, \rightarrow \sigma\ = 0;\ \dot{\varepsilon} = \sigma/\eta$) and strain increases without limit.

8.3.2.1 Generalized Maxwell Model

A generalized Maxwell model is composed of n Maxwell models connected either in series or in parallel. A generalized Maxwell model (in series) is presented in Figure 8.6.

In accordance with Equation (8.10), the constitutive equation may be written in the form:

$$\dot{\varepsilon} = \dot{\sigma}\sum\frac{1}{E_i} + \sigma\sum\frac{1}{\eta_i} \tag{8.14}$$

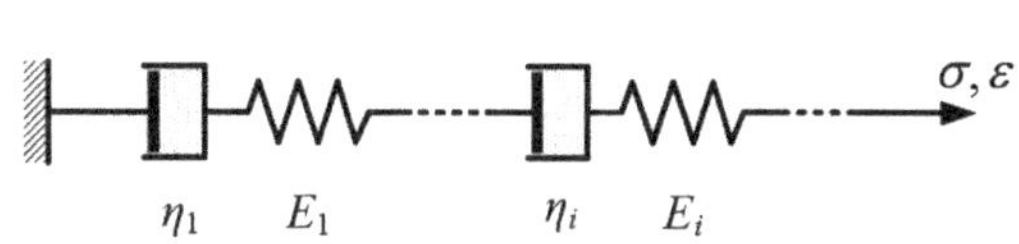

Figure 8.6 A generalized Maxwell model.

8.3.3 Voigt-Kelvin Model (K = H | N)

Two members of the model (the spring representing Hooke's (H) model and the linear viscous dashpot representing Newton's (N) model) which were in series in Maxwell's model, are now connected in parallel, and this model is known as the Voigt–Kelvin model (K), see Figure 8.7.

Let the K model at time $t_0 = 0$ be subjected to constant stress ($\sigma = \sigma_0$). The spring tends to deform immediately for σ/E but this is prevented by the simultaneous action of the viscous dashpot [6]; thus, there is no initial strain (deflection) since the spring and the dashpot are in parallel. However, deformation increases gradually as we asymptotically approach the value σ/E. If at moment t_1 the model is unloaded, this unloading will equally not happen immediately because of the viscous dashpot, as shown in Figure 8.6 (b). The level of stress in Hooke's model, as a part of the Kelvin model, is different from that in Newton's model ($\sigma_1 \neq \sigma_2$), but deformation in both parts (Hooke's and Newton's) is the same ($\varepsilon_1 = \varepsilon_2 = \varepsilon$). As such, for the Voigt–Kelvin model we have:

$$\sigma_1 = E\varepsilon \tag{8.15}$$

$$\sigma_2 = \eta\dot{\varepsilon} \tag{8.16}$$

The total stress is:

$$\sigma = \sigma_1 + \sigma_2. \tag{8.17}$$

If the quantities in Equations (8.15) and (8.16) are included in Equation (8.17), the following equation is obtained:

$$\dot{\varepsilon} + \frac{E}{\eta}\varepsilon = \frac{\sigma}{\eta}. \tag{8.18}$$

Equation (8.18) is a linear differential equation of the first order. Its solution is of the form:

$$\varepsilon = e^{-\int P\mathrm{d}x}\left[\int Q\cdot e^{\int P\mathrm{d}x}\mathrm{d}x + \mathrm{C}\right]. \tag{a}$$

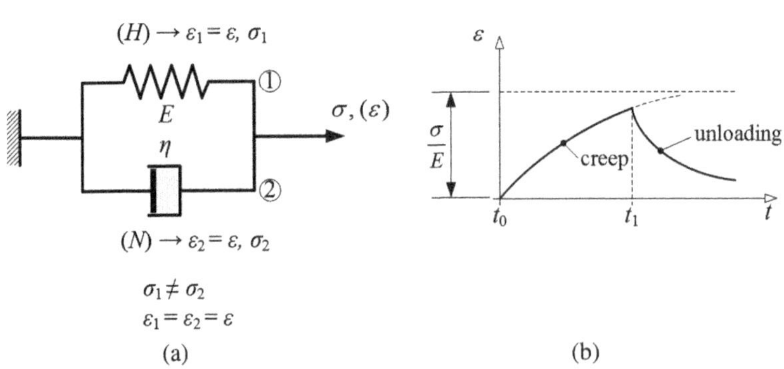

Figure 8.7 A Voigt–Kelvin model (spring and linear viscous dashpot in parallel). (a) Model; (b) strain–time diagram.

Comparing Equations (8.18) and (a), it follows that:

$$\varepsilon = e^{-\int \frac{E}{\eta} \mathrm{d}t} \left[\int \frac{\sigma}{\eta} \cdot e^{\int \frac{E}{\eta} \mathrm{d}t} \mathrm{d}t + \mathrm{C} \right]. \tag{b}$$

Since (in general):

$$\int e^{at} \mathrm{d}t = \frac{1}{a} e^{at}, \tag{c}$$

the solution of Equation (b) is:

$$\varepsilon = e^{-\frac{Et}{\eta}} \left[\frac{\sigma}{E} \cdot e^{\frac{Et}{\eta}} + \mathrm{C} \right]. \tag{d}$$

On the basis of the boundary condition:

$$(t = 0, \varepsilon = 0) \rightarrow \mathrm{C} = -\frac{\sigma}{E}, \tag{e}$$

Equation (d) becomes:

$$\varepsilon = \frac{\sigma}{E} \left(1 - e^{-Et/\eta} \right). \tag{8.19}$$

Equation (8.19) is the solution of Equation (8.18) and represents the behavior of the Voigt–Kelvin model under constant stress; that is, this equation simulates the change in strain over time at constant stress ($\varepsilon = \varepsilon(t)$, $\sigma = \mathrm{const.}$). When this equation is plotted in the form of a curve (Figure 8.7(b)), it is clear that this curve can model creep (the so-called transient creep stage, that is the first stage of creep). In contrast to Maxwell's model, the Voigt–Kelvin model cannot simulate the relaxation process.

8.3.3.1 Generalized Voigt–Kelvin Model

A generalized Voigt–Kelvin model can also be composed of n Kelvin models connected in series or in parallel. A generalized Kelvin model composed of single models connected in parallel is presented in Figure 8.8.

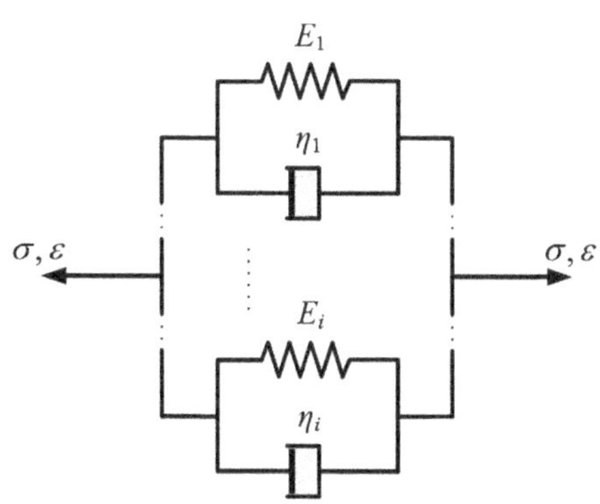

Figure 8.8 A generalized Voigt–Kelvin model.

For simplicity, the constitutive equation of this model is written in accordance with Equation (8.17), that is, with respect to stress:

$$\sigma = \varepsilon \sum E_i + \dot{\varepsilon} \sum \eta_i \tag{8.20}$$

8.3.4 Standard Linear Solid Model (SLS)

A model consisting of Maxwell's model and Hooke's model in parallel is known as a Standard Linear Solid model (see Figure 8.9) [7]. If the H_2 model is treated as infinite then this model reduces to the K model, whereas if H_1 is zero then this model reduces to the M model. Let the SLS model be subjected to constant stress (σ).

Since the SLS model consists of Maxwell and Hooke models, the rheological equation may be derived as follows. For this model we have:

$$\varepsilon = \varepsilon_1 = \varepsilon_2; \quad \sigma = \sigma_1 + \sigma_2 \tag{a}$$

The boundary condition is:

$$t = 0 \rightarrow \varepsilon_{t=0} = \frac{\sigma}{E_1 + E_2} \tag{b}$$

For the Maxwell model we have (from Equation (8.10)):

$$\dot{\varepsilon} = \frac{\sigma_2}{\eta} + \frac{\dot{\sigma}_2}{E_2} \tag{8.21}$$

while for Hooke's model we have:

$$\sigma_1 = E_1 \varepsilon_1 = E_1 \varepsilon \tag{8.22}$$

If Equations (a) and (8.22) are incorporated into Equation (8.21), it follows that:

$$\dot{\varepsilon} = \frac{\sigma - E_1 \varepsilon}{\eta} + \frac{\dot{\sigma} - E_1 \dot{\varepsilon}}{E_2} \tag{c}$$

$$\dot{\varepsilon} = \frac{\sigma}{\eta} - \frac{E_1}{\eta} \varepsilon + \frac{\dot{\sigma}}{E_2} - \frac{E_1}{E_2} \dot{\varepsilon} \tag{d}$$

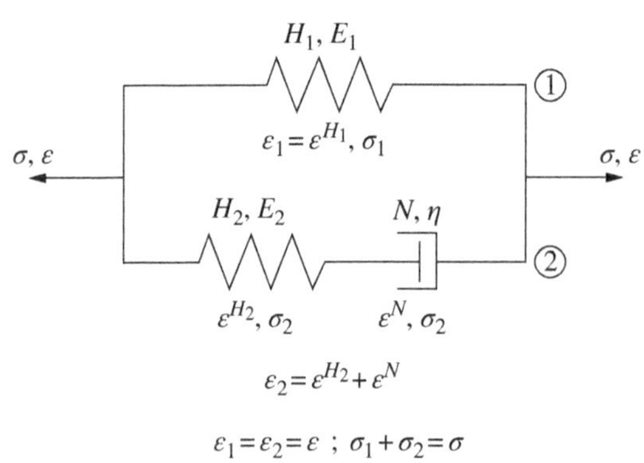

Figure 8.9 A Standard Linear Solid (SLS) model.

After further mathematical calculation, the following differential equation can be obtained:

$$\dot{\varepsilon} = (E_1 + E_2)^{-1}\left[\dot{\sigma} + \frac{E_2}{\eta}(\sigma - E_1\varepsilon)\right] \tag{8.23}$$

The solution of this differential equation (with respect to strain, as stress is constant) can be obtained as follows. Based on Equation (8.23), it follows that:

$$\dot{\varepsilon} + \frac{E_1E_2}{\eta(E_1 + E_2)}\varepsilon = \frac{\sigma E_2}{\eta(E_1 + E_2)} \tag{e}$$

The solution of Equation (e) may be presented in the form given earlier in the chapter (after Equation (8.18)):

$$\varepsilon = e^{-\int P\mathrm{d}x}\left[\int Q \cdot e^{\int P\mathrm{d}x}\mathrm{d}x + \mathrm{C}\right] \tag{f}$$

On the basis of Equation (f), now Equation (e) can be written in the form:

$$\varepsilon = e^{-\frac{E_1E_2t}{\eta(E_1+E_2)}}\left[\int \frac{\sigma E_2}{\eta(E_1 + E_2)} e^{\int\frac{E_1E_2\mathrm{d}t}{\eta(E_1+E_2)}}\mathrm{d}t + \mathrm{C}\right] \tag{g}$$

$$\varepsilon = \frac{\sigma}{E_1} + \mathrm{C}e^{-\frac{E_1E_2t}{\eta(E_1+E_2)}} \tag{8.24}$$

In accordance with the boundary condition (b), the constant C is:

$$\mathrm{C} = \sigma\left[-\frac{E_2}{E_1(E_1 + E_2)}\right] \tag{8.25}$$

Finally, if Equation (8.25) is inserted into Equation (8.24), it follows that:

$$\varepsilon = \varepsilon(t) = \frac{\sigma}{E_1(E_1 + E_2)}\left(E_1 + E_2 - E_2 \cdot e^{-\frac{E_1E_2 \cdot t}{(E_1+E_2)\eta}}\right) \tag{8.26}$$

or:

$$\varepsilon = \varepsilon(t) = \frac{\sigma}{E_1} - \frac{\sigma E_2}{E_1(E_1 + E_2)}e^{-\frac{E_1E_2 \cdot t}{(E_1+E_2)\eta}} \tag{8.26a}$$

Equation (8.26) defines the constitutive (rheological) equation of the Standard Linear Solid model and can simulate creep (the transient/primary creep stage).

8.3.5 Voigt–Kelvin–Hooke's Model (K–H)

A model consisting of a K model in series with an H model is presented in Figure 8.10. For this model, the following are valid:

- Stress in the Kelvin model is equal to the stress in "Hooke's 1" model, that is:

$$\sigma^K = \sigma^{H_1} \tag{a}$$

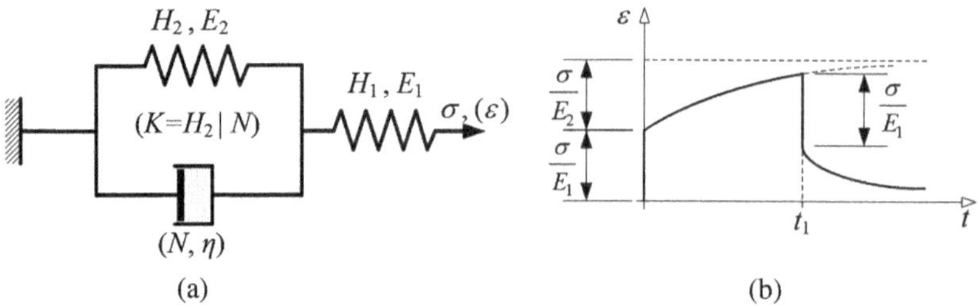

Figure 8.10 A Voigt–Kelvin model in parallel with Hooke's model. (a) Model; (b) strain–time diagram.

- Strain overall is the strain in the Kelvin model plus the strain in "Hooke's 1" model:

$$\varepsilon = \varepsilon^K + \varepsilon^{H_1} \tag{b}$$

Since strain versus time in the Voigt–Kelvin model is defined by Equation (8.19), the strain overall (total strain) is:

$$\varepsilon = \varepsilon(t) = \frac{\sigma}{E_1} + \frac{\sigma}{E_2}\left(1 - e^{-E_2 \cdot t/\eta}\right). \tag{8.27}$$

8.3.6 Burgers' Model

A model consisting of two Hooke models (springs) and two Newton models (linear viscous dashpots) assembled as shown in Figure 8.11 is known as Burgers' model [8–10]. In short, Burgers' model consists of a Maxwell model (M = H – N) and a Voigt–Kelvin model (K = H | N) connected in series.

Consider the case when the model is subjected to constant stress, σ, at $t = t_0$.

If the considered model is divided into three parts (part 1: Hooke 1 → describes elastic strain; part 2: Hooke 2 and Newton 1 in parallel → describes primary or transient creep stage; part 3: Newton 2 → describes secondary or steady-state creep stage), as shown in Figure 8.7(a), the following equations relating to strain and stress may be written:

- Strain:

$$\varepsilon = \varepsilon_1 + \varepsilon_2 + \varepsilon_3; \quad \dot{\varepsilon} = \dot{\varepsilon}_1 + \dot{\varepsilon}_2 + \dot{\varepsilon}_3$$
$$\varepsilon_1 = \varepsilon_0 = \varepsilon^{\mathrm{H}_1} \rightarrow \text{(strain in part 1, i.e., instantaneous strain in spring H}_1\text{)}$$
$$\varepsilon_2 = \varepsilon_{\mathrm{pc}} = \varepsilon^{\mathrm{H}_2} = \varepsilon^{\mathrm{N}_1} \rightarrow \text{(strain in part 2, i.e., strain in primary creep stage)}$$
$$\varepsilon_3 = \varepsilon_{\mathrm{sc}} = \varepsilon^{\mathrm{N}_2} \rightarrow \text{(strain in part 3, i.e., strain in dashpot N}_2\text{, secondary creep stage)} \tag{a}$$

- Stress:

$$\sigma = \sigma_1 = \sigma_2 = \sigma_3 \tag{b}$$
$$\sigma_1 = \sigma^{\mathrm{H}_1} = E_1 \varepsilon_1$$
$$\sigma_2 = \sigma^{\mathrm{N}_1} + \sigma^{\mathrm{H}_2} = \eta_1 \dot{\varepsilon}_2 + E_2 \varepsilon_2 \tag{c}$$
$$\sigma_3 = \sigma^{\mathrm{N}_2} = \eta_2 \dot{\varepsilon}_3$$

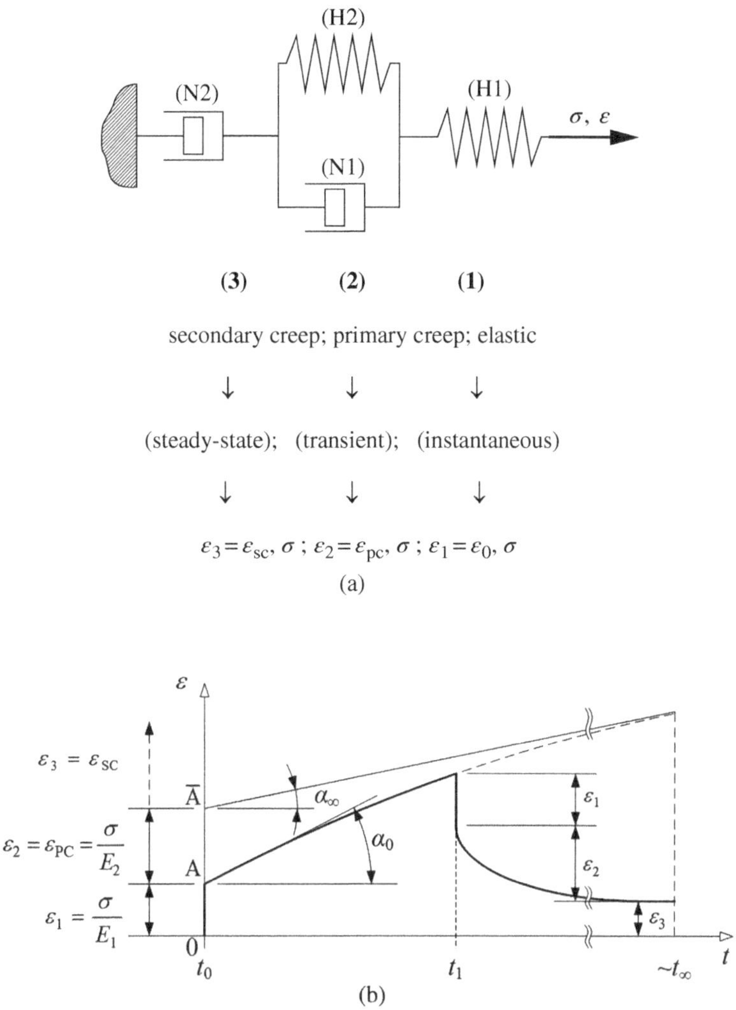

Figure 8.11 Burgers' model. (a) Model; (b) strain–time diagram ($\varepsilon_1 = \varepsilon_0, \sigma$; $\varepsilon_2 = \varepsilon_{pc}$; $\varepsilon_3 = \varepsilon_{sc}$).

The boundary conditions are (according to Figure 8.7(a) and Equations (a), (b) and (c)):

$$t = t_0 = 0 \rightarrow$$
$$\varepsilon = \varepsilon_1 = \frac{\sigma}{E_1}, \quad \varepsilon_2 = \varepsilon_3 = 0 \tag{d}$$

and

$$\dot{\varepsilon} = \frac{\sigma}{\eta_1} + \frac{\sigma}{\eta_2}.$$

After mathematical calculations, based on Equations (a), (b), (c) and (d), the differential equation of Burgers' model can be obtained. The procedure is as follows:

$$\varepsilon_1 = \frac{\sigma}{E_1};\ \varepsilon_2 = \frac{\sigma - \eta_1 \cdot \dot{\varepsilon}_2}{E_2};\ \varepsilon_3 = \frac{\sigma \cdot t}{\eta_2}$$

$$\dot{\varepsilon}_2 = \dot{\varepsilon} - \dot{\varepsilon}_1 - \dot{\varepsilon}_3 = \dot{\varepsilon} - \frac{\sigma}{\eta_2};$$

$$\varepsilon = \varepsilon_1 + \varepsilon_2 + \varepsilon_3 \longrightarrow$$

$$\dot{\varepsilon} + \frac{E_2}{\eta_1}\varepsilon - \sigma\frac{E_2}{\eta_1\eta_2}t - \sigma\left(\frac{E_2}{\eta_1 E_1} + \frac{\eta_1 + \eta_2}{\eta_1\eta_2}\right) = 0 \tag{8.28}$$

Equation (8.28) represents the differential equation of Burgers' model.

The solution of Equation (8.28) with respect to strain is:

$$\varepsilon = \varepsilon(t) = \sigma\left[\frac{1}{E_1} + \frac{1}{E_2}\left(1 - e^{-(E_2/\eta_1)t}\right) + \frac{t}{\eta_2}\right]. \tag{8.29}$$

In Equation (8.29), $\varepsilon = \varepsilon(t)$ is the strain consisting of instantaneous strain (elastic or elastic–plastic) and creep strain; σ is the constant stress of the considered creep process; E_1 is the modulus of elasticity of the considered material at the considered creep temperature; t is the duration of the creep process; E_2, η_1 and η_2 are material parameters which are to be determined from the creep curve.

Equation (8.29) is the constitutive (rheological) equation of Burgers' model and can describe creep behavior (primary and secondary creep stages) under constant stress. Comparison of Equation (8.29) with Equations (8.9) and (8.19) indicates that the creep behavior described by Burgers' model is summmarized by the creep behavior described by the Maxwell and Voigt–Kelvin models, Figures 8.5(c) and 8.7(b). Thus, this model can describe primary and secondary creep stages. In addition, the quality of modeling (that is, the correspondence between the real creep curve and the modeled curve) depends on the length and shape of the first creep stage. Further, the first derivative of Equation (8.29) satisfies the last of the equations, Equation (d).

Data relating to α_0 and α_∞ shown in Figure 8.11(b), may be determined as follows.

By differentiating Equation (8.29), the creep rate, ε, is:

$$\dot{\varepsilon} = \frac{\sigma}{\eta_2} + \frac{\sigma}{\eta_1}e^{-\frac{E_2}{\eta_1}t} \tag{8.30}$$

The creep rate starts at $t = 0$, with a finite value:

$$\dot{\varepsilon}_{t=0} = \frac{\sigma}{\eta_1} + \frac{\sigma}{\eta_2} = \tan\alpha_0, \tag{8.31}$$

while as the process progresses, it approaches asymptotically (at time $t = \infty$):

$$\dot{\varepsilon}_{t=\infty} = \frac{\sigma}{\eta_2} = \tan\alpha_\infty \tag{8.32}$$

Burgers' model may be used, as stated above, to simulate creep behavior of a material in its primary (transient) and secondary (steady-state) stages.

8.4 Differential Form of Constitutive Equations

In the case of rate-sensitive, linear viscoelastic materials, in general, two types of stress/strain states may be distinguished. Constitutive equations can be expressed as a linear function of stress, strain…, and so on, as follows:

$$f(\sigma, \dot{\sigma}, \ldots.\varepsilon, \dot{\varepsilon}\ldots.) = 0 \tag{8.33}$$

In Equation (8.33), the dots represent the derivatives of the variables with respect to time. The commonly used form of the constitutive equation is:

$$P\sigma(t) = Q\varepsilon(t), \tag{8.34}$$

where the linear differential operators with respect to time are:

$$P\sigma(t) = p_1\sigma + p_2\dot{\sigma} + \ldots \tag{8.35a}$$

$$Q\varepsilon(t) = q_1\varepsilon + q_2\dot{\varepsilon} + \ldots \tag{8.35b}$$

If the coefficients (material constants) p_i, q_i are selected properly, then, using Equation (8.34), the behavior of viscoelastic models, for example, the response of the Kelvin or Maxwell models, can be described and the process is better described as more coefficients are selected. In spite of this, some particular effects cannot be modeled appropriately, for example, instantaneous elastic strain or similar effects. Although in uniaxial stress, material behavior can be modeled with some degree of efficiency, in a multiaxial state of stress, usually another type of equation is used. This is because the behavior (response) of viscoelastic materials is somewhat different with respect to dilatation and shear effects. In this case, the constitutive equations are written in such a way that the deviatoric stress (S_{ij}) and strain (e_{ij}) parts are separated from the dilatation parts ($\sigma_{kk}, \varepsilon_{kk}$), as follows:

$$P' S_{ij}(t) = Q' e_{ij}(t) \tag{8.36a}$$

$$P'' \sigma_{kk}(t) = Q'' \varepsilon_{kk}(t) \tag{8.36b}$$

In Equations (8.36), P', P'', Q', Q'' are operators similar to those in Equation (8.34). In addition, p_i', p_i'', q_i', q_i'' in these equations are material constants determined by experiments.

Using some of the rheological models presented in this chapter, such as the Burgers and SLS models, modeling of the creep behavior for the materials 42CrMo4 and AISI 316Ti, subjected to creep at defined temperatures and stress levels, was carried out based on experimentally obtained creep curves. Table 8.1 shows the creep modeling data related to steel 42CrMo4 (1.7225) and steel AISI 316Ti (1.4571), while Figures 8.12 and 8.13 show the modeled creep curves.

Example 8.1.

Using the Standard Linear Solid model and Burgers' model, simulate the creep process of 42CrMo4 (1.4571, AISI 4140) steel. The creep process* occurs at 400 °C/240 MPa = 0.75 $\sigma_{0.2}$/1000 min. The chemical composition of the steel (mass %) is as it usually is for this alloy. (This material is used in the manufacture of statically and dynamically stressed components in engines and machines.)

Table 8.1 Creep modeling data.

Material and creep process data	Material	42CrMo4 (1.7225)	AISI 316Ti (1.4571)
	Temperature (°C)	400	700
	Time (min)	1000	1000
	σ (Pa)	$240{\cdot}10^6$	$72{\cdot}10^6$
Burgers' model	E_1 (Pa)	$1.57{\cdot}10^{11}$	$1.7{\cdot}10^{11}$
	E_2 (Pa)	$6.2005278{\cdot}10^8$	$9.0036808{\cdot}10^8$
	η_1 (Pa min)	$1.0738040{\cdot}10^{10}$	$3.1752396{\cdot}10^8$
	η_2 (Pa min)	$3.4607635{\cdot}10^{11}$	$3.8320398{\cdot}10^{11}$
SLS model	E_1 (Pa)	$2.2074021{\cdot}10^8$	$2.0324095{\cdot}10^8$
	E_2 (Pa)	$9.0010355{\cdot}10^8$	$7.4914857{\cdot}10^8$
	η (Pa min)	$8.6232436{\cdot}10^{10}$	$1.6634030{\cdot}10^{11}$

Solution:

The modeled creep curve, using Burgers' and SLS models, is presented in Figure 8.12.

Example 8.2.

Using the Standard Linear Solid model and Burgers' model, simulate the creep process of AISI 316Ti (1.4571, X6CrNiMoTi17-12-2) steel. The creep process** occurs at 700 °C/72 MPa = 0.3 $\sigma_{0.2}$/1000 min. The chemical composition of the steel (mass %) is as it usually is for this alloy. (Applications of this material are in the construction of equipment in the chemical and pharmaceutical industries, pressure vessels, and so on.)

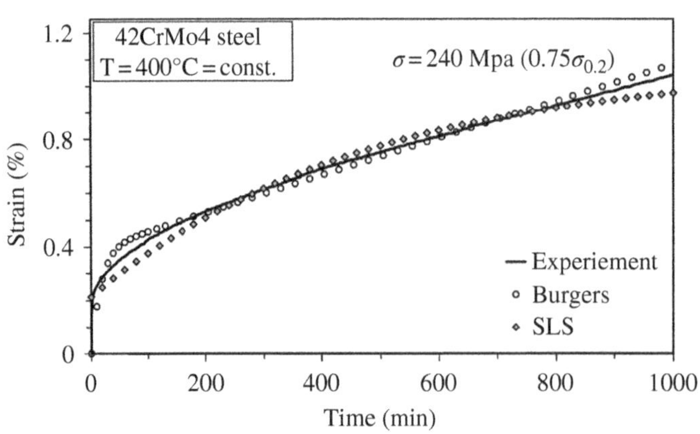

Figure 8.12 Modeled creep curve – material 42CrMo4 (1.7225)/400 °C/240 MPa/1000 min.

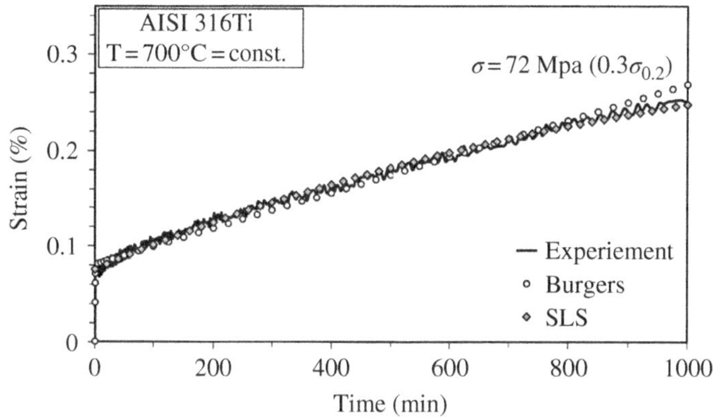

Figure 8.13 Modeled creep curve – material AISI 316Ti (1.4571)/700 °C/72 MPa/1000 min.

Solution:
The modeled creep curve, using Burgers' and SLS models, is presented in Figure 8.13.

*, ** Note: The testing of these materials was carried out at the Laboratory of Materials Strength Testing, Department of Enginering Mechanics, University of Rijeka – Faculty of Engineering. Experimental data related to the behavior of the materials mentioned in Examples 8.1 and 8.2 can be found in [11, 12].

References

1 Eley, R. R. (1995) Rheology and Viscometry. In J. V. Koleske (Ed.) *ASTM Paint and Coating Testing Manual, The Gardner–Sward Handbook*, 14th edition, Ann Arbor, pp. 333–350.
2 Collins, J. A. (1993) *Failures of Materials in Mechanical Design*, 2nd edition, John Wiley & Sons, New York.
3 Dowling, N. E. (2013) *Mechanical Behavior of Materials*, Pearson, New York.
4 Findley, W. N., Lai, J. S. and Onaran, K. (1989) *Creep and Relaxation of Nonlinear Viscoelastic Materials*, Dover Publications, Inc., New York.
5 Drozdov, A. D. (1996) *Finite Elasticity and Viscoelasticity*, World Scientific Publishing, Co., New Jersey.
6 Alfirević, I. (1975) *Advanced Strength of Materials* (in Croatian), Faculty of Mechanical Engineering and Naval Architecture, Zagreb.
7 Plaseied, A. and Fatemi, A. (2008) Deformation response and constitutive modeling of vinyl ester polymer including strain rate and temperature effects, *Journal of Materials Science*, **43**, 1191–1199.
8 Betten, J. (2002) *Creep Mechanics*, Springer, New York.
9 Brnic, J., Canadija, M., Turkalj, G. and Lanc, D. (2010) 50CrMo4 Steel – Determination of Mechanical Properties at Lowered and Elevated Temperatures, Creep Behavior and

Fracture Toughness Calculation, *Journal of Engineering Materials and Technology*, **132**(2), 021004-1–021004-6.

10 Brnic, J., Turkalj, G., Niu, J., Canadija, M. and Lanc, D. (2013) Analysis of experimental data on the behavior of steel S275JR – Reliability of modern design, *Materials and Design*, **47**, 497–504.

11 Brnic, J., Turkalj, G., Canadija, M., Lanc, D. and Brcic, M. (2015) Study of the Effects of High Temperatures on the Engineering Properties of Steel 42CrMo4, *High Temperature Materials and Processes*, **34**(1), 27–34.

12 Brnic, J., Turkalj, G., Canadija, M. and Lanc, D. (2011) AISI 316Ti (1.4571) steel – Mechanical, creep and fracture properties versus temperature, *Journal of Constructional Steel Research*, **67**, 1948–1952.

9

Creep in Metallic Materials

9.1 Introduction

In recent times, the main goal of structural design has been to achieve the best design and to manufacture the best end product. The "best" product is achieved as the result of a compromise made between the demands and the available opportunities. In other words, the best product is a compromise between the required "product quality" and restrictions that prevent the achievement of this required quality. The product must, first and foremost, comply with functionality, safety and low cost requirements, but at the same time, this should be accompanied by satisfying a series of constraints such as allowable stress, allowable strain, appropriate manufacturing technology, suitable material selection, and so on that arise during the creation of the product.

During its service life, various failures may occur within a structure. This issue was discussed in Section 1.4. The number of disasters that are caused by such failures in comparison to the number of successful designs is minimal, but the costs of such failures are enormous and are often paid with human lives. The great advances in technology, unfortunately, do not completely eradicate the occurrence of failures, which are often accompanied by large losses. However, the analysis of structural failures, no matter their type or the reasons for their occurrence, helps to reduce these failures to a minimum. The analysis of failures which have occurred in many types of structures has resulted in better designs and design procedures, better maintenance and more appropriate selection of the materials used as well as in increased attention to the uses of the structures.

In many fields of application within engineering practice, many types of failures can occur. Yielding, an excess deformation, is probably one of the most recognized and analyzed deformation modes. On the other hand, mechanical failures can involve complex interactions between load, environment and time. Temperature and corrosion are included in environmental conditions. Materials such as metals, polymers and so on subjected to loads in special environmental conditions exhibit time-dependent behavior. The time-dependent phenomenon which occurs in materials subjected to a certain level of constant stress (load) at high enough temperature and results in a continuous increase in deformation is known as *creep*.

Creep is one of the commonly observed failure modes in engineering practice. The term "high temperatures" mentioned above does not automatically mean elevated temperatures. In this case, "high temperatures" refers to a temperature that is close to the melting point. In some materials such as polymers (and others), creep may occur at room temperature. In addition, in engineering practice, some structures or machines, such as chemical reactors, automobile engines, aircraft engines, nuclear plants and so on are subjected to fatigue at elevated temperatures. As such, mechanical fatigue load is associated in action with creep, metallurgical aspects and environmental oxidation, and here, the fatigue behavior and life assessment of a considered member are complicated. In such situations, possible creep–fatigue interaction becomes the subject of discussion. The design of such structures in terms of their life predictions is usually undertaken on the basis of a permissible amount of creep strain [1–8]. In engineering applications, it is common practice to accept creep strain of 1–2% as an allowable level [9]. A design for a structure in which creep strain is deemed a critical factor is usually known as a creep-sensitive design.

Creep analysis includes the avoidance of failure due to creep from excessive deformation and creep rupture. Creep of a particular engineering element subjected to load at high temperature can be analyzed and monitored regardless of whether the element contains or does not contain a crack. In the case where the element does contain a crack, under consideration is creep crack growth, that is, crack propagation in the presence of creep [9]. This implies a need to find a solution for the prediction of crack propagation life at elevated temperatures in the presence of creep. Usually, in many processes, metals are modeled as isotropic, homogenous and linearly elastic. If mechanical properties are determined experimentally, that is, a stress–strain diagram is determined experimentally, or an S–N curve (stress versus life curve: S – stress, N – cycles to failure) is determined experimentally, then microstructural effects are inherently included. If the data related to material properties (tensile strength, yield strength, endurance limit and so on) are not obtained experimentally, then heat treatment, cold working, anisotropy and chemical composition as well as inclusions, voids and many other imperfections may have a significant impact on the properties, and the modeling mentioned is not correct at the microscopic level. This is of importance when metal fatigue is considered and especially when creep crack growth is being monitored, as it is known that material properties depend on chemical composition, processing path and microstructure. Since most mechanical properties depend on microstructure, and the strength of materials, for example, influences the fatigue limit, this means that fatigue in metal alloys is influenced by microstructure. The properties mentioned are known as *structure-sensitive properties* [10].

Failures can originate at the surface of the engineering element or inside the element. For example, the cyclic impact load (pressure) that arises between two elements in engagement may cause yielding below the contact surfaces, that is, inside the element, and a crack that is initiated in the plastic zone will spread to the contact surface. However, most fatigue failures originate at the surface, which means that the surface condition or the effects of the surface caused by chemical composition, differences in surface roughness, microstructure and so on will affect the behavior of the material under fatigue. It is known that the surface finish of an engineering element is different from that of experimental specimens where the surface is polished and smooth. These conditions may cause degradation in fatigue strength; in other words, they will have an influence on fatigue behavior (fatigue resistance). On the other hand, the effect of test frequency related to the *S–N* curve, if its value is up to 200 Hz may be neglected. The fracture surface that is a result of creep fracture is normal to the direction of maximum normal stress. A fracture surface caused by creep can be different from a fracture surface caused by other factors since different mechanisms cause these fractures (the fracture surface topographies are different).

9.2 Plastic Deformation – General

An engineering structure or a considered engineering element subjected to load may be deformed elastically, plastically or may be fractured. In Section 5.1, the elastic and inelastic responses of a solid were discussed. Depending on the level of stress as well as on environmental conditions, plastic deformation can lead to fracture, for example, separation of the element into two (or more) parts. Plastic deformation in metallic materials may be caused by a sufficiently high level of stress at "low" temperatures but also by sufficiently constant stress at high temperatures. "Low" temperatures can be assumed to mean temperatures below creep temperatures. In general, plastic deformation or plastic flow may refer to either time-independent or time-dependent behavior. Several processes in which plastic deformation can occur may be considered: *slip, cleavage, twinning, grain boundary sliding, void coalescence* and *creep*. Creep, as a phenomenon, will be discussed separately. When plastic deformation leads to material separation (fracture of the element), it is of interest to find which process – referred to as the *fracture micromechanism* – is the cause of the fracture. This issue is discussed later in Section 9.3.1.2.

9.2.1 Slip

As the principal mode of plastic deformation, *slip* means the gliding (parallel movement) of one plane of atoms (material layer) over another (see Figure 9.1(a)), and it depends on external load as well as the level of shear stress caused by this load, the crystal structure geometry, the active slip plane and the active slip direction. Within a crystal lattice, some crystallographic planes and directions, notably the most dense atomic planes and directions, are more susceptible to slip than others. The creation of the slip direction and the plane containing it is known as the *slip system*.

Slip is initiated when the shear stress on the active slip plane in the active slip direction exceeds the critical resolved shear stress. The critical resolved shear stress represents the influences of various factors affecting the appearance of slip. It is known that single

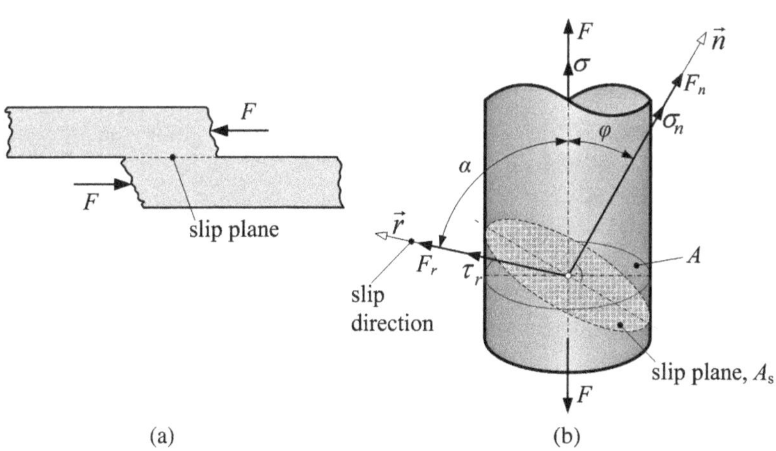

Figure 9.1 Slip as a plastic deformation mechanism. (a) Plastic deformation; (b) slip plane and slip direction.

crystals in different orientations require different stress levels to initiate slip. The plastic deformation of a polycrystalline specimen may be comparable with the distortion of the individual grain(s) with respect to slip. In this sense, let the single crystal (a cylindrical element) be subjected to an axial force, F, as shown in Figure 9.1(b) [4]. Let A be the cross-sectional area of the element (F is normal to the cross-sectional area, A) and A_s be the area of the active slip plane in the crystal defined by its normal, $\vec{n}$. The normal $\vec{n}$ forms an angle φ with respect to the axial axis of the element (direction of F) and finally, the active slip direction $\vec{r}$ in the mentioned active slip plane is defined by angle α. The normal stress σ due to the force F acting on the cross-sectional area, A, of the element is $\sigma = F/A$. It is easy to show that the resolved shear stress on the active slip plane in the active slip direction, $\vec{r}$, is:

$$\sigma = \frac{F}{A}, A_s = \frac{A}{\cos\varphi}, F_r = F\cos\alpha, \tau_r = \frac{F_r}{A_s} = \frac{F\cos\alpha}{\frac{A}{\cos\varphi}}$$

$$\tau_r = \frac{F}{A}\cos\alpha\cos\varphi = \sigma\cos\alpha\cos\varphi \tag{9.1}$$

and similarly, the resolved normal stress in the slip plane is:

$$\sigma_n = \frac{F_n}{A_s} = \frac{F\cos\varphi}{\frac{A}{\cos\varphi}}$$

$$\sigma_n = \frac{F}{A}\cos^2\varphi = \sigma\cos^2\varphi. \tag{9.2}$$

The slip direction changes at the grain boundaries, and the planes and directions on which slip occurs depend on the material's crystal structure. After a crystal is loaded, since shear stress increases for all slip systems, slip is initiated on that system for which the critical shear stress first appears. This system depends on the orientation of the crystal (grain). When any of the mentioned angles is 90°, the resolved shear stress will be zero and slip will not occur; the crystal (specimen) will tend to fracture rather than slip.

9.2.2 Cleavage

It is also possible for the critical resolved normal stress to occur, which causes cleavage of the planes. This means that the planes will separate (cleave). However, as the external load on a crystal increases, both of the mentioned stresses (shear and normal) also increase, and depending on which critical stress is reacheds first, the appropriate phenomenon will occur – slip or cleavage. Both of these stresses are temperature and strain rate dependent.

Cleavage as a plastic deformation phenomenon or fracture mechanism can occur in materials which may exhibit a negligible level of plastic deformation but also in materials with significant plastic deformation, for example, BCC structures. This type of fracture is treated as a transgranular, low-energy fracture characterized primarily by separation of atomic bonds. Since a cleavage fracture occurs along defined crystallographic planes within each grain, its directions will be changed by crossing through a grain. A Mode I brittle fracture may occur without or with a limited degree of plastic deformation

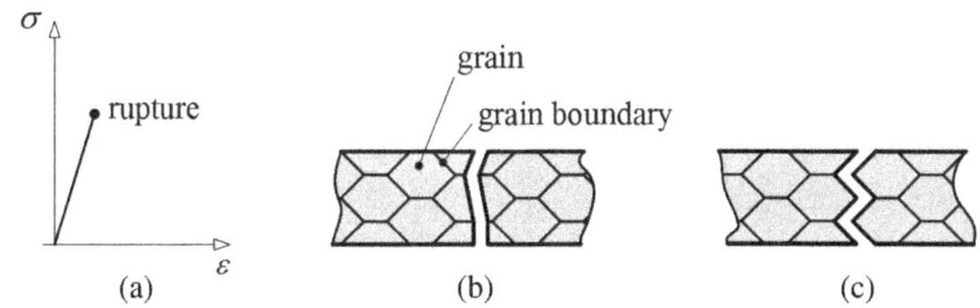

Figure 9.2 Brittle fracture, Mode I. (a) Mode I; (b) transgranular propagation (cleavage); (c) intergranular separation (intergranular brittle fracture, decohesion rupture).

and can propagate transgranularly or intergranularly (decohesion rupture), as shown in Figure 9.2.

In Mode II brittle fracture, fracture precedes microscopic plastic deformation, and a nucleated crack may propagate by cleavage or intergranularly. In Mode III brittle fracture, macroscopic deformation occurs before fracture.

It should be noted that brittle fracture can be divided into three groups: Modes I, II and III, where these modes are different from the I, II and III used to describe crack opening modes (and growth) later in Section 9.3.2.2. The intergranular fracture or decohesion rupture mentioned above occurs before slip or cleavage at the grain boundaries when the boundaries are the weakest locations. This process is associated with low toughness and low ductility. Cleavage fracture is also possible in metals under conditions of stress-corrosion cracking.

9.2.3 Twinning

This type of deformation is different from slip deformation. When, due to the action of shear stress, one portion of a crystal lattice takes an orientation that is the mirror image of the lattice orientation in the parent crystal, the process is called *twinning* (see Figure 9.3).

9.2.4 Grain Boundary Sliding

Grain boundary sliding may be treated as a phenomenon associated with high temperatures (low strain rates), where sliding in polycrystalline materials occurs along the grain

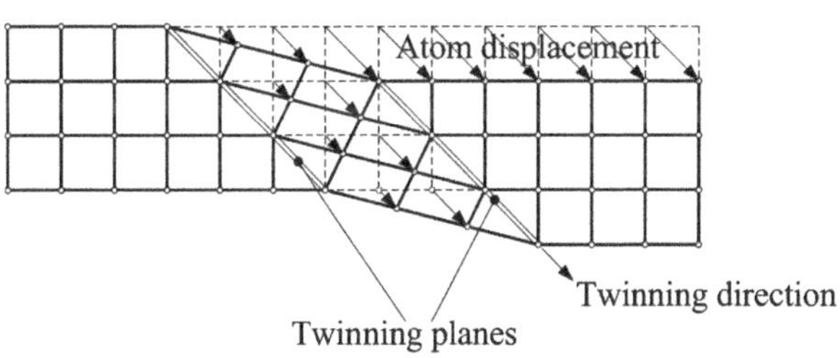

Figure 9.3 Schematic representation of plastic deformation by twinning.

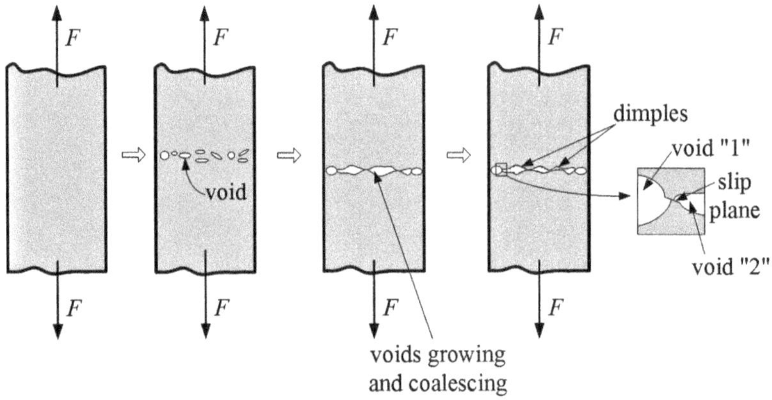

Figure 9.4 Void coalescence mechanism (or mode of dimpled rupture).

boundaries. So, in creep processes taking place at low stresses in pure metals, some percentage of the strain may be due to grain boundary sliding.

9.2.5 Void Coalescence

The process or mechanism of ductile fracture, usually called *void coalescence*, is a common mechanism associated with slip and is due to the separation of a solid (material) by the formation of internal voids. After the voids are formed, the material continues to deform by slip. This is a ductile fracture mechanism, for example, for BCC and FCC metal structures and is usually described and known as *dimpled rupture mode* (see Figure 9.4).

9.3 The Creep Phenomenon and Its Geometrical Representation

Creep belongs to the category of high-temperature failures. In addition to the aforementioned definition of creep, there are some other definitions of this phenomenon. A material can be deformed slowly and continuously over time under constant stress and such a thermally assisted and time-dependent process of deformation is defined as creep. In short, creep is time-dependent, inelastic strain under constant stress at elevated temperature. In this section, creep in metallic materials will be considered.

When a structural member operates under a creep regime, its life may be limited even if the applied load is less than the design load. It may be said that creep is appreciable at temperatures above 0.4 T_m, where T_m is the melting temperature measured in degrees Kelvin [11]. In some applications (for example, an aircraft's gas turbine engine), the use of materials is limited by the creep properties of the material (creep resistance). In such a structure, the design is usually based on a maximum 1% permissible amount of creep strain during the expected service life of the component. It is clear that the operating function of an engineering member can be limited by creep, which is directly related to the melting point, in contrast to yield strength, which is not directly related to the melting point. In general, a temperature sufficient to initiate creep is 205 °C for aluminum alloys and 370 °C for low-alloy steels [12]. Since creep deformation depends on temperature and time, creep is a thermally activated process and this fact may serve as a supplement to our previously mentioned explanation of the complex interaction between creep, metallurgical aspects and environmental oxidation. Many engineering components such as nuclear power plants, reactors, aircraft engines and so on are operating at service temperatures in excess of the creep threshold temperature. In general, for a considered metallic material in an air environment, its fatigue resistance decreases with an increase in temperature. Creep failure in an engineering element occurs when accumulated creep strains exceed the design limit, while creep rupture means that a considered element is separated into two parts and is unable to serve the purpose for which it was designed. A uniaxial creep process in metallic materials is usually represented by a curve (A, B, C, D) consisting of three distinct parts: I – primary (transient) creep stage; II – secondary (steady-state) creep stage; III – tertiary (accelerating) creep stage (see Figure 9.5). This creep curve represents the stress–strain response of a material subjected to constant stress (σ_1) at elevated constant temperature (T). However, schematically similar responses are obtained when the temperature is kept constant but stress increases from test to test and is kept constant for a considered test, or stress is kept constant but temperature increases from test to test, or both temperature and stress increase from test to test.

Consider the creep process represented by creep curve (A, B, C, D). At the beginning, after the load (stress) is applied, an instantaneous deformation (OA) appears, which can be elastic or elastic–plastic depending on the level of applied stress at the considered temperature. After that, the primary (transient) stage of the creep (AB) occurs; that is, the region of the creep in which the creep strain rate decreases with time. At the moment when the creep strain rate reaches the minimum, the secondary (steady-state) creep stage (BC) begins. The duration of this phase will depend on the temperature and the applied level of constant stress. The tertiary (accelerating) creep stage (CD) is the region of the

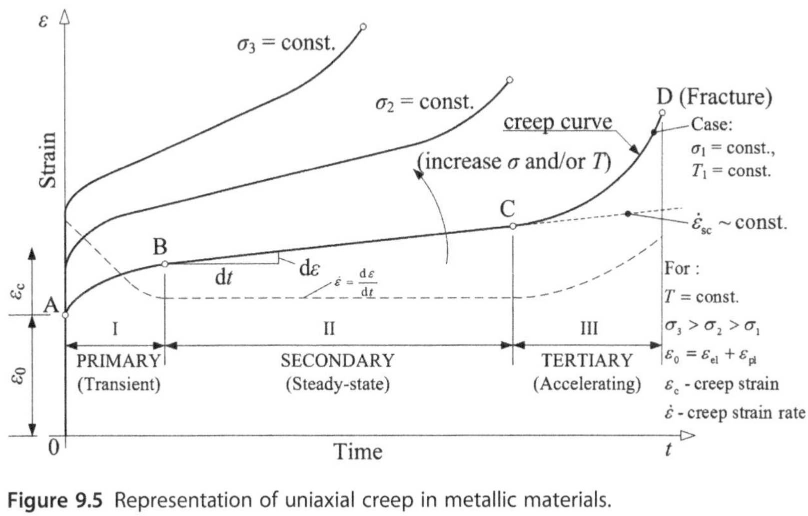

Figure 9.5 Representation of uniaxial creep in metallic materials.

creep process in which the creep strain accelerates with time. Two categories of material degradation due to creep are usually mentioned: mechanical and environmental. The first (mechanical) refers to inelastic deformation and changes in size with time due to applied stress. The second category covers environmental degradation that is due to the material's reaction with the environment.

9.3.1 Creep Deformation Maps and Fracture Mechanism Maps

Metal alloys and the engineering ceramics used in engineering applications are composed of crystalline grains which are separated by grain boundaries. A material with such a structure is said to be a *polycrystalline material*; in other words, a polycrystalline material consists of crystals (grains) which are interconnected along their boundaries. Grains individually have anisotropic properties. The orientation of the grains' crystallographic planes and axes differs from grain to grain.

Metal alloys and ceramics have similar physical mechanisms for creep deformation. Different classes of materials have different physical mechanisms causing creep. Even the same considered material may have different mechanisms depending on the particular combination of stress and temperature levels. In accordance with the theory of optimal structural design, special attention must be paid to the selection of a material that can be exposed to high temperatures during its expected service life. Therefore, knowledge of materials' behavior under creep conditions is of great importance. A great advance in the development of materials and knowledge of their creep resistance when used under high-temperature conditions has been the introduction of so-called *creep deformation maps* and *fracture mechanism maps* [3, 13]. For a given material (alloy), these maps indicate the *creep deformation* and *fracture mechanisms* as a function of the given combination of temperature and stress levels. In both of the mentioned map types (deformation and fracture), the ratios (σ/E) are placed on the ordinate while the ratios (T/T_{m}) are placed on the abscissa. The maps can be made for a given alloy as well as for some groups of metals. From a creep deformation map, one can observe the kind of deformation (elastic, plastic) that arises in the vicinity of the temperature 0.4 T_{m} but also the kind of creep mechanisms (power-low creep, viscous creep) that cause creep deformation at temperatures above 0.4 T_{m}. Fracture mechanism maps, which are similar to creep deformation maps and are also shown as a function of the given combination of temperature and stress levels, indicate the mode of failure. By examining the creep fracture surface in combination with these fracture mechanism maps, some information about operating conditions at the failure time may be obtained.

9.3.1.1 Creep Deformation Mechanisms

Creep is usually considered at temperatures above 0.4 T_{m} (the melting temperature), while below the temperature of 0.4 T_{m} deformation is either elastic or plastic. The modes of creep failure may be compared to the modes of failure that occur below creep temperatures. Creep deformation maps indicate creep deformation mechanisms for a considered alloy as a function of the different combinations of stress and temperature levels [3]. Similar considerations are possible for fracture mechanism maps. During deformation at temperatures above 0.4 T_{m}, that is, at higher temperatures, the motion of molecules and atoms, dislocations, vacancies and so on that occur within a considered solid material are intensive. Such motions form part of the behavior known as diffusion.

Physical mechanisms for creep deformation may be classified into several categories and, according to different authors, such mechanisms are:

1) Diffusional flow (diffusion creep, viscous creep)
2) Dislocation creep (power-law creep) [7];

or

A) Power-law creep
B) Viscous creep [3];

or

a) Vacancy diffusion
b) Dislocation climb
c) Grain boundary sliding [11].

Diffusion Creep

The deformation of a crystalline solid by the diffusion of vacancies through a crystal lattice is known as *diffusion creep*. This diffusion of vacancies through the crystal (often called a grain) can happen in several ways. The diffusional flow that includes the movement of vacancies through the crystal (grain), that is, through the grain itself, is known as *Nabarro–Herring creep*, while the diffusion of vacancies/atoms in a material along the grain boundaries is known as *Coble creep* (see Figure 9.6). Although in both of these mechanisms, the strain rate ($d\varepsilon/dt$) is proportional to the applied stress (σ), the mechanisms differ in their relationships between the strain rate and grain size "d". In Nabarro–Herring creep, the strain rate is proportional to d^{-2}, while in Coble creep, it is proportional to d^{-3}. These relationships are commonly used to determine which mechanism is dominant in a material [14].

The occurrence of the mechanism known as diffusional flow may be expected in the field of low stresses but relatively higher temperatures. This mechanism includes the movement of vacancies (missing atoms/holes) in the crystal lattice.

Since both types of diffusional flow, that is, the movement of vacancies through the crystal lattice and the movement along grain boundaries, are viscous processes, diffusional flow or diffusion creep is also called *viscous creep*.

Dislocation Creep

Dislocation creep is a deformation mechanism that takes place due to movements of dislocations through the crystal lattice and is also known as *power-law creep*. It involves the motion of dislocations and it is dominant at high stresses and relatively low temperatures. Turbine blades, steam pipelines and other applications may be mentioned as areas where this type of creep can be expected. Many further details on creep and the mechanisms of deformation can be found in [15].

Basic Information from the Theory of Dislocation Our discussion will only cover basic information related to the problem of dislocation. More details about this problem can be found in specialized literature, and some suggested books are listed in the references at the end of this chapter.

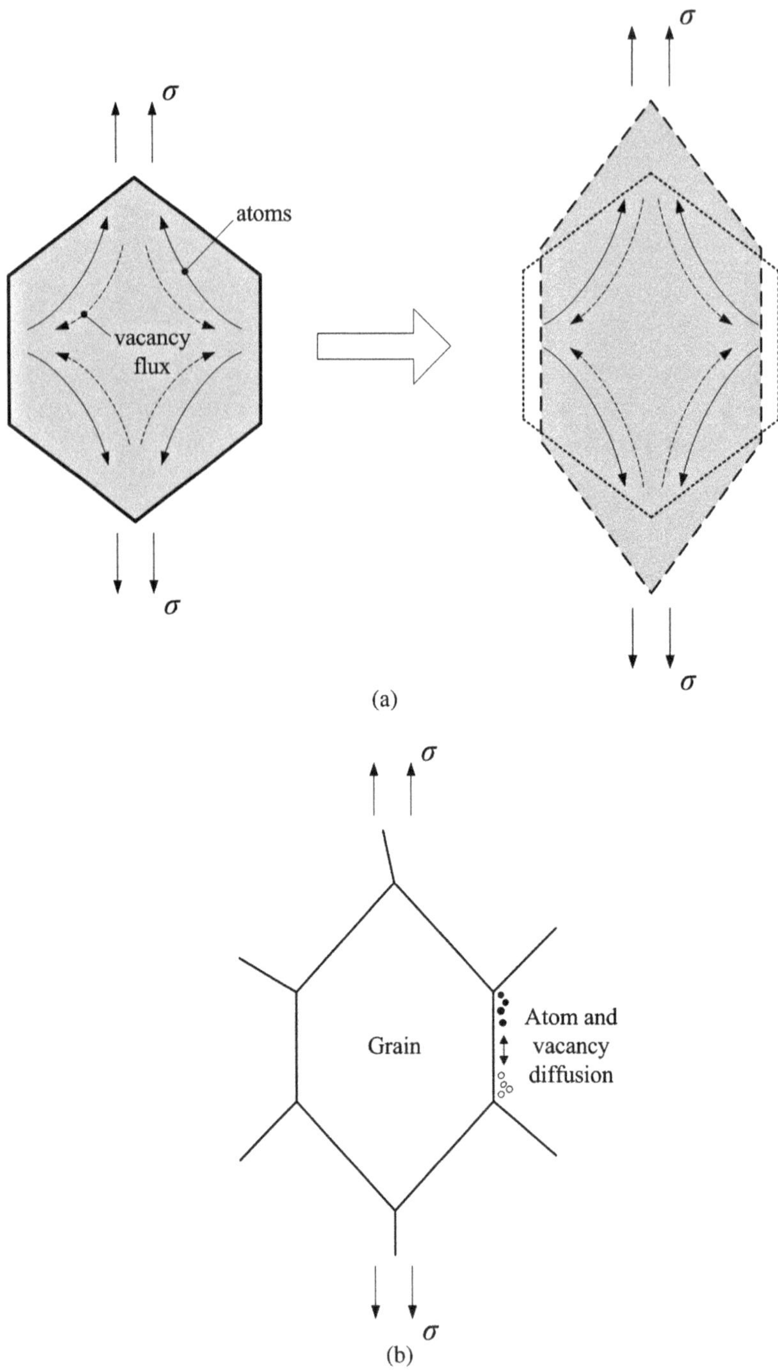

Figure 9.6 Diffusion creep. (a) Nabarro–Herring creep; (b) Coble creep.

Dislocation is a crystallographic defect (or irregularity) within a crystal structure. Crystals (grains) make up polycrystalline materials that may have some irregularities but also some defects. Defects (inclusions, microcracks) may be induced by the manufacturing process and can be observed on the macro scale. In addition, some faults on the micro scale are treated as defects in the crystal lattice. As such, we may distinguish *point defects* (vacancies/missing atoms, impurity atoms and so on), *line defects* (dislocations), *surface defects* (stacking faults, twin boundaries and so on) and *volume defects* (voids, bubbles and so on).

Two types of line defect (dislocation) can be mentioned: edge and screw. In real materials, dislocations are typically mixed, that is, they have the characteristics of both types of dislocation. To explain or visualize edge dislocation, first, let a cubic lattice structure be cut half-way through, as shown in Figure 9.7, and spread. When an extra half-plane of atoms is introduced mid-way through the crystal, distorting the planes of atoms, this is treated as an edge type of dislocation (I). In Figure 9.7, the inserted half-plane of atoms is shown as the plane with white circles. Edge dislocations may be taken as positive (⊥) or as negative (⊤). Positive edge dislocation occurs when there are $(n + 1)$ atoms above the line

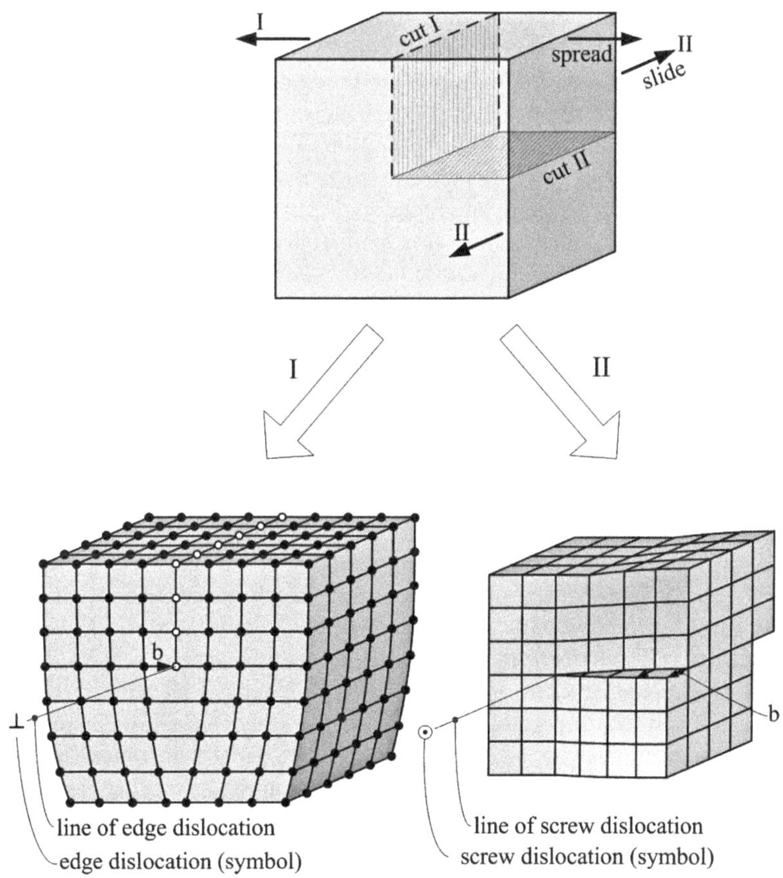

Figure 9.7 Representations of edge and screw dislocations.

and (n) atoms below the line. If sufficient force is applied in direction I (I-spread), the extra plane passes through the planes of atoms, breaking bonds with them. A line direction running along the bottom of the mentioned extra half-plane and the so-called Burgers' vector are two properties of dislocation. The magnitude and direction of the displacement that occurs when considering a dislocation are described by the so-called Burgers' vector (**b**). Burgers' vector is perpendicular to the dislocation line in edge dislocation.

Lattice imperfections (or defects) called dislocations have been postulated and explained by several authors. Although we mentioned two types of dislocation above, dislocations are often classified into three categories: (a) edge dislocations, (b) screw dislocations and (c) mixed dislocations. A representation of edge and screw dislocations using a cubic lattice structure is shown schematically in Figure 9.7.

9.3.1.2 Fracture Micromechanisms and Macromechanisms

In engineering practice, one encounters different types of: *structures* (plane construcions, spatial constructions), *engineering elements* (rods, beams, plates, rings and so on), *loads* (uniaxial, torsion, bending, nonuniform bending, excentrical and so on), *states of stress and strain* (one-, two- or three-dimensional) and *environmental conditions* (elevated temperatures, low temperatures, corrosive media and so on). Many overloads, especially when combined with unfavorable environmental conditions, lead to the fracture of the element (material separation). Based on findings from previous designs, the intention is to make newer designs better and more reliable. This means that knowledge of how and why an engineering element has failed is of huge importance. In order to obtain this information, or to analyze the fractured element, it is necessary to know what type of material was used (brittle, ductile), what type of fracture occurred (brittle, ductile), what type of load was applied and, in accordance with the review, which process occurred when the material was subjected to overload that led to material separation. Generally, it is said that macroscopic orientation of the fracture surface can be correlated with the loading conditions. Some considerations on these problems are also given in Section 3.2.

Let's consider a brittle material subjected to uniaxial tension overload. When the fracture did not show plastic deformation, that is, the fracture is also brittle, it is expected that the macroscopic fracture surface will be flat and perpendicularly oriented to the principal stress. A material is treated as brittle when elongation is less than 5% [4]. Fracture is also, generally, treated as brittle if there is no noticeable significant plastic deformation on the macroscopic plane. In addition, since the human eye cannot observe the process which leads to material separation, in engineering practice, a microscope is used. This process is referred to as a fracture micromechanism (fracture mechanism); it is a fine-scale process and is correlated with the topology of the fracture surface. As such, it is necessary to distinguish between fracture mechanisms (fine-scale processes) that lead to material separation and fracture macromechanisms (coarse-scale processes).

Fracture as a mechanism that separates the element (specimen) into two (or more) parts can occur in a specimen made of a ductile or a brittle material, regardless of whether the specimen possesses a notch or not. Some confusion can arise when fracture is classified as either brittle or ductile. The correct use of these terms (brittle and ductile) relating to fracture depends on whether the term is referring to a fracture micromechanism or the macroscopic work of fracture. Consider two fracture processes. Let both fracture processes be performed on specimens made from ductile material, but one specimen is manufactured as unnotched while the other is manufactured as notched. In the case of the unnotched specimen, fracture will occur under fully plastic deformation and that type of fracture will be ductile, while in the case of the notched specimen, the plastic region will be restricted to the tip of the crack while overall plastification will be substantially reduced. From an engineering point of view, this case is treated as brittle fracture. This is the reason why the terms "brittle" and "ductile" fracture may have an appropriate meaning on the macroscopic level. The main characteristics with which a brittle fracture is described are: "small" energy consumption prior to fracture, little or no noticeable plastic deformation before fracture and a fracture surface that is practically flat and perpendicular to the axial axis of the element. On the other hand, ductile fracture is described as a process that is accompanied by a large amount of plastic deformation before fracture, a fracture cross-section that is reduced by necking and slow crack growth. The fracture micromechanisms indicate the processes that caused the fractures, and each fracture may be classified as brittle or ductile depending on whether or not it required plastic flow for the material separation into two parts. In short, brittle fracture is usually described as a phenomenon which is characterized by rapid crack propagation, much less expenditure of energy in comparison with ductile fracture and as a process without noticeable plastic deformation.

Some of the fracture deformation mechanisms were discussed in Section 9.2. So-called fracture deformation maps (fracture mechanism maps) are constructed for BCC, FCC and other types of structures [16]. Brittle fracture mechanisms are usually associated with transgranular cleavage (transcrystalline cleavage) or intergranular separation (intergranular brittle fracture or intercrystalline fracture), while ductile fracture is associated with dimpled rupture. Transgranular and intergranular fracture processes are macroscopic and brittle. Brittle fracture can be divided into three groups: Mode I, II and III, where these modes are different from those that described crack opening modes (and growth) earlier. Mode I includes brittle fracture with a little plastification that can occur at low normalized stress (σ/E) and may originate from a flaw or crack. In a Mode II brittle fracture, fracture is preceded by microscopic plastification (that is, it is not significant). During a Mode III brittle fracture, macroscopic deformation precedes the fracture.

9.3.1.3 Creep Fracture Mechanisms

As noted earlier, a material may be deformed in the elastic, elastic–plastic, elastic–viscoplastic and so on range. A brittle material was distinguished from a ductile material in the sense that it is not able to withstand larger deformations before fracture, although in both types of material, an elastic phase can be observed. In those other chapters, material behavior was discussed practically only on the basis of stress–strain diagrams. Some of the material response processes caused by subjected load lead to material separation (fracture). As stated in the previous section, examination of the topology of the fracture surface to determine the mechanism that caused fracture can be done using low magnification, that is, on the basis of so-called coarser-scale processes or on the basis of fine-scale processes. In the first category are fracture macromechanisms (processes) which cause fracture, while in the second category are fracture micromechanisms (usually simply called fracture mechanisms). Since the fine topology of the fracture surface is determined by the fracture mechanism, this means that careful observation of the fracture surface topology can give clues and possible confirmation as to which fracture mechanism has caused fracture. At this point, the fracture mechanism can be discussed on a more appropriate basis and this requires the use of higher levels of magnification. A comparison of fracture mechanisms between processes conducted at temperatures below the creep temperature (low-temperature failure) and those sufficient to cause creep may be made. More precise explanations can be found in specialized literature [3, 4, 16] dealing with such issues. The following sections will discuss various creep fracture mechanisms.

Intergranular Creep Fracture

This mode or mechanism of creep fracture occurs in structures that have been exposed to long-time creep processes and may be manifested as nucleation of voids along grain boundaries (creep voids), as shown in Figure 9.8(a), or as wedge cracks which appear in joint points, as shown in Figure 9.8(b), that is, cracks which are formed at the intersection of the grain boundaries. As the failure increases, its result is fracture accompanied by a small decrease in the cross-sectional area.

Transgranular Creep Fracture

In contrast to intergranular creep fracture, this mode of failure occurs in short-time creep processes and, similar to tensile rupture, results in a large decrease in cross-sectional area; voids can grow transgranularly or intergranularly, as shown in Figure 9.8(c) and Figure 9.8(d).

Fracture as Rupture in a Point

In this failure mode, necking occurs at a point. It is possible that this failure appears at high temperatures and low stresses.

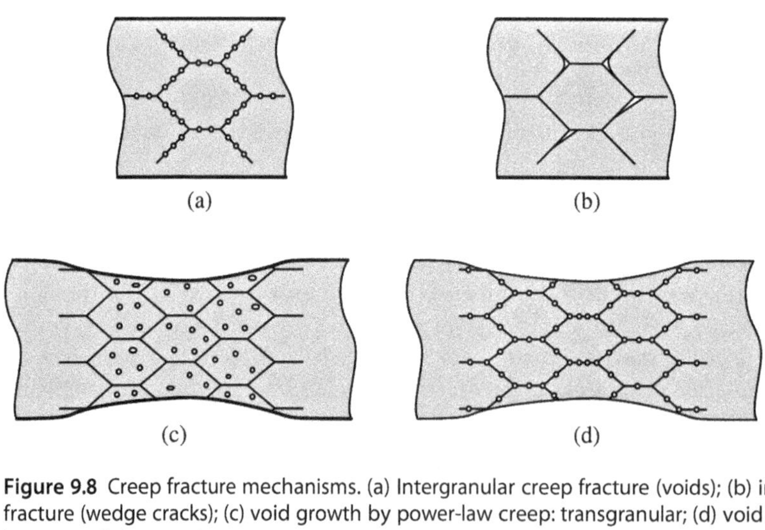

Figure 9.8 Creep fracture mechanisms. (a) Intergranular creep fracture (voids); (b) intergranular creep fracture (wedge cracks); (c) void growth by power-law creep: transgranular; (d) void growth by power-law creep: intergranular.

9.3.2 Short-time Uniaxial Creep Tests, Creep Modeling and Microstructure Analysis

9.3.2.1 Short-time Uniaxial Creep Tests

Uniaxial creep tests were already shown schematically in Figure 9.5, but experimentally obtained (real) creep tests are presented in this section (see Figure 9.10). Uniaxial tensile testing, testing of creep behavior (resistance), impact energy testing and microstructure examination can provide an insight into material behavior under prescribed environmental conditions and also some possibilities for a material's applications.

In structural design, one should choose materials with properties that correspond to the purpose of the structure and its operating conditions. The basic material properties are determined by uniaxial tests. The situation is similar to the determination of the material properties at high/low temperatures. The reason for choosing uniaxial tests in many situations is that these tests are simple in comparison with other types of test. Using uniaxial tests at high temperature it is possible to obtain tensile/compression strength, yield strength, the modulus of elasticity and so on. In accordance with this, uniaxial tests performed at constant stress and at constant high temperature may give an insight into material creep resistance.

Short-time uniaxial creep tests are often performed and these tests may be treated as appropriate for those applications where materials may be exposed to high temperatures in short periods of time, such as where there is a fire hazard situation or similar.

Only some special types of materials are intended for use at high temperatures over a long period. Tests that are intended to be performed at high temperatures over long periods require the use of special testing equipment. A basic testing machine and a furnace with a high-temperature extensometer is not the best combination for these measurements.

Later in this chapter, we present the results of research relating to the behavior of some materials at elevated temperatures. Engineering stress–strain diagrams as well as creep curves relating to the selected structural and stainless steels are shown. On the basis of these stress–strain diagrams it is possible to gain an insight into the level of strength and the decrease in strength with an increase in temperature. It is also possible to assess creep resistance for considered materials and for given conditions. Figure 9.9 presents some engineering stress–strain diagrams at prescribed temperatures for some considered materials, while Figure 9.10 shows creep curves from some actual creep tests. To complete the data on the tested materials, first information is given on the chemical elements of which the particular material is composed. Stress–strain diagrams at prescribed temperatures are shown in order to see the difference in the level of strength of materials as well as in their extension and creep resistance. The percentage of a particular chemical element and also the final state of the microstructure of a particular material may vary depending on the shipment of material. In these investigations the chemical compositions (in %) are not tested, they are displayed only with respect to the chemical elements in accordance with experiments previously performed on similar materials [17–20]. However, approximate values for the chemical composition of considered materials can be found in the literature. In addition, the state of the as-received material can usually be found in the certificate from the supplier of the material. Some tested materials are listed in Table 9.1. Figure 9.9 shows engineering stress–strain diagrams corresponding to room temperature (23 °C) in Figure 9.9(a), and high temperatures in Figure 9.9(b), for the

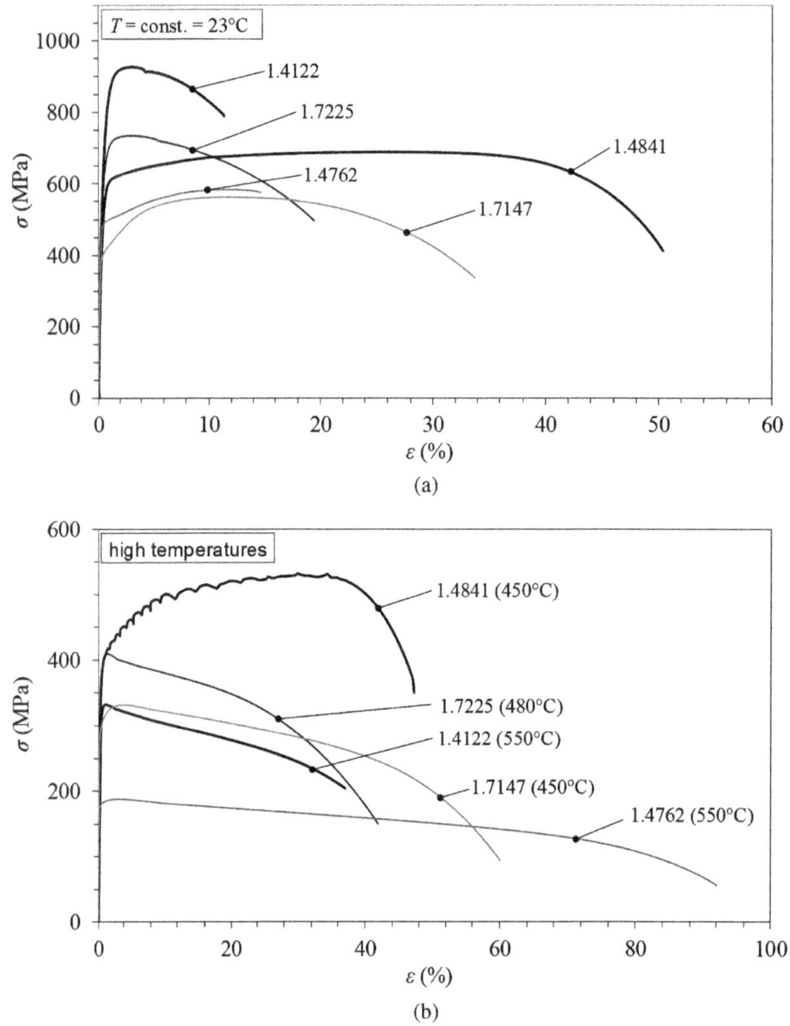

Figure 9.9 Engineering stress–strain diagrams at room temperature (~23 °C) and high temperatures for structural steels (1.7147/20MnCr5; 1.7225/42CrMo4) and stainless steels (1.4122/X39CrMo17-1; 1.4762/X10CrAlSi25; 1.4841/X15CrNiSi25-20). (a) Room temperature (~23 °C); (b) high temperatures.

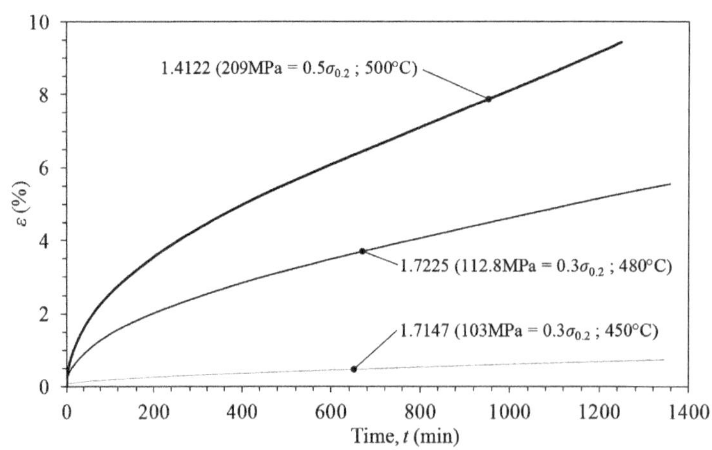

Figure 9.10 Short-time uniaxial creep tests for selected materials from Table 9.1.

Table 9.1 Tested materials: the constitutive chemical elements.

Material	Chemical elements
20MnCr5/**1.7147**/AISI SAE 5120 Round bar, structural steel	C, Cr, Si, Ni, Mn, Mo, Al, Cu, P, S, Rest/Balance
42CrMo4/**1.7225**/AISI 4140 Round bar, construction steel	C, Cr, Si, Mn, Mo, P, S, Rest/Balance
X39CrMo17-1/**1.4122**/AISI 420RM Round bar, martensitic stainless steel	C, Cr, Si, Ni, Mn, Mo, Cu, P, S, Rest/Balance
X10CrAlSi25/**1.4762**/AISI 446 Round bar, ferritic heat-resistant steel	C, Cr, Si, Ni, Mn, Mo, Al, Cu, P, S, Rest/Balance
X15CrNiSi25-20/**1.4841**/AISI 314 Round bar, austenitic heat-resistant steel	C, Cr, Si, Ni, Mn, Mo, Al, Cu, P, S, Rest/Balance

materials in Table 9.1, while Figure 9.10 shows creep curves for some of these materials. These creep curves were obtained on the basis of creep tests carried out at prescribed temperatures and stress levels. Tests were carried out in the Department of Engineering Mechanics at the University of Rijeka.

9.3.2.2 Creep Modeling

The responses of structures subjected to loads and technological processes are real processes in engineering practice. Often, the real process, for example, the response of a structure under a given load and defined operating conditions (service life conditions), may be modeled/simulated. This is done in such a way that the curve that describes the real response is as authentic a match as possible to the actual response curve.

The design of a new structure in terms of its performance and service life conditions can be very expensive. If the designer does not have enough previous knowledge about the responses of such a structure, the design may be prone to many failures. This is the reason why the predicted behavior (response) of a new design based on the simulated response of a previously known model can be of huge importance. This simulated response, based on some previously known facts about such a structure, does not require the creation of a real structure or even a geometric model of it.

As we know, in principle, two types of technological process, structural response or similar process can be distinguished: real and modeled processes. A real process (a process that is happening in reality) can also be modeled (simulated, predicted), and this means that this real process is presented using some appropriate tools. On the other hand, in many cases, a new, imaginary process for which the result is not known in advance, can also be replaced by a model, and now it can be said that the process is being simulated, modeled or predicted. The difference between these two cases is as follows: in the first case, a real process is modeled or simulated, while in the second case, the real process does not exist but it is imagined or predicted by the model.

The same may apply to the creep process. Some already-performed real creep processes are modeled using well-known rheological models or with proposed analytical equations.

As mentioned in Chapter 8, some relationships based on rheological models have been proposed [6] to relate factors such as stress, strain and time at a considered, constant temperature. There are also some analytical expressions. For some metals or other engineering applications involving relatively short times of loading at a prescribed temperature, it is necessary (and is of interest) to consider the transient (first/primary) stage of creep. As mentioned previously (Section 8.3.3), the Voigt–Kelvin model can be used for this purpose (see Equation (8.19)) but there is no initial strain, as explained there. Another rheological model that can be used in the modeling of the primary creep stage (but also in the secondary stage) of short-time uniaxial creep is the well-known Burgers' model [6, 21], shown in Figure 8.11 and represented by Equation (8.29):

$$\varepsilon = \varepsilon(t) = \sigma\left[\frac{1}{E_1} + \frac{1}{E_2}\left(1 - e^{-(E_2/\eta_1)t}\right) + \frac{t}{\eta_2}\right], \tag{a}$$

where $\varepsilon = \varepsilon(t) = \varepsilon_0 + \varepsilon_c$, is the strain consisting of instantaneous strain (ε_0), which can be elastic or elastic–plastic, arising after the specimen is subjected to stress and creep strain (ε_c), σ is the constant stress of the considered creep process, E_1 is the modulus of elasticity of the considered material at the considered creep temperature, t is the duration of the creep process, while E_2, η_1 and η_2 are parameters which are to be determined from the real or similar response. A further rheological model that is used in the modeling of the primary creep stage is the Standard Solid Linear model (SLS) [6, 21], as depicted in Figure 8.9 and represented by Equation (8.26):

$$\varepsilon = \varepsilon(t) = \frac{\sigma}{E_1(E_1 + E_2)}\left(E_1 + E_2 - E_2 \cdot e^{-\frac{E_1 E_2 \cdot t}{(E_1 + E_2)\eta}}\right), \tag{b}$$

where $\varepsilon = \varepsilon(t) = \varepsilon_0 + \varepsilon_c$, is the strain, also consisting of the instantaneous strain (ε_0), which can be elastic or elastic–plastic strain and creep strain (ε_c), σ is the constant stress of the considered creep process, t is the duration of the creep process and E_1, E_2 and η are material parameters which are to be determined from the creep curve. The possibilities for modeling using these rheological models were discussed in Chapter 8.

There are also a number of analytical equations that serve the purpose of modeling the primary stage of short-time creep. One formula that was suggested in [21] is:

$$\varepsilon(t) = D^{-T}\sigma^p t^r, \tag{c}$$

where ε represents the strain, T represents the temperature, σ represents the stress, t represents time and D, p and r are parameters which are to be determined. For this formula, an instantaneous strain is included when the formula is applied to the modeling of the overall creep curve (that is, when it comprises the deformation of the specimen at the beginning of its exposure to stress). In this case, for strain $\varepsilon = \varepsilon(t) = \varepsilon_0 + \varepsilon_c$, is valid, as mentioned above for Equation (b). Equation (c) can be used as an analytical model for consideration of the primary creep stage and it is given in the time–temperature–stress dependence form. This formula may be used to model an existing process (that is, an already-performed process) but also to model (or to predict) a new one. For the modeling of an already-performed process, the parameters will be adopted in such a way that they fit the considered curve. It is clear that an isothermal process is being considered.

For a new process at the same temperature but different stress, the same equation (Equation (c)) can be used. Generally speaking, there are several types of creep process that may be modeled: $\varepsilon(t) = \varepsilon(t)$, $\sigma = \text{const.}$, $T = \text{const.}$; $\varepsilon(t) = \varepsilon(\sigma, t)$, $T = \text{const.}$; $\varepsilon(t) = \varepsilon(\sigma, T, t)$. For the first case, the creep process is modeled for the prescribed constant stress at the prescribed constant temperature. In the second case, creep behavior can be modeled for a required constant stress within the range of constant stresses but at a prescribed constant temperature. Finally, in the third case, the creep process can be modeled for a required constant stress level and a required constant temperature within both of the mentioned ranges.

Equation (c) can also be used for the case where $\varepsilon(t) = \varepsilon(\sigma, T, t)$, that is, when a range of temperatures and the appropriate range of stresses are considered: $[(T_1 = \text{const.}: \sigma_{1(T_1)} = \text{const.}, \sigma_{2(T_1)} = \text{const.}, \ldots); (T_2 = \text{const.}: \sigma_{1(T_2)} = \text{const.}, \sigma_{2(T_2)} = \text{const.} \ldots) \ldots]$. This model is known as the time–temperature–stress dependence model. Some other known models of this type can be found in [12]. For a new, let's say desirable, process that involves a desirable temperature and a desirable stress level, data modeling will be based on data known for the above-mentioned combination of temperature ranges and the ranges of stress to which the desirable process belongs. In both mentioned cases, strain may be treated as being predicted, since it cannot be exactly the same as in an actual test. One well-known and widely used formula for the isothermal condition (creep process with known constant temperature) intended for use in modeling the primary creep stage is the Baily–Norton equation [12]:

$$\varepsilon_c = k \cdot \sigma^p \cdot t^r \ (\text{c} = \text{creep}) \tag{d}$$

Equation (c) above is of similar form to Equation (d). The proposed rheological and analytical models (Voigt–Kelvin, SLS, Burgers' and Equations (c) and (d) presented above) can be applied to model the first stage of the considered creep process (Burgers' model may also be used outside of the primary stage). The proposed analytical equations belong to the most appropriate and simple models when the primary creep stage is being considered. The most appropriate model is one where the material's creep behavior can be modeled for a certain field of temperatures and for a certain field of stress levels, as mentioned above. In this case it is: $\varepsilon = \varepsilon(T,\sigma,t)$

To demonstrate creep modeling, some materials were tested for creep under given environmental conditions and the results are shown in Table 9.2, where the creep modeling data for the materials are also given. Figure 9.11 shows the modeled creep curves.

9.3.2.3 Microstructure Analysis – Basic

To complete the insight into material behavior under different conditions, the material is usually uniaxially tested at room and at high temperatures; it is then tested to creep and to fatigue. Moreover, after completion of the scheduled creep process, the microstructure of the investigated material is also analyzed. The microstructure of the as-received material is compared with the microstructure of the specimen that was previously subjected to creep. Figure 9.12 (a), (b) and (c) show, for several selected materials (some of which are listed in Table 9.1), the microstructures for as-received materials as well as for the specimens that were previously subjected to creep.

Table 9.2 Creep modeling data.

Model	**Material**	**1.4122**	**1.7225**	**1.4122**
	Temperature (°C)	500	480	450
	Time (min)	1200	1200	1200
	σ **(Pa)**	$209 \cdot 10^6$	$112.8 \cdot 10^6$	$103 \cdot 10^6$
***Formula** (c)*	D	0.9261179	1.5087304	0.6568385
	p	−2.0838488	10.5352343	−10.4861582
	r	0.5269094	0.5217963	0.5568786
SLS	E_1	$1.8592207 \cdot 10^7$	$1.760162 \cdot 10^7$	$1.0146676 \cdot 10^8$
	E_2	$1.7735108 \cdot 10^8$	$1.4773960 \cdot 10^8$	$9.8430326 \cdot 10^8$
	η	$1.3972717 \cdot 10^{10}$	$1.3202841 \cdot 10^{10}$	$1.0420645 \cdot 10^{11}$

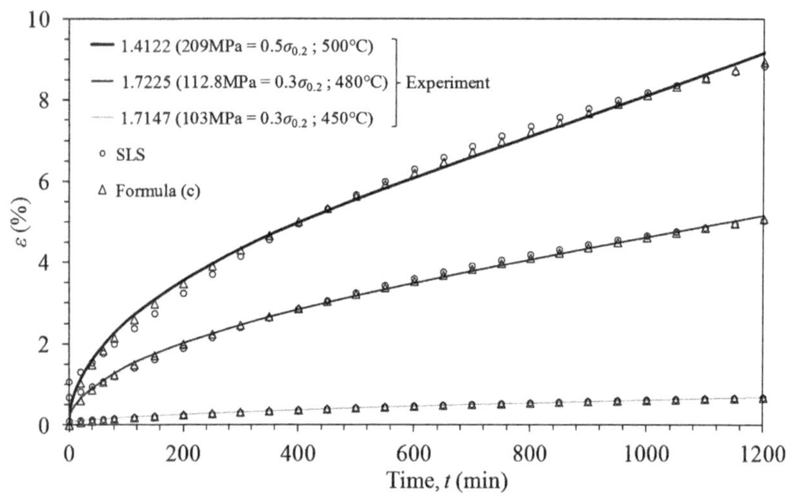

Figure 9.11 Modeled creep curves for: X39CrMo17-1 (1.4122); 42CrMo4 (1.7225); 20MnCr5 (1.7147).

9.3.3 Long-term Creep Behavior Prediction Based on the Short-time Creep Process

In engineering practice, in addition to short-time creep processes, long-term creep processes can also be of interest. In the earlier section describing creep deformation mechanisms (Section 9.3.1.1), two creep mechanisms were mentioned – diffusion creep and dislocation creep – and both of them are diffusion controlled. Creep occurs by diffusion along the grain boundaries at low temperatures while at high temperatures, the diffusion path occurs through the bulk of the material. As such, in general, it may be said that creep failure can result from bulk damage or from localized damage. Creep fracture mechanisms may be recognized on the basis of surface topology, as mentioned in Section 9.3.1.3. However, intensive creep deformation, which may be caused by crack development or other damage, can result in a rupture point in the material (element).

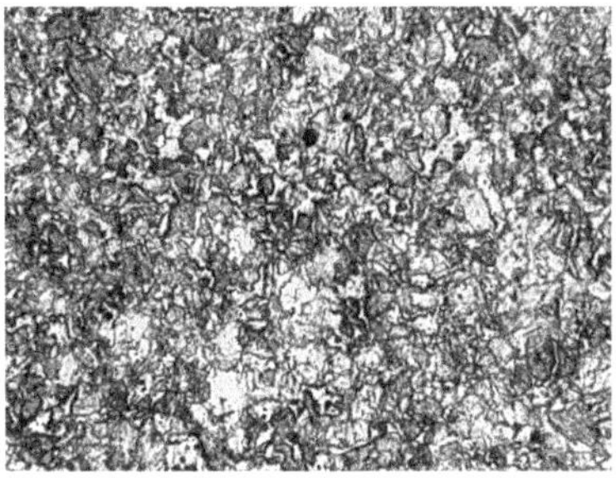

20MnCr5 (1.7147) steel

- As received material:
  Room temperature, 23 °C
- Lengthwise section (500×)
- Microstructure: mixture of ferrite and cementite

20MnCr5 (1.7147) steel

- After creep process:
  103 MPa, 450 °C, 1200 min
- Lengthwise section (500×)
- Microstructure: no significant changes before and after creep, except some elongations in the grains.

(a)

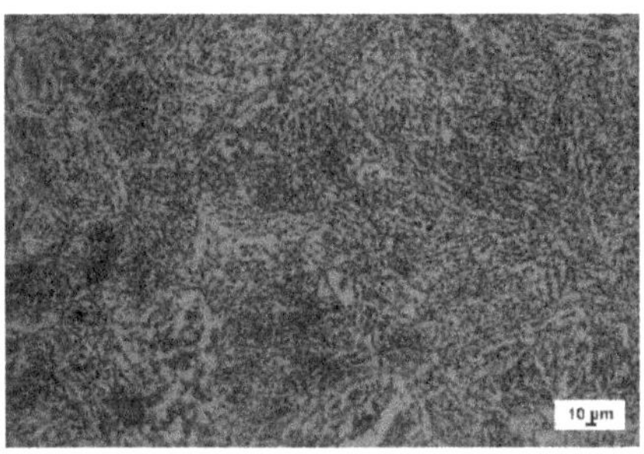

42CrMo4 (1.7225) steel

- As received material:
  Room temperature, 23 °C
- Cross-section (1000×)
- Microstructure: Thin pearlite, a few ferrite and smaller particles of cementite.

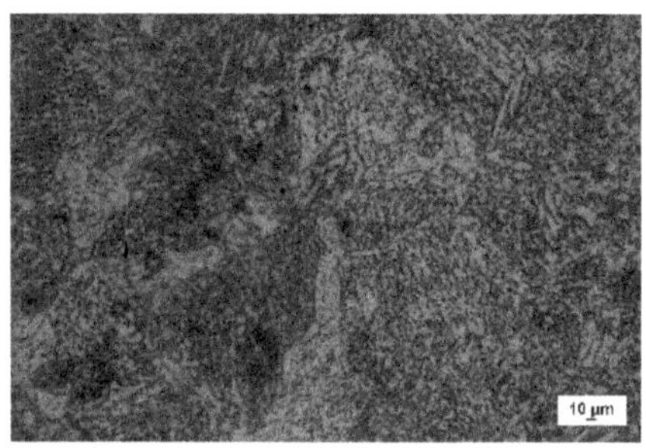

42CrMo4 (1.7225) steel

- After creep process:
  113.6 MPa, 500 °C, 1200 min
- Cross-section (1000×)
- Microstructure: No noticeable changes in relation to room temperature, except for a slight increase in grain size.

(b)

Figure 9.12 Optical micrographs of investigated materials (4% nital). (a) 20MnCr5 (1.7147) steel; (b) 42CrMo4 (1.7225) steel); (c) X10CrAlSi25 (1.4762) steel.

In metals, for example, the "other damage" referred to above may include the appearance of voids along the grain boundaries or at other places of stress concentration, such as places of precipitate particles, and the enlarging and joining of voids causes cracks. A crack can progress to the point of fracture, and this is known as *creep rupture*. When creep is under consideration, several facts can be emphasized:

- In engineering practice, most of the laboratory-performed creep tests/processes are short-time processes;

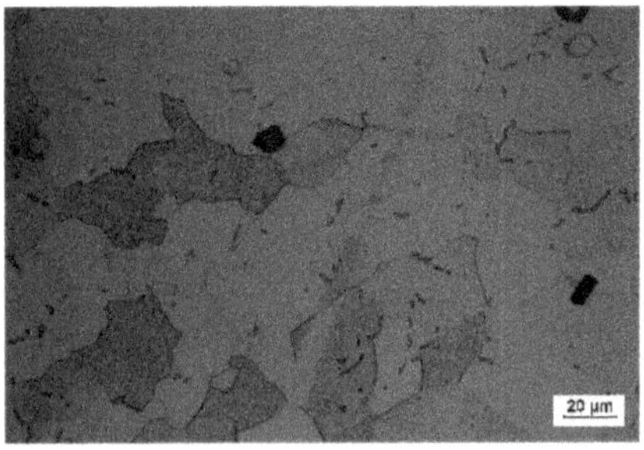

X10CrAlSi25 (1.4762) steel

- As received material:

  Room temperature, 23 °C
- Cross-section (500×)
- Microstructure: In general, the base phase with light color and dark gray color with different orientations is ferrite. Light gray color is the austenite.

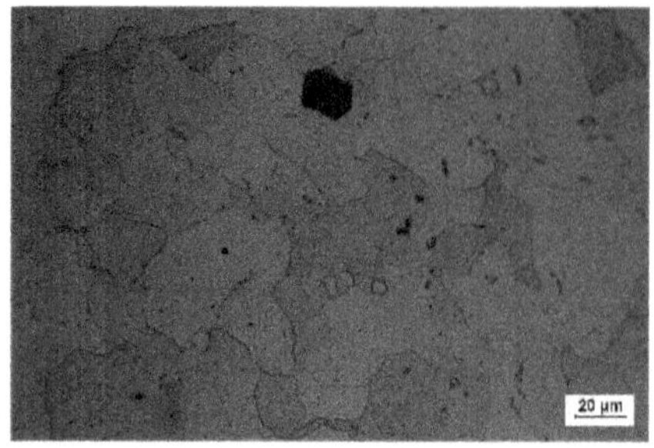

X10CrAlSi25 (1.4762) steel

- After creep process:

  155 MPa, 500 °C, 170 min
- Cross-section (500×)
- Microstructure: No significant changes, grains are elongated.

(c)

Figure 9.12 *(Continued)*

- Most of the performed creep processes are uniaxial processes;
- In assessing a material's creep behavior, it is desirable to establish a correlation between short-time and long-term creep processes;
- It is necessary to consider the choice of material for use in processes at high temperatures.

Since real data related to creep processes can be obtained by creep tests, and it is impossible to wait for years to obtain long-term creep data, the question that arises is how is it possible to obtain reliable creep data related to long-term creep, which may serve as input design data, without performing long-term creep tests? The need to establish a correlation between short-time creep and long-term creep is important here; the need arises because there is a discrepancy between the required structural design life and the impossibility of conducting long-term creep processes in order to obtain realistic data on the creep behavior of a material. Some attempts relating to the establishment of a correlation between short-time and long-term creep behavior will be documented below.

9.3.3.1 Extrapolation Methods

One of the known methods is the *Abridged method* [22].

Here, several laboratory short-time creep tests are performed at different stress levels but at the expected temperature, that is, at a temperature that is constant and is expected in the structure's service life (see Figure 9.13). Since the structure is designed for a desired life, creep curves are extrapolated to that required boundary. It needs to be said that these extrapolations do not include the possibility of creep rupture occurrence before reaching the (creep) design life.

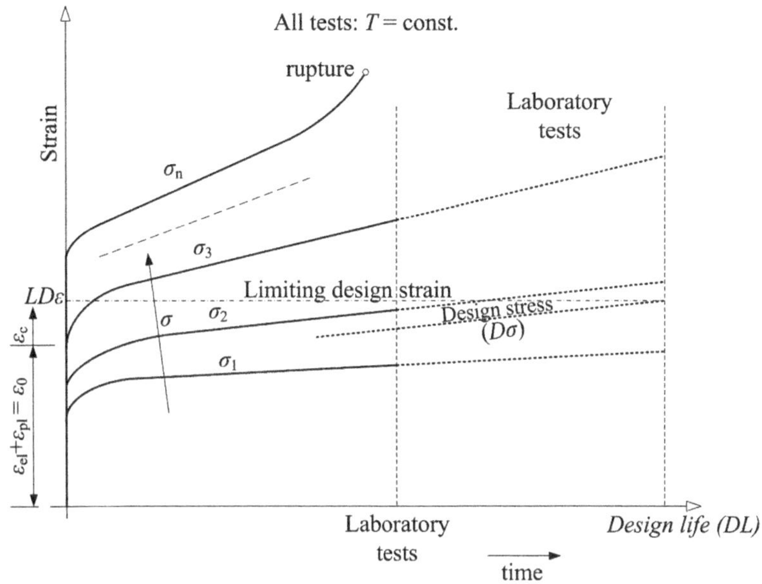

Figure 9.13 Graphical representation of the Abridged method.

Let the design be defined by limited design strain (LDε); the intersection of the lines of limiting design strain and required design life (DL) determines the allowable (design) stress (Dσ) that may be involved in the structure's service life.

Some other methods include the *Thermal acceleration method* and the *Mechanical acceleration method.*

9.3.3.2 Time–Temperature Parameters

Unlike the extrapolation methods, which are based on short-time creep tests conducted at constant temperature and different constant stress levels pertaining to the particular material, there are theories that relate to the so-called *time–temperature parameters.* One such theory is the *Larson–Miller theory.* Using this theory, on the basis of a short-time combination of temperature and time, it is possible to find a temperature and time equivalent to any desired long-term combination that can occur in the structure's service life. The theory and experiments have shown a good match for a wide range of materials. This theory is usually given in such a way that the temperature is expressed in °*F*. The Larson–Miller parameter, which is related to temperature and time for the considered material, is written in the following form [22]:

$$P = (T + 460)\left(C + \log_{10} t\right), \tag{9.3}$$

where P is the Larson–Miller parameter (constant for a considered material and stress level), T is the temperature (°F), t is the time (hours) and it relates to the time to rupture or the time to reach a specific level (for example, a limited level) of creep strain, C is a constant (usually adopted as 20).

As an example, let the test for the material under consideration be carried out at a temperature of 1400 °F and for 1500 hours (operating conditions). Let the achieved level of deformation be "ε". An equivalent test, which will be conducted at a temperature of 1500 °F to achieve the same level of deformation, will run for 98.55 hours. The result is obtained on the basis of the same parameter, P, in accordance with Equation (9.3). So, the equivalent test conditions are: 98.55 hours at 1500 °F. Similar to the existence of different extrapolation methods, there are also different time–temperature parameters, such as, for example, the Manson–Hafered parameter [22] and the Sherby–Dorn parameter [7].

9.3.4 Multiaxial Creep

Just as uniaxial creep is used as the name for creep that takes place under a uniaxial state of stress, multiaxial creep is used as the name for creep that takes place under a multiaxial state of stress. Consideration of multiaxial creep is more complicated than the consideration of uniaxial creep. Under uniaxial creep, only creep strain and the creep strain rate are involved and in this case, perhaps, only the determination of the start of the creep process (the end of the instantaneous deformation) may be a problem. By contrast, a multiaxial creep test includes a creep strain tensor and a creep strain rate tensor. Despite the fact that the multiaxial creep process is very complex, many engineering applications, such as turbine rotors, machines and so on, may involve multiaxial creep conditions since they operate under multiaxial stress state conditions. As has been mentioned, the yield criterion for a multiaxial stress state is based on the uniaxial stress state (tension test). Let a linear elastic isotropic material be considered. In this case, the principal axes of stresses and strains coincide and Hooke's Law with respect to principal strains (dilatations) is (from Equation (7.29a)):

$$
\begin{aligned}
\varepsilon_1 &= \frac{1}{E}\left[\sigma_1 - \nu(\sigma_2 + \sigma_3)\right] \\
\varepsilon_2 &= \frac{1}{E}\left[\sigma_2 - \nu(\sigma_1 + \sigma_3)\right] \\
\varepsilon_3 &= \frac{1}{E}\left[\sigma_3 - \nu(\sigma_1 + \sigma_2)\right]
\end{aligned}
\tag{9.4}
$$

In this theory, the (linear elastic) history of loading is neglected. However, creep belongs to the group of inelastic deformations where loading history affects the deformations. This means that creep strain rate–stress relations need to be included in the consideration of creep deformation. Total deformation in the creep process includes instantaneous deformation (elastic or elastic–plastic) and creep deformation. Let's consider strain as elastic and inelastic (creep). The creep deformation process is mostly inelastic and, in accordance with experimental observations, does not involve a change in volume. Therefore, we can assume that volumetric change due to creep deformation is zero, and this can be written as:

$$\varepsilon_{kk_c} = 0, \quad (\varepsilon_{1_c} + \varepsilon_{2_c} + \varepsilon_{3_c} = 0), \quad \mathrm{c} = \mathrm{creep} \tag{9.5a}$$

We can also write:

$$\dot{\varepsilon}_{kk_c} = 0, \quad (\dot{\varepsilon}_{1_c} + \dot{\varepsilon}_{2_c} + \dot{\varepsilon}_{3_c} = 0) \tag{9.5b}$$

Another assumption is proportionality between the maximum values of shear strain rates and shear stresses, similar to the theory of plasticity [12]. Maximum shear stresses were defined in Chapter 2 (Equations (2.31a)), while maximum shear strains were given in Equations (2.31b). The mentioned proportionality is written as:

$$\frac{\dot{\varepsilon}_{1c}-\dot{\varepsilon}_{2c}}{\sigma_1-\sigma_2}=\frac{\dot{\varepsilon}_{2c}-\dot{\varepsilon}_{3c}}{\sigma_2-\sigma_3}=\frac{\dot{\varepsilon}_{3c}-\dot{\varepsilon}_{1c}}{\sigma_3-\sigma_1}=C. \tag{9.6a}$$

On the basis of this equation, we have:

$$\begin{aligned}\dot{\varepsilon}_{1c}-\dot{\varepsilon}_{2c}&=C(\sigma_1-\sigma_2)\rightarrow\dot{\varepsilon}_{2c}\\ \dot{\varepsilon}_{1c}-\dot{\varepsilon}_{3c}&=C(\sigma_1-\sigma_3)\rightarrow\dot{\varepsilon}_{3c}\end{aligned} \tag{9.6b}$$

If the obtained values for $\dot{\varepsilon}_{2c}$ and for $\dot{\varepsilon}_{3c}$ are inserted into Equation (9.5b), the first of Equations (9.7) can be calculated. On the basis of Equations (9.6a) and (9.5b), using similar procedures, the remaining equations in the set of Equations (9.7) can be calculated:

$$\begin{aligned}\dot{\varepsilon}_{1c}&=\frac{2}{3}C\left[\sigma_1-\frac{1}{2}(\sigma_2+\sigma_3)\right]\\ \dot{\varepsilon}_{2c}&=\frac{2}{3}C\left[\sigma_2-\frac{1}{2}(\sigma_3+\sigma_1)\right]\\ \dot{\varepsilon}_{3c}&=\frac{2}{3}C\left[\sigma_3-\frac{1}{2}(\sigma_1+\sigma_2)\right]\end{aligned} \tag{9.7}$$

The constant C is changed within the transient creep stage while it remains constant in the second creep stage. This constant can be determined on the basis of known creep behavior for the considered material at a given strain rate and stress level. The creep rate equations (Equations (9.7)) can be adapted to creep behavior under uniaxial (tensile) creep. If the principal stresses σ_2 and σ_3 are considered to be zero, this yields:

$$\dot{\varepsilon}_{1c}=\frac{2}{3}C\sigma_1 \tag{9.8}$$

From the distortional energy density criterion (von Mises criterion) in Equation (5.16), an equivalent stress for a multiaxial state of stress is given by:

$$\sigma_e=\frac{1}{\sqrt{2}}\sqrt{(\sigma_1-\sigma_2)^2+(\sigma_2-\sigma_3)^2+(\sigma_3-\sigma_1)^2} \tag{9.9}$$

If Equation (9.9) is adapted to the uniaxial stress case, then it becomes:

$$\sigma_e=\sigma_1, \tag{9.10}$$

and, by combining Equation (9.8) and Equation (9.10), it follows that:

$$\sigma_e=\sigma_1=\frac{3}{2}C\dot{\varepsilon}_{1c}. \tag{9.11}$$

When a creep test is slightly longer and primary creep is quite short, then the creep process may be approximated by the steady-state stage of creep. In this case, the creep rate depends on the stress level only. This dependence was proposed by Bailey, and it is generally written as (creep rate) [12]:

$$\dot{\varepsilon} = B\sigma^N. \tag{9.12}$$

Now, Equation (9.11) becomes:

$$\sigma_1 = \frac{3}{2}CB\sigma_1^N \tag{9.13}$$

Solving this equation with respect to "C", it follows that:

$$\frac{2}{3}C = B\sigma_1^{N-1} = B\sigma_e^{N-1}. \tag{9.14}$$

Based on the value in Equation (9.14), Equation (9.7) can be written in the following form:

$$\begin{aligned} \dot{\varepsilon}_{1c} &= B\sigma_e^{N-1}\left[\sigma_1 - \frac{1}{2}(\sigma_2 + \sigma_3)\right] \\ \dot{\varepsilon}_{2c} &= B\sigma_e^{N-1}\left[\sigma_2 - \frac{1}{2}(\sigma_3 + \sigma_1)\right] \\ \dot{\varepsilon}_{3c} &= B\sigma_e^{N-1}\left[\sigma_3 - \frac{1}{2}(\sigma_1 + \sigma_2)\right] \end{aligned} \tag{9.15}$$

Thus, the principal creep strain rates are defined on the basis of the principal creep stresses and parameters (B, N) that are determined experimentally on the basis of uniaxial tensile creep. Creep strains may also be determined using further mathematical calculations [22], and the prediction of creep behavior in multiaxial creep can be made by equations based on the result of the uniaxial creep test.

9.4 Relaxation Phenomenon and Modeling

Creep, relaxation and recovery of a material are known as material phenomena. Creep, as time-dependent behavior of a material has been discussed previously. Another type of behavior that may be considered is called recovery. This can be described as the gradual disappearance of creep strain that may occur after stress removal. In metals, the amount of recoverable strain in comparison with creep strain is (generally) small. Creep and relaxation phenomena are associated with viscoelastic materials and metals. In other words, metals subjected to certain loadings and environmental conditions may demonstrate the properties of both viscous and elastic materials. The relaxation phenomenon can be most simply described as a gradually decrease in stress over a time under constant strain; that is, this is the phenomenon of stress relief while the strain is kept constant, as shown in Figure 9.14. It may be said that relaxation in metals is predominant at elevated temperatures, and it may be responsible for loss of external load, especially in bolted engineering elements. As such, material relaxation behavior can be of importance to

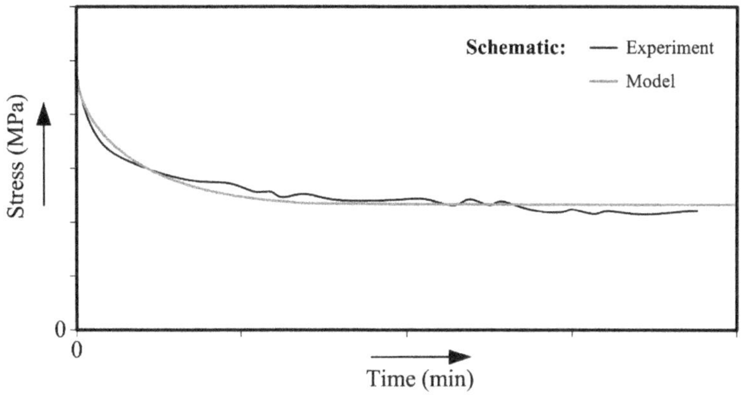

Figure 9.14 General representation of the relaxation phenomenon.

structural designers who are involved, for example, with sealing (tightening) problems. Similar to the problem of creep, the relaxation problem can be determined experimentally but also modeled.

Information about creep and relaxation modeling using the Standard Linear Solid model can be found in [23]. As was discussed in Chapter 8, some rheological models may be used to describe the process of relaxation. The solution of the differential equation (Equation (8.23)) of the Standard Linear Solid model (refer back to Figure 8.9) with respect to stress gives:

$$\sigma(t) = E_1\varepsilon_0 + e^{-\frac{E_2}{\eta}t}(\sigma_0 - E_1\varepsilon_0). \tag{9.16}$$

Equation (9.16) describes the change in stress under constant strain. In other words, this equation can be obtained as the solution of Equation (8.23) if one takes $\dot{\varepsilon} = 0$ (strain is constant) and applies Equation (a) from Section 8.3.3 to Equation (8.23). In Equation (9.16), σ represents stress, E_1, E_2 and η are parameters that are to be determined, t represents time, σ_0 represents the level of stress at the beginning of the relaxation process, ε_0 represents constant strain (that is, strain at the beginning and throughout the relaxation process).

On the other hand, if relaxation modeling is performed using Burgers' model (see Figure 8.11), it can be described by the following formula [24]:

$$\sigma(t) = \varepsilon_0\left[E_1 + E_2 e^{-\frac{E_2.t}{\eta_1}}\right] \tag{9.17}$$

In Equation (9.17), σ represents stress, E_1, E_2 and η_1 are parameters that are to be determined, t represents time, ε_0 represents constant strain (that is, strain at the beginning and throughout the relaxation process).

On the basis of both mentioned equations it is clear that stress (corresponding to strain at the beginning of the relaxation process) decreases exponentially. Strain is kept constant. To confirm the validity of applying these models, several experimental tests of relaxation processes were undertaken and then modeled.

References

1 Gross, D. and Seeling, T. (2011) *Fracture Mechanics*, Springer-Verlag, Berlin/Heidelberg.
2 Stephens, R. I., Fatemi, A., Stephens, R. R. and Fuchs, H. O. (2001) *Metal Fatigue in Engineering*, John Wiley & Sons, New York.
3 McEvily, A. J. (2002) *Metal Failures*, John Wiley & Sons, New York.
4 Brooks, C. R. and Choundhury, A. (2002) *Failure Analysis of Engineering Materials*, McGraw-Hill, New York.
5 Saxena, A. (1998) *Nonlinear Fracture Mechanics for Engineers*, CRC Press, New York.
6 Findley, W. N., Lai, J. S. and Onaran, K. (1989) *Creep and Relaxation of Nonlinear Viscoelastic Materials*, Dover Publications, Inc., New York.
7 Dowling, N. E. (2013) *Mechanical Behavior of Materials*, Pearson, New York.
8 Solecki, R. and Conant, R. J. (2003) *Advanced Mechanics of Materials*, Oxford University Press, New York.
9 Kanninen, M. F. and Popelar, C. H. (1985) *Advances in Fracture Mechanics*, Oxford University Press, New York.
10 Bramfitt, B. L. (1997) Effect of Composition, Processing and Structure on Properties of Irons and Steels. In G. E. Dieter (Ed.) ASM Handbook, Vol. **20**, *Materials Selection and Design*, ASM International, USA.
11 Raghavan, V. (2004) *Materials Science and Engineering*, Prentice- Hall of India, New Delhi.
12 Boresi, A. P., Schmidt, R. J. and Sidebottom, O. M. (1993) *Advanced Mechanics of Materials*, John Wiley & Sons, New York.
13 Frost, H. J. and Asby, M. F. (1982) *Deformation-Mechanism Maps: The Plasticity and Creep of Metals and Ceramics*, Pergamon Press.
14 Meyers, M. A. and Chawla, K K. (1999) Mechanical Behavior of Materials, Prentice-Hall.
15 Kassner, M. E. (2015) *Fundamentals of Creep in Metals and Alloys*, 3rd edition, Butterworth-Heinemann, Oxford.
16 Gross, T. S. (1997) Micromechanisms of Monotonic and Cyclic Crack Growth. In S. R. Lampman (Ed.) ASM Handbook, Vol **19**, *Fatigue and Fracture*, ASM International, USA.
17 Brnic J., Turkalj, G. and Canadija, M. (2014) Mechanical Testing of the Behavior of Steel 1.7147 at Different Temperatures, *Steel and Composite Structures*, **17**(5), 549–560.
18 Brnić, J., Turkalj, G., Lanc, D., Čanadija, M., Brčić, M., Vukelic, G. and Munjas, N. (2013) Testing and Analysis of X39CrMo17-1 Steel Properties, *Construction and Building Materials*, **44**, 293–301.
19 Brnić, J., Turkalj, G., Krscanski, S., Niu, J. and Li, Q. (2015) Changes in the Material Properties of Steel 1.4762 Depending on the Temperature, *High Temperature Materials and Processes*, DOI 10.1515/htmp-2015-0075.
20 Brnić, J., Turkalj, G., Krscanski, S., Vukelić, G. and Čanađija, M. (2017) Uniaxial Properties versus Temperature, Creep and Impact Energy of an Austenitic Steel, *High Temperature Materials and Processes*, **36**(2), 135–143.
21 Brnic, J., Turkalj, G., Niu, J., Canadija, M. and Lanc, D. (2013) Analysis of experimental data on the behavior of steel S275JR – Reliability of modern design, *Materials and Design*, **47**, 497–504.
22 Collins, J. A. (1993) *Failure of Materials in Mechanical Design*, 2nd edition, John Wiley & Sons, New York.

23 Almagableh, A., Mantena, P. R. and Alostaz, A. (2010) Creep and stress relaxation modeling of vinyl ester nanocomposites reinforced by nanoclay and graphite platelet, *Journal of Applied Polymer Science*, **115**(3), 1635–1641.

24 Eley, R. R. (1995) Rheology and Viscometry. In J. V. Koleske (Ed.) *Paint and Coating Testing Manual*, fourteenth edition of the Gardner–Sward Handbook, Ann Arbor, pp. 333–350.

10

Fracture Mechanics

10.1 Introduction

Although this chapter is written as an independent chapter, a matter covered within it is closely related to Chapters 1 (Section 1.4), 6 and 9. We know that products such as vessels, shafts and so on should be optimal. Furthermore, the product, let's call it a structure or a set of engineering elements, should be maintained and controlled, including control of the possible occurrence of failures. Analysis of engineering failures, as a basis for determining and understanding the causes of failures and the modes of their manifestations, can give the answers as to why and how an engineering element has failed. This is important since, on the basis of this knowledge, it may be possible to avoid similar failures in the future.

In many respects, the analysis of a structure can be compared with the analysis of the human body; we can compare, for example, the control of structural failure with the control of disease in the human body. Just as a structural failure can result in the inability of the structure to perform its function, so, too, can a particular disease lead to the inability of the human body to perform a particular function. In addition, the fracture of an engineering element related to any type of failure can be compared with the termination of human life caused by any disease. The fracture of an engineering element, that is, the separation of the element into two (or more) parts, depends on the element's loading, the element's service life conditions including environmental conditions, and so on.

As already mentioned, in engineering practice there are many failure modes as well as many reasons why they appear and many different modes of failure manifestation. For example, creep as a possible failure mode can be caused by a short- or long-term constant load to which the considered element is exposed at high temperature, and its mode of manifestation is usually recognized as large deformation. Creep in its final stage may result in fracture of the element. Creep deformation and creep fracture mechanisms were described in the previous chapter. Other types of failure mentioned earlier may also lead to the fracture of the element. The yielding failure that occurs when plastic deformation becomes large enough may lead to separation or, for example, ductile rupture caused by enormous plastic deformation, or impact failure, which may be caused by a very high magnitude of stresses and strains, can also lead to separation of the considered member into two parts. In addition, fatigue failure as a consequence of the application of fluctuating loads can lead to separation as a result of the initiation and growth of a crack.

Analysis of Engineering Structures and Material Behavior, First Edition. Josip Brnić.

Some answers to the questions "why" and "how" some element has failed may be obtained on the basis of the failure manifestation mode and by consideration of the topology of the fracture surface on the coarse scale (macroscopic) or fine scale (microscopic).

Many types of failure in their final phase lead to fracture based on crack propagation. A crack in the structure may be nucleated and may grow under fatigue loading; that is, fatigue is a cumulative failure under repeated loading where a crack can grow from the microscopic level. Further, a crack may be caused by the manufacturing process or it may stem from defects in the material [1].

An understanding of the causes of failure as well as an awareness of the damage that they cause, gives hope for design improvements and future prevention of these failures. Fracture mechanics, as a relatively new engineering discipline, has become a powerful tool in the analysis of failures and their prevention.

10.2 Fracture Classification

Engineering design and manufacture of the structure precede its use. Design is an iterative, decision-making process for the creation and optimization of the structure, while manufacture of the structure should include the best choice of technological process. Types of possible structural failures that can be encountered in engineering practice were discussed in Section 1.4 and were also mentioned in the introduction to this chapter.

Fracture mechanisms (fracture deformation mechanisms) in general, and creep fracture mechanisms and modes of fracture manifestation were discussed in Chapter 9. Fracture can be classified into several categories with respect to typical phenomena involved in the fracture process. The classification can be made: *with respect to the amount of plastification, then with respect to the loads that cause it and, for example, with respect to the dominant stress in the fracture surface orientation.* The fracture process is complete when the considered element has separated into two (or more) parts.

As discussed in Section 9.3.1.2, the difference between *ductile* and *brittle* fracture can be determined from the macroscopic point of view. *Ductile fracture* is, in general, characterized by a large amount of plastic flow (large plastic deformations) before fracture. A fracture of a notched (cracked) specimen at the microscopic level can be treated as a brittle fracture, even if it is made of a ductile material. Further, a fracture is usually called a *brittle fracture* if, at the macroscopic level, only a small amount of plastic deformation is noticeable.

A fracture that occurs as a result of creep crack growth (a creep process) is called a *creep fracture*, and a fracture that is the result of crack growth due to repeated (cyclic) loading is known as a *fatigue fracture*.

It can be said that the fracture of an engineering element can occur in a brittle or in a ductile manner and can be caused by overload, cyclic loading, creep, different types of defects associated with load actions and so on. Since, at the end of the fracture process, the member is separated into two (or more) parts, fracture surfaces are created and they often reveal the loading condition. If, for example, a *tension overload* is considered, we can say that if the fracture surface plane coincides with the cross-section that is perpendicular to the principal stress, that is a *normal-stress-dominated fracture*; if the fracture surface

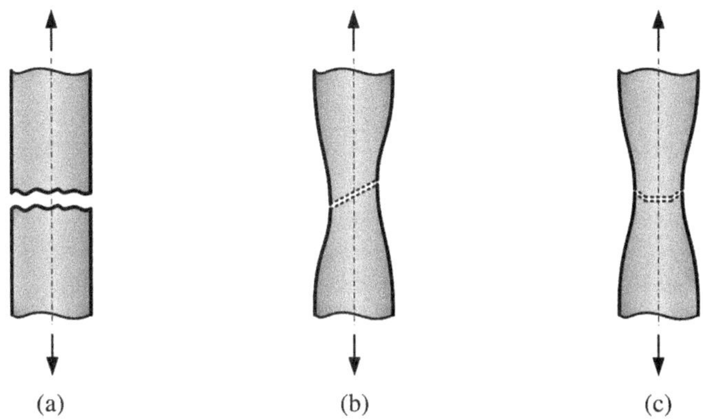

Figure 10.1 Types of fracture classified with respect to fracture surface orientation (tensile-loaded specimens). (a) Normal-stress-dominated fracture; (b) shear-stress-dominated fracture; (c) mixed (normal with shear lips).

takes a position that is oriented as certain critical shear stress (that is, the fracture surface coincides with critical shear stress), that is a *shear-stress-dominated fracture*. There is also a mixed form, as shown in Figure 10.1 [2, 3]. The fracture surface that occurs after fracture depends of the material and its microstructure as well as the loading conditions. Under the above-mentioned tension overload, brittle and ductile fracture can be recognized. Although the amount of plastic deformation is somewhat arbitrary, the material is usually considered to be brittle if elongation is less than 5% [3]. In accordance with the above, in the case of an axially tensile overloaded brittle material, since it is known that the maximum normal stress occurs in the axial direction (Equations (6.11) and (6.12)), the brittle fracture surface is perpendicular to the principal stress direction, as shown in Figure 10.1 (a). In an axially tensile overloaded ductile material (the fracture clearly involves larger plastic deformation), the fracture surface orientation can take several positions. First, the fracture plane orientation should be in accordance with the maximum shear stress (Equation (6.13) and Figure 10.1(b)), that is, the fracture surface lies at 45° to the loading axis, or it may be a *cup-and-cone* type of fracture (Figure 10.1(c)), that is, the plane of the central rough region of the fracture surface is perpendicular to the loading axis and on the periphery there is a fracture surface inclined at 45° to the central plane. This phenomenon is known as *shear lips*, where the lips on the periphery are not parallel but they are on the same side of the central rough region/plane). Shear lips on the periphery can also be parallel but placed on different sides of the central plane (one above, the other down).

In the case of a torsionally overloaded cylinder specimen, fracture can occur in a brittle or in a ductile manner. In a torsionally overloaded brittle material, it is expected that the fracture surface will take a position along the planes sustaining the maximum normal stress, that is, at 45° to the longitudinal axis (pure (shear) torsion: maximum normal stress $\sigma_{\max,\min} = \pm\tau$ at 45° to the longitudinal axis). In a torsionally overloaded ductile material, the fracture surface is at 90° to the longitudinal axis.

In the case of bending overload, it can be said that the fracture surface is similar to that caused by tensile overload, whereby the crack is formed at the tensile side and propagates through the section to the rupture. Some explanations of these issues can be found, for example, in [3].

10.3 Fatigue Phenomenon

10.3.1 Known Starting Points

Many years ago structures were designed in a simple way without powerful computers and numerical methods of analysis, and the manufacture of structures was completed without the use of modern technologies. Although many of those processes would now be considered outdated, engineering practice at that time achieved acceptable solutions. Technological development has brought new methods of computing, new materials and new production processes. Despite this progress, there is still a large number of failures which reduce both the quality and safety of products. Recent design and manufacture of a product are based on new developments in the creation and production technologies, that is, those that are supported by powerful computers and modern production facilities. It can be said that the idea of a new product is based on an optimal design which avoids failures and fracture and has programmed its required lifetime. In this regard, the design criteria for a construction undergoing cyclic load must also fit in the mentioned design philosophy.

Fatigue design criteria, in accordance with the required durability of the structure, include a time range from infinite life to damage tolerance, depending on the application. Known fatigue life models, such as stress–life: σ–N (S–N); deformation-life: ε–N; and models related to the crack propagation rate ($\mathrm{d}a/\mathrm{d}N - \Delta K$), are used. Let's consider the fatigue process as a known and usually observed mechanical failure mode which can cause cracks. In a higher stage of development of the mentioned failure, that is, in a higher stage of crack propagation, deformation growth can lead to fracture of the element. At the microscopic level, when the applied stress is sufficiently high to break the bonds between the atoms, molecules and so on, the material (or element) fractures. In this case, under consideration is the so-called cohesive strength at the atomic level.

From an engineering point of view, it is of importance to study the field of the causes of cracks and their propagation, that is, the field of fatigue in materials. Fatigue is a failure due to repeated loading and is a common cause of fracture. As a very simple example, we may consider the breaking of a narrow tin strip by bending it back and forth a number of times.

10.3.2 Stress versus Life Curves (σ–N/S–N), Endurance Limit

Most loads to which engineering structures are subjected are dynamic in character. Engineering machines and structures are frequently subjected to loading that can be approximated by known laws. The investigation of the mechanical behavior of materials leads to knowledge of their resistance to failures that can be caused by different conditions. This may include failures that occur under so-called static imposed loads and dynamic loads but also at room as well as high temperatures. To these failures belong, among others, fatigue. Fatigue, like many other failures, can result in the appearance of a crack. The resistance to crack propagation is called *fracture toughness*. On the other hand, it is known that when materials (consider metals) are subjected to repeated (fluctuating) loads, fatigue failure may result in fracture of the element at a stress level that is much lower than the fracture stress corresponding to a monotonic tension load. Monotonic loading means loading that proceeds in only one direction.

To gain some details, knowledge and understanding of material fatigue behavior, that is, of the fatigue failure phenomenon, so-called fatigue tests are conducted. There are machines that allow the implementation of fatigue tests in the field under axial loads, torsion and rotational bending. In engineering practice, fatigue is often considered in cases of tensile loading and also for unnotched (smooth) specimens, that is, specimens that have no stress raiser. Attention should be paid to ensure that the test specimen is compatible with the testing machine that generates the test data. There are several types of specimen. Both the geometry of the specimen (including sizes/dimensions) and the method of performing the fatigue test are recommended by standards (for example, the ASME standard, the ISO standard). For a considered (specific) material, for each fatigue test performed at a prescribed fatigue stress and a prescribed stress ratio, one piece of data related to the number of the cycle when the specimen has failed is obtained and this is recorded in the *S–N* system as one point; that is, each fatigue test generates one point. It is possible that for the prescribed number of cycles (let's say, 10^7 cycles), the specimen does not break. Usually, several tests are performed for the same level of fatigue stress (at the same stress ratio) since a different number of cycles to failure can be obtained for the same stress level; that is, scatter related to the number of cycles for the same stress level occurs. On the basis of obtained test data, the dependence relating to fatigue stress versus cycles to failure is usually displayed in the form of a diagram known as a *Wöhler* (or σ–*N*) curve. This curve is often also designated an *S–N* curve (or diagram) and it is a curve showing stress versus the number of cycles to failure (a stress–life curve).

The cyclic loads that occur in engineering structures and machines are almost random in nature and vary in intensity over time [4], as shown in Figure 10.2(a). Repeated loads (cyclic loads) in engineering structures induce cyclic stresses within the material. This results in microscopic damage that accumulates over time as the cyclic stressing continues. Finally, a macroscopic crack (or some other damage) can become evident. As such, *fatigue* may be defined as the process of damage accumulation due to cyclic loading. Let cyclic loading be described by the shape of a known function, for example, by a sinusoidal function. Consider the case where the mean stress and stress amplitude remain constant, as shown in Figure 10.2(b). Let us introduce terms such as maximum stress ($\sigma_{\max}$), minimum stress ($\sigma_{\min}$), mean stress (σ_{m}), stress amplitude (σ_{a}) and the stress ratio (R), which is sometimes called the factor of asymmetry of the cycle. Using two of the mentioned quantities, each type of cyclic load can be defined. Depending on the stress ratio (R), it is possible to distinguish several types of cyclic process, as shown in Figure 10.2(b).

The relationships between the above-mentioned characteristics of a cyclic load are defined as follows:

$$\begin{aligned} \sigma_{\mathrm{m}} &= \frac{\sigma_{\max} + \sigma_{\min}}{2} \\ \sigma_{\mathrm{a}} &= \frac{\sigma_{\max} - \sigma_{\min}}{2} \\ R &= \frac{\sigma_{\min}}{\sigma_{\max}} \end{aligned} \tag{10.1}$$

In Equation (10.1), σ_{m}= mean stress, σ_{a}= the amplitude of the stress, R = the stress ratio. The mean stress is characterized by the maximum and minimum stress levels but also by

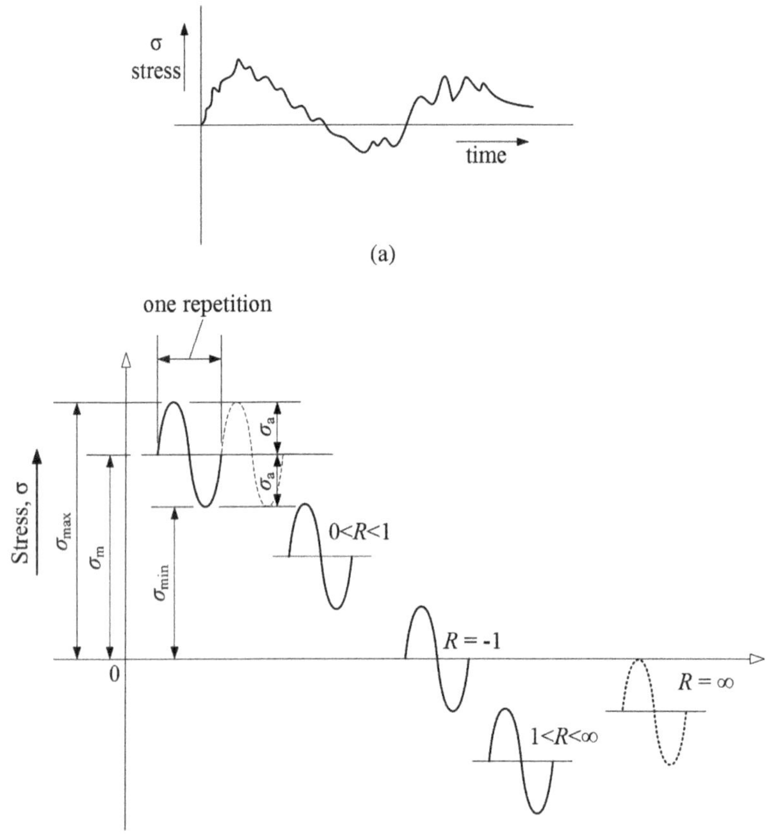

σ_{max} = maximum stress
σ_{min} = minimum stress
σ_m = mean stress
R = factor of asymmetry (stress ratio)

(b)

Figure 10.2 Random and constant-amplitude load cycles. (a) Random load; (b) types of sinusoidal variable stress (cycles with different stress ratios).

the stress ratio. In accordamce with the definition of the stress ratio, $R = -1$ refers to fully reversed loading, $R = 0$ to zero tension fatigue and $R = 1$ to a static load. All cycles with $R \neq 1$ are called asymmetrical cycles, while a cycle with $R = -1$ is called a symmetrical or alternate cycle.

A fatigue test can be presented in such a way that the ordinate shows cyclic maximum stress or cyclic stress amplitude or cyclic mean stress, while the abscissa shows the number of cycles to failure (fracture) of the specimen. In terms of the definition of one load cycle, the most appropriate solution, from the fatigue point of view, is that which contains the display of the maximum and minimum stresses, since they are the stress levels at which the loading direction is reversed (this means that cyclic slip is reversed and crack propagation stops at maximum stress) [4]. The abscissa (fatigue life – that is, the number of cycles (N) to failure) is usually plotted on a logarithmic scale, while the ordinate (fatigue stress test data – that is, cyclic maximum stress or cyclic mean stress or amplitude stress) is plotted on a linear scale or on a logarithmic scale. In general, on both axes any scale may be used.

These fatigue tests are a useful tool for predicting the life of a considered engineering element made of a given material that is subjected to constant amplitude cyclic load. When, regardless of the number of cycles, a test specimen remains unbroken at the given conditions of the test (stress ratio and maximum stress or stress amplitude), the fatigue limit has been reached and the stress associated with this limit is called the *endurance limit* (or the *fatigue limit*) [4–6]. The curve appears to become practically horizontal (although this is not the case for all materials). This means that the number of cycles to failure becomes practically infinite, see Figure 10.3, or it may be said that for given test conditions, a distinct stress level (endurance limit) appears, under which fatigue failure does not occur. In engineering practice, the number of cycles that a specimen must withstand without rupture for a considered material is usually prescribed, and in this case, the information related to maximum fatigue stress can be used as the fatigue limit. Figure 10.3 shows two curves. One curve refers to a material with a clearly expressed

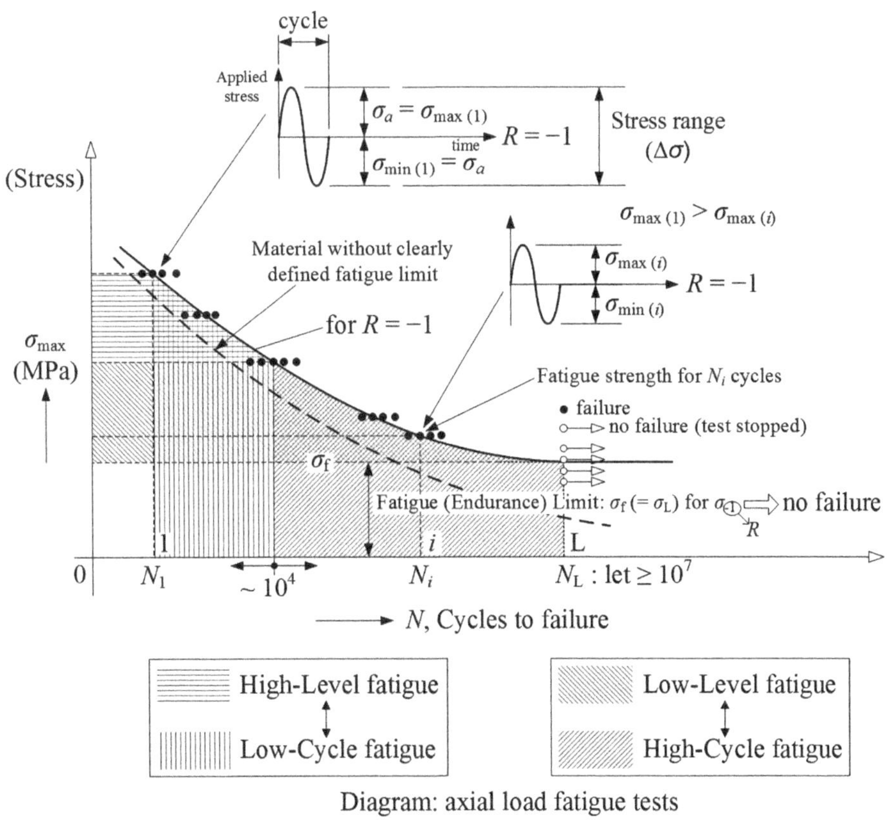

Figure 10.3 Schematic representation of Wöhler σ–N curve for $R = -1$, obtained on the basis of axial load fatigue test data. (The abscissa is usually shown on a log scale; the ordinate on a log or linear scale).

fatigue limit and the other to a material without a clear fatigue limit. Two regions in the diagram can be recognized. One region belongs to the finite fatigue region (life) and the second belongs to the infinite fatigue region.

If dependences related to stress versus the number of the cycles to failure are shown (modeled) as straight lines, then the first straight line is an inclined line while the second is a horizontal line. The *S–N* diagram shows which level of fatigue strength corresponds to the number of cycles to failure.

As previously stated, the number of cycles a specimen must withstand without rupture is usually prescribed, and since the scatter related to the number of cycles in this case may also arise, it is important to determine the fatigue limit. Various methods exist to determine the fatigue strength that corresponds to the number of cycles to failure (in this case, the fatigue limit). These tools include the conventional *S–N* test method and quantal response tests. To each of these tools belong some particular methods. For example, quantal response tests may involve the staircase method, the Probit method and so on.

The shape of the curve depends on the considered material and the scale used. For a given material, let's consider a metal alloy, factors that affect the shape of the σ–N curve are: the stress ratio, types of cyclic load (tension, torsion and so on), the material conditions (heat treatment and so on) and environmental conditions such as temperature, corrosion and so on. The endurance limit is also affected by stress concentration, surface finish and so on.

Common values of the threshold number of cycles at which the test specimen remains unbroken (a smooth specimen) are: for steel alloys $\rightarrow N_L = 10^7$; for copper alloys $\rightarrow N_L = 5 \cdot 10^5$; for light metals and their alloys $\rightarrow N_L = 10^8$. Data about a comparison between the fatigue limit (at a mentioned number of cycles) and the ultimate tensile strength for a considered group of materials can be found in [6]. In the case of a completely reversed bending test using a polished specimen, the above ratios may be given as: 0.5 for low and intermediate strength steels and 0.4 for aluminum alloys.

If the curve representing fatigue testing at $R = -1$, under a finite fatigue regime (Figure 10.3) is now displayed as a logarithmic line, then this line will appear as an inclined straight line.

Types of specimens (geometry and dimensions) are defined by standard, for example, ASTM E466-96.

Testing is typically carried out in such a way that, for example, six to eight stress levels are selected and for each of them two to three specimens are used; all the tests are performed at the same stress ratio [1]. Figure 10.4 shows tests for different stress ratios.

The endurance limit is often considered to be a material property. In a case when, for example, a specific stress level for a particular "life" (that is, a particular number of cycles) from the *S–N* curve is of interest, then the term *fatigue strength* is usually used. In engineering practice, two types of fatigue test are usually discussed: *high-cycle fatigue* (low-level fatigue/low stress level) and *low-cycle fatigue* (high-level fatigue/high stress level) [4]. High-cycle fatigue is characterized by a relatively large number of cycles, let's say over 10^4 cycles, and it refers to elastic behavior (stress and strain within the elastic range). Low-cycle fatigue is usually associated with a number of cycles to failure less than 10^4 cycles and is associated with macroscopic deformation in every cycle. In engineering practice, most known data about fatigue behavior of a considered material are based on fatigue tests carried out by applying fully reversed loading tests, that is, on fatigue tests where the stress ratio, $R = -1$, and the mean stress, $\sigma_m = 0$. When data obtained by these

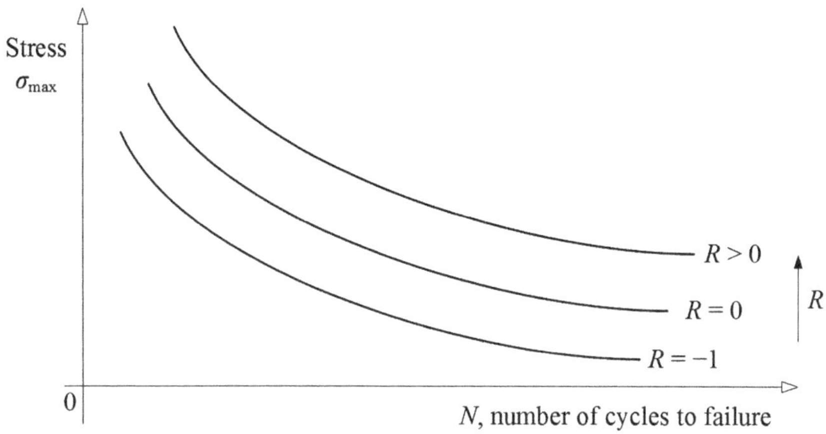

Figure 10.4 A schematic representation of fatigue tests carried out at different stress ratios.

tests are plotted in the $\sigma - N$ system, then the $\sigma - N$ (also known as the S–N) curve is developed. The second known case is $R = 0$.

However, sometimes for a particular purpose it will be very useful to know some data relating to a specific stress ratio, for example, the specific mean stress that corresponds to it. As such, Figure 10.5(a) shows some S–N curves based on fatigue tests carried out at different mean stresses (σ_m). Based on data from S–N curves it is possible to construct a *fatigue diagram* consisting of curves that describe constant fatigue life. The principle of construction of a fatigue diagram is shown in Figure 10.5(b). It is clear that stress amplitude has a larger effect on fatigue than the mean stress. A diagram of the type presented

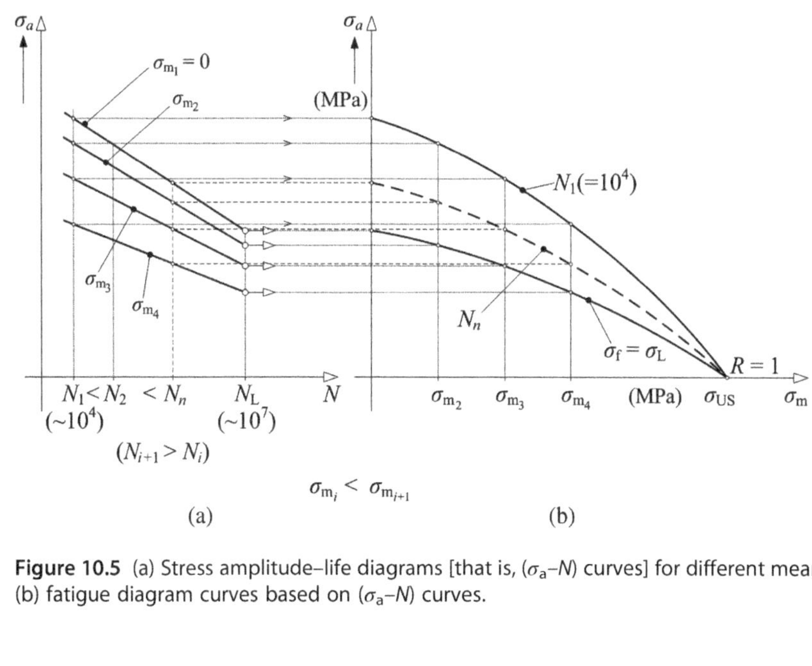

Figure 10.5 (a) Stress amplitude–life diagrams [that is, (σ_a–N) curves] for different mean stresses (σ_m); (b) fatigue diagram curves based on (σ_a–N) curves.

in Figure 10.3 with lines of constant fatigue life (usually constructed for some alloy) is usually referred to as a Goodman diagram. In engineering, similar data can also be presented by a so-called Smith diagram.

Unlike tests that are performed on unnotched specimens, in engineering practice, tests are also performed on notched specimens. Many engineering elements have some discontinuities such as keyways, holes and so on that cause stress to be locally higher. These discontinuities are called *stress raisers* or *notches* and cracks start to propagate in these places. For a notched specimen, that is, a specimen containing a stress raiser, the fatigue strength at long lives (the fatigue limit) is reduced because of the effect of the fatigue notch factor [6].

10.4 Linear Elastic Fracture Mechanics (LEFM)

10.4.1 Basic Consideration

To return to our earlier question of "why" and "how" some engineering element has failed, one common, general answer is the growth of crack(s). The origins and phenomena of failure that lead to the fracture of an element are manifold. As is well known, a crack may be a pre-existing defect in a material or it may be initiated by some known mechanical failure. Loads on the element may cause the appearance of a crack, but may also cause crack propagation.

There are different mechanisms of microcrack formation. Let's look at the mechanism of dislocation at obstacles. In this case, a high stress concentration occurs, which can lead to bond breaking at preferred lattice planes and finally to cleavage that can be transcrystalline or intercrystalline, as was shown earlier in Figure 9.2(b) and (c) and is shown again in Figure 10.6(a) and (b). Dislocation may also cause the formation and coalescence of voids (see Figure 10.6(c)). Cyclic slip is considered to be the cause of the crack [1] and this is shown in Figure 10.6(d).

A fatigue crack can be initiated in an element subjected to repeated load (cyclic load). In general, two phases of crack "life" may be discussed: the initiation phase and the growth phase. The initiation phase refers to crack nucleation and its microgrowth, while the growth phase includes stable and unstable crack propagation periods (see Figure 10.7 (a)). A similar representation is shown (Figure 10.7(b)) on the cross-section of a cylindrical specimen which was subjected to tension and to bending [1].

In the first phase, that is, in nucleation and microgrowth, the microcrack size is of the order of a single grain. Since the grain is of crystalline structure, in the beginning there is a large number of slip systems and any of them may be activated. In this early stage of the formation of a crack, fatigue is a surface phenomenon of the material and this ceases to be the case when the crack penetrates the material and becomes an integral part. Microcrack growth depends on so-called cyclic plasticity, so there is a relationship between threshold cyclic plasticity and crack growth (by slip).

In reality, if we consider materials at a sufficiently small size (microlevel), all of them are anisotropic. Engineering materials, let's consider metals, consist of small grains (crystals) and, due to crystallographic planes within each grain, the behavior is anisotropic.

With respect to the crack growth rate (more commonly alluded to as crack propagation), there are some considerations. Various explanations exist for this phenomenon [2].

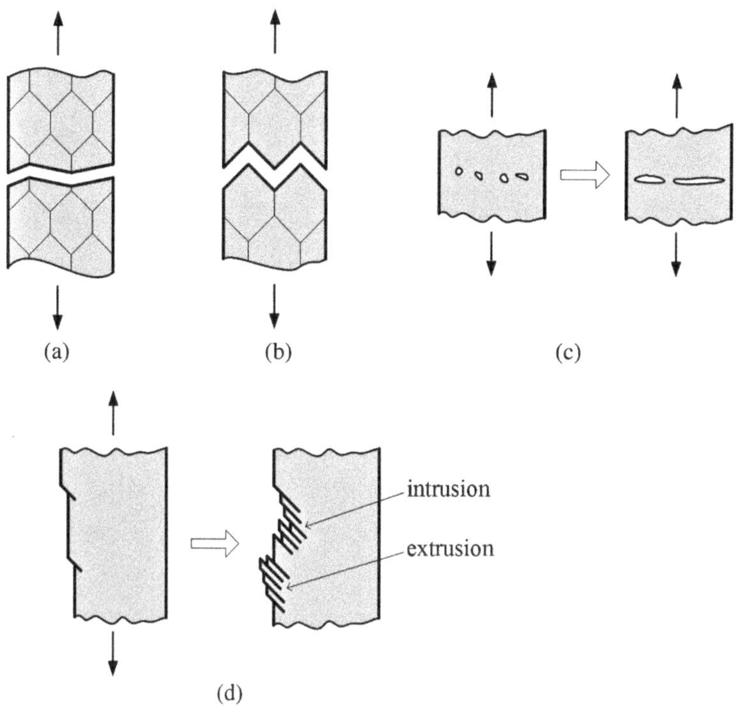

Figure 10.6 Crack formation and fracture. (a) Dislocation mechanism causes transcrystalline cleavage; (b) dislocation mechanism causes intercrystalline cleavage; (c) formation and coalescence of voids; (d) slip process.

If a crack propagates (increases) under constant load at a velocity of 1 mm/s, this is treated as subcrititical. Crack growth under a cyclic load at incremental steps (about 10^{-6} mm/cycle) is called fatigue crack growth. Sometimes the term "crack arrest" is mentioned and this means rapid crack growth, that is, the growth rate is very high.

Cracks as a phenomenon have been studied theoretically and examined experimentally. The causes of the occurrence of cracks and their propagation, stress field developments at the tips of cracks and service life predictions for members possessing a crack based on the crack growth rate have been studied and many of the findings verified experimentally. As mentioned earlier, discontinuities (notches) in structural elements are among the sites where cracks may grow. Comparison in stress field distribution is usually made for two cases: around the discontinuity and at the crack tip. However, an infinite value of stress in a real material does not exist. Thus, for applied stress of an average value, a relatively sharp crack tip accommodates another, less sharp, one from plastic zone development that is caused by yielding. In further development of the crack, the tip becomes blunted, that is, a very small radius at the tip of the crack occurs (the crack opens near its tip by a small amount "δ" – called *crack-tip opening displacement*). The sharp tip of the crack and the theoretical field stress around it as well as the real crack, the plastic zone around the crack tip and the appropriate stress field for a metallic material are presented in Figure 10.8.

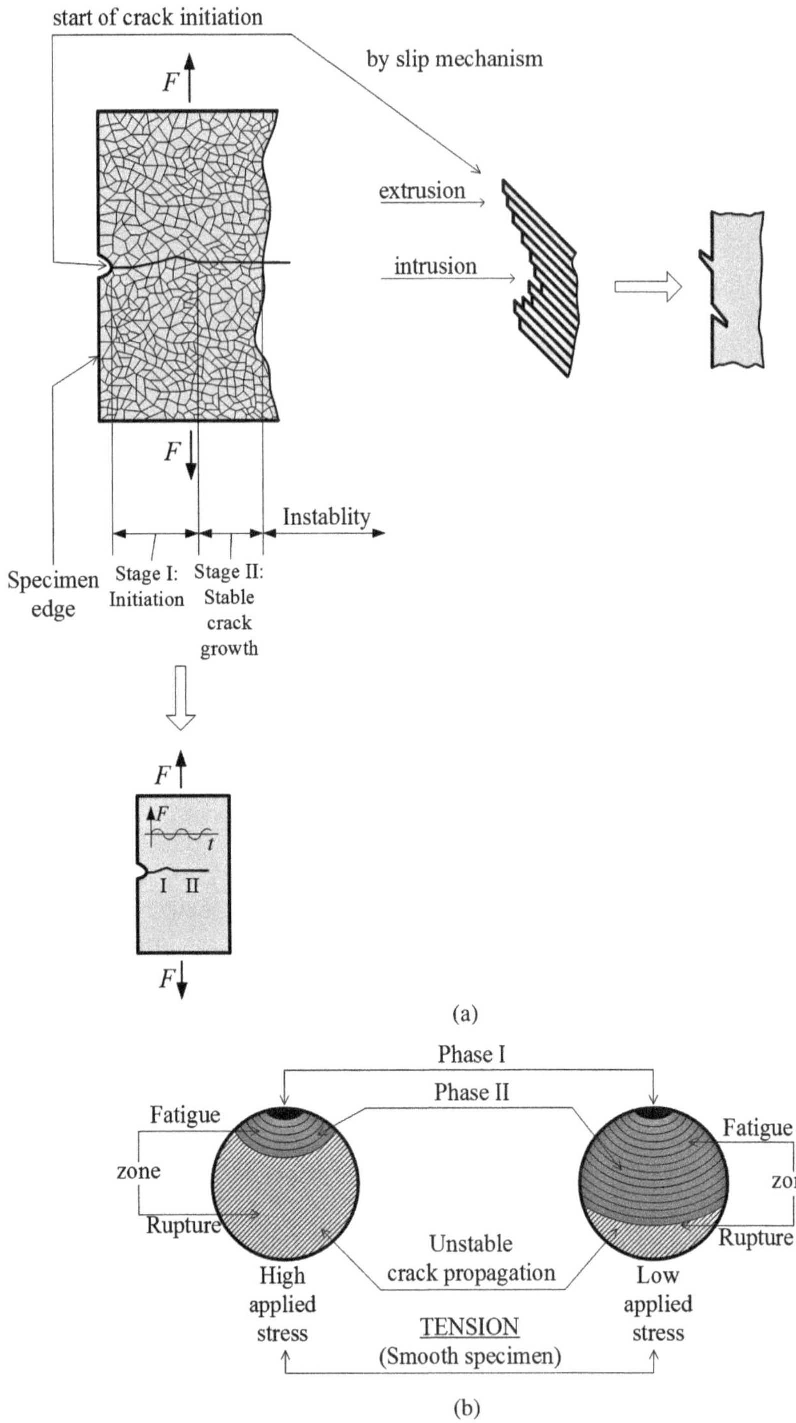

Figure 10.7 A schematic representation of crack growth phases. (a) Crack growth phases; (b) areas on the cross-section of a specimen related to crack phases.

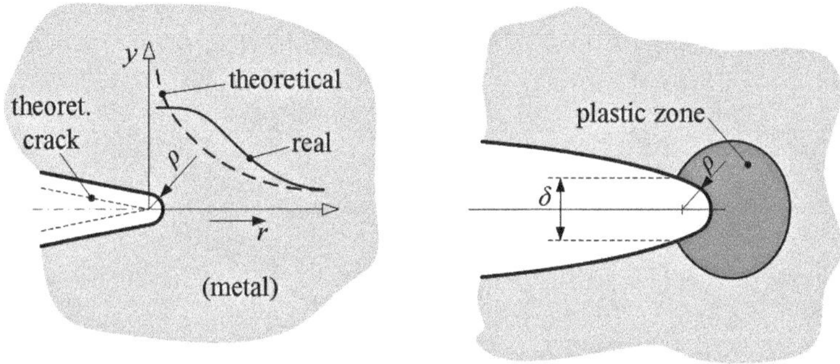

Figure 10.8 Tip of the crack and stress field in materials: theoretical and real.

In the case where the stress applied to an element containing a crack is high, the element may suddenly rupture because of sudden growth of the crack. In this case, only a small amount of plastification occurs; that is, this is a brittle fracture.

10.4.2 Crack Opening Modes

In an engineering structure, a crack may exist or it may be formed over time and propagate further regardless of the thickness of the structure. We mention the thickness of the element here since it is a factor that defines a particular condition at the tip of the crack: the plane strain state or the plane stress state. This problem will be discussed later.

Imagine a body (an engineering element) subjected to stress in which a crack is formed. Consider the behavior of the crack; in other words, consider "deformation" of the crack or, let's say, consider the possible crack-opening modes that indicate crack propagation. The crack "deformation" mode indicates the mode of the applied load that causes the crack to open (that is, crack propagation). As such, the crack "deformation" mode is usually known as the crack opening mode. There are three types of crack opening mode, as shown in Figure 10.9. There also exists a mixed mode.

The crack opening modes are sometimes called: Mode I (*opening* mode), Mode II (*shear*/in-plane shear/edge or forward sliding mode) and Mode III (*tearing*/out-of-plane shear/side sliding/parallel shear mode). Mode I is caused by tensile loading, while the other two modes are caused by shear loading. However, it is known that most materials are more suspectible to fracture by normal stress (produced by tensile loading) than by shear stress.

Most engineering problems relating to cracks involve Mode I. When linear elastic behavior is under consideration, this also implies that force versus strain (displacement) behavior is linear. In the process of crack propagation, the *strain energy release rate* is usually introduced and designated $\mathcal{G}$. In fact, the strain energy release rate introduced by Irwin, as used in this context, does not refer to the derivative of energy with respect to time, but defines the rate of change in potential energy of the body (total potential) with an increase in crack area (surface) during crack propagation; that is, it is a measure of the energy available for an increment of crack extension [8]:

$$\mathcal{G} = -\frac{\mathrm{d}\Pi}{\mathrm{d}A} = -\frac{\mathrm{d}\Pi}{t \cdot \mathrm{d}a}. \tag{10.2}$$

In Equation (10.2): Π is the total potential of the body, $\mathrm{d}A$ is the increase in the surface of the crack, t is the thickness of the specimen and $\mathrm{d}a$ is the increase in the crack whose

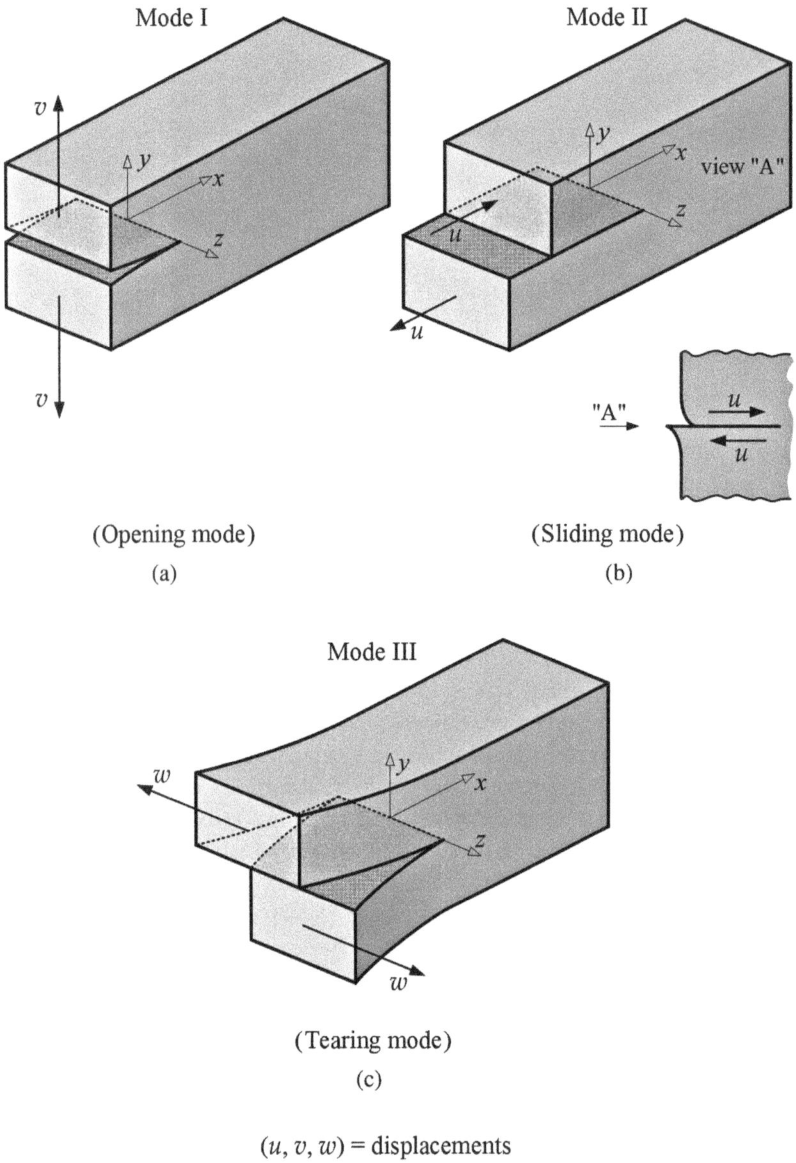

Figure 10.9 Crack opening modes. (a) Mode I (opening mode); (b) Mode II (sliding mode); (c) Mode III (tearing mode).

length previously was a (shown, for example, in Figure 10.12). Since strain energy release follows as a derivative of the potential energy, it is also called the *crack extension force* or the *crack driving force.*

10.4.2.1 Stress Intensity Factor (*K/SIF*)

A notch is a raiser and it is unfavorable (even dangerous) for an engineering member that is subjected to load, since it reduces its carrying capacity. As such, analysis of the stress field around the tip of the raiser is of importance. Although unfavorable, many raisers are necessary in engineering practice and each of them has a certain function. The stress distribution around the tip of the notch shows that the stress concentration is present there and the stress has its peak value. The stress concentration factor (a dimensionless factor), usually designated α_{kt} (or K_t) is defined as the ratio between the maximum stress (peak value) at the tip of the raiser and nominal (average) stress that is calculated without the existence of the raiser.

So why mention the stress concentration factor now? The reason is that the crack tip (or crack front) is also a type of notch (raiser) which has zero tip radius or has a radius of minimal value. By considering an infinite sheet with an elliptical hole [4] subjected to tension, an exact solution referring to the stress concentration factor was obtained. In this case, by decreasing the minor axis to zero, the elliptical hole became the crack (raiser/notch), and since the tip radius of the crack became equal to zero, the stress concentration factor reached an infinite value regardless of the semi-crack length; that is, for any crack length. This result does not offer the possibility of describing the severity of the stress distribution around the crack tip. On the other hand, examination of the photo-elastic model with different cracks shows similar isochromatic pictures around the crack tips, and this suggests similar stress distributions in those places. However, in this case, the stress concentration factor is not an appropriate (meaningful) quantity with which to define the severity of the stress distribution around the crack tip. As stated above, it would become infinite in a case involving a very sharp crack tip, as it is in reality, regardless of the crack length. To define the intensity/severity of the stress distribution around the crack tip, a so-called *stress intensity factor* (designated *SIF* or *K*) is introduced. The stress distribution around the crack tip will be described as a linear function of the stress intensity factor, *K*, as written later in Equation (10.7). It can be seen that, in its calculation, the loading stress, crack length and dimensionless factor are used. As will be discussed later, based on this factor a new material property is derived that defines resistance of the material to crack propagation, while the stress concentration factor is a dimensionless quantity only.

Fatigue crack growth occurs after the crack initiation period. Stresses around the crack tip in the cracked body, as shown in Figure 10.10 (an infinitesimal element of the cracked body around the point A in the vicinity of the crack tip), assuming linear elastic material behavior, may be described by the following equation:

$$\sigma_{ij} = \frac{k}{\sqrt{r}} f_{ij}(\varphi) + \ldots \tag{10.3}$$

where: σ_{ij} is the stress tensor, k is a constant, f_{ij} is a dimensionless function of (φ). Angle φ is defined as shown in Figure 10.10. It is clear that a polar coordinate system is used.

Since the leading term in Equation (10.3) depends on the factor ($1/\sqrt{r}$), stress in the vicinity of the crack tip varies with it and is proportional to it. Each type of loading

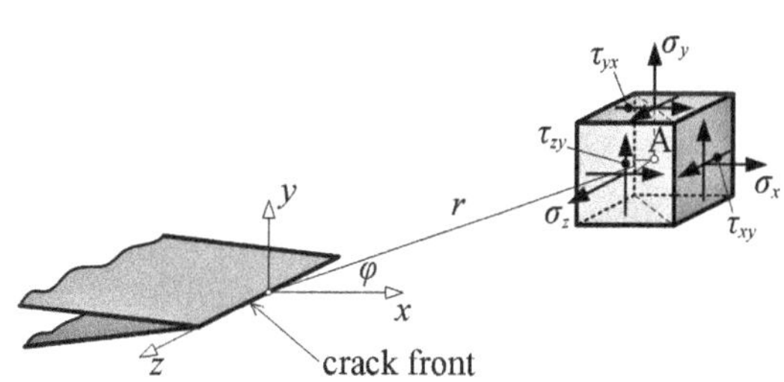

Figure 10.10 Stress field around the crack tip.

(that is, each type of crack opening mode) produces $1/\sqrt{r}$ singularity at the tip of the crack. However, singularity occurs in the stress distribution at the crack tip since all stress components tend to infinity for $r \to 0$ for each angle φ. Other high-order terms in Equation (10.3) that occur after the (+) sign, that is, stress components parallel to the crack (those relating to x and z) such as σ_{x0}, τ_{xz0}, as well as terms of the order $[O]r^{1/2}$ are usually neglected in comparison with the leading term, which is dominant in a small region surrounding the crack tip. This region is the so-called *K-dominant region*, where K is the stress intensity factor. As stated, based on Equation (10.3), it is clear that when $r \to 0$, this produces singularity; that is, stress becomes asymptotic for $r = 0$. Constant k and function f_{ij} depend on the crack opening mode. As such, it is of interest to introduce the so-called stress intensity factor in the form: $K = \mathrm{k}\sqrt{2\pi}$, and then three stress intensity factors, K_I, K_{II} and K_{III}, related to the three opening modes may be distinguished. (The subscript here refers to the type of opening mode.) Modes II and III are treated as less important for engineering practice, and generally for testing. In accordance with this, using the above-mentioned value for K as well as Equation (10.3), the stress field ahead of the crack tip in an isotropic linear elastic material, for a Mode I opening mode, can be written as:

$$\sigma_{ij}^{I} = \frac{K_I}{\sqrt{2\pi r}} f_{ij}(\varphi). \tag{10.4}$$

Without displaying the principles of the determination (the basis is the theory of linear elasticity), for a Mode I crack opening mode, the stress components in the crack-tip stress field (the region surrounding the crack tip) may be written in the form [6–14]:

$$\begin{aligned}
\sigma_x &= \frac{K_I}{\sqrt{2\pi r}} \cos\frac{\varphi}{2}\left[1 - \sin\frac{\varphi}{2}\sin\frac{3\varphi}{2}\right] \\
\sigma_y &= \frac{K_I}{\sqrt{2\pi r}} \cos\frac{\varphi}{2}\left[1 + \sin\frac{\varphi}{2}\sin\frac{3\varphi}{2}\right] \\
\tau_{xy} &= \frac{K_I}{\sqrt{2\pi r}} \sin\frac{\varphi}{2}\cos\frac{\varphi}{2}\cos\frac{3\varphi}{2} \\
\sigma_z &= 0 \,(\text{Plane stress conditions}) \\
\sigma_z &= \nu(\sigma_x + \sigma_y)\ldots.(\text{Plane strain conditions}: \varepsilon_z = 0) \\
\tau_{zx} &= \tau_{zy} = 0
\end{aligned} \tag{10.5a}$$

For Mode II (shear mode or forward sliding mode), in the crack tip stress field there are the following stress components:

$$\begin{aligned}
\sigma_x &= -\frac{K_{II}}{\sqrt{2\pi r}} \sin\frac{\varphi}{2}\left[2 + \cos\frac{\varphi}{2}\cos\frac{3\varphi}{2}\right] \\
\sigma_y &= \frac{K_{II}}{\sqrt{2\pi r}} \sin\frac{\varphi}{2}\cos\frac{\varphi}{2}\cos\frac{3\varphi}{2} \\
\tau_{xy} &= \frac{K_{II}}{\sqrt{2\pi r}} \cos\frac{\varphi}{2}\left[1 - \sin\frac{\varphi}{2}\sin\frac{3\varphi}{2}\right] \\
\sigma_z &= 0\ldots..(\text{Plane stress conditions}); \\
\sigma_z &= \nu(\sigma_x + \sigma_y)\ldots.(\text{Plane strain conditions}: \varepsilon_z = 0) \\
\tau_{zx} &= \tau_{zy} = 0.
\end{aligned} \tag{10.5b}$$

For Mode III (tearing or side sliding mode), in the crack tip stress field there are the following stress components:

$$\begin{aligned} \tau_{xz} &= \frac{K_{\mathrm{III}}}{\sqrt{2\pi r}} \sin\frac{\varphi}{2} \\ \tau_{yz} &= \frac{K_{\mathrm{III}}}{\sqrt{2\pi r}} \cos\frac{\varphi}{2} \\ \sigma_x &= \sigma_y = \sigma_z = \tau_{xy} = 0 \end{aligned} \tag{10.5c}$$

In general, the stress intensity factor K, designated K_{I} when Mode I is being considered, defines the intensity of the stress around (in the vicinity of) the sharp crack tip and it is a measure of the severity of the stress distribution around the crack tip. It defines the crack tip conditions. It may also be said that it represents the stress field in this area and in the physical sense this means the intensity of the load transmittal through the crack tip. When $\varphi = 0$, the shear stress is zero and the crack plane (the x–z plane when the z axis is a crack front) in Figure 10.11 or Figure 10.15 becomes a principal plane, for example, for Mode I. In this case, in accordance with Equation (10.5a), we have:

$$\sigma_x = \sigma_y = \frac{K_{\mathrm{I}}}{\sqrt{2\pi r}}, \tag{10.6}$$

and this value, when displayed as a curve, represents the σ_y stress distribution (perpendicular to the crack plane) at the tip of the crack. In accordance with Equation (10.4), the dimension of the stress intensity factor is: stress$\cdot\sqrt{\text{length}} \to \text{MPa}\sqrt{\text{m}}$, $(\text{MPa}\sqrt{\text{m}} = 31.62\,\text{Nmm}^{-3/2})$. Consider a semi-infinite, tensile-loaded (σ) plate (Figure 10.11) that contains a crack of length $2a$ through its thickness, and dimensional analysis of the quantities in Equations (10.5). It is convenient for the stress intensity factor to use the expression in the form:

$$K = \beta\sigma\sqrt{\pi a} \tag{10.7}$$

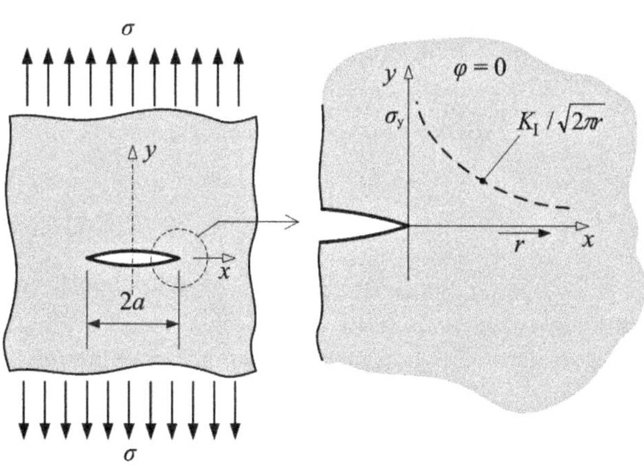

Figure 10.11 Defining the form of the stress intensity factor.

where σ is the remote loading stress, a is the crack length, β is a dimensionless factor/function that depends on the type of loading configuration and the geometry of the specimen/structural element; that is, $\beta = \beta$ (loading configuration, geometry, a/b) [6, 12]. See, for example, Figure 10.12.

Closed form Solutions for the Stress Intensity Factor

Some of the most common solutions [6, 12–13] that are used for the stress intensity factor, for the cases shown in Figure 10.12, are displayed below each part of the figure. For all shown cases:

$$K_{I_a} = \beta_{Ia}\sigma\sqrt{\pi a},\ (\sigma = \sigma_a,\ \text{a...axial}) \tag{10.8}$$

The subscript "I" refers to the opening mode (that is, Mode I). Subscript "a" refers to the normal stress that is a consequence of the axial load, which, in fact, causes Mode I. In Figure 10.12, the subscript "a" that refers to the load is omitted. In engineering applications, cracks can be found with special shapes, such as circles, semi-circles and so on [4, 6]. An embedded circular crack (internal crack) is shown in Figure 10.12(d). This is a penny-shaped crack in a solid under tension. If the solid (member) is considered to be of finite size, then the given solution is valid when $\left(\frac{a}{t} < 0.5,\ \frac{a}{b} < 0.5\right)$.

In the above-mentioned references [6, 12], factor β (from Equation (10.9)), is presented in the form of a curve: $\beta = f\ (a/b)$. In [12], a similar curve is also given for the case in Figure 10.12(b) when, instead of stress, a bending moment is applied. It should be noted that the presented value of the factor β is defined for a small ratio a/b while for higher ratios, this constant (factor β) is additionally a function of the ratio h/b.

Effect of Assembled Loads

For linear elastic materials (that is, in the field of linear elastic behavior of materials), the basic assumption of the superposition of the individual components of strain, stress and displacement can be applied. Similarly, when the total (combined) stress intensity factor induced by several types of loads that cause the same type of opening mode is considered, the same assumption can be applied. A very simple example, which represents a load that is a combination of an axial load and a bending moment that simultaneously induce Mode I, is presented in Figure 10.13.

As such, for example, considering Mode I deformation at the crack tip, we may write [1, 6]:

$$K_{i,\ \text{tot}} = \sum K_{i_j} \rightarrow K_{\text{I,tot}} = K_{\text{I}_a} + K_{\text{I}_b} + \ldots \tag{10.9a}$$

In Equation (10.9a), i refers to the crack opening mode (for example, Mode I) and j to the applied load that produces the same opening mode (for example, axial load, bending). This means that stress intensity factors are additive as long as the crack opening mode is the same. Also, we may write:

$$K_{\text{tot}} \neq K_{\text{I}} + K_{\text{II}} + K_{\text{III}} \tag{10.9b}$$

and this means that normal and shear stresses cannot be summed.

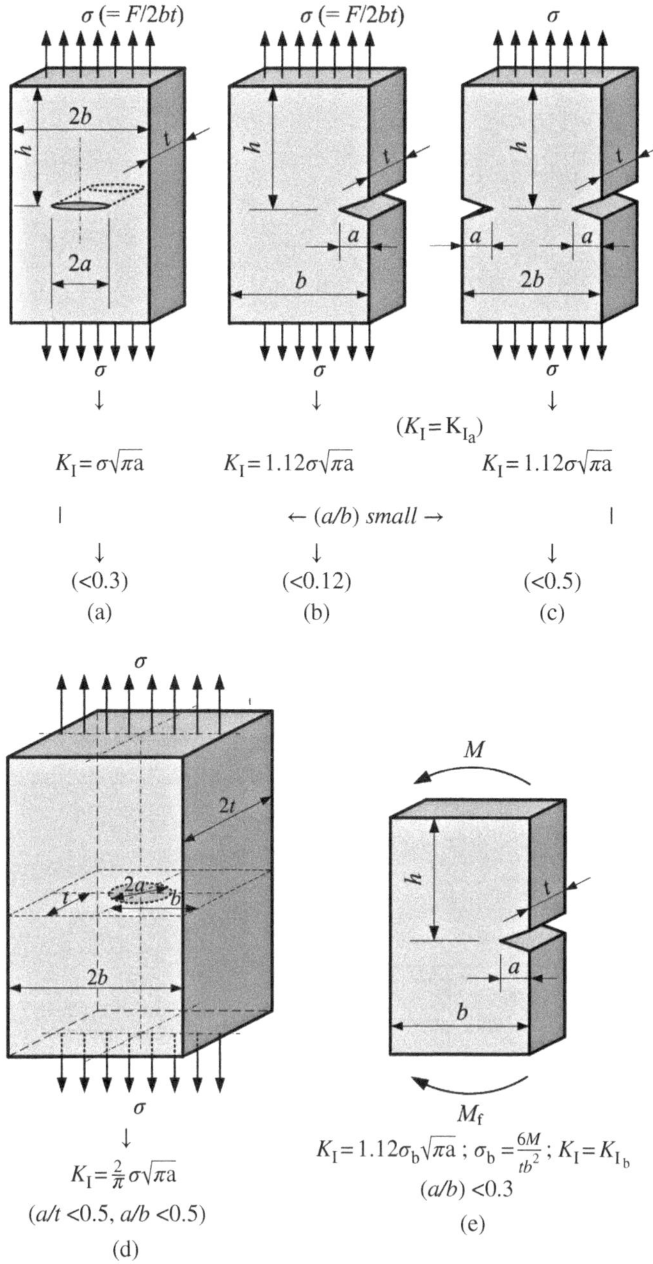

Figure 10.12 Solutions for the stress intensity factor in closed form. (a) A through-cracked plate (specimen) – tensile load; (b) edge-cracked plate (specimen) – tensile load; (c) double-edge-cracked plate (specimen) – tensile load; (d) penny-shaped circular crack embedded in a solid – tensile load; (e) edge-cracked plate (specimen) – bending moment.

Based on Equation (10.9a), for the case presented in Figure 10.13(a), it follows that

(a)

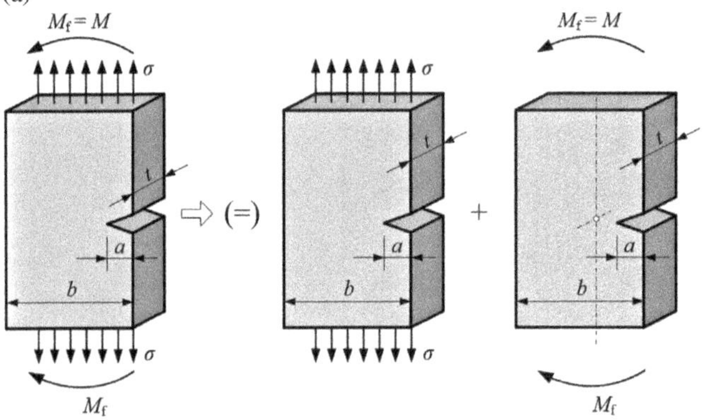

↓ (axial + moment) = axial (tension) + moment (bending)

$$K_{I_{tot}} = K_{I_a} + K_{I_b} \quad (10.10)$$

$$K_{I_{tot}} = \beta_{I_a}\,\sigma_a\sqrt{\pi a} + \beta_{I_b}\,\sigma_b\sqrt{\pi a} \quad (10.10a)$$

$$\sigma = \sigma_a\,;\ \sigma_b = \frac{6M}{tb^2} \quad (10.10b)$$

Figure 10.13(a) Cracked plate with a single edge through-the-thickness crack simultaneously subjected to both tensile loading and pure bending. Tensile load plus bending moment.

Subscript "b" refers to the bending moment that causes normal stresses which induce Mode I. Factors $(\beta_{Ia}, \beta_{Ib}) = f(\text{loading}, a/b)$ can be found in the form of a curve in [6, 12].

For the case presented in Figure 10.13(b), it follows that:

(b)

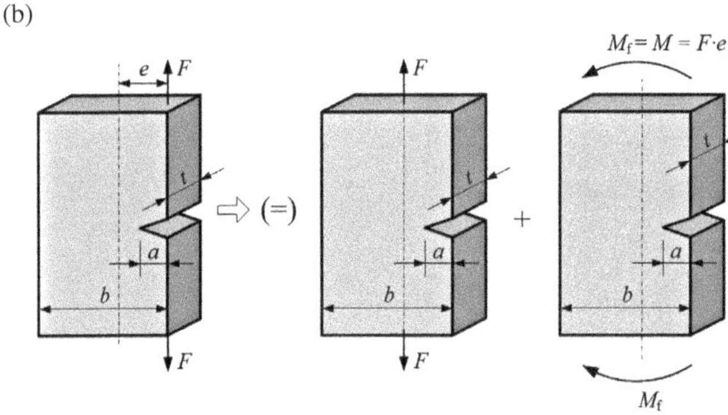

↓ (Eccentric tensile load) = axial (tension) + moment (bending)

$$K_{I_{tot}} = K_{I_a} + K_{I_b}$$

$$K_{I_{tot}} = \beta_{I_a}\,\sigma_a\sqrt{\pi a} + \beta_{I_b}\,\sigma_b\sqrt{\pi a} \quad (10.10c)$$

$$\sigma = \sigma_a = \frac{F}{bt};\quad \sigma_b = \frac{6M}{tb^2};\quad M = F \cdot e \quad (10.10d)$$

Figure 10.13(b) Cracked plate with a single edge through-the-thickness crack simultaneously subjected to both tensile loading and pure bending. Eccentric load (combination of axial load plus bending moment).

Critical Stress Intensity Factor

As discussed, three different types of crack opening, that is, three different crack opening modes or crack deformation modes, are recognized: Mode I, Mode II and Mode III. Each of these modes is associated with the corresponding stress intensity factor: K_I, K_{II} or K_{III}.

Fracture of a structural element may occur when any of these stress intensity factors reaches its critical value, that is, K_{I_c}, K_{II_c} or K_{III_c}. Let, for example, the critical value of the stress intensity factor (the *critical stress intensity factor*) be denoted K_c and assume we are dealing with Mode I. K_c generally varies with the loading mode (the type of opening mode), so the mentioned critical values are mutually different. Engineering materials are more susceptible to fracture by normal stress; that is, Mode I is of practical importance and consequently K_{Ic} needs to be considered. Solutions in fracture mechanics that relate to fracture failure of a cracked element are usually reduced to two plane-dominated states at the crack tip: plane stress and plane strain conditions. For the considered material, investigations indicate that the critical stress intensity factor, K_c, decreases as the thickness of the specimen increases and tends to the limiting (minimum) value that is reached when the state of plane strain at the tip of the crack is ensured (see Figure 10.14). A test specimen with "small" thickness (compared to the plastic zone size) corresponds to plane stress fracture conditions at the tip of the crack. When the thickness of the specimen reaches a sufficient plateau that ensures plane strain conditions at the tip of the crack, further increases in thickness of the specimen do not cause any change in the level/value of the K_I stress intensity factor. This level of the stress intensity factor that remains practically constant in spite of an increase in the thickness of the specimen is the critical value of the stress intensity factor, K_I and it is associated with plane strain conditions at the crack tip. It is known as the *plane strain fracture toughness* and is usually denoted K_{Ic}; as already noted, it is considered thickness independent.

It should be noted that the term "toughness" was mentioned in Chapter 4 (Section 4.5.1.1) in the context of *tensile toughness*. The definition and an explanation of the meaning of tensile toughness are given there.

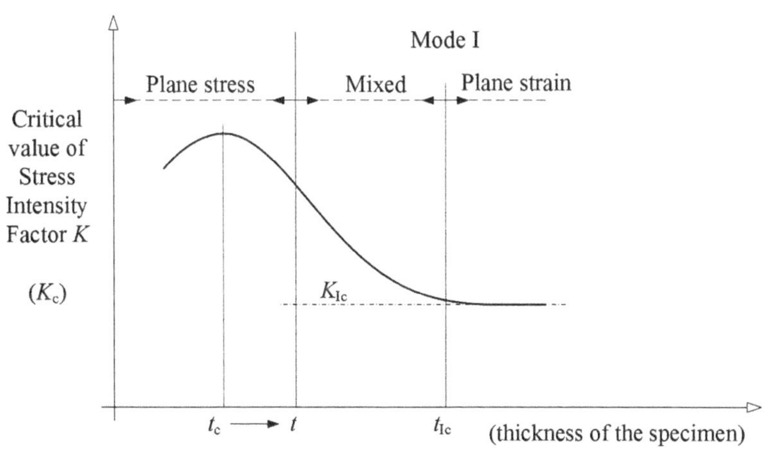

Figure 10.14 Critical stress intensity factor (K_c) versus thickness of the specimen.

Critical values of K_I for specimen thicknesses less than those corresponding to the plane strain constraint are not known as K_{Ic} and may be designated as K_c, although this designation may cause some confusion because of other modes. In any case, fracture toughness that may be considered at the plane strain or plane stress state at the crack tip is defined as the resistance of the material to unstable crack growth; that is, the material's resistance to crack propagation. Usually, the thickness of the specimen associated with plane strain conditions is also designated t_{Ic} and that associated with plane stress conditions as t_c. The relationship between the stress intensity factor, K_I, and the critical value, K_{Ic} (the fracture toughness or critical stress intensity factor for Mode I), is similar to the relationship between stress, σ and the ultimate tensile strength, denoted by σ_{US} (or sometimes σ_m). Therefore, K_I represents the "stress intensity" at the tip of the crack and K_{Ic} represents the highest value of stress intensity under plane strain conditions. If the thickness of the specimen does not meet the plane strain conditions, the critical stress intensity factor, K_c, is usually associated with plane stress conditions. In this case, the critical stress intensity factor, K_c, for failure prediction under plane stress conditions may be estimated using the following equations (Irwin's equation) [1, 12]:

$$K_c = K_{Ic}\left[1 + \frac{1.4}{t^2}\left(\frac{K_{Ic}}{\sigma_{YS}}\right)^4\right]^{1/2}; \quad K_c = K_{Ic}\left[1 + \frac{1.4}{t}\left(\frac{K_{Ic}}{\sigma_{YS}}\right)^2\right] \tag{10.11}$$

In Equations (10.11), t is the thickness of the specimen (see, for example, Figure 10.12), while σ_Y is the yield strength (variously denoted by σ_{YS} or σ_Y or by $\sigma_{0.2}$). As mentioned in [1], K_c is thickness and crack length dependent. Assessment of K_c based on K_{Ic} is quite conservative and it may be possible to use it only as an orientation value.

Fracture is predicted to occur when (from Equation (10.7)):

$$\beta\sigma\sqrt{\pi a} \geq K_c. \tag{10.12}$$

As has been stated, of greatest interest is the critical stress intensity factor that relates to Mode I, and this is denoted K_{Ic}. Standards governing the determination of K_{Ic} set specific requirements for the specimen sizes to obtain a valid result in metals. One such requirement is:

$$t \geq 2.5\left(\frac{K_{Ic}}{\sigma_{YS}}\right)^2. \tag{10.13}$$

The requirement defined in Equation (10.13) means that the specimen must be thick enough to ensure plane strain conditions at the tip of the crack.

10.4.2.2 Plastic Zone Size around the Crack Tip

Although theoretical considerations in linear stress analysis in the case of a sharp crack indicate that stress may be infinite at the crack tip, the behavior of real materials indicates that stress at the crack tip is of a finite level (see, for example, Figure 10.8). Two facts contribute to this: the crack tip has a finite radius, and during the deformation process, a plastic zone is formed that yields to stress relaxation. This means that there is some discrepancy between linear elastic fracture mechanics based on linear stress analysis, which practically ignores the existence of the plastic zone around the crack tip (or assumes it to be very small) and reality. This discrepancy can be corrected by taking into account the existence of the small plastic zone at the crack tip.

In the case of significant plastification around the crack tip, an alternative parameter must be applied. As stated, linear elastic fracture mechanics starts from the view that, eventually, a very small plastic zone (r_p) at the crack tip exists compared to the K-dominated region (Figure 10.15), or we may describe this as small-scale yielding. In a simple version, we assume that the yielding (that is, the plastic zone) is without significant impact. However, if we consider a sufficiently small plastic zone (r_p), Equations (10.5a), representing the elastic stress field, may be applied in the region outside of this very small plastic zone where the so-called K_I-dominated zone (or K_I field) is. This K_I-dominated zone is represented by r_e, and Equations (10.5a) may represent the stress field until $r = r_e$. The existence of this zone is important for the application of the LEFM.

The normal stress, σ_y, on the crack plane is given by Equation (10.6). The first assessment of the size of the plastic zone at the crack tip was made by Irwin (Irwin's approach) in his first approach on the basis that the boundary between the elastic and the plastic zone occurs when stress σ_y, given by Equation (10.6) satisfies a yield criterion. As mentioned above and shown in Figure 10.11, the stress σ_y on the crack plane ($\varphi = 0$) at the crack tip is [1, 8–9, 11–18]:

$$\sigma_y = \frac{K_I}{\sqrt{2\pi r}}. \tag{10.14}$$

If we consider that, in a case involving *plane stress* conditions, yielding occurs when stress σ_y reaches yield stress level σ_Y of uniaxial stress, that is, $\sigma_y = \sigma_Y$ then, from Equation (10.14), the plastic zone ($r_y = r_p$) may be estimated (see Figure 10.15) as:

$$r_p = \frac{1}{2\pi}\left(\frac{K_I}{\sigma_{YS}}\right)^2. \quad (\varphi = 0°) \tag{10.15}$$

In this case, the stress distribution in the area $r \le r_p$ can be represented by a horizontal line ($\sigma_Y - \sigma_y$), as illustrated in Figure 10.15.

This analysis does not seem to be entirely correct. It is based on an elastic crack tip. After yielding occurs, in order to satisfy equilibrium, the stress must be redistributed. However, the stress distribution in the area above the estimated plastic region (the area above the horizontal line σ_Y and between the vertical axis, σ_y, and the curve representing the elastic stress distribution, σ_y) that was obtained on the basis of an elastic stress distribution cannot be ignored. As such, Irwin's second approach states that the mentioned elastic stress distribution around the crack tip above the estimated plastic zone needs to be replaced by an elastic–plastic stress distribution [8]. This consideration leads to a plastic zone that must be larger – the plastic zone must increase since stress must be redistributed after yielding occurs in order to satisfy the equilibrium condition. In other words, we must have:

$$\sigma_Y \cdot R_p = \int_{r_p} \frac{K_I}{\sqrt{2\pi r}} dr. \tag{10.16}$$

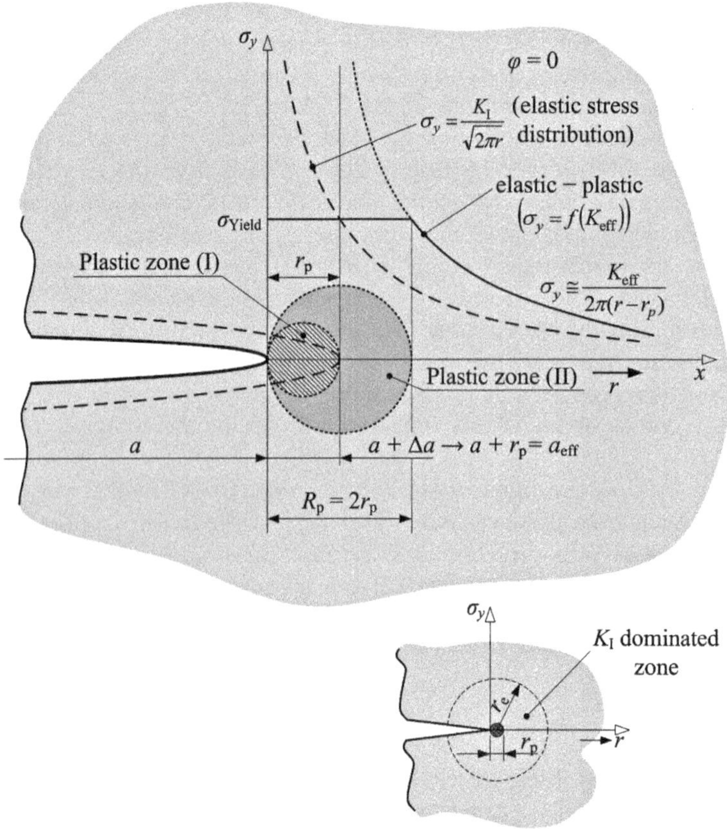

Figure 10.15 Crack tip plasticity (elastic and inelastic stress distribution around the crack tip).

Solving this equation with respect to R_p, it follows that:

$$R_p = \frac{1}{\pi}\left(\frac{K_I}{\sigma_{YS}}\right)^2, \left(R_p = 2r_p\right) \left(\varphi = 0^\circ\right), \tag{10.17}$$

which is the new estimation of the plastic zone size (Figure 10.15) relating to plane stress. In this case, the so-called effective stress intensity factor, K_{eff}, which corresponds to the effective crack length ($a_{eff} = a + \Delta a = a + r_p$) can be considered. Since the opening mode (Mode I), especially under *plane strain* conditions, is significant, it is necessary to define an assessment of the plastic zone size at the crack tip under such conditions. It is known that the triaxial stress state (three-dimensional stress state/multiaxial stress state) corresponds to the plane strain state. Von Mises's yielding criterion (octahedral shear stress yield criterion, distortional energy density criterion) corresponds to the multiaxial stress state and is (from Equation (5.15)):

$$(\sigma_1 - \sigma_2)^2 + (\sigma_2 - \sigma_3)^2 + (\sigma_3 - \sigma_1)^2 = 2\sigma_{YS}^2. \tag{a}$$

When considering Mode I, in the case of plane strain conditions, the stress components in the crack tip stress field are defined by Equations (10.5a). If the principal plane (for example, the crack plane, $\varphi = 0$) is considered (Figure 10.11), then we have:

$$\sigma_1 = \sigma_y, \sigma_2 = \sigma_x, \sigma_3 = \sigma_z = \nu(\sigma_1 + \sigma_2), \sigma_x = \sigma_y. \tag{10.18}$$

If the values listed in Equations (10.18) are entered into Equation (a), and Poisson's ratio is adopted as $\nu = 0.3$, it follows that:

$$\begin{aligned}
&0 + [\sigma_x - \nu(\sigma_x + \sigma_x)]^2 + [\nu(\sigma_x + \sigma_x) - \sigma_x]^2 = 2\sigma_{YS}^2 \\
&[\sigma_x(1 - 2\nu)]^2 + [\sigma_x(2\nu - 1)]^2 = 2\sigma_{YS}^2 \\
&\sigma_x = \sqrt{6.25}\,\sigma_{YS} \\
&\sigma_x = \sigma_y = 2.5\sigma_{YS}.
\end{aligned} \tag{10.19}$$

The size of the plastic zone determined on the basis of Equation (10.14), when the stress that causes yielding (defined by Equation (10.19)) is entered into it, is no longer appropriate in the opinion of Irwin. The constraint on lateral contraction introduced by the plane strain condition, for example, by a triaxial state of stress, elevates the stress that is required to initiate yielding. This means that a triaxial stress state suppresses the plastic zone size and the plastic zone radius is smaller. As such, Irwin estimated that the cause of the yielding may be defined as [6]:

$$\sigma_y = \sqrt{3}\,\sigma_{YS}. \tag{10.20}$$

On the basis of this value, the plastic zone size (from Equation (10.14)) is now:

$$r_p = \frac{1}{6\pi}\left(\frac{K_I}{\sigma_{YS}}\right)^2. \tag{10.21}$$

10.4.2.3 Plastic Zone Shape around the Crack Tip

In the previous section, our consideration was directed to defining the size of the plastic zone in the principal plane, that is, in the plane of the crack ($\varphi = 0$) in accordance with Irwin's approach. In accordance with this, the plastic zone is designed under the assumption that it is of circular shape, taking only the size r_p (or R_p) in the direction of crack propagation. However, the shape of the plastic zone around the crack tip may be designed depending on the yield criterion and taking the considered size as a function of the angle φ. As such, the plastic zone can be discussed in terms of the most common yield criteria such as the von Mises and Tresca criteria. If the von Mises criterion is under consideration, then the following size and shape of the plastic zone relating to Mode I may be listed.

Von Mises Criterion

The von Mises criterion is defined by Equation (5.15):

$$(\sigma_1 - \sigma_2)^2 + (\sigma_2 - \sigma_3)^2 + (\sigma_3 - \sigma_1)^2 = 2\sigma_e^2 \tag{a}$$

where σ_e is equivalent stress, while σ_1, σ_2 and σ_3 are the principal stresses, which, in this case, refer to the stress components shown in Equations (10.5a). Yielding occurs when equivalent stress exceeds the yield stress under a uniaxial stress state.

Plane Stress Conditions The principal stresses in a biaxial stress state are (from Equation (2.21)):

$$\sigma_{1,2} = \frac{1}{2}\left(\sigma_x + \sigma_y\right) \pm \frac{1}{2}\sqrt{\left(\sigma_x - \sigma_y\right)^2 + 4\tau_{xy}^2} \tag{b}$$

Note that:

$$\sigma_z = 0\ldots\ldots(\text{Plane stress conditions});$$

$$\sigma_z = \nu\left(\sigma_x + \sigma_y\right)\ldots.(\text{Plane strain conditions}: \ \varepsilon_z = 0)$$

$$\tau_{zx} = \tau_{zy} = 0.$$

If components σ_x, σ_y and τ_{xy} defined by Equation (10.5a) are entered into Equation (b), then the principal stresses become [1, 18]:

$$\sigma_1 = \frac{K_I}{\sqrt{2\pi r}}\cos\frac{\varphi}{2}\left(1 + \sin\frac{\varphi}{2}\right) \tag{10.22a}$$

$$\sigma_2 = \frac{K_I}{\sqrt{2\pi r}}\cos\frac{\varphi}{2}\left(1 - \sin\frac{\varphi}{2}\right) \tag{10.22b}$$

$\sigma_3 = 0$ (plane stress)

$$\text{and}\,\sigma_3 = \frac{2\nu K_I}{\sqrt{2\pi r}}\cos\frac{\varphi}{2}\ (\text{plane strain}) \tag{10.22c}$$

Inserting the values of the principal stresses defined by Equations (10.22) into the von Mises criterion (Equation (a)) and taking $\sigma_e = \sigma_Y$, $\sigma_3 = 0$ (for plane stress), the plastic zone size (r_p) as a function of φ can be written in the form:

$$r_p(\varphi) = \frac{1}{4\pi}\left(\frac{K_I}{\sigma_{YS}}\right)^2\left(1 + \cos\varphi + \frac{3}{2}\sin^2\varphi\right). \tag{10.23}$$

Comparing this result with that obtained in Equation (10.17), that is, with R_p, which also relates to the plane stress state, it follows that there is a dimensionless ratio [1, 8, 18–19]:

$$\frac{r_p(\varphi)}{R_p(\varphi = 0)} = \frac{1}{4}\left(1 + \cos\varphi + \frac{3}{2}\sin^2\varphi\right). \tag{10.24}$$

This dimensionless ratio is shown in Figure 10.16(a), and, for example, when $\varphi = 0°$ (the crack plane), it amounts to 0.5. In these considerations, the angle φ is in degrees.

Plane Strain Conditions Taking into consideration the values given by Equation (10.18), and using a similar procedure to the one used above, in the case of plane strain state conditions, the size of the plastic zone can be obtained as:

$$r_p(\varphi) = \frac{1}{4\pi}\left(\frac{K_I}{\sigma_{YS}}\right)^2\left[(1 - 2\nu)^2(1 + \cos\varphi) + 3/2\,\sin^2\varphi\right]. \tag{10.25}$$

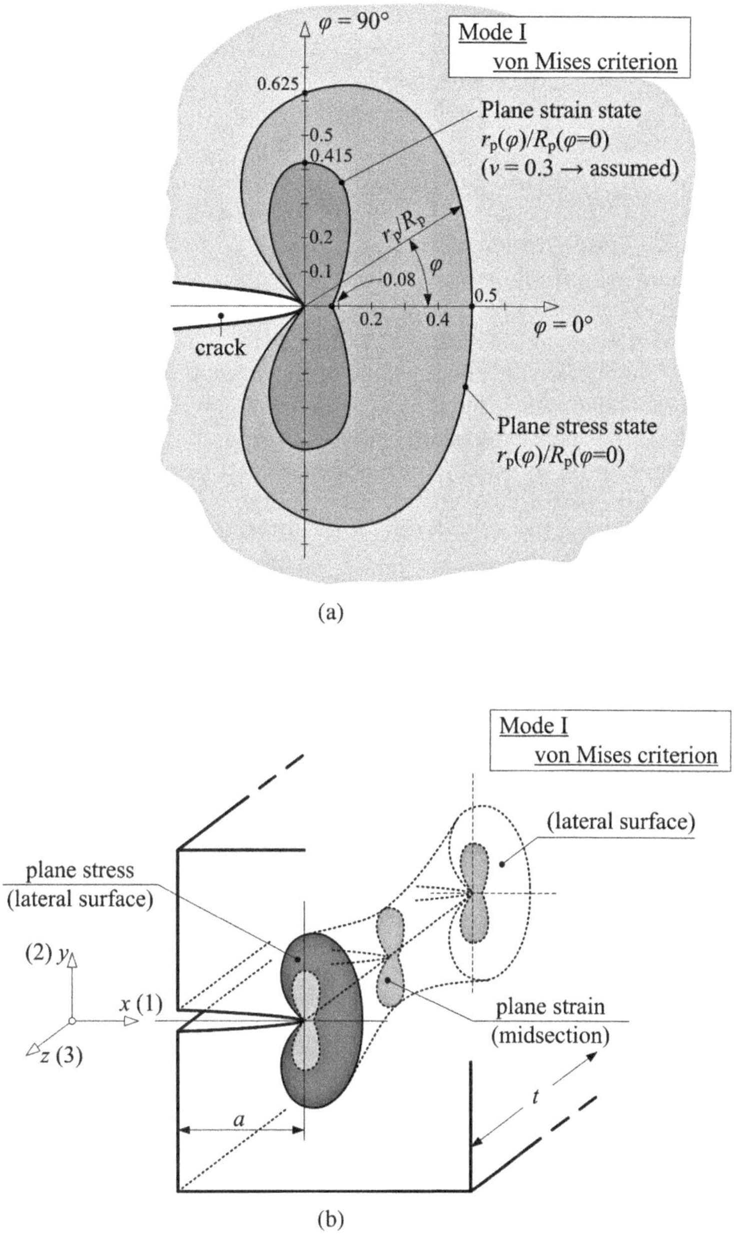

Figure 10.16 Plastic zone shape around the crack tip in the case of a plane stress state and in the case of a plane strain state corresponding to Mode I and the von Mises criterion, and transition of the plastic zone shape through the thickness of the specimen. (a) The shape of the plastic zone around the crack tip (Mode I, von Mises); (b) transition of the plastic zone from the lateral surface to the middle of the specimen.

Comparing this result with that obtained by Equation (10.17), the dimensionless ratio/quantity is:

$$\frac{r_p(\varphi)}{R_p(\varphi=0)}=\frac{1}{4}\left[(1-2\nu)^2(1+\cos\varphi)+\frac{3}{2}\sin^2\varphi\right]. \tag{10.26}$$

This dimensionless ratio is also shown in Figure 10.16(a) and, for the crack plane ($\varphi = 0°$), it amounts to 0.08 if Poisson's ratio is assumed to be 0.3. A similar analysis could also be drawn up in the case of the Tresca criterion.

Although the concept and the definition of fracture toughness have been mentioned several times so far, as a new material property, fracture toughness deserves special attention since it is very important in terms of fracture mitigation in the design of structures. As discussed, the fracture toughness (the critical value of the stress intensity factor) is dependent on the thickness of the wall (specimen), as shown in Figure 10.14. The shape and size of the plastic zone at the crack tip depend on whether plane strain or plane stress (or mixed) conditions exist at that point; that is, the type of the stress field at the crack tip defines the shape and size of the plastic zone. Plane stress conditions are associated with thin sections, while plane strain conditions are associated with thick sections. However, in thick sections, although fracture toughness is treated as being independent of thickness, plane stress conditions exist on the edge surfaces and plane strain conditions exist in the interior. The variation in the plastic zone's shape and size according to the thickness of the specimen (the specimen is thick enough to ensure plane strain conditions) based on crack opening Mode I and for the von Mises yield criterion is illustrated in Figure 10.16(b). As can be seen, the plane strain state ($\varepsilon_3 \sim 0$) dominates in the middle of the specimen and the plane stress state dominates at the lateral (edge) surfaces ($\sigma_3 \sim 0$). This means that the shape and size of the plastic zone correspond to the plane strain state in the middle section, while the shape and size of the plastic zone correspond to the plane stress state at the edges. There is a gradual transition between these two states. In accordance with the form of the gradual transition from the plane stress state to the plane strain state and back to the plane stress state, as shown in Figure 10.16, this model is usually known as the *dog bone model*. It should also be noted that, although we have considered plane strain fracture toughness (K_{I_c}) to be independent of thickness, it can vary with temperature and strain rate.

10.5 Elastic–Plastic Fracture Mechanics (EPFM)

The application of linear elastic fracture mechanics in the analysis of the stress field around the crack tip, and consequently in the assessment of fracture toughness which is a very important material property, has some limitations. The application of linear elastic fracture mechanics is valid as long as plasticity around the crack tip may be considered very small; that is, as long as the small-scale yielding assumption is valid. Also, as explained later, an experimentally determined fracture parameter, fracture toughness K_{Ic}, is valid if the requirements described by Equations (10.44) are fulfilled. By contrast, a consequence of excessive yielding is that the stress intensity factor can no longer represent the stress field around the crack tip, or we might say that this factor does not correctly represent the severity of the crack. So, in a case when the assumption of small-scale

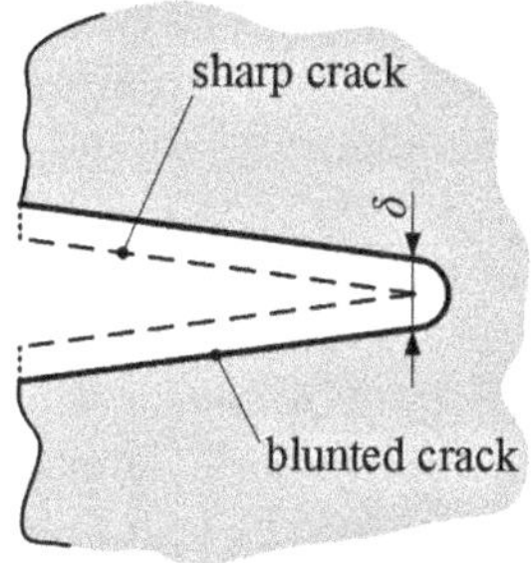

Figure 10.17 A crack tip: sharp and blunted.

yielding is not valid, that is, when plastic deformation is present to a greater extent, other elastic–plastic parameters are introduced into the analysis of the material's behavior.

Let's mention two parameters – the *J-contour integral* and *CTOD* (the *crack tip opening displacement*) – using which the crack tip conditions in elastic–plastic material behavior can be described. These two parameters may be used as fracture criteria. They give some measure of the fracture toughness of a considered material, although both of these parameters have some limitations which can be manifested when crack growth becomes significant as well as when plasticity is excessive. In general, several measures of material fracture toughness may be mentioned and used. We can list the *plane strain fracture toughness*, K_{Ic}, the fracture toughness in the area of thinner specimens (thin sheets, thin plates), which is designated K_c (the *critical stress intensity factor* for thinner specimens), as well as J_{Ic} (the critical value of the J integral) and the critical value of CTOD, both independent measures of fracture toughness. As is known and experimentally verified, the fracture toughness associated with thin specimens of minimal thickness, t_c, is the *plane stress fracture toughness*, K_c, while the *plane strain fracture toughness*, K_{Ic}, is associated with thicker specimens of thickness t_{Ic}.

Experimental investigations relating to fracture toughness have indicated that some metals exhibit low levels of fracture toughness while some exhibit quite high levels of fracture toughness; that is, they are tough. As already mentioned, if the plastic zone is no longer small compared to the in-plane dimensions of the specimen, LEFM is no longer an appropriate tool for indicating the fracture toughness behavior of metals. On the basis of experimental investigations in such cases, it was found that an initially sharp crack tip becomes a blunted one, and the crack tip opening displacement (CTOD) was established as a new measure of fracture toughness (see Figure 10.17). In engineering practice, there are many applications where low-strength materials with high toughness are used, such as pressure vessels, ships, aircraft and so on. Here, the assumption of small-scale yielding at a crack tip ceases to be valid and linear elastic fracture mechanics lose their validity. Crack tip opening displacement would be useful in these scenarios but we will not discuss the subject further.

10.5.1 The *J* Integral

In further research relating to fracture problems, some attempts were made to find solutions for the analysis of crack behavior (propagation) in the field of nonlinear material behavior; that is, to extend the analysis of cracks beyond the limits of linear elastic fracture mechanics. The idea was that the description of nonlinear elastic material behavior could be used as an idealization for the description of elastic–plastic material behavior, and in that way it would be possible to find a solution for the description of the stress intensity around the crack tip in cases involving greater plasticity.

Finding the solution may be based on the consideration of uniaxial stress–strain behavior of nonlinear elastic and elastic–plastic materials. The loading path, that is the response of the material subjected to stress, for nonlinear elastic and elastic–plastic behavior is considered to be the same. They differ in their unloading paths. The unloading path in a case of linear elastic behavior is the same as its loading path, while in a case of elastic–plastic behavior, the unloading path has a slope that is equal to the direction of Young's modulus. As such, and in order to crack the analysis, Rice introduced the so-called J-integral [8], which is a path-independent contour integral whose value is equal to the energy release rate in a nonlinear elastic body containing a crack. The same definition of energy release rate (Equation (10.2)) for linear and nonlinear elastic materials is valid when $\mathcal{G}$ is replaced by J. So the J integral becomes, at the same time, an energy parameter and a stress intensity parameter. However, Rice suggests that nonlinear elastic material behavior and elastic–plastic material behavior may be considered the same if unloading is not carried out. In order to analyze the state of stress around the crack tip, the J-integral is taken to be a line integral enclosing the crack tip. In the case of an elastic–plastic material, it no longer has meaning as a physical interpretation with respect to the total potential of the considered body, but its significance as a measure of stress intensity around the crack tip is retained. It is important that for nonlinear elastic behavior, J is path-independent and has the same value for all choices of considered contour, C. Rice stated that a planar integral, designated $J\,(J_C, J_\phi, J_\Gamma)$ for a nonlinear elastic, homogeneous, isotropic body in an equilibrium state along a closed path is always equal to zero. The same investigation also refers to the problem of the body containing a crack [7, 8]. If the problem is enlarged to three dimensions, the loading paths for nonlinear elastic and elastic–plastic behavior of materials do not have to be the same, but a good assumption may exist. Consider a closed contour "c" that surrounds a region A, as shown in Figure 10.18.

The J_c-integral, in the Cartesian coordinate system (x_1, x_2) or in (x, y), is defined in the form [7, 14]:

$$J_c = \int_c \left(W n_1 - T_i \frac{\partial u_i}{\partial x_1} \right) \mathrm{dc}, \tag{10.27a}$$

which may be transformed into the form:

$$\mathrm{J_c} = \int_c \left(\mathrm{W n_1} - (\sigma_{ij} \mathrm{n_j}) \frac{\partial \mathrm{u_i}}{\partial \mathrm{x_1}} \right) \mathrm{dc}, \tag{10.27b}$$

The terms in Equation (10.27a) are explained below.

Finally, for a cracked body, in the x, y coordinate system, the J integral may be written in the form:

$$J = J_c = \int_c W \mathrm{d}y - T_i \frac{\partial u_i}{\partial x} \mathrm{dc}. \tag{10.27c}$$

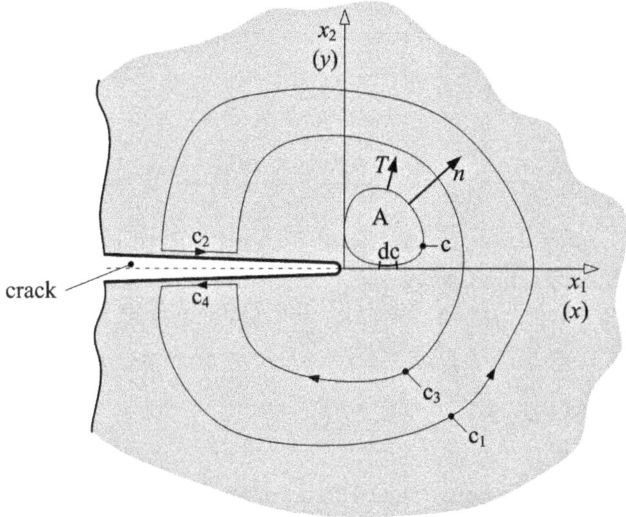

Figure 10.18 A *J* integral (a path-independent contour integral).

In the above equations, "c" is the contour of the considered curve surrounding the considered region, while the other designations are as follows:

Elastic strain energy density:

$$W = \int_0^{\varepsilon_{ij}} \sigma_{ij} \mathrm{d}\varepsilon_{ij}, \tag{10.28}$$

Stress components:

$$\sigma_{ij} = \frac{\partial W}{\partial \varepsilon_{ij}}, \tag{10.29}$$

The traction vector, taken as the outward normal **n** to the contour "c", written in index notation is:

$$T_i = \sigma_{ij} n_j, \tag{10.30}$$

Displacement vector:

$$\mathrm{u} = \mathrm{u_i} + \mathrm{u_j}. \tag{10.31}$$

In a state of static equilibrium of the body, and neglecting body forces, we have:

$$\frac{\partial \sigma_{ij}}{\partial x_j} = 0, \tag{10.32}$$

and, considering small deformation at the same time, the following relationship is valid (from Equation (3.17)):

$$\varepsilon_{ij} = \varepsilon_{ji} = \frac{1}{2}\left(\frac{\partial u_i}{\partial x_j} + \frac{\partial u_j}{\partial x_i}\right) \tag{10.33}$$

Now we have:

$$\frac{\partial W}{\partial x_1} = \sigma_{ij}\frac{\partial \varepsilon_{ij}}{\partial x_1} = \sigma_{ij}\frac{\partial}{\partial x_j}\left(\frac{\partial u_1}{\partial x_1}\right), \tag{10.34}$$

or, applying Equation (10.32), it follows that:

$$\frac{\partial W}{\partial x_1} = \frac{\partial}{\partial x_j}\left(\sigma_{ij}\frac{\partial u_i}{\partial x_i}\right). \tag{10.35}$$

Considering all of the mentioned terms, the J_c integral may be written in the form shown in Equation (10.27b). Further, using Green's theorem, the J_c-integral in Equation (10.27b), referring to the closed area A, may be writen as [7]:

$$J_c = \int_A \left(\frac{\partial W}{\partial x_1} - \frac{\partial}{\partial x_j}\left(\sigma_{ij}\frac{\partial u_i}{\partial x_1}\right)\right)\mathrm{d}x_1\mathrm{d}x_2 \tag{10.36}$$

and finally, introducing the quantity from Equation (10.35) into Equation (10.36), it follows that:

$$J_c\,(=J_\Phi) = 0. \tag{10.37}$$

It can be shown that the J-integral is an independent integral of the path "c" surrounding the crack tip; that is, the value of this integral is the same for any path originating at a point on the lower crack surface and finishing at any point on the upper crack surface surrounding the crack tip. In accordance with this, consider the J integral along the closed path consisting of four segments, displayed in Figure 10.18. In this case we have:

$$\sum J_{C_i} = 0 \rightarrow J_{C_1} + J_{C_2} + J_{C_3} + J_{C_4} = 0 \tag{10.38}$$

$$\left(\text{or written}: J_{C_1} + J_{C_2(+)} - J_{C_3} + J_{C_4(-)} = 0\right) \tag{10.38a}$$

The crack surfaces are traction free and therefore $T_i = 0$ along C_2 and C_4. Also, it can be noted that: $n_1 = \mathrm{d}y/\mathrm{d}c$ and $\mathrm{d}y = 0$ along C_2 and C_4. With regard to the foregoing, it follows that $J_{C_2} = J_{C_4} = 0$ (the "y" direction is taken normal to the crack line). However, on the basis of Equation (10.38), we have:

$$J_{C_1} = -J_{C_3}, \tag{a}$$

$$\left(\text{or written in accordance with Equation (10.38a)}: J_{C_1} = J_{C_3}\right) \tag{a1}$$

Thus, the value of the J integral is the same for any counterclockwise path around the crack tip. The dimension of J may be expressed, for example, in: MPa·m = kN/mm or in N/mm, kP·m or similar. Information about the application possibilities of the J integral in crack analysis can be found in [20].

10.6 Experimental Determination of Fracture Toughness

It does not hurt to mention once again that the best data with which to characterize a material are obtained experimentally. The same is true for fracture toughness, which characterizes a material's resistance to crack propagation. Fracture toughness is a material property that is used to design structures that are resilient to fracture, in the same way that yield strength is a material property used to design structures that are resilient to plastic deformation.

The result of a fracture toughness test may be either a single value of fracture toughness or a resistance curve. Depending on which toughness parameter (*K*, *J*, etc.) is plotted against crack extension, a *K–R*, *J–R* or *CTOD–R* curve can be considered. However, standards prescribe the procedures for obtaining fracture toughness or an appropriate curve; for example, to obtain K_{Ic}, a *K–R* curve; for J_{Ic}, a *J–R* curve; and so on. There are several standards prescribing fracture toughness (K_{Ic}, J_{Ic}, ...*CTOD*) and resistance curve (*K–R*, *J–R*) testing procedures. The standards are produced by various organizations, among which are theASTM (American Society for Testing and Materials), the BSI (British Standards Institution) and the ISO (International Organization for Standardization).

Although experimentally obtained data are the most useful data from which to draw conclusions about using a specific material, some comments need to be made. The value or utility of experimentally obtained data depends on whether the experiment is a credible interpretation of the real process. For example, tensile tests are quite close to the actual events in the element, while investigations which relate to fracture toughness can represent an actual event only to a certain degree of coincidence due to the existence of differences between a crack in the specimen and a crack in a real structure.

10.6.1 Test Specimens: Shapes, Dimensions, Orientations and Pre-cracking

10.6.1.1 Shapes and Dimensions of the Specimens

Some types of specimens typically used in fracture toughness testing are presented in Figure 10.19 [21]. There are three dimensions that characterize a specimen: crack length (a), width (b) and the specimen's thickness (t).

Note that the tensile force acting on the opening of a crack is always the force designed to act in the middle of the thickness of the specimen, despite the depiction shown in the figure. In this sense, in Figure 10.19, tensile force F is displayed for a 2-D image, but its real position is in the middle of the specimen. A similar situation is the case in Figure 10.29.

Apart from standardized specimens, there also exist other types of specimen that can give some information on material toughness, but such information needs to be taken with a certain level of skepticism. As discussed, the possibilities of crack propagation in real structures may differ compared to those that occur in manufactured specimens. First, any specimen is designed and manufactured to contain a crack that differs from a real crack in a real structural element. These differences can be manifested in the crack size but also in the shape of the crack tip, although the crack tip in the specimen is sharp, as it is also assumed to be in a real structural element.

Our discussion of considered specimens is oriented and based on the existing standards, but the designations we've used in relation to a specimen's geometry and sizes differ somewhat from those used in the standards. This, however, should not be an obstacle to understanding. Generally, it is accepted that the dimensions are as follows: $b = 2t$ and $0.45 \le a/b \le 0.55$.

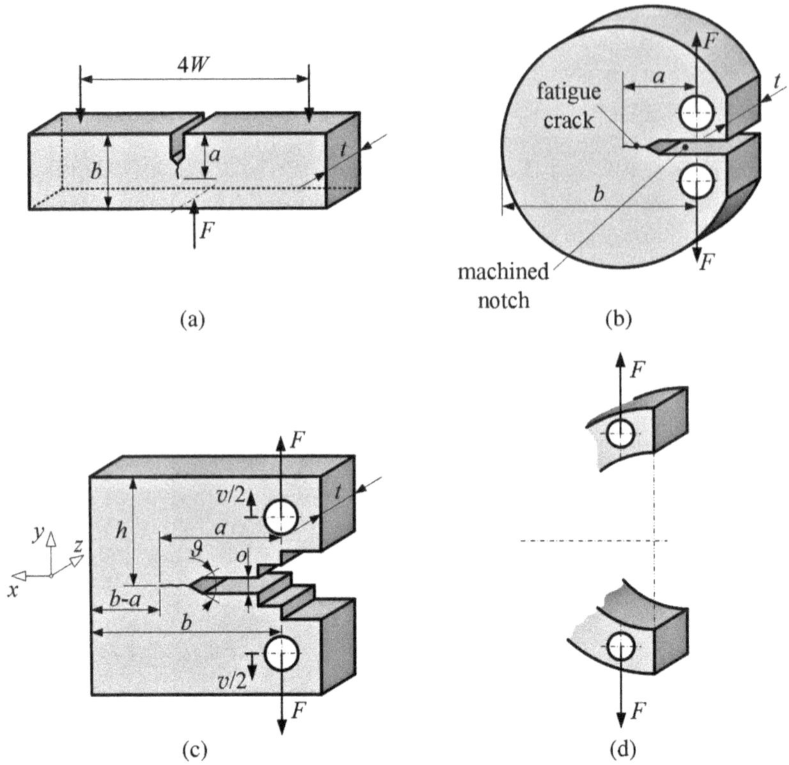

Figure 10.19 Geometry of the specimens – sketches. (a) SENB (single-edge notched bend) specimen; (b) DCT (disc-shaped compact type/tension) specimen; (c) CT (compact type/tension) specimen; (d) arch-shaped specimen.

10.6.1.2 Orientation of a Specimen Made from Base Material

As was discussed in Section 1.3, material properties that depend on the microstructure are called *structure-sensitive properties*. A material's microstructure is often sensitive to the direction of the final product forming. Although, in accordance with considerations in mechanics, a material is mostly assumed to be isotropic, it is not always so in engineering practice and the microstructure may differ depending on orientation. This means that a certain microstructure orientation may facilitate the progress of a crack due to the occurrence of planes of weakness where crack propagation is easier. As such, the standards require that the resulting report which relates to fracture toughness testing contains information on the crack orientation (that is, the specimen orientation extracted from the material). Figure 10.20 shows possible types of specimen orientation.

The specimen (crack) orientation presented in Figures 10.20(a) and 10.20(b) may be found, for example, in ASTM Standard E399-90 [21]. As can be seen from Figure 10.20, the first letter indicates the direction of the principal tensile stress (always perpendicular to the crack plane for Mode I), while the second letter indicates the direction of crack propagation.

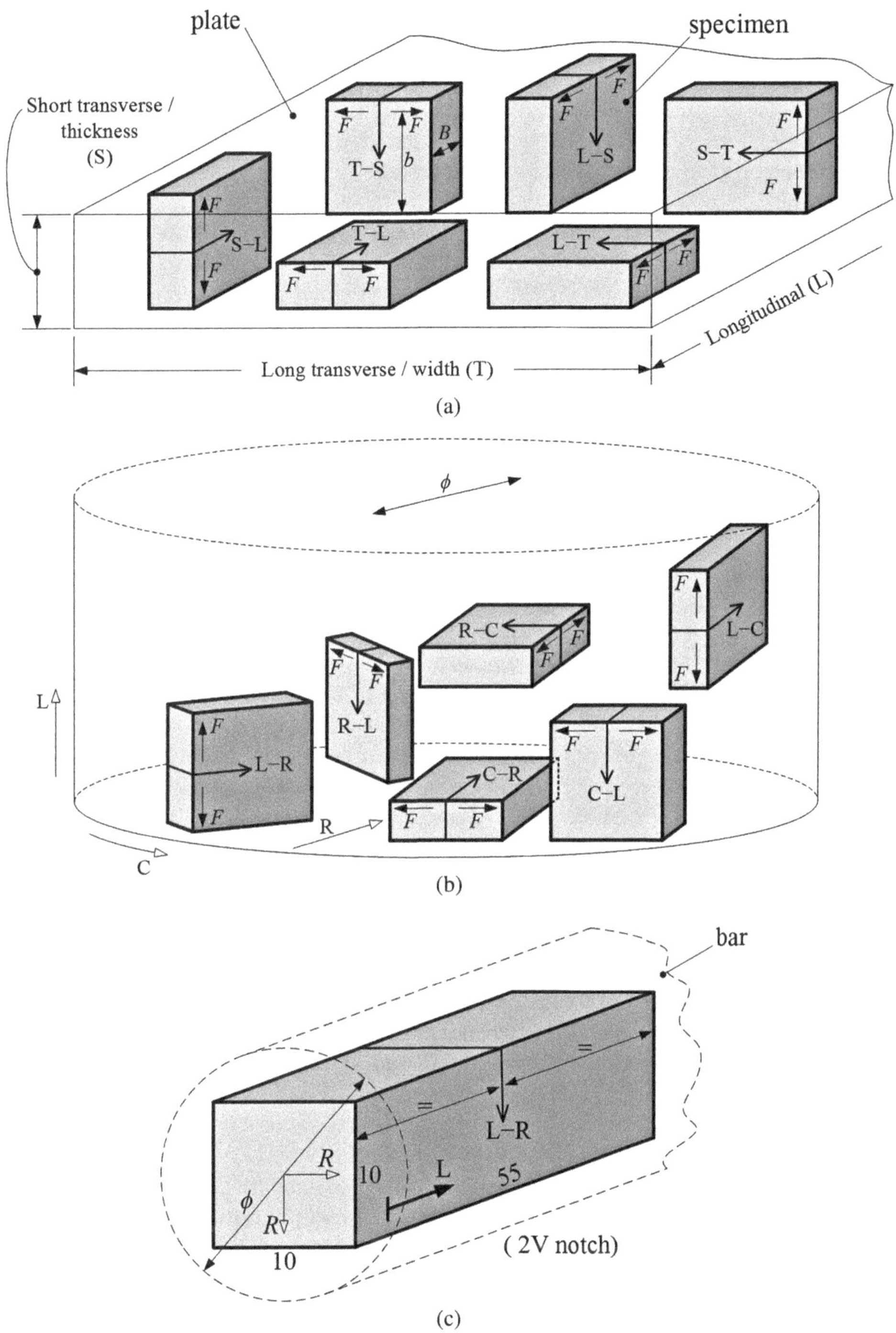

Figure 10.20 Notation for a specimen's orientation in the manufacturing process. (a) Specimen manufactured (extracted) from a rolled plate (rolling direction coincides with the longitudinal direction); (b) specimen manufactured (extracted) from a disc and cylinder; (c) specimen manufactured for Charpy impact testing from a bar. (L-longitudinal, T-long transverse, S-short transverse, C-circular, R-radial).

10.6.1.3 Fatigue Pre-cracking

In fracture mechanics it is assumed that the tip of the crack is sharp. Since the notch on the specimen is manufactured for laboratory investigation, its sharpness does not have to comply with a real scenario. To achieve a sufficiently sharp tip of the crack, a pre-cracking procedure is used. A fatigue crack is initiated at the tip of a machined notch and developed to the desired size through controlled cyclic loads. In accordance with ASTM standard E 399-90 [21], the width of the notch is defined as $o < b/10$ (but not less than 1.6 mm), while the angle at the end of the notch is $\vartheta < 90°$ (see Figure 10.19(c)). The required length of the fatigue crack (pre-cracking) is defined as: at least 0.025 b or a minimum of 1.3 mm, whichever is larger. When experimental investigations relate to the fatigue crack growth rate (ASTM standard E 647-00 [21]), some details relating to pre-cracking are somewhat different.

10.6.2 Fracture Toughness, K_{Ic} and the *K–R* Curve

As mentioned earlier, the comparison between stress and tensile stress ($\sigma : \sigma_{UTS}$) is similar to that between the stress intensity factor and the critical stress intensity factor ($K_I : K_{Ic}$). The critical stress intensity factor is known as fracture toughness. Tensile stress and fracture toughness are material properties.

10.6.2.1 *R*-curve (*K–R* Curve)

As discussed in some references, for example, [2, 7–8], resistance to crack propagation in a tested material that behaves in a linear elastic manner prior to failure, and if the plane strain condition is considered, may be defined by a single value of fracture toughness (K_{Ic}). In addition, the curve that describes a material's resistance to crack extension is known as the *R*-curve (or *K–R* curve, crack resistance curve, crack resistance, crack resistance force, crack extension resistance curve, crack growth resistance curve) or, similarly, the *K*-resistance curve, K_R. Based on the force–displacement ($F - v$) data, the curves representing driving forces ($\sigma - a$) are displayed. From this, the resistance curve (*K–R*) is constructed. The critical stress intensity, K_c, is obtained as a point where the driving force is tangential to the (*K–R*) curve. This is the so-called instability point and it is not treated as a material property since this point depends on the shape of the driving force curve. This is the reason why this critical value is not transferable to a real structure.

The *R*-curve for a brittle material (this material is treated as being linear elastic) is flat. For some linear elastic materials as well as for nonlinear elastic (and elastic–plastic) materials, the *R*-curve may have different, rising shapes. The ASTM E561-10e2 standard describes a procedure for determining a rising *R*-curve (*K–R* curve) for material that exhibits linear elastic behavior. The ASTM standard E399-90 [21] describes the procedure for K_{Ic} testing. In contrast to this standard, the *K–R* standard (such as the one mentioned above – ASTM E561-10e2) is not limited to a minimum thickness of specimen. This means that this standard may be applied in cases involving plane stress conditions (thinner specimens) but it may be appropriate only when the plastic zone is small; that is, fracture toughness is plane stress linear elastic. Usually these tests are applied to thin sheets (or plates) made of high-strength materials and to materials that exhibit linear elastic fracture behavior. It has been shown that a material exhibiting a single value of K_{Ic} under plane strain conditions may exhibit a rising *R*-curve under plane stress conditions. Characterization of a material with a rising *R*-curve is not determined uniquely by a single toughness property.

A material's resistance to crack propagation starts from the initial crack length, a_0, and increases during the process of crack growth, as shown in Figure 10.21. Determination of the crack growth is based on the equilibrium:

$$\mathcal{G}(\sigma,a) = R(\Delta a). \tag{10.39a}$$

It is clear that crack growth is defined by the above equilibrium condition between the crack driving force (crack extension force) and the crack resistance force. The equilibrium state is stable (loading is fixed) when the crack resistance (R) increases faster than the driving force ($\mathcal{G}$):

$$\left.\frac{\partial \mathcal{G}}{\partial a}\right|_{\sigma=\text{const}} < \frac{\mathrm{d}R}{\mathrm{d}a} \tag{10.39b}$$

and it becomes unstable when:

$$\left.\frac{\partial \mathcal{G}}{\partial a}\right|_{\sigma=\text{const}} > \frac{\mathrm{d}R}{\mathrm{d}a} \tag{10.39c}$$

while the limit of stable crack growth occurs when the critical condition occurs [2, 8, 19]:

$$\left.\frac{\partial \mathcal{G}}{\partial a}\right|_{\sigma=\text{const}} = \frac{\mathrm{d}R}{\mathrm{d}a}. \tag{10.39d}$$

In cases of experimentally determined R-curves, tests are usually carried out under displacement control. The shape of the R-curve depends on the material's behavior and the configuration of the cracked specimen. The point of instability can be seen in Figure 10.21 for tests conducted under load (stress) control. Instability occurs when the driving force curve becomes tangential to the resistant curve (R-curve, K–R curve). The critical load (stress) σ_c and the corresponding value $\mathcal{G}^*$ depend on the loading type,

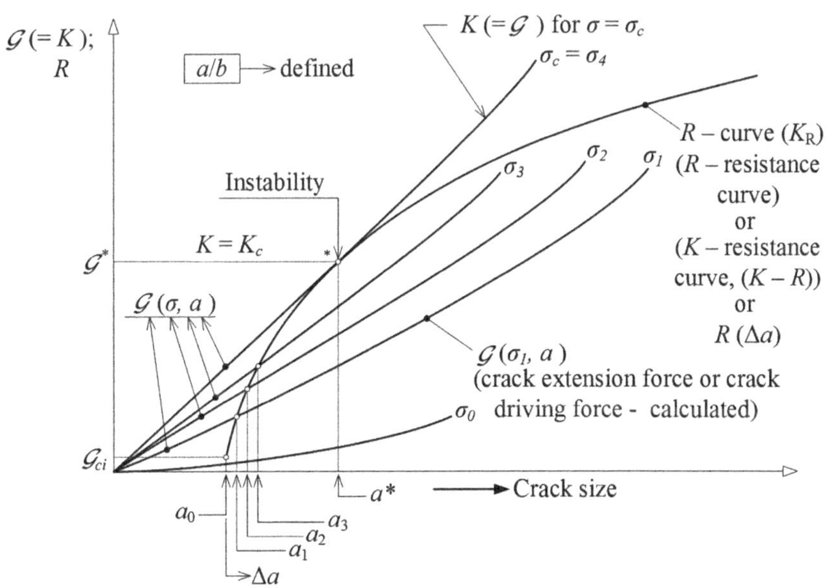

Figure 10.21 Point of instability → R curve/K–R curve ($R-\Delta a$) (for a given a/b).

the R-curve and the crack geometry. Now, $\mathcal{G}^*$ is a critical energy release rate that is equivalent to K and J in linear fracture mechanics. So-called *crack resistance* or the *crack resistance force* (let's denote this $\mathcal{G}_c$) or the material's resistance to crack propagation has an initial value $\mathcal{G}_{ci}$ at initial crack length a_0. Fracture is initiated if, during crack growth, the energy release rate is equal to the required energy. Since, for the plate in Figure 10.11, the strain energy release rate $\mathcal{G}$ and stress intensity factor K_I are:

$$\mathcal{G} = \frac{\pi\sigma^2 a}{E}, \quad K_I = \sigma\sqrt{\pi a}, \tag{10.40}$$

And, by comparing the two formulas, it follows that:

$$\mathcal{G} = \frac{K_I^2}{E^*}. \tag{10.41}$$

In Equation (10.41), for plane stress we have:

$$E^* = E, \tag{10.42a}$$

while for plane strain, we have:

$$E^* = \frac{E}{1-\nu^2} \tag{10.42b}$$

The previously defined *energy release rate* (crack extension force, crack driving force) from Equation (10.2), for example, for the CT specimen taken as double cantilever beam in Figure 10.19(c), may be written as:

$$\mathcal{G} = -\frac{d\Pi}{dA} = -\frac{d\Pi}{t \cdot da}. \tag{a}$$

If the total potential is written in the form:

$$\Pi = U + V \tag{b}$$

where the strain energy (for a linear relationship between force and displacement) is:

$$U = \frac{1}{2}\int_v F dv = \frac{Fv}{2} \tag{c}$$

and the potential of the external load is:

$$V = -Fv, \tag{d}$$

It follows that:

$$\Pi = -U, \tag{e}$$

and then:

$$\mathcal{G} = \frac{1}{t}\left[\frac{dU}{da}\right]. \tag{f}$$

Introducing compliance:

$$C = \frac{v}{F}, \tag{g}$$

and on the basis of known values from beam theory, we have:

$$\frac{v}{2} = \frac{Fa^3}{3EI} \text{ and } I = \frac{th^3}{12} \tag{h}$$

It follows that:

$$\mathcal{G} = \frac{F^2}{2t}\frac{\mathrm{d}C}{\mathrm{d}a} = \frac{12F^2a^2}{t^2h^3E}. \tag{i}$$

10.6.2.2 Plane Strain Fracture Toughness (K_{Ic}) Testing

Experimentally obtained fracture toughness may be treated as an appropriate fracture parameter with which to define resistance of the material to fracture (that is, to crack propagation) if the material's behavior prior to failure is in the linear elastic domain, that is, the plastic zone around the crack tip is small compared to the crack size and the specimen dimensions. The K_{Ic} testing procedure and some other demands (specimen size, configuration/geometry of the specimen, preparation, results interpretation and so on) are defined, for example, by the ASTM E 399-90 standard [21]. At least three tests need to be made for each material condition. A specimen is manufactured in such a way that it is fatigue pre-cracked and the crack length compared to the specimen width (a/b) is in the range between 0.45 and 0.55 (Figure 10.19(c)). Designations of the dimensions used here and in the standard are as follows: $b = W$, $t = B$. Dimension "a" includes fatigue pre-cracking. However, the plastic zone must be small compared to these dimensions.

Generally speaking, the fracture toughness (K_{Ic}) of engineering metals is in the range 20–200 MPa$\sqrt{\text{m}}$ and it can be affected by a very small change in the chemical composition or processing. So, the impact of sulfide inclusions in steels is reflected in increased rates of fracture: with an increase in sulfur, there is a decrease in fracture toughness.

The test procedure for K_{Ic} determination includes monitoring the load and displacement of a pre-cracked specimen to failure. Load–displacement curves are shown in Figure 10.22. Displacement relating to the crack opening can be monitored in accordance, for example, with Figure 10.19(c).

In Figure 10.22, three types of load–displacement curve are displayed. A load–displacement curve is obtained in such a way that a pre-cracked test specimen is loaded

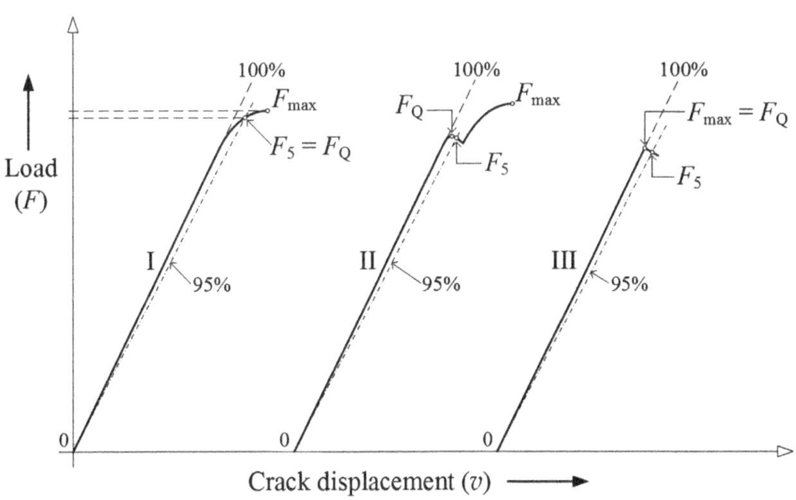

Figure 10.22 Load displacement (F–v) curve in K_{Ic} testing.

to failure and the load and displacement are monitored. Depending on the type of curve, the critical load, F_Q, is determined. As can be seen from the first diagram (I) in Figure 10.22, the direction of 100% slope is an original elastic loading slope (relative to a horizontal line), while F_5 is a 5% secant line taken from the origin (that is, its slope is equal to 95%). In this diagram, a small nonlinearity that occurs can be caused by plasticity. In the second diagram (II), unstable crack growth occurs before the deviation of 5% from linearity occurs, and, finally, from the third diagram it is clear that the specimen fails before achieving 5% nonlinearity. Determination of the critical load, F_Q, can be seen. The crack length has to be measured on the basis of the fracture surface as the average value based on three evenly spaced measurements. So, the crack length may vary through the thickness of the specimen. After determination of F_Q and the crack length, the fracture toughness, $K_Q = K_{Ic}$ can be determined as [8, 21–22]:

$$K_Q = K_{Ic} = \frac{F_Q}{t\sqrt{b}} f(a/b). \tag{10.43}$$

The value obtained from Equation (10.43) may be valid as the plane strain fracture toughness, K_{Ic} (in accordance with the ASTM standard) if the following requirements are met:

$$0.45 \le a/b \le 0.55 \tag{10.44a}$$

$$t, a \ge 2.5\left(\frac{K_Q}{\sigma_{YS}}\right)^2; \ \sigma_{YS} = \sigma_{0.2} \tag{10.44b}$$

$$F_{max}/F_Q \le 1.1. \tag{10.44c}$$

Usually, the overall requirements for meeting plane strain conditions are given in the form:

$$\text{a, t, (b} - \text{a), h} \ge 2.5\left(\frac{K_{Ic}}{\sigma_{YS}}\right)^2, \ \text{b} \ge 5\left(\frac{K_{Ic}}{\sigma_{YS}}\right)^2 \tag{10.45}$$

In Equation (10.43), function $f(a/b)$ is a dimensionless function which is given in polynomial form depending on the type of specimen. As such, if the SENB specimen from Figure 10.19(a) is used, then this function is given in the following form [5]:

$$f\left(\frac{a}{b}\right) = \frac{3\left(\frac{a}{b}\right)^{1/2}\left[1.99 - \left(\frac{a}{b}\right)\left(1 - \frac{a}{b}\right)\left(2.15 - 3.93\frac{a}{b} + 2.7\left(\frac{a}{b}\right)^2\right)\right]}{2\left(1 + 2\frac{a}{b}\right)\left(1 - \frac{a}{b}\right)^{3/2}} \tag{10.46a}$$

or, if the CT specimen from Figure 10.19(c) is used, then this function may be written as:

$$f\left(\frac{a}{b}\right) = \frac{\left(2 + \frac{a}{b}\right)\left[0.866 + 4.64\frac{a}{b} - 13.32\left(\frac{a}{b}\right)^2 + 14.72\left(\frac{a}{b}\right)^3 - 5.6\left(\frac{a}{b}\right)^4\right]}{\left(1 - \frac{a}{b}\right)^{3/2}} \tag{10.46b}$$

Generally, fracture toughness increases with an increase in temperature,and this change can be especially visible in metals with a BCC (body-centered cubic) crystal structure, while in metals with an FCC (face-centered cubic) structure and in high-strength steels, these changes are less noticeable. It is important to pay attention to the so-called DBTT point (ductile to brittle transition temperature) and NDT point (nil ductility temperature). Since these may be visible in Charpy impact tests, they will be discussed later.

10.6.3 Fracture Toughness J_{Ic} and the *J–R* Curve

As is known, under small-scale yielding, any of the mentioned single fracture toughness parameters, such as K, J and so on, can correctly represent the crack tip (yield field) conditions and thus may be used as an appropriate fracture criterion. The limitations of their use occur when excessive yielding takes place. With respect to K, its use is no longer appropriate when the conditions of linear elastic fracture mechanics are not valid. Also, excessive yielding is one reason why J may not be an appropriate parameter. Namely, crack conditions also need to be such that the strain region is small compared to the in-plane dimensions of the cracked specimen; in other words, crack growth must not be significant in comparison to the mentioned dimensions. Usually, the term *J-controlled fracture* is used to describe the crack tip conditions where this parameter may be treated as an appropriate parameter. In elastic–plastic conditions, the J parameter may be used while the K parameter may not, and it may be said that J is of significance in EPFM just as K is in LEFM. The J application is oriented to stationary cracks under monotonous load. The J integral is equal to the strain energy release rate for nonlinear elastic material behavior, although the part relating to plastic strain energy is no longer available as the crack driving force.

LEFM deals with linear elastic material behavior and EPFM with nonlinear elastic and elastic–plastic material behavior. Perhaps it would be useful to make some explanations relating to the *driving* and *resistance forces* that are often used in engineering. These terms are intended to be used to differentiate between the mathematical side and the materials side of fracture mechanics. The driving force is, for example, K_I in Equations (10.7) and (10.8), or, in fatigue, $\Delta K = K_{Imax} - K_{Imin}$ (Equation (10.61)), and the driving force is the calculated value and depends upon the specimen, the applied load and so on. For its calculation, material properties are not needed. As such, it can be compared to the calculation of the applied stress in an elastic stressed structure. On the other hand, the resistance force, K_c, (K_{Ic}, J_{Ic}) is a material property and can be obtained only by testing, similar to the measurement of yield strength.

10.6.3.1 *R*-curve (*J–R* Curve)

To characterize the state around the crack tip in a cracked body when LEFM cannot be applied, J may be determined as indicated in Section 10.5.1. The J integral, as the path-independent integral around the crack tip, is defined by Equation (10.27). For its determination in a case involving a cracked body and a larger plastic zone, numerical methods are generally required; for example, the finite element method, which is practically indispensable. Experimental methods may be used to determine the critical value of J, that is, J_{Ic} at ductile fracture, using the same types of specimens (Figure 10.19) as for the K_{Ic} determination in LEFM.

When talking about the use of the finite element method in determining the J integral, since this integral is independent of the chosen line around the crack tip, it is advantageous to choose a contour line far away from the crack tip, that is, in the region where the material behaves elastically. As we know from linear elastic fracture mechanics (LEFM), crack propagation (fracture) occurs when the stress intensity factor K_I reaches the critical value, that is, K_{Ic} (fracture toughness), where subscript I refers to the opening Mode I.

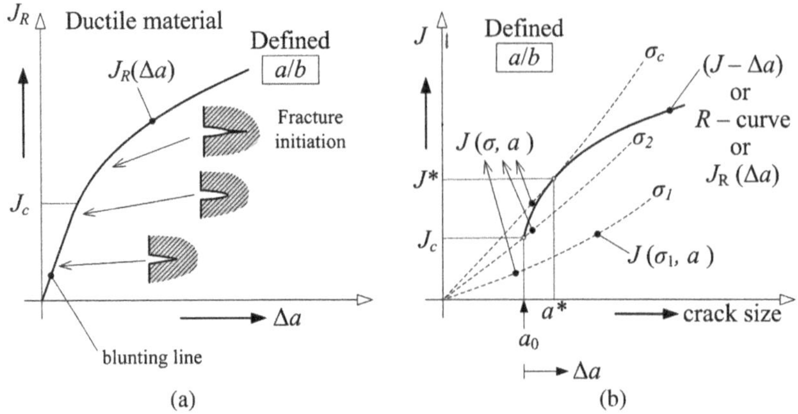

Figure 10.23 Point of instability, *R*-curve (*J*–*R* curve, (J_R)). (a) curve stages, (b) curve construction.

In the presence of the critical value of the stress intensity factor, the necessary condition for starting the fracture process is fulfilled and, at the end of the process, this leads to element separation. A similar situation occurs for elastic–plastic material behavior, and crack propagation is initiated when J reaches the critical value, J_c (J_{Ic}).

We previously described the *R*-curve, which can also be designated the *K*–*R*, K_R or *R* (Δa) curve, and is known as the *K*-resistance curve. Now, we can also show the *R*-curve (alternatively the *J*–*R*, $J_R(\Delta a)$curve), known as the *J*-resistance curve (see Figure 10.23 (a)). In Figure 10.23(a), a general presentation of the $J_R - \Delta a$ curve is shown to give an insight into the crack blunting before J_c is reached for a ductile material. The figure shows the crack growth resistance behavior under small-scale yielding.

In terms of stability, *J*-controlled crack propagation may be considered in the same way as Equations (10.39). The equilibrium condition (Figure 10.23(b)), is:

$$J(\sigma, a) = J_R(\Delta a). \tag{10.47a}$$

Under constant load, the equilibrium state is stable when:

$$\left.\frac{\partial J}{\partial a}\right|_{\sigma = \text{const}} < \frac{dJ_R}{da}, \tag{10.47b}$$

and the limit for stable crack growth is:

$$\left.\frac{\partial J}{\partial a}\right|_{\sigma = \text{const}} = \frac{dJ_R}{da}. \tag{10.47c}$$

The curves designated $J(\sigma, a)$ are crack driving forces, while $J_R(\Delta a)$ is the *R*-curve. The problem that needs to be solved is how to determine J_{Ic}, the fracture toughness parameter.

10.6.3.2 Fracture Toughness (J_{Ic}) Determination/Testing

The procedure for fracture toughness (K_{Ic}) determination was previously explained and it was based on the force versus displacement (F–v) diagram. The force versus

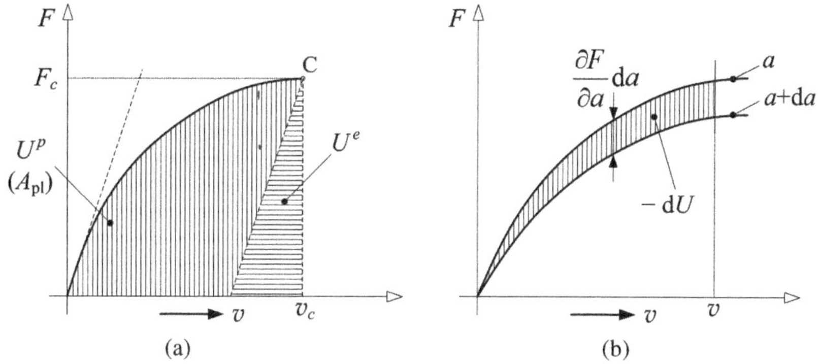

Figure 10.24 Explanation of J_c determination. (a) load-displacement curve, (b) change in strain energy.

displacement (F–ν) diagram for the J_{Ic} (or J_c) determination is now highly nonlinear as a result of crack growth and plastic deformation. Consider the point C (F_c, ν_c), shown in Figure 10.24(a). The area under the F–ν curve corresponds to the strain energy, that is, to the external work, W (the work of the external load acting on the specimen) that is transformed to the strain energy ($U = W$). Since strain energy is a function of the crack length, at fixed displacement (ν = const), and for unit thickness (t = 1), we have:

$$J = -\frac{dU}{da} = -\left.\frac{\partial U}{\partial a}\right|_{\nu = \text{const}}, \tag{10.48}$$

and thus, this equation shows J (energy release rate) for a crack extension that corresponds to the area between the curves for a crack of length "a" and "a + da" (Figure 10.24(b)).

This is the basic idea for determination of J_{Ic} using a technique with multiple specimens with different crack lengths. A correlation between J and the deformation work (strain energy under the F–ν curve) can be written as follows:

$$J = J_{el} + J_{pl}, \tag{10.49}$$

where the elastic and plastic parts may be presented as:

$$J_{el} = \frac{K_I^2(1-\nu^2)}{E}; \quad J_{pl} = \frac{\eta A_{pl}}{t(b-a)}. \tag{10.50}$$

The elastic part, J_{el}, can be derived on the basis of Equations (10.41) and (10.42b), that is, on the basis of the elastic stress intensity. So, for a linear elastic material, $J = \mathcal{G}$.

In the plastic part (J_{pl}), η is a dimensionless constant and may be taken as [8]:

$$\eta = 2 \text{ for an SENB specimen,} \tag{10.51a}$$

and

$$\eta = 2 + 0.522\,(1a/b) \text{ for a CT specimen.} \tag{10.51b}$$

In Equation $(10.50)_2$, A_{pl} is the plastic area under the F–ν curve (Figure 10.24(a)).

Explanations of both J and J_{Ic} testing can be found in ASTM standard E1829-01. As we know, in K_{Ic} testing, the ratio between the crack length, a, and the width, b, of the specimen was defined as $0.45 \le a/b \le 0.55$ (from Equation (10.44a)), while in J_{Ic} testing, this

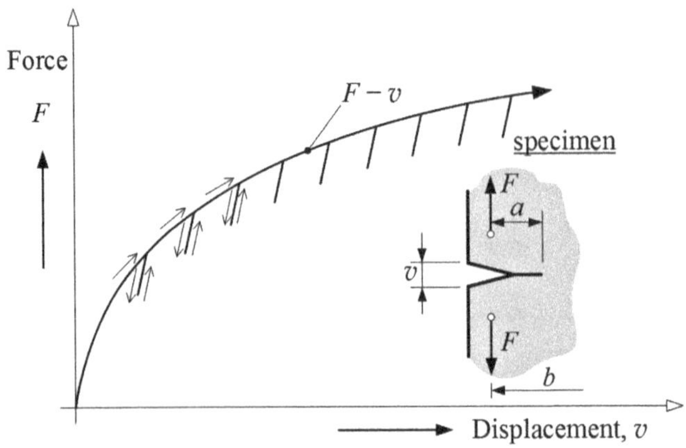

Figure 10.25 Force–displacement behavior with periodic unloading.

ratio is usually $0.5 \leq a/b < 0.7$ because the method used (the unloading compliance method) is not sensitive enough when a/b is less than 0.5.

Fracture toughness, J_{Ic}, is determined on the basis of the resistance curve (the *J*–*R* curve). To obtain the necessary *J*–*R* curve, the values of the *J* integral and crack extension Δa data along the *F*–*v* curve are needed. Some special means are needed to measure crack growth. A *single-specimen technique* or a *multiple-specimen technique* can be used to obtain the mentioned *J*-crack extension (J–Δa) data needed to form the *J*–*R* curve. The commonly used method for a single-specimen test technique is the so-called *unloading compliance method*, where the specimen is periodically unloaded (see Figure 10.25) by a small amount to measure *F*–*v* behavior, that is, to determine *J* and crack extension, Δa. The slopes observed in Figure 10.25 are measures of the specimen's elastic stiffness (it decreases with an increase in crack length).

When the single-specimen technique is used, instead of the unloading compliance method, the so-called *potential drop method* can also be applied.

As regards the *multiple-specimen technique*, several identical specimens are tested in such a way that they are loaded to different load levels and then unloaded. This produces different amounts of crack extension in the tested specimens. The samples under consideration are heated to tint the new fracture toughness to aid observation, and the amounts of crack growth (extension) are measured under a microscope.

In the so-called *basic test method*, the crack growth (extension), Δa, is approximated to be equal to the initial crack length; that is, $a = a_{\mathrm{i}} + \Delta a \approx a_{\mathrm{i}}$. In the *resistant curve test method*, which may be used with the single-specimen technique when the unloading compliance method is applied, the *J* values are calculated by updating the value at each point along the *F*–*v* diagram where crack lengths have been determined. The calculated *J* values are plotted against crack extension Δa to obtain the curve referred to as the resistance curve or *R*-curve (*J*–*R* curve, J_{R}) for stable crack growth (Figure 10.26), from which the critical value of the *J* integral (that is, J_{Ic}) can be obtained.

The procedure for determining J_Q (the provisional J_{Ic}) from the resistance curve is shown in Figure 10.26. In this case, σ_Y represents the yield stress (flow stress),

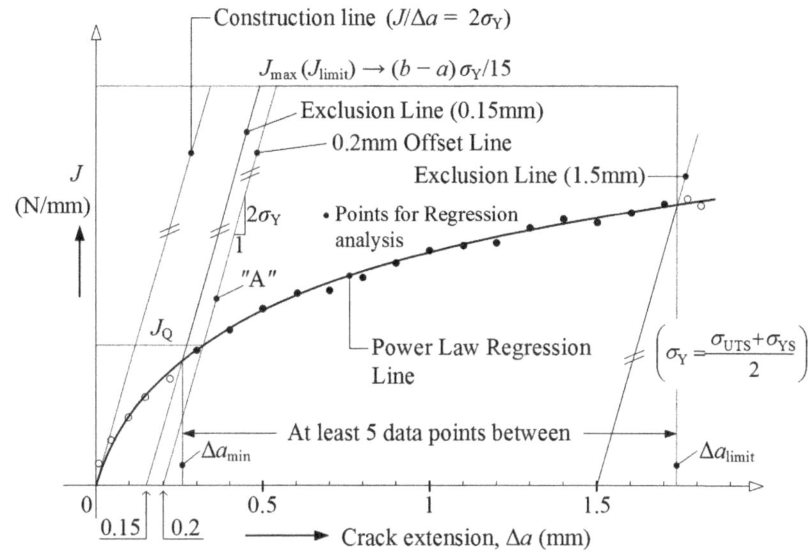

Figure 10.26 *J–R* curve (*J* integral versus Δa). Determination of J_Q (the provisional value of J_{Ic}).

which is now defined as an average value between the ultimate tensile strength (sometimes denoted σ_m or *UTS* or *US* or σ_{US}) and the yield strength (sometimes denoted $\sigma_{0.2}$, or *YS* or σ_{YS}). In Figure 10.26 can be seen four straight lines that are parallel to each other and have a slope of 2 σ_Y. The first of these plotted straight lines is known as the construction line and has a slope of 2 σ_Y. Two further straight lines, also with the same slope, are drawn at crack extensions of $\Delta a = 0.15$ mm and $\Delta a = 1.5$ mm, and these are called the *exclusion lines*. The intersection of the straight line at a crack extension $\Delta a = 0.2$ mm (line "A", known as the *0.2 offset line*) and the fitted J_R curve between the two exclusion lines gives J_Q (the provisional J_{Ic}). Further, a horizontal exclusion line, J_{max}, is taken as in [8]. The measurement capacity of the specimen is governed by this value to ensure *J*-controlled crack growth. Only data between the two mentioned exclusion lines, that is, between 0.15 mm and 1.5 mm (points falling within the qualifying area) and that also fall below the horizontal exclusion line can be used in regression analysis. The curve J_R is a power law regression line. Because of the plastic blunting effect, the tip of the crack moves forward and the *J* versus Δa curve has an initial/blunting line, mentioned earlier as the construction line. J_Q may be used as the fracture toughness, J_{Ic} as long as the following requirements are fulfilled (according to [21]):

$$t, (b-a) > 25\frac{J_Q}{\sigma_Y}, \tag{10.52}$$

where t is the thickness of the specimen while $(b-a) = b_0$ is the initial ligament.

As an equivalent, the quantity K_{Ic} can be calculated from Equation (10.41) ($\mathcal{G} = J$):

$$K_{Ic} = \sqrt{J_{Ic} \cdot E^*}, \tag{10.53}$$

where E^* is defined by Equations (10.42).

10.7 Charpy Impact Energy Testing

Fracture toughness is the most important property of a material when resistance to fracture is the main consideration in structural design. This property was discussed previously through the plane strain fracture toughness parameters, K_{Ic} and J_{Ic}. Both of these parameters can be determined experimentally but it needs to be stressed that these types of investigations are performed under laboratory conditions. We mention this because behavior in laboratory tests and the real behavior of a structure are not necessarily identical to each other. Specimen crack geometry in a laboratory and that in real engineering elements is not necessarily equal either. Furthermore, specimen preparation and testing procedures, particularly for quite large specimens with complicated geometry, also require a large expenditure of money and time.

To avoid these problems, some "indirect" experimental methods exist to determine the fracture resistance of a structure. One such method is the Charpy impact test, using which one may determine the impact energy, which can be correlated with fracture toughness. The Charpy impact test procedure is described in ASTM standard E 23-05. This method also has several limitations such as very small specimen size, loading rate and a blunt machined tip of the crack (notch acuity). In spite of this, however, the results obtained by Charpy tests and those obtained by standard tests can be compared and correlated. Research results in the field of impact energy using a Charpy impact machine (small-scale test) and research results in the field of fracture toughness (large- scale test) established a particular correlation between these two quantities [23–25]. Some of the correlations between the Charpy impact energy and the fracture toughness (in short: CVN versus K_{Ic}) are shown in Equations (10.54)–(10.57), in accordance with the mentioned references:

$$\begin{aligned} &K_{Ic} = 8.47\,(CVN)^{0.63}(\text{MPa}\sqrt{\text{m}}) \\ &CVN \rightarrow (\text{J}) \end{aligned} \tag{10.54}$$

Equation (10.54) is known as the *Roberts–Newton* equation and may be applied independent of temperature level and impact energy range.

$$\begin{aligned} &K_{Ic} = 14.6\,(CVN)^{1/2}(\text{MPa}\sqrt{\text{m}}) \\ &CVN \rightarrow (\text{J}) \rightarrow (7-68) \end{aligned} \tag{10.55}$$

Equation (10.55) is known as the *Sailors–Cortens* equation and may be applied if CVN energy is in the range 7–68 J.

$$\begin{aligned} &(K_{Ic}^2/E) = 0.22CVN^{3/2}, \rightarrow K_{Ic};(\text{MPa}\sqrt{\text{m}}) \\ &CVN \rightarrow (\text{J}) \rightarrow (3-82) \end{aligned} \tag{10.56}$$

Equation (10.56) is known as the *Barson–Rolf* equation and may be applied if CVN energy is in the range 3–82 J.

$$K_{Ic}^2 = \frac{1}{f}(E\bar{K})(1+\nu), \rightarrow K_{Ic};\ (\text{MPa}\sqrt{\text{m}}) \tag{10.57}$$

$$\begin{aligned} &E \rightarrow (\text{MPa});\ \bar{K} = CVN/A \rightarrow (\text{J/mm}^2) \\ &CVN \rightarrow (\text{J});\ A(\text{mm}^2) \end{aligned} \tag{10.57a}$$

$$f = \frac{E(1+\nu)}{71.74 \cdot CVN^{0.26}} \tag{10.57b}$$

Equation (10.57) is proposed in [26]. In this equation, E is the modulus of elasticity, $\bar{K}$ is a toughness, ν is a Poisson ratio and A is the cross-sectional area of the specimen at the notch.

From the view that the effects of loading rates and notch acuity are not critical at the upper shelf (see Figure 10.28), as in the transition-temperature region, it is interesting to present a reasonable expected correlation in this region in the form [11]:

$$\left(\frac{K_{Ic}}{\sigma_{YS}}\right)^2 = \frac{5}{\sigma_{YS}}\left(CVN - \frac{\sigma_{YS}}{20}\right). \tag{10.58}$$

The form of Equation (10.58) requires the input of data as follows: K_{Ic} $\left(\text{ksi}\sqrt{\text{in}}\right)$; $\sigma_{YS} = \sigma_{0.2}$ (ksi) at the upper shelf temperature; CVN (ft-lb) is the impact test value at the upper shelf. On the basis of experimentally obtained impact energy, when it is divided by the area of the cross-section of the specimen at the notch, the toughness of the considered material can be assessed.

A specimen intended for use in a Charpy test may have, for example, a V-notch of 2 mm or a U-notch of 2, 3 or 5 mm. The choice of specimen with a particular type and size of notch depends on the material being investigated but also on the standard used.

Factors that influence the toughness of the material (also the impact energy) include: temperature, microstructure, type and size of the notch, specimen dimensions, impact velocity and so on.

A Charpy impact machine is shown in Figure 4.3(a). One of the typically used specimens (containing a V-notch) in Charpy impact testing is shown in Figures 10.20(c), 10.27 and 4.3(b) and (c).

The CVN impact energy versus temperature is shown in Figure 10.28.

Unless specifically defined, the test relating to impact energy determination is usually conducted at a temperature of 23 °C ± 5 °C. As well as the aforementioned Charpy test, other tests, such as the Izod test or the Brugger test, may also be used to determine the impact energy.

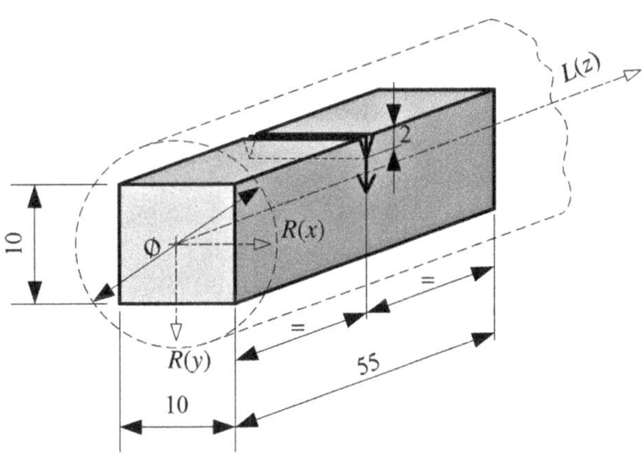

Figure 10.27 Charpy V-notch specimen.

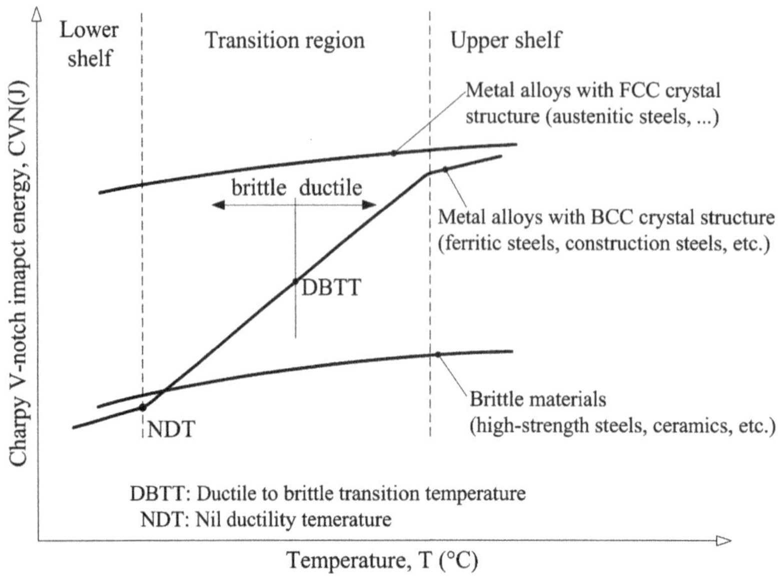

Figure 10.28 Impact energy versus temperature (schematic view).

As noted previously, a structure is designed in accordance with its purpose and the conditions under which it will operate. Both the structure's purpose and its operating conditions call for a careful choice of material. The chosen material(s) must have the appropriate properties such as strength, ductility, fracture toughness and so on. A particular material that is ductile at high temperatures may become brittle at low temperatures (the comparison may be similar at room temperature and low temperatures). This transition is known as a *ductile-to-brittle transition*. The point at which this transition occurs is called the *DBTT point* (see Figure 10.28) and it is dependent on the chemical composition of the metal. Steel is one material that exhibits such behavior. In short, ductility may be taken as the material's ability to deform under tensile stress. Some structural failures may be caused primarily by the brittleness of the steel used in their design. As temperature decreases, a material's ability to deform in a ductile manner decreases. The DBTT is the minimal temperature that ensures the ability of a given material to absorb a specific amount of energy without fracturing and is characterized by a sudden drop in the absorbed energy by a metal subjected to an impact load. If the ductility suddenly decreases almost to zero at some temperature, the transition is often more abrupt, and this temperature is known as the *nil-ductility temperature* (NDT) or the nil-ductility transition temperature. Body-centered cubic structures do exhibit DBTT characteristics; face-centered cubic structures exhibit no DBTT characteristics and remain ductile at all temperatures. For low-temperature designs, the designer wishes to have the material exhibit a DBTT as low as possible. Steel that belongs to the BCC group evidently changes its fracture toughness over a small temperature range. Some of the experimentally obtained data relating to a Charpy V-notch (2 V) impact energy test for particular materials at room temperature (20–23 °C) are listed in Table 10.1.

Table 10.1 Charpy V-notch impact energy (CVN).

Material	CVN impact energy (J); (2 V- notch), (Room temp.)	Ultimate tensile strength, σ_m (MPa)	0.2 Offset yield strength, $\sigma_{0.2}$ (MPa)	Modulus of elasticity, E (GPa)
	Approximated values*			
S275 JR / 1.0044/ St-44-2	210	453	323	210
20MnCr5/ 1.7147/ AISI 5120	178	562	398	218
42CrMo4/ 1.7225/ AISI 4140	168	732	592	201
50CrMo4/ 1.7228/ AISI 4150	70	1147	1089	204
X6CrNiMoTi17-12-2/ 1.4571/ AISI 316Ti	157	633	459	185
X15CrNiSi25-20/ 1.4841/ AISI 314	311	687	497	218

* Experimental data obtained in the Laboratory for Structural Strength Testing, Department of Engineering Mechanics, Faculty of Engineering, University of Rijeka.

10.8 Crack Propagation

10.8.1 Introduction

Let's go back briefly to the core of the theory of design. In accordance with the requirements of the optimal structural design, it is necessary to carry out an assessment of the life of the structure. Two main principles of structural design were mentioned: "design against plasticity" and "design against fracture". Since fracture can occur not only as a result of excessive loads but also as a result of fatigue, it is reasonable to mention the principle of "design against fatigue". Undesirable fatigue failure, such as crack, may lead to fracture. Any design must bring together a combination of experience, experimental research and engineering estimates. Engineering elements, as well as engineering structures in general, may be exposed to static or dynamic loads, depending on their intended use.

As stated above, any structural element should have an assessed and guaranteed lifetime. The life assessment of the structural member can be based on the use of a $\sigma-N$ (or $S-N$) diagram. As discussed previously, this is constructed on the basis of laboratory tests of cyclic loading of specimens simulating service conditions. These tests are conducted under the assumption that the tested specimens are initially free from defects. In addition, tested specimens may be manufactured as unnotched or notched. The fatigue regime may be low-cycle fatigue or high-cycle fatigue.

On the other hand, linear elastic fracture mechanics in the life assessment of a structural element assume that a crack has already initiated in the material, and this is often the case. It is known that the static strength of a material differs from the dynamic strength. In structural design, it is important that the used data on the behavior of the material are as close as possible to those that will be present in the structure's service life. These considerations must also include special effects/factors that can reduce the measured data (experimentally obtained under laboratory conditions) relating to the strength of the material. Any larger defect may reduce the bearing capacity even for a statically loaded structure, but this should not be too dangerous for this type of loading as long as the mentioned defect and the load level remain below a critical level.

Our consideration here is, primarily, focused on the occurrence of cracks in the material. The occurrence of a crack in the material has a significant influence when a structural element operates under repeated (cyclic) loading because of fatigue effects. A crack may already exist in the material or it may be initiated due to cyclic loading and can grow to a critical size which is sufficient to cause brittle fracture. It would be unusual (and unexpected) for a crack of dangerous size to exist initially in a material. The material as received from the producer (or vendor) may contain defects of a small size such as porosities (small holes) in the base metal or in welded parts, or inclusions or small microcracks. During repeated load (cyclic loading), that is, due to fatigue, a microcrack can propagate to a large enough size to induce fracture, and such a crack propagation is called *fatigue crack growth*. This is typically characterized by the crack growth rate ($\mathrm{d}a/\mathrm{d}N$), where a is the crack length and N is the number of cycles. An adverse environmental impact, even under a steady load, may cause environmental crack growth, and this kind of crack growth can be especially supported by cyclic loading. Of special interest are two cases of crack growth: fatigue crack growth and creep crack growth. Fatigue crack in metals involves growth through the grains of the material (that is, transgranularly) and its growth is a consequence of cyclic slip; creep crack as well as corrosion crack involve intergranular growth (that is, growth between grain boundaries).

10.8.2 Fatigue Crack Growth

The fatigue fracture process is usually deemed to consist of three phases: crack initiation, crack propagation and final fracture. After an engineering element has broken, the fracture surface can be analyzed. In Section 10.2, we discussed the fracture surface orientation [3, 6]. Separation of the element into two parts is caused by crack propagation through the whole of the sectional area of the element. Crack initiation and its propagation can be caused by overload, fatigue, creep and so on. A crack can also grow from existing defects in the material associated with the load. Since some information relating to the fracture surface under fatigue, and thus with crack growth, were discussed in Section 10.2, here we will mention only a few details.

Crack initiation and its growth in the fatigue process are the consequence of cyclic slip. If fatigue is considered, for example, from the viewpoint of cyclic tension or cyclic torsion, as discussed in the aforementioned section, cyclic slip is considered the influential factor for microcrack initiation and its early growth. The maximum shear stress under tension occurs at 45° with respect to the longitudinal axis (Sections 2.4.1 and 6.1), but at the same time, in this plane occurs normal stress of the same intensity $(\varphi = 45^\circ \rightarrow \sigma = \tau = (1/2)\sigma_z = (1/2)F/A)$. In terms of torsional load, the maximum shear stress occurs in the planes that are parallel and perpendicular to the longitudinal axis (Figure 6.8(h)) and the normal stress component on that slip plane is zero. Therefore, these two cases (tension and torsion) are different when microcrack initiation is considered to be a matter of cyclic slip in a single surface grain. The cases of brittle and ductile fracture for both of these loads were discussed in Section 10.2. In terms of a tensile load, it is said that the fracture surface, that is, the growth of the crack, can take several directions: perpendicular to the principal stress, at 45° with respect to the longitudinal axis, the cup and cone type (shear lips at the same side of the central part of the surface), or shear lips parallel but on different sides with respect to the central part of the fracture surface. This depends on the material's classification as either brittle or ductile. When a material is of the brittle type (when the crack tip plastic zone is small), the crack surface is usually flat and perpendicular to the principal stress. In a tensile-loaded ductile material, a different situation can be encountered relating to the fracture surface. For this case, some of the positions of fracture surfaces are shown in Figure 10.29. As the crack grows (Figure 10.29(a)), the plastic zone around the crack tip increases and, although the crack plane propagation may start perpendicular to the principal stress, for example, as a horizontal plane, completion of its propagation takes place in such a way that the crack plane turns to a 45° shear (slant) mode. This shear mode can be singular (Figure 10.29(b)) or double shear (shear lips), as shown in Figure 10.29(c). Figure 10.20(a) shows a cracked specimen under tensile stress. The crack propagation can start immediately at 45° if the plastic zone related to the specimen thickness is high enough (plane stress conditions). Fatigue crack growth is a common mode of failure in aircraft, machines and other structures operating under cyclic loads.

Fatigue crack growth (or fatigue crack propagation) implies crack extension in the presence of, for example, cyclic loading. Knowledge of the fatigue crack growth rate is at the core of understanding fatigue fracture caused by crack propagation. Consider the

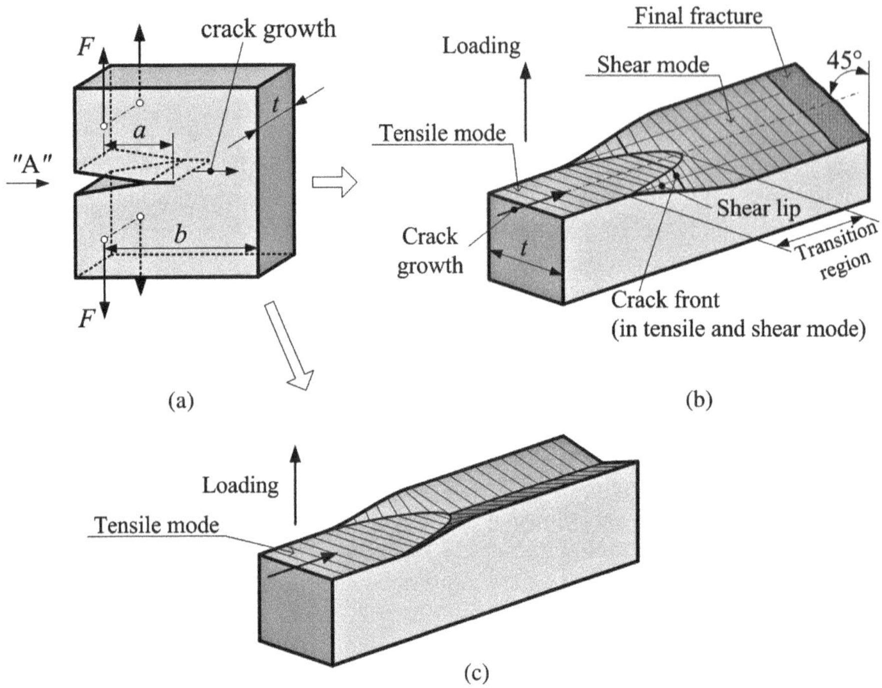

Figure 10.29 Tensile loading: crack growth transition from tensile mode to shear mode. (a) Specimen; (b) single shear mode; (c) double shear mode.

case of crack propagation under fatigue (cyclic) tensile stress loading of constant amplitude, as shown in Figure 10.30(a). At the crack tip, a plastic zone is formed and, because of crack propagation, a so-called plastic wake follows behind (Figure 10.30(b)).

In the case of a sufficiently small plastic zone at the crack tip, the stress intensity factor K defines the state/conditions around the crack tip. So, in our crack growth analysis, K may be used. This means that crack growth, and thus structure life to failure, may be controlled by K. As discussed previously, K depends on loading, crack length and geometry (from Equation (10.7)):

$$K = \beta\sigma\sqrt{\pi a}. \tag{a}$$

Let the applied cyclic loading of constant amplitude, σ_a, be defined with maximum stress, σ_{max}, and minimum stress, σ_{min}, where $(\Delta\sigma = 2\sigma_a)$. The stress ratio is defined as $R = \sigma_{min}/\sigma_{max}$. Since, in accordance with Equation (a), K and σ are proportional for a given crack length, the stresses and the stress ratio are as follows:

$$K_{max} = \beta\sigma_{max}\sqrt{\pi a}, \tag{10.59a}$$

$$K_{min} = \beta\sigma_{min}\sqrt{\pi a}. \tag{10.59b}$$

Further, we have:

$$\Delta K = \beta\Delta\sigma\sqrt{\pi a}, \tag{10.60a}$$

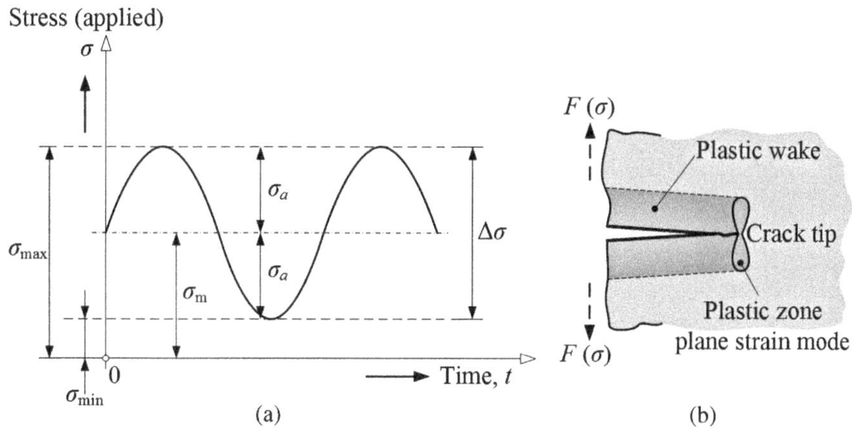

Figure 10.30 (a) Fatigue tensile loading with constant amplitude; (b) fatigue crack growth under small-scale yielding.

where:

$$\Delta\sigma = \sigma_{max} - \sigma_{min}. \tag{10.60b}$$

Finally, it follows that:

$$\Delta K = K_{max} - K_{min}, \quad R = \frac{K_{min}}{K_{max}}. \tag{10.61}$$

In general, many factors that affect the crack growth may be enumerated and the fatigue lifetime of a structure can be determined. Let the aforementioned cyclic loading of constant amplitude cause growth in a crack's length by an amount Δa due to a number of cycles ΔN. If a very small time interval is considered, these values can be written as da and dN. As such, the crack growth per cycle (or crack propagation per cycle, crack extension per cycle) may be written as a function of some parameters:

$$\frac{da}{dN} = f(P_i), \tag{10.62}$$

where P_i represents parameters such as the geometry of the specimen and the crack, the stress amplitude and stress ratio, the influence of the environment, mechanical material properties and so on.

Throughout history, some recommendations have been proposed for solving the problem described by Equation (10.62). The development of fracture mechanics, a very powerful tool in engineering failure analysis, has led to the knowledge that the stress intensity factor enables stress analysis in the vicinity of the crack tip, that is, it enables us to know the stress around the crack tip. Therefore, crack growth may be characterized using the K concept, that is, by using K_{min} and K_{max}.

Consider two fatigue processes that are carried out on specimens with the same properties, including the same initial crack length (see Figure 10.31(a)). One fatigue process is defined by $\Delta\sigma_1 = \sigma_{max1} - \sigma_{min1}$ and the second one by $\Delta\sigma_2 = \sigma_{max2} - \sigma_{min2}$, where

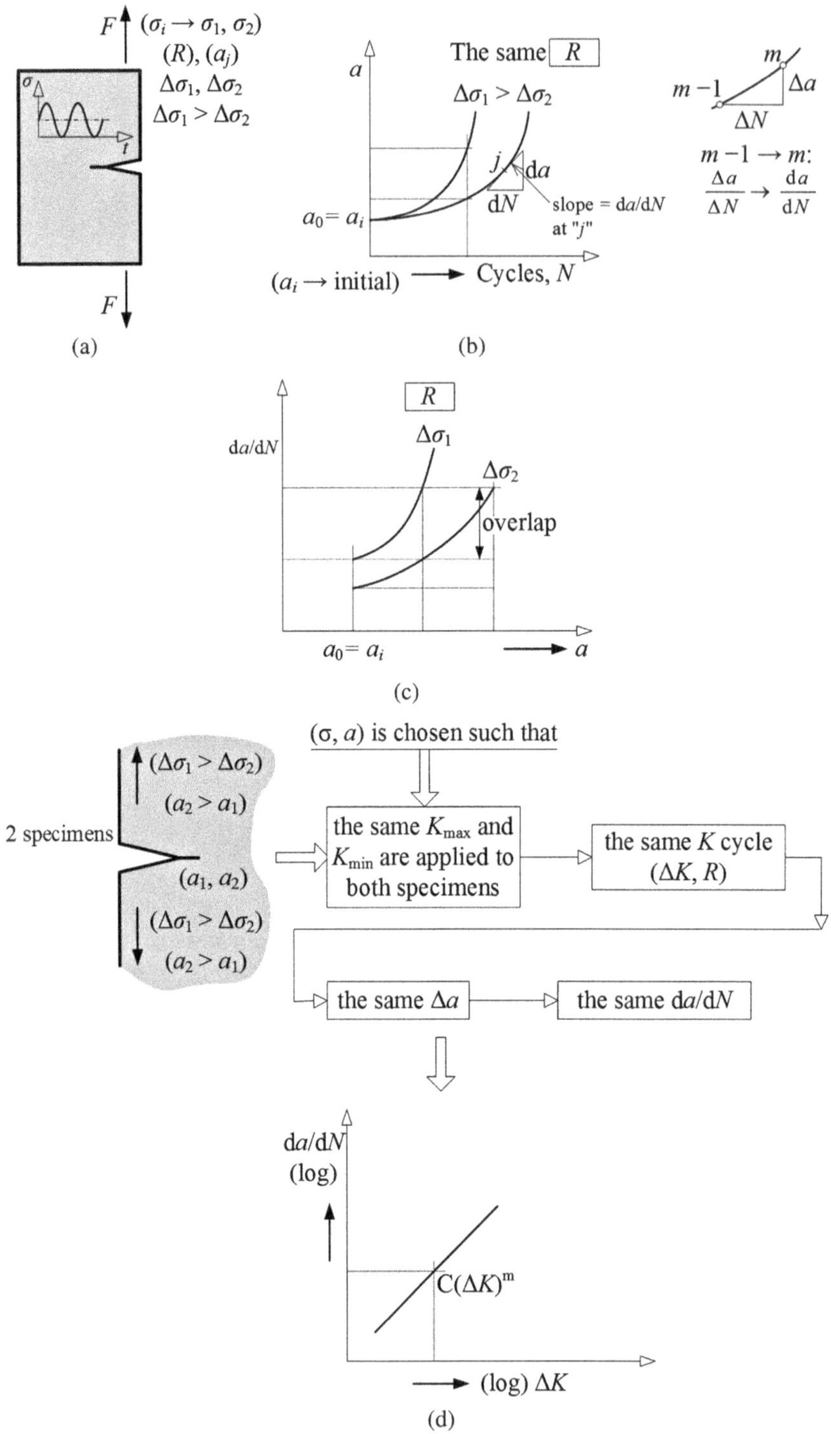

Figure 10.31 Fatigue crack growth and the principle of similarity. (a) Specimen; (b) fatigue processes with different $\Delta\sigma$; (c) overlapping; (d) the same K cycle.

$\Delta\sigma_1 > \Delta\sigma_2$ while the applied stress ratio, R, is the same for both processes (Figure 10.31 (b)). In accordance with that setup, we have:

$$\Delta\sigma_1/\sigma_{\max 1} = 1 - R = \Delta\sigma_2/\sigma_{\max 2} \text{ and } \sigma_{\max 1} > \sigma_{\max 2}$$

On the right-hand side of Figure 10.31(b), the designation ($\rightarrow$) means tends to. As can be seen from Figure 10.31(b), the curve representing crack growth versus cycle shows the slope da/dN of the crack growth which represents the crack growth rate. Crack growth simply means its extension, while the crack growth rate means the crack extension in a single cycle. It is evident that the crack propagation in the first case ($\Delta\sigma_1$) is faster; that is, for the same number of cycles, in the process characterized by $\Delta\sigma_1$, crack extension is more significant. Here, two different crack lengths occur. If the slopes (da/dN) for both curves are presented in terms of crack length, it is possible to display the curves representing crack growth rates shown in Figure 10.31(c), where a partial overlap can be seen. Based on this, it is expected that similar crack growth rates exist under two tests, although consider the different values for the crack length. This was first observed by Paris, and this leads to the principle of similarity based on the stress intensity factor. To clarify the principle of similarity, imagine that two similar specimens are being considered. One is a specimen with a large crack and this is cyclic loaded by low stress, while the other specimen with a small crack is cyclic loaded by high stress. The cyclic loading and lengths of the cracks of these specimens are chosen in such a way that they correspond to the condition that both specimens are subjected to the same K cycle, which also includes the same stress ratio, and the same crack extension should be expected [4] (see Figure 10.31(d)).

So, the crack growth rate, based on this principle of similarity, may be written as a function of $K_{\min}$ and $K_{\max}$. Equation (10.62) can now be written to describe the cyclic crack growth rate (crack growth/extension per cycle) in the form:

$$\frac{da}{dN} = f(K_{\max}, K_{\min}), \tag{10.63a}$$

and since the K cycle may be defined using Equation (10.61), that is, as a function of ΔK and stress ratio R, we can now write:

$$\frac{\mathrm{da}}{\mathrm{dN}} = \mathrm{f}(\Delta \mathrm{K}, \mathrm{R}) \tag{10.63b}$$

or simply:

$$\frac{da}{dN} = f(\Delta K) \tag{10.63c}$$

For a selected material, this equation characterizes the resistance of the material to fatigue crack growth. For the function f many expressions have been recommended and most of them were experimentally based. To estimate the fatigue life of a considered element – that is, to determine the number of cycles from an initial crack length ($a_0 = a_i$) to failure crack length (a_f) – Equation (10.63) needs to be integrated. In a case of excessive plasticity, when K can no longer characterize the condition around the crack tip, some researchers have applied the J integral to the fatigue process assuming the same form as in Equation (10.63). However, applying the J integral for the characterization of fatigue crack growth in a case of large-scale plasticity becomes questionable.

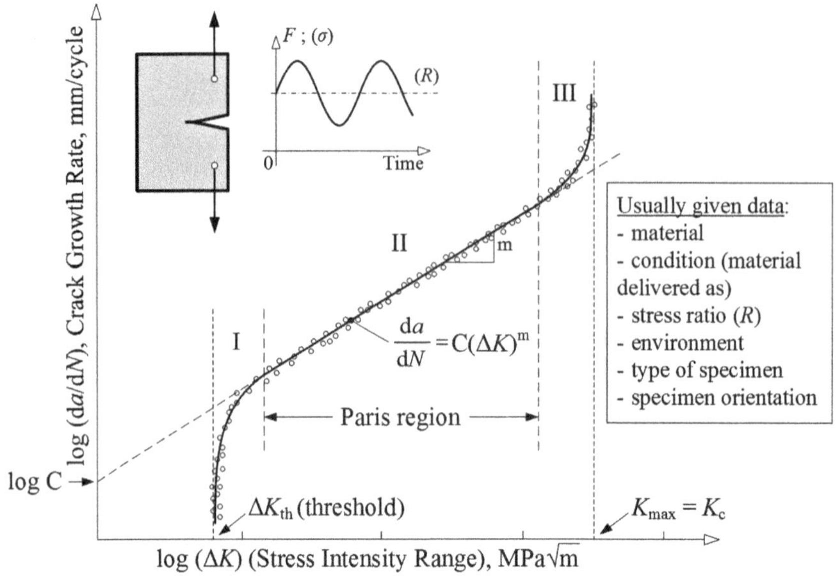

Figure 10.32 Schematic representation of the fatigue crack growth rate.

Equation (10.63) represents the crack growth rate in a general form. It is known that experimental results for crack growth rates in tested specimens made from a considered material (metal) and tested at a prescribed stress ratio (and Mode I) look similar to the representation in Figure 10.32, where a fitted curve is also shown [6, 14, 27]. This sigmoidal curve represents typical fatigue crack growth behavior in metals and it is a schematic log–log plot of da/dN versus ΔK. Paris (and Erdogan) was the first to research the crack growth rate in Region II (Figure 10.32) [8]. Figure 10.32 represents a general illustration of crack growth rate in metallic materials and is very similar to the illustration of the collected experimental data (at room temperature) by Paris *et al.* for one kind of steel at the prescribed stress ratio [18].

As can be seen, the fitted sigmoidal curve contains three distinct regions. Region I relates to the beginning of the crack growth, that is, to low growth rates, when the curve becomes very steep, and its asymptote is practically vertical. This quantity of ΔK is denoted ΔK_{th} and is called the *fatigue crack growth threshold*. It is the lowest limiting value of ΔK below which crack growth does not occur. The threshold value is associated with macro fatigue crack growth and not with micro cracks nucleated at the surface. An intermediate phase – Region II – is usually approximated by a straight line (on a log–log scale). Finally, in Region III, at high growth rates, the curve representing the crack growth rate may again become very steep. This can happen due to rapid, unstable crack growth that leads to the final failure of the specimen. The asymptote to the curve corresponds to $K_{max} = K_c$, that is, to the fracture toughness that is associated with the considered material and thickness of the specimen.

10.8.2.1 The Paris Equation

For the description of the linear part of the curve (Region II), shown in Figure 10.32, Paul Paris, in line with his opinion that fatigue crack growth increment da/dN depends only on the stress intensity factor range ΔK, has suggested the following power-law relationship [1, 4, 6, 14, 27]:

$$\frac{da}{dN} = C(\Delta K)^{m}, \tag{10.64}$$

in which m and C are experimentally determined material constants. As can be seen, m is the slope on a log–log plot in the intermediate region. On the basis of the shape of the central part of the curve, we can state [8] that the fatigue crack growth rate depends only on ΔK and is insensitive to the stress ratio, R, in this part. However, Equation (10.64) is applicable for the prescribed value of the stress ratio [5]. In accordance with experimental investigations, as shown in [6], the stress ratio, R, affects the crack growth rate versus ΔK. In general, it can be said that the mentioned constants are influenced by material microstructure, temperature, the environment and the stress ratio, R. Equation (10.64) is known as *the Paris equation*. In this double log plot, this relationship is represented as: $\log(da/dN) = \log C + m \log(\Delta K)$. The C and m values (Paris constants) for different types of steel may be found, for example, in [6, 11, 27], and are shown in Table 10.2 using the SI system for a case when $R = 0$. The values given in Table 10.2 are also publicly accessible (Fatigue Crack Growth – Paris Law) and correspond to those in the aforementioned references.

A lot of research [28–37] has been carried out concerning the determination of the coefficients C and m (m is the so-called strain hardening coefficient) in the Paris equation as well as for other equations that define crack growth rates. As for experiments dealing with fatigue processes in materials under constant but different stress ratios (S–N curves), where the stress ratio (R) affects the maximum applied stress (or stress amplitude or mean stress) and material endurance (see Figure 10.4), here, one would also expect a similar influence of the stress ratio on the crack growth rate. Experiments have indicated, as stated above, that an increase in the R ratio (which also means increasing the mean stress, see Equations (10.1) and (10.61)) of the fatigue loading causes an increase in crack growth rates in all parts of the sigmoidal curve for the considered ΔK. However, the increase in Region II may be small and is recognized as material dependent. So, similar to the mentioned effect of the R ratio observed in $S-N$ curves, the effect of the R ratio is also observed in the crack growth rate, as schematically presented in Figure 10.33 [6]. This effect is more pronounced in materials that are more brittle.

Since the Paris equation has some limitations regarding the R ratio and also Regions I and III are not included herein, some alternative equations have been proposed.

Table 10.2 Constants C and m for some types of steel (Paris equation: $da/dN = C(\Delta K)^{m}$, $R = 0$).

Material: Steels	$C\left[\frac{\text{m/cycle}}{(\text{MPa}\sqrt{\text{m}})^{m}}\right]$	m
Austenitic	$5.6 \cdot 10^{-12}$	3.25
Ferritic – Pearlitic	$6.89 \cdot 10^{-12}$	3
Martensitic	$1.36 \cdot 10^{-10}$	2.25

(Note: "m" in the numerator of "C" means meters)

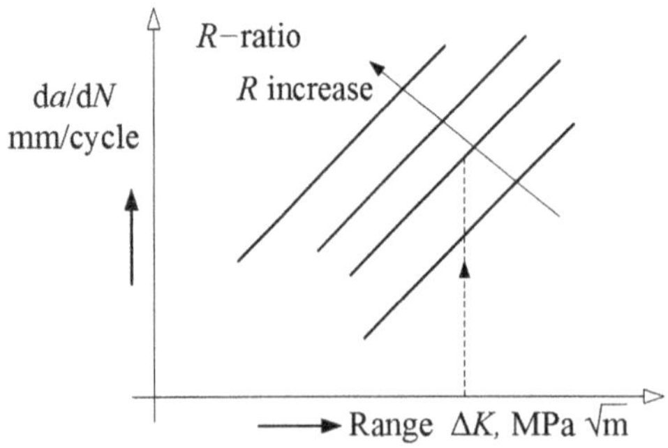

Figure 10.33 Schematic representation of the effect of stress ratio R on crack growth rates.

10.8.2.2 The Walker Equation

The Walker equation is an attempt to include the R effect in the da/dN versus ΔK curves. Similar to the Walker equation given for the equivalent zero-to-maximum stress range [6] (for the case $R = 0$),

$$\Delta\sigma_{eq,R} = \sigma_{max}(1-R)^{\gamma}, \tag{b}$$

where the use of γ as a constant of the material is limited to the range zero to 1.0, and may be treated as an inverse measure of the sensitivity to R or σ_m (the stress ratio, mean stress). When γ increases, sensitivity decreases. In accordance with Equations (10.59a)–(10.61), we can write:

$$\Delta K_{eq,R} = K_{max}\,(1-R)^{\gamma}. \tag{10.65}$$

The value of the constant γ depends on the material and varies between 0.3 and 1, but it is usually 0.5. It is understood that $\overline{\Delta K}$ causes the same crack growth rate as the K_{max}, R combination. Since, from Equation (10.61), we have:

$$K_{max} = \frac{\Delta K}{1-R}, \tag{c}$$

Equation (10.65) may now be writen as:

$$\Delta K_{eq,R} = \frac{\Delta K}{(1-R)^{1-\gamma}} \tag{10.66}$$

If, for the special case when $R = 0$ in Equation (10.64), constant C is designated C_0, then it can be written in the form:

$$\frac{da}{dN} = C_0(\Delta K)^{m} \text{ for } (R=0) \tag{d}$$

If the quantity given by Equation (10.66), as an equivalent ΔK for $R = 0$, is inserted into Equation (d) then, after mathematical rearranging, it follows that [6, 38]:

$$\frac{da}{dN} = \frac{C_0}{(1-R)^{m(1-\gamma)}}(\Delta K)^{m}. \tag{10.67}$$

On a log–log plot, this equation represents parallel straight lines of slope "m" (the family of da/dN versus ΔK curves), similar to those shown in Figure 10.33. Now, it is clear that constant C, which is the first term on the right-hand side of Equation (10.67), becomes a function of R while "m" is not affected by R. Constants (C_0, m and γ (for $R > 0$ and $R < 0$)) for the Walker equation may be found for some materials, for example, in [6].

10.8.2.3 The Forman Equation

To turn on the R effects on da/dN versus ΔK curves, Forman has recommended the following relation for Regions II and III [1, 4–6, 8, 18]:

$$\frac{da}{dN} = \frac{C_F(\Delta K)^{m_F}}{(1-R)K_c - \Delta K} \tag{10.68a}$$

and, using Equation (10.61), it follows that:

$$\frac{da}{dN} = \frac{C_F(\Delta K)^{m_F}}{(1-R)(K_c - K_{max})}. \tag{10.68b}$$

In Equations (10.68), K_c is the critical value of the fracture toughness of the material and thickness considered (K_c – plane stress or K_{Ic} – plane strain), while C_F and m_F are constants in the Forman equation. Equations (10.68) can be used to fit the data in intermediate and accelerated (high-growth rate) regions because when K_c tends to K_{max}, the denominator tends to zero and da/dN becomes large. Constants C_F and m_F may be found in the literature, for example, in [6].

10.8.2.4 The Forman–Newman–de Koning Equation

The equation that can include all of the crack growth rate regions of the crack growth curve is the Forman–Newman–de Koning equation, usually written in the form [1, 39–40]:

$$\frac{da}{dN} = C\left[\frac{1-f}{1-R}\Delta K\right]^n \frac{[1-(\Delta K_{th}/\Delta K)]^p}{[1-(K_{max}/K_c)]^q}. \tag{10.69}$$

This expression was jointly introduced by NASA and ESA, and their fatigue crack growth computer program, *NASGRO*, can be used in crack propagation analysis. In Equation (10.69), R is the stress ratio, ΔK and ΔK_{th} are the stress intensity range and the stress intensity range threshold, while C, n, p and q are experimentally derived constants. Constant C differs from the Paris constant. The factor f is the crack opening parameter and it refers to the effect of closure behavior on the crack growth rate under constant-amplitude cyclic loading [1].

10.8.3 Creep Crack Growth

Since the creep phenomenon was discussed in Chapter 9, here, failure due to creep will be mentioned only briefly. Chapter 9 outlines many elements that relate to the details of creep fracture mechanisms, creep deformation and so on. Engineering components such as pressure vessels, boiler tubes and so on are subjected to uniform loading at elevated temperatures, and they fail by creep rupture. On the other hand, turbine blades that are subjected to stress and temperature gradients often experience creep crack growth as a result of crack propagation starting from a critical location (stress raisers, stress concentrations) [41]. In structures subjected to conditions which allow the occurrence of creep,

creep fracture after an incubation period may be followed by a period of crack growth and a loss of structural integrity can be expected. Creep crack growth, as a time-dependent phenomenon, can be induced by creep effects, environmental effects or both.

10.8.4 Life Assessment of Engineering Components

10.8.4.1 Constant Amplitude Loading

In accordance with many experimental results mentioned in the literature [1, 6, 11, 14], it is clear that an initial crack propagates with the number of cycles (N), and the crack growth rate ($\mathrm{d}a/\mathrm{d}N$) depends on the stress intensity range (ΔK), as is generally described by Equation (10.63b):

$$\frac{\mathrm{da}}{\mathrm{dN}} = \mathrm{f}(\Delta \mathrm{K}, \mathrm{R}). \tag{a}$$

A quantity that defines the crack growth rate in Equation (a) may be any of the previously mentioned equations, for example, the Paris equation (Equation (10.64)). If the initial defect size (crack size) is known, the fatigue life, or, more accurately, the structural service life, can be determined/estimated on the basis of the crack propagation law when the parameters needed for the crack propagation law are known. The lifetime of an engineering element relating to the period from an initial crack size, a_i, to the considered/required crack size, a_f, that is achieved by crack propagation, calculated by the number of cycles is:

$$\int_{a_\mathrm{i}}^{a_\mathrm{f}} N = \int_{a_\mathrm{i}}^{a_\mathrm{f}} \frac{\mathrm{d}a}{f(\Delta K, R)} \tag{10.70}$$

The integral boundaries are: N_i, N_f and a_i, a_f. The crack growth goes from an initial size a_i to the final size a_f (which may be the crack size at fracture or at a certain required size of the crack, or the crack size at a required number of cycles). N_i corresponds to a_i and N_F to a_f. The number of cycles in the considered period of crack propagation, that is, in the period between the two crack sizes (a_i and a_f) is the difference between the number of cycles N_f and N_i. Let this difference be designated by $\Delta N = N_\mathrm{f} - N_\mathrm{i}$. Since the similitude of the crack tip condition implies a unique relationship between the crack growth length ($\mathrm{d}a/\mathrm{d}N$), the stress intensity range (ΔK) and the stress ratio, R, it is valid only for constant amplitude loading ($\mathrm{d}a/\mathrm{d}K = 0$). When the constant amplitude loading assumption is used, it means that excessive changes in stress between the initial and final crack lengths are excluded/disallowed.

10.8.4.2 Variable Amplitude Loading

In contrast to the ideal situation where constant amplitude loading is applied to a structure, over its service life a real structure is usually subjected to a spectrum of stress. The crack growth rate of such a loaded structure at any moment depends not only on current stress conditions but also the loading history. In this case, the estimation of crack growth, Δa, for each cycle of variable amplitude loading, is based on the crack growth rate versus stress intensity range curve. If all the crack extensions for all cycles are summed, a life estimation can be made (the procedure is equivalent to numerical integration using variable a, which is the crack length).

Let Δa_k be a crack extension for one cycle. The starting crack length is a_k, and the crack length for the next cycle is a_{k+1}, so we have:

$$a_{k+1} = a_k + \Delta a_k, \Delta a_k \left(\frac{da}{dN}\right)_k. \tag{10.71}$$

10.8.5 Crack Closure

Although the Paris equation (Equation (10.64)) has become a powerful tool in the modeling of crack propagation behavior, that is, in describing da/dN versus ΔK curves, it is not acceptable as a general solution. Experimental data for a wide range of tests shows the shape of the curve is not fully covered by this equation and, as discussed, it is insensitive to the effect of the stress ratio, R.

10.8.5.1 Elber Crack Closure Phenomenon

Under consideration is crack closure phenomenon at constant amplitude loading for various positive stress ratios (R). The fatigue crack growth rate in linear elastic fracture mechanics is usually predicted under the assumption of an idealized fatigue crack that looks like a saw-cut of zero width. This implies that the field condition ahead of the crack tip provides crack propagation in small-scale yielding. A tensile stress keeps the crack open. Based on elastic material behavior, crack opening displacements return to zero after the tensile stress is removed. However, Elber [1, 4–5, 8, 42] discovered that a fatigue crack under tensile unloading was already closed before the stress reached zero, that is under existing positive tensile stress. This has resulted in the belief that crack propagation and crack growth rate are influenced by the conditions ahead of the crack tip as well as by the nature of the crack surfaces behind the crack tip; that is, the crack growth rate depends on prior history.

However, although small-scale yielding is being considered, a plastic zone is created and it is proportional to $(K_{max}/\sigma_{YS})^2$ (from Equation (10.15)). The plastic zone that is created is also plastically elongated in the loading direction and is, at any time, longer than before. As a consequence, the zone is loaded under compression during the process of unloading, since the tensile load did not reach zero. As a result of tensile deformation occur plastically deformed material and residual compressive stresses. In all stages, the crack tip plastic zone forms an envelope containing all previous plastic zone sizes, and this behavior causes crack closure (contact realized between the crack surfaces).

This explanation refers to the so-called crack closure effect of constant fatigue amplitude loading for different positive R values. In this sense, Elber's discovery provides certain clarification relating to the threshold and the effect of the stress ratio, R. In accordance with his discovery and explanation of the crack closure process, Elber compares so-called compliance (dv/dF) at high and low loads. He states that, at high loads, compliance corresponds to the standard formulas for specimen behavior description, but at low loads, the compliance is close to that of an uncracked specimen and thus the change in compliance is due to contact between the crack surfaces. It is concluded that only a part of the stress intensity range, the so-called effective stress intensity range, ΔK_{eff}, is responsible for the fatigue crack growth rate and this is a result of crack closure, as shown in Figure 10.34(a). As can be seen, a

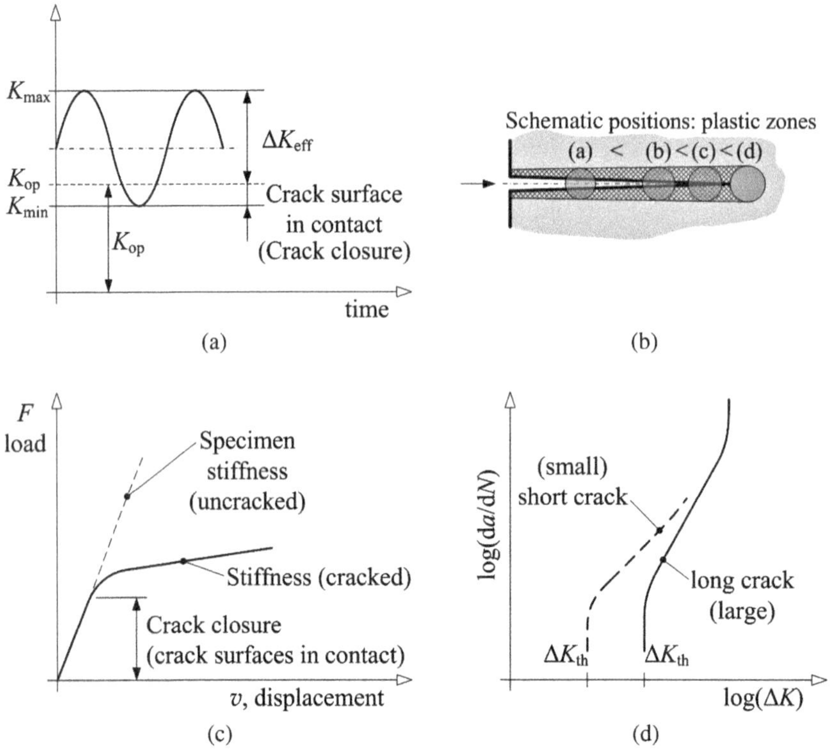

Figure 10.34 Crack closure interpretation. (a) Effective stress intensity range; (b) schematic representation of plastic zones; (c) compliance – crack closure representation; (d) ΔK_{th}: short–long crack behavior.

crack opens at K_{op}, while below that, the crack surfaces are in contact (crack closure), which means that this portion of ΔK is not involved in crack growth. The effective stress intensity range ratio is introduced:

$$U = \frac{\Delta K_{eff}}{\Delta K} = \frac{K_{max} - K_{op}}{K_{max} - K_{min}} = \frac{1 - K_{op}/K_{max}}{1 - R}, \tag{10.71a}$$

and the Paris (Paris–Erdogan) equation (Equation (10.64)) is modified as follows:

$$\frac{da}{dN} = C\Delta K_{eff}^{m}. \tag{10.72}$$

A schematic representation of the plastic zones during crack propagation as well as an explanation regarding compliance of specimen stiffness are shown in Figures 10.34(b) and (c). Since the plastic deformation during the fatigue process affects the partial closure of the crack surfaces, the crack will be closed and opened at a stress level higher than zero tensile stress.

Other researchers have also dealt with this issue and additional crack closure mechanisms have been identified and discussed [43]. The effect of the crack length on the threshold stress intensity factor ΔK_{th} has also been studied. If a crack is classified as being long (sometimes also described as large)or short (sometimes also described as small), its behavior towards the threshold value may be shown as in Figure 10.34(d). The two terms (short/small and long/large) that relate to the crack length are used interchangeably and have the same meaning. Usually (using ASTM standard E 647-00), small fatigue cracks may be classified as [21]: mechanically small ($a \leq r_y$); microstructurally small ($a \leq 5-10d_g$), physically small ($a \leq 1$mm) or chemically small ($a < 10$ mm), where a is the crack length, r_y is the plastic zone size of the notch and d_g is the grain size.

10.8.6 A Brief Review of Testing of Unnotched, Axially Loaded Specimens

In Chapter 4, and also in this chapter, some details relating to the behavior of an unnotched specimen subjected to tensile loading have been presented. Based on static tensile tests and cyclic tests, some material properties can be obtained. Classifications relating to tensile tests can be made in terms of:

1) **The nature of loading:**
   A) Static tensile tests, the results of which can be:
      A_1 – Engineering stress–strain diagrams;
      A_2 – True stress–strain diagrams;
      A_3 – Force–displacement diagrams for plane strain fracture toughness determination.

      Static tensile tests can be performed at low, room or high temperatures.
      Properties that can be obtained by performing static tensile tests are:

      a_1 – Ultimate tensile strength;
      a_2 – Yield strength;
      a_3 – Modulus of elasticity;
      a_4 – Fracture toughness.

   B) Cyclic (fatigue) loading tests, the results of which can be:
      B_1 – S–N curve (fatigue curve, Wöhler curve).

A property that can be obtained on the basis of this curve is:
   b_1 – the fatigue limit (endurance limit), which is associated with the fatigue life.

Cyclic loadings can be:
- Low-cycle fatigue (high-level fatigue): up to 10^4 cycles;
- High-cycle fatigue (low-level fatigue): large number of cycles, 10^4 or more.

Cyclic loading may refer to constant amplitude loading or variable amplitude loading, as well as to different stress ratios. However, the boundary between low-cycle fatigue and high-cycle fatigue is not strictly defined in terms of the number of cycles. The difference between these two types of fatigue loadings can be distinguished in terms of deformation and is associated with macroplastic deformation in any cycle under low-cycle fatigue and with elastic behavior on a macro scale under high-cycle fatigue.

2) **The specific phenomena:**
   A) The Bauschinger effect;
   B) Hysteresis loop;
   C) Causes of crack initiation and propagation, which include:

- Fatigue loading,
- Creep,
- Other effects.

D) Crack propagation properties, which include:
- Crack growth rates,
- Threshold value,
- Crack closure.

Stress–strain tension diagrams (engineering diagrams and true/real, actual diagrams) were presented in Chapter 4 (see Figure 4.8). True stress is obtained by dividing the load to which the specimen is currently subjected at a particular point by the actual cross-sectional area that corresponds to that particular point instead of the original cross-sectional area. Just as a note, in engineering analysis, one usually uses engineering stress–strain data instead of the true stress–strain data, since engineering design is based on the elastic range with the intention of avoiding permanent deformation and maintaining structural integrity. These two types of tension diagram (engineering and true) are presented once again in Figure 10.35(a). The first under consideration is the engineering stress–strain diagram obtained by conducting a monotonic tensile test (loading proceeds in only one direction) on smooth, unnotched specimens. Part of a compression test is also shown.

Let's now consider a standard true stress–true strain diagram, as shown in Figure 10.35 (b). It is clear that, upon unloading of the tensile-loaded specimen at the point N in the inelastic (nonlinear) part of the considered stress–strain diagram, the total strain is composed of an elastic part and a plastic (permanent set) part. As can be seen, under consideration is the tensile test with $R = 0$, which means that unloading is going from the tensile stress (force) at the considered point, N to the point at which stress equals zero. The unloading path is a straight line equal to the slope of the elastic modulus when it is presented as an ideal line; however, in reality, it is modified as shown on the right of the figure. The plastic strain is a part of the total strain that is not recovered upon unloading, and it forms a larger part of the total strain when reloading is done at a later stage of the tension loading process – that is, as the point N is moved to the right. When the specimen is reloaded, the reloaded stress–strain path will be the same "straight" line as at unloading. It can be seen that, upon loading and unloading to a zero stress level, the reloading path will always follow the stress–strain curve that is the original curve associated with the monotonic true stress–strain curve.

Let's consider now the part of the stress–strain curve between the yield point and the ultimate tensile strength point (that is, the data between Y and M). Here, Y represents the yielding point and M represents the ultimate tensile strength point. This curve is called the yield curve (or flow curve) and may be represented by the power law equation (Equation (4.3)) [1, 43–44]:

$$\sigma = K(\varepsilon_p)^n \tag{a}$$

which is usually known as the Holloman–Ludwig equation, as mentioned in Chapter 4, and is presented as an inclined straight line. So, for most metals, a logarithmic plot of true stress versus true plastic strain, that is, a plot of the so-called flow curve, is a straight line, as shown in Figure 10.35(b), which may be expressed as indicated in Equation (a). ε_p is the true plastic strain, K is the strength coefficient (the intercept on a log scale of true

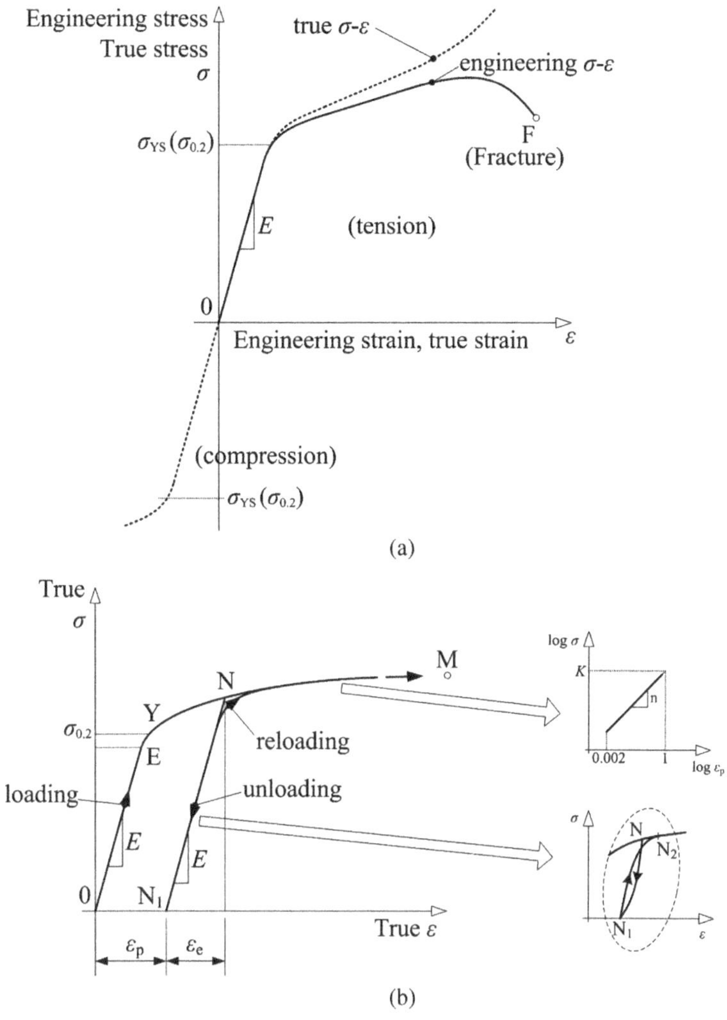

Figure 10.35 Cyclic phenomena. (a) Engineering and true stress–strain diagrams; (b) total strain; (c) Bauschinger effect; (d) cyclic hardening under strain-controlled amplitude cycling; (e) stable stress–strain hysteresis loop (completely reversed, $R = -1$); (f) the cyclic stress–strain curve.

stress versus true strain, at plastic strain equal to 1), while n is the strain-hardening exponent (slope). Coefficient K is not the earlier-mentioned stress intensity factor.

When total true strain is considered, that is, when we also include the elastic part, it may be given by [1, 43]:

$$\varepsilon = \varepsilon_e + \varepsilon_p \longrightarrow \varepsilon = \frac{\sigma}{E} + \left(\frac{\sigma}{K}\right)^{1/n} \tag{10.73}$$

Equation (10.73) is known as the Ramberg–Osgood relationship. When the actual (true) unloading path from the point N to the stress equal to zero is presented, and a reloaded

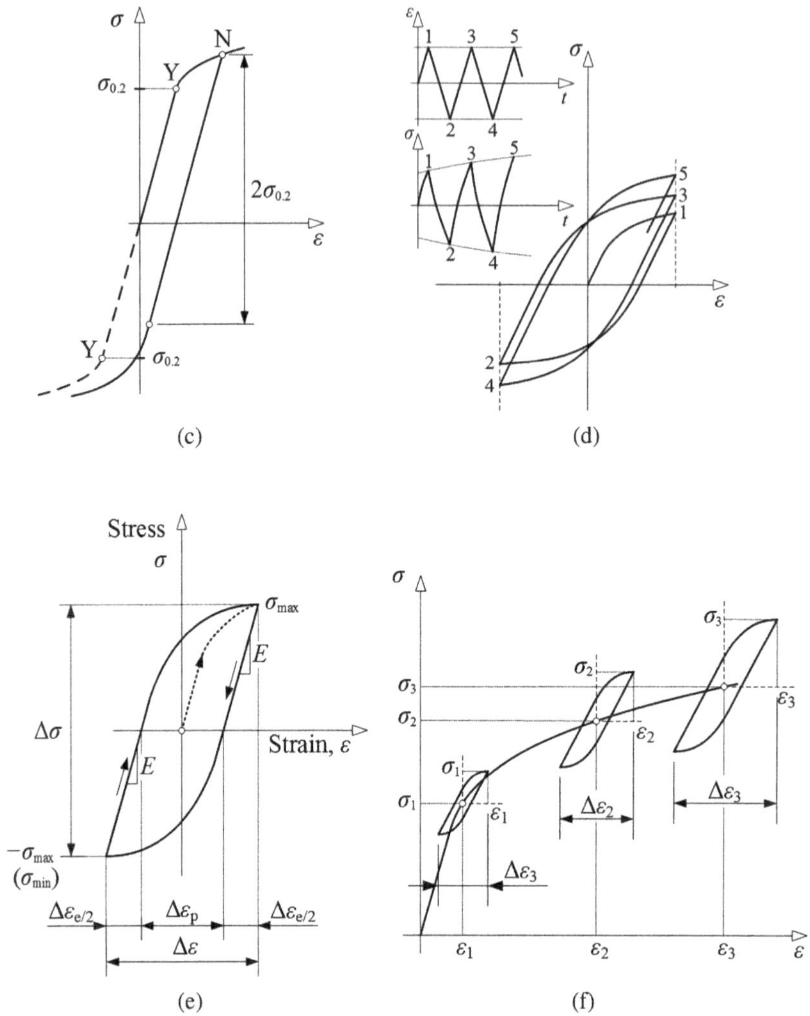

Figure 10.35 *(Continued)*

path is also plotted, then it can be observed that neither of the mentioned paths follows the ideal linear elastic unloading/reloading path. The difference between these curves (N, N_1 and N_1 N_2), that is, between the unloading path and the reloading path, is a phenomenon known as *hysteresis* (see Figure 10.35(b)). It can also be observed that the new position of the point N (that is, N_2) is increased. So, if a specimen is subjected to stress above the elasticity limit (E), for example, to stress that corresponds to the point (N) (Figure 10.35(c)), and after that it is unloaded, it is observed that in the reloading, the levels that correspond to (E) and (N) will both be increased. This situation can be seen when a specimen is subjected to cyclic loading and reloading of $\pm\sigma$, as presented in Figure 10.35(d), where a generated hysteresis loop for a complete, fully reversed cycle $(R = -1)$ is visible.

In addition, a new phenomenon is observed when fully reversed fatigue cyclic (fatigue) loading is performed. The difference between a monotonic tensile (compression) test and a cyclic test was first observed by Bauschinger. In Figure 10.35(c), the unloading path starts from the stress that corresponds to the point N and moves to a point where the stress equals zero; after that follows a reversed path which refers to compression loading. Here, it was observed that the stress–strain path followed (that is, the compression path) differed from the initial monotonic one. It ican be seen that yielding on the unloading path occurs (at point defined by 2Y from N) prior to the yield strength that corresponds to the monotonic compression. Since this early yielding was first noted by Bauschinger, it is known as the *Bauschinger effect* (Figure 10.35(c)). Later it was recognized that the quantity of $2\sigma_{0.2}$ (that is, $2\sigma_{YS}$) takes place from the reached level $\sigma_{\max}$ (or $\sigma_{\min}$).

Depending on the metal's initial state (annealed, quenched and tempered, cold worked, normalized and so on) and the test conditions, a metal may be cyclically stable or may cyclically harden or soften or may exhibit mixed behavior [44, 45]. Some experimental research results relating to low-cycle fatigue at room and elevated temperatures for a well-known material (steel 9Cr-1Mo) can be found in [46]. Under cyclic loading, the material response is usually characterized by strain-controlled conditions rather than stress-controlled conditions. Annealed materials, in general, harden under strain-controlled conditions. A case of cyclic hardening under strain-controlled amplitude cycling is shown in Figure 10.35(d). Cyclic stabilization may be treated as complete within 40% of the fatigue life. A hysteresis loop achieved at about 50% of the fatigue life may represent stable (steady-state) cyclic stress–strain behavior (Figure 10.35(e)), which is schematically presented in the figure for completely reversed ($R = -1$) strain-controlled conditions with zero mean strain. A similar situation may also be shown for cyclic softening under controlled strain amplitude cycling. Finally, let it be assumed that three specimens were tested to failure at three different controlled strain amplitudes, and a cyclic stress–strain curve (based on stabilized hysteresis loops) was made (see Figure 10.35(f)).

To model the true cyclic stress–strain curve, we can use a similar model to the one used for the monotonic, true tensile stress–strain curve. As such, similar to Equation (10.73), we can now write:

$$\varepsilon = \frac{\sigma}{E} + \left(\frac{\sigma}{K*}\right)^{1/n^*}, \tag{10.74}$$

where $K*$ is the cyclic strength coefficient and $n*$ is the cyclic hardening exponent (the value is between 0.1 and 0.2). As discussed earlier, the fatigue limit (endurance limit) has been experimentally determined for many steels using Wöhler's curves (S–N curves) obtained from the results of cyclic constant stress–amplitude loadings for different stress ratios. The fatigue limit as a material property is of paramount importance for designing structural elements subjected to cyclic loading, where fatigue takes a leading role in the selection of a suitable material and in the assessment of the structural service life. This property (the fatigue limit) does not exist for high-strength steels and nonferrous metals (aluminum alloys) [44]. As such, for structures made of such materials, obtaining a cyclic stress–strain curve may be useful and very important, since the cyclic stress–strain curve can be compared to the monotonic or tensile stress–strain curve and may serve directly in the quantitative assessment of cyclically induced changes in the mechanical behavior of a structural element. A cyclic stress–strain curve for a material can be obtained based

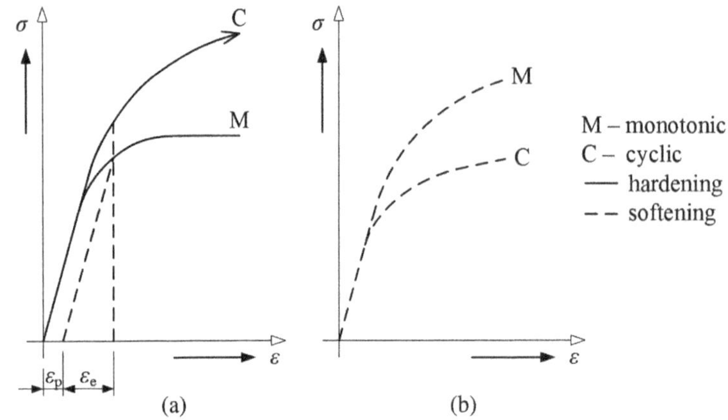

Figure 10.36 Comparison of monotonic and cyclic stress–strain curves. (a) Cyclic hardening; (b) cyclic softening.

on several fatigue tests performed on smooth, highly polished specimens. These fatigue tests are performed under strain-controlled conditions with a stress ratio $R = -1$ and different strain amplitudes. The stress–strain curve is constructed via the tensile tips of stabilized hysteresis loops of different strain amplitudes (see Figure 10.35(f)).

The material behavior under strain-controlled conditions depends on the nature of the heat treatment. Some materials (annealed materials) will harden; some materials (cold worked materials) have the tendency to become soft. Figure 10.36 shows a schematic representation of monotonic and cyclic stress–strain curves for cases of cyclic hardening and cyclic softening.

Example 10.1.

Consider a center-cracked specimen subjected to cyclic stress of constant amplitude, as shown in Figure 10.37.

Data:

Let the data relating to the specimen and loading be known. The specimen is made of steel. Cyclic loading is of constant amplitude with stress ratio $R = 0$. The crack growth curve and stress intensity ranges are as shown in Figure 10.37. For this center-cracked specimen with initial crack length of $a_i = 2.3$ mm and maximum applied stress of $\sigma = 80$ MPa, it is necessary to determine: a) the Paris and Forman constants and b) the number of cycles that correspond to crack growth of $\Delta a = 20$ mm using the Paris equation.

Solution:

a) The Paris and Forman constants

On the basis of two points from the linear portion of the crack growth curve, the mentioned constants may be determined as detailed below.

∗Paris constants: C, m

On the basis of the given data (Figure 10.37), we have:

$$da/dN = 0.00001\,\text{mm/cycle}, \Delta K = 6.8\,\text{MPa}\sqrt{\text{m}};$$

$$da/dN = 0.0001\,\text{mm/cycle}, \Delta K = 37\,\text{MPa}\sqrt{\text{m}};$$

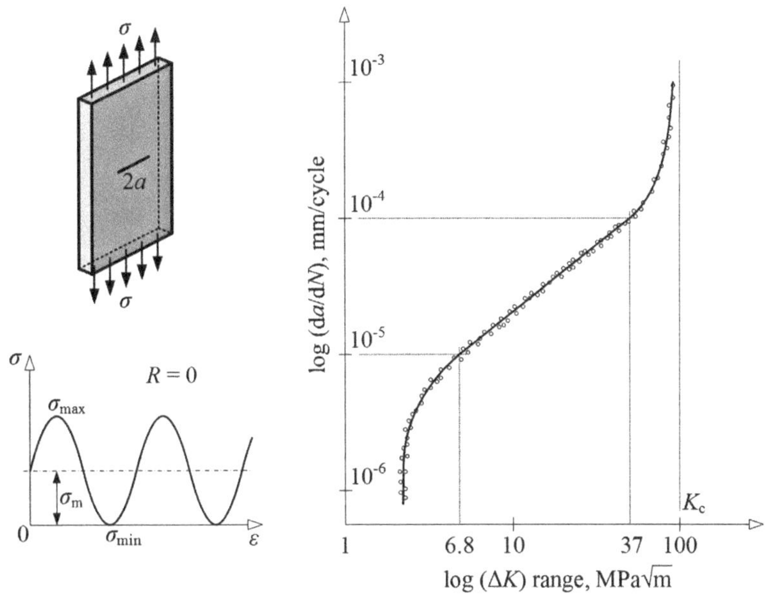

Figure 10.37 A center-cracked specimen subjected to a cyclic load of constant amplitude (Example 10.1).

Using Equation (10.64), it follows that:

$$\begin{aligned} \frac{da}{dN} &= C(\Delta K)^m, \\ 0.00001 &= C(6.8)^m \\ 0.0001 &= C(37)^m \end{aligned} \tag{a}$$

Solving these equations with respect to C and m, it follows that:

$$10 = 5.441^m$$

$$m = 1.3593;\ \ C = 7.3848 \cdot 10^{-7}\left[\frac{mm/cycle}{(MPa\sqrt{m})^m}\right].$$

∗Forman constants: C_F, m_F

Using Equation (10.68a), it follows that:

$$\begin{aligned} \frac{da}{dN} &= \frac{C_F(\Delta K)^{m_F}}{(1-R)K_c - \Delta K},\ (R=0) \\ 0.00001 &= [C_F(6.8)^{m_F}]/(100-6.8) \\ 0.0001 &= [C_F(37)^{m_F}]/(100-37) \\ 10 &= 5.441^{m_F} \cdot (63/93.2) \\ m_F &= 1.5905;\ \ C_F = 2.0189 \cdot 10^{-5}. \end{aligned} \tag{b}$$

The Forman equation requires data on the critical stress intensity factor. Sometimes this property can be experimentally determined or it can be calculated on the basis of known impact energy using the Roberts–Newton formula. Data relating to the impact energy for one very widely used material in pressure vessel applications (ASTM A 516 steel) can be found in [47]. Descriptions of some impact tests can be found in [48].

b) The number of cycles corresponding to the growth of the crack $\Delta a = a_f - a_i = 20\,\text{mm}$.
Consider crack growth rates as defined by Equation (10.64):

$$\frac{da}{dN} = C(\Delta K)^m \tag{c}$$

Let ΔK be defined as (from Equation (10.61)):

$$\Delta K = K_{max} - K_{min}, \tag{d}$$

and (from Equation (10.60 a)):

$$\Delta K = \beta \Delta \sigma \sqrt{\pi a}. \tag{e}$$

Assume that $\beta = \beta(a/b)$ can be approximated as constant over the range of crack growth from a_i to a_f. This means that $\Delta\sigma$ and R may also be treated as constants. As such, integration of Equation (10.70) results in the number of cycles that need to be applied for crack growth from the initial (a_i) to the requested (a_f) crack length, that is, for the requested crack extension (Δa):

$$\int dN = \int \frac{da}{f(\Delta K, R)} \tag{f}$$

$$\Delta N = N_{if} = \int_{a_i}^{a_f} \frac{da}{C(\Delta K)^m} = \int_{a_i}^{a_f} \frac{da}{C(\beta \Delta \sigma \sqrt{\pi a})^m} \tag{10.75}$$

For the considered case, $\beta = 1$ (assumed constant).

Since, C, $\Delta\sigma$, β and m are all constants, integration of Equation (10.75) gives:

$$\Delta N = N_{if} = \frac{a_f^{1-m/2} - a_i^{1-m/2}}{C(\beta \Delta \sigma \sqrt{\pi})^m \left(1 - \frac{m}{2}\right)} \tag{10.76}$$

For the case m = 2, this equation becomes indeterminate. For the considered case, we have:

$$\Delta\sigma = \sigma_{max}; \quad (R = 0)$$

After the inclusion of the data in Equation (10.76), it follows that:

$$\Delta N = N_{if} = \frac{0.0223^{1-1.3593/2} - 0.0023^{1-1.3593/2}}{7.3848 \cdot 10^{-7}(80\sqrt{\pi})^{1.3593}\left(1 - \dfrac{1.3593}{2}\right)}$$

$$\Delta N = N_{if} = 4183.9\ \text{cycles}$$

We recommend further study of experimental investigations relating to material fatigue behavior and the determination of the fatigue limit (endurance limit). Information about such investigations for a defined/prescribed stress ratio, R, in the field of the finite fatigue region and the infinite fatigue region, as well as the fatigue limit calculation can be found in [49, 50].

References

1 Farahmand, B., Bockrath, G. and Glassco, J. (1997) *Fatigue and Fracture Mechanics of High Risk Parts*, Chapman & Hall, New York.
2 Gross, D. and Seeling, T. (2011) *Fracture Mechanics*, 2nd edition, Springer, New York.
3 Brooks, C. R. and Choudhury, A. (2002) *Failure Analysis of Engineering Materials*, McGraw-Hill, USA.
4 Schivije, J. (2009) *Fatigue of Structures and Materials*, Springer, Delft, the Netherlands.
5 Suresh, S. (2003) *Fatigue of Materials*, Cambridge University Press, UK.
6 Dowling, N. E. (2013) *Mechanical Behavior of Materials*, Pearson, New York.
7 Saxena, A. (1998) *Nonlinear Fracture Mechanics for Engineers*, CRC Press, New York.
8 Anderson, T. L. (1995) *Fracture Mechanics*, CRC Press, New York.
9 Hertzberg, R. W. (1996) *Deformation and Fracture Mechanics of Engineering Materials*, John Wiley & Sons, New York.
10 Zhang, L. (2010) *Failure Assessment of Thin-Walled Structures with Particular Reference to Pipelines*, WIT Press, Southampton.
11 Barson, J. M. and Rolfe, S. T. (1999) *Fracture and Fatigue Control in Structures*, Butterworth-Heinemann, Woburn.
12 Collins, J. A. (1993) *Failure of Materials in Mechanical Design*, 2nd edition, John Wiley & Sons, New York.
13 Blinn, M. P. and Williams, R. A. (1997) Design for Fracture Toughness. In G. E. Dieter (Volume Chair) *ASTM Handbook*, Vol. **20**, Materials Selection and Design, ASM International, pp. 534–544.
14 Recho, N. (2012) *Fracture Mechanics and Crack Growth*, John Wiley & Sons, Chichester.
15 Freiman, S. W. and Mecholsky, J. J. (2012) *The Fracture of Brittle Materials*, John Wiley & Sons, USA.
16 Pommier, S., Gravouil, A., Combescure, A. and Moes, N. (2011) *Extended Finite Element Method for Crack Propagation*, John Wiley & Sons, New Jersey.
17 Zhang, L. (2010) *Failure Assessment of Thin-Walled Structures with Particular Reference to Pipelines*, WIT Press, Boston.
18 Kanninen, M. F. and Popelar, C. H. (1985) *Advanced Fracture Mechanics*, Oxford University Press, New York.
19 Kuna, M. (2013) *Finite Element in Fracture Mechanics*, Springer Science + Business Media, New York.
20 Vukelić, G. (2011) *Numerical Analysis of Crack Propagation in Structures* (in Croatian), PhD thesis, Faculty of Engineering, University of Rijeka, Rijeka.
21 Annual Book of ASTM Standards (2005) *Section 3*, Metals Test Methods and Analytical Procedures, Vol. 03.01, Metals – Mechanical Testing; Elevated and Low-Temperature Tests; Metallography, ASTM International, West Conshohocken, PA.
22 Franc, M. (1998) Testing of mechanical and technological properties of materials (in Croatian). In *Engineering Manual IP4*, Školska knjiga Zagreb, Zagreb.

23 Roberts, R. and Newton, C.(1981) Interpretive report on small scale test correlation with K_{Ic} data, *Welding Research Council Bulletin*, **265**, 1–16.
24 Shekhter, A., Kim, S., Carr, D. G., Crocker, A. B. L. and Ringer, S. P. (2002) Assessment of temper embrittlement in an ex-service 1Cr–1Mo–0.25 V power generating rotor by Charpy V-notch testing; KIc fracture toughness and small punch test, *International Journal of Pressure Vessels and Piping*, **79**(8–10), 611–615.
25 Chao, Y. J., Ward, J. D. and Sands, R. G. (2007) Charpy Impact Energy, Fracture Toughness and Ductile–Brittle Transition Temperature of Dual-Phase 590 Steel, *Materials and Design*, **28**(2), 551–557.
26 Brnić, J., Turkalj, G., Čanađija, M. and Lanc, D. (2011) AISI 316Ti (1.4571) steel – Mechanical, creep and fracture properties versus temperature, *Journal of Constructional Steel Research*, **57**, 1948–1952.
27 Krupp, U. (2007) Fatigue Crack Propagation in Metals and Alloys, Microstructural Aspects and Modelling Concepts, Wiley-VCH Verlag Gmbh & Co, KGaA, Weinheim.
28 Farahmand, B. and Nikbin, K. (2008) Predicting fracture and fatigue crack growth properties using tensile properties, *Engineering Fracture Mechanics*, **75**, 2144–2155.
29 Pugno, N., Ciavarella, M., Cornetti, A. and Carpinteri, A. (2006) A generalized Paris' law for fatigue crack growth, *Journal of the Mechanics and Physics of Solids*, **54**, 1333–1349.
30 Branco, R., Antunes, F. V., Costa, J. D., Yangc, F. P. and Kuang, Z. B. (2012) Determination of the Paris law constants in round bars from bench marks on fracture surfaces, *Engineering Fracture Mechanics*, **96**, 96–106.
31 Carpinteri, A. and Paggi, M. (2007) Self-similarity and crack growth instability in the correlation between the Paris constants, *Engineering Fracture Mechanics*, **74**, 1041–1053.
32 Taheri, F., Trask, D. and Pegg, N. (2003) Experimental and analytical investigation of fatigue characteristics of 350WT steel under constant and variable amplitude loadings, *Marine Structures*, **16**, 69–91.
33 Cortie, M. B. and Garret, G. G. (1998) On the correlation between the C and m in the Paris equation for fatigue crack propagation, *Engineering Fracture Mechanics*, **30** (1), 49–58.
34 Li, Y., Wang, H. and Gong, D. (2012) The interrelation of the parameters in the Paris equation of fatigue crack growth, *Engineering Fracture Mechanics*, **96**, 500–509.
35 Cameron, D. W. and Hoeppner, D. W. (1997) Fatigue Properties in Engineering. In S. R. Lampman (Ed.) *ASM Handbook*, Vol. **19**, Fatigue and Fracture, ASM International, USA.
36 Saxena, A. and, Muhlstein, C. L. (1997) Fatigue Crack Growth Testing. In S. R. Lampman (Ed.) *ASM Handbook*, Vol. **19**, Fatigue and Fracture, ASM International, USA.
37 Antolovich, S. D. and Antolovich, B. F. (1997) An Introduction to Fracture Mechanics. In S. R. Lampman (Ed.) *ASM Handbook*, Vol. **19**, Fatigue and Fracture, ASM International, USA.
38 Rosenfield, A. R. (1997) Fracture Mechanics in Failure Analysis. In S. R. Lampman (Ed.) *ASM Handbook*, Vol. **19**, Fatigue and Fracture, ASM International, USA.
39 Liu, A. F. (2005) *Mechanics and Mechanisms of Fracture*, ASM International, USA.
40 Raduta, A. M., Locovei, C., Nicoara, M. and Cucuruz, L. R. (2010) On The Influence of Residual Stresses on Fatigue Fracture of Railway Axles, *WSEAS Transactions on Applied and Theoretical Mechanics*, **5**(3), 197–207.
41 Norris, R. H., Grover, P. S., Hamilton, B. C. and Saxena, A. (1997) Elevated Temperature Crack Growth. In S. R. Lampman (Ed.) *ASM Handbook*, Vol. **19**, Fatigue and Fracture, ASM International, USA.

42 Elber, W. (1970) Fatigue Crack Closure Under Cyclic Tension, *Engineering Fracture Mechanics*, **2**, 37–45.

43 Stephens, R. I., Fatemi, A., Stephens, R. R. and Fuchs, H. O. (2001) *Metal Fatigue in Engineering*, John Wiley & Sons, New York.

44 Mitchell, M. R. (1997) Fundamentals of Modern Fatigue Analysis for Design. In S. R. Lampman (Ed.) *ASM Handbook*, Vol. **19**, Fatigue and Fracture, ASM International, USA.

45 Christ, H. J. (1997) Cyclic Stress–Strain Response and Microstructure. In S. R. Lampman (Ed.) *ASM Handbook*, Vol. **19**, Fatigue and Fracture, ASM International, USA.

46 Guguloth, K., Sivaprasad, S., Chakrabarti, D. and Tarafder, S. (2014) Low-cyclic fatigue behavior of modified 9Cr–1Mo steel at elevated temperature, *Materials Science & Engineering A*, **604**, 196–206.

47 Begley, J. A. and Clarke, G. A. (1983) *Elastic–Plastic Fracture*, ASM International, Baltimore.

48 Bowman, K. (2004) *Mechanical Behavior of Materials*, John Wiley & Sons, New York.

49 International Standard ISO 12107, 2nd edition, Metallic materials – fatigue testing – Statistical planning and analysis data. Reference number ISO 12107:2012(E).

50 Brnić, J., Turkalj, G., Čanađija, M., Lanc, D., Kršćanski, S., Brčić, M., Li, Q. and Niu, J. (2016) Mechanical Properties, Short Time Creep and Fatigue of an Austenitic Steel, *Materials* **9**(4), 298-1–298-19.

11

The Finite Element Method and Applications

11.1 The Finite Element Method (FEM) in the Analysis of Engineering Problems

11.1.1 Applications of FEM

In engineering practice, designers encounter various issues associated with the design of structures, technological processes, procedures for the heat treatment of materials and many other problems. In order to control the quality of the final product, for example, the structure, in the design procedure it is necessary to make an analysis of its response to an applied load under given environmental conditions as well as controlling the manufacturing process. These controls are complicated and cannot be implemented by simple methods. As such, the finite element procedure (FEP) is an important and very powerful tool in engineering analysis related to structural design and manufacturing. Its application has been shown to be reliable in structural design, the simulation of structural responses and manufacturing and it is based on the finite element method (FEM). The basic concept of the finite element method, when applied to structural (continuum/domain) analysis, is that the structure is divided into simple regions, called *finite elements* (or subdomains). The behavior of each of the regions is monitored by a finite element equation consisting of assumed functions representing the stress or displacement in that region. A finite element consists of finite element nodes and a finite element field placed within the finite element nodes. A significant increase in the use of FEM is expected in the coming years [1].

The use of the finite element method in structural analysis requires the idealization of the considered system (structure) to some degree so that it can be analyzed. This idealization includes descriptions of the geometry, load and environmental conditions. It is possible to make an analysis of discrete and continuous systems, and for different types of problems such as steady-state, propagation and eigenvalue problems. In steady-state problems, the response of the structure does not change with time (time is not a variable), while in dynamic or propagation problems, the response of the structure changes with time. In eigenvalue problems, no unique solution exists to the structural response, so the analysis includes calculations of various solutions. In general, the analysis covers some steps such as the idealization of a system as an assemblage of structural components including their interconnections regulating load transfer (that is, joint constraints), setting element equilibrium equations and their transfer to a global coordinate system

Analysis of Engineering Structures and Material Behavior, First Edition. Josip Brnić.

(forming a global equation, that is, an equation of the structure) and a solution for the response as well as interpretation of the results.

11.1.2 The Advantages of Using the FEM

It is necessary to emphasize several advantages of the application of the finite element method (FEM) in comparison with analytical methods, such as the possibility of its use in structural modeling independent of: the constraints related to the linearity or nonlinearity of the considered problem; the shape of the structure; the applied load; environmental conditions; boundary conditions and so on. In other words, the effectiveness and general applicability of the FEM may be treated as its advantages. The finite element method can be used to solve physical problems in engineering practice. The finite element procedure will solve/consider only the selected mathematical model and all the assumptions contained within it will be reflected in the expected response. The analysis of engineering structures using the finite element method is based on matrix algebra and highly capacitive computers [2]. Dealing with the finite element analysis of a structure requires some familiarity with the theory of elasticity and matrix structural analysis [3].

11.1.3 A Brief Overview of the Historical Development of the FEM

The design of structures throughout history was based primarily on experience and analytical methods. In earlier times, optimal design was not explicitly required. Solutions based on analytical methods were primarily directed towards frames, trusses and similar structures.

Idealization of the geometry of a structure and the assumption of a stress distribution were used to provide a structural response and the determination of forces as unknowns. However, limitations on the solvability of algebraic equations with more than one unknown as well as excessive idealization of the structure, which leads to large variations in the results obtained compared to a real response, gave rise to the need for new solutions.

Numerous authors, based on their own knowledge, contributed to the historical development of the finite element method. In general, the concept of representing a considered domain as an assemblage of a finite number of subdomains (finite elements/regions) has been known since ancient times. Mathematicians used this idea to solve the problem of the determination of the perimeter (circumference) of a circle of known radius. They approximated the circle by an inscribed polygon of a finite number of sides. These sides of the polygon represent an assemblage of finite elements, and the perimeter of the polygon as the sum of its individual lengths is equal to the perimeter of the circle. In 1941, in the field of solid mechanics, Hrenikoff represented a plane elastic medium (the elastic continuum problem) as an assemblage of bars and beams. In 1943, Courant used triangular elements and the principle of total potential to study the Saint Venant torsion problem. Further presentations involving the significant use of the finite element method may be attributed to Argyris and Kelsey (1960), then to Turner, Clough and others (1956), but many others have also made a great contribution in the development and application of the finite element method [4–8].

In terms of the first applications of finite elements in engineering analysis, we may mention the analyses of trusses and frames, for example, in the aerospace industry.

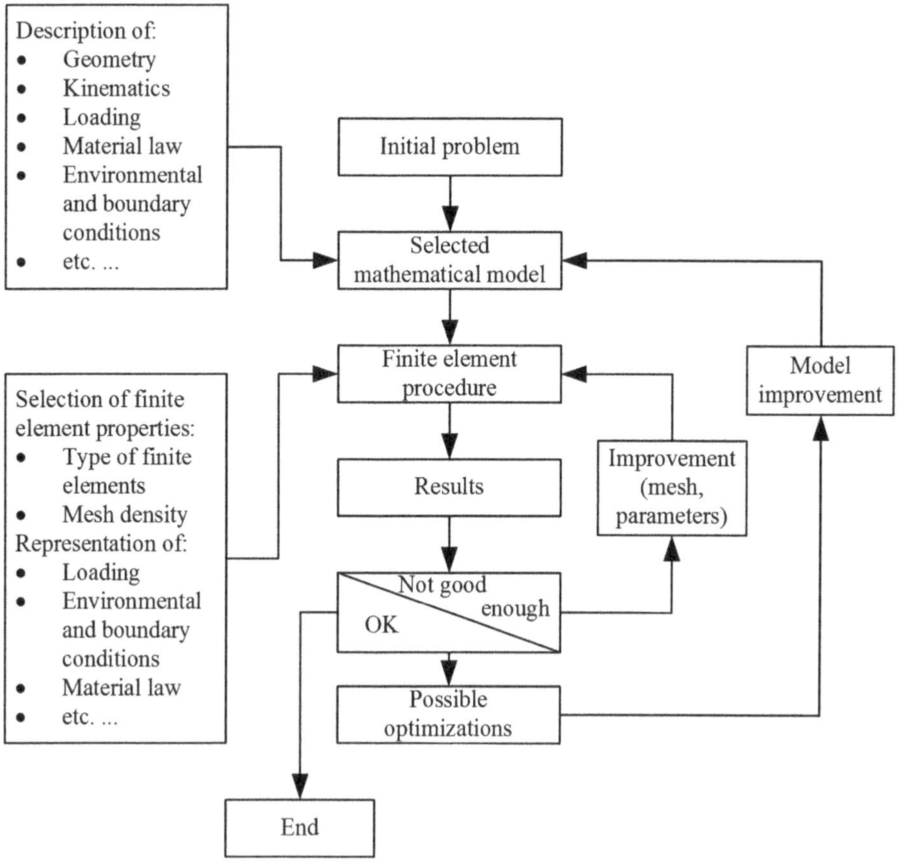

Figure 11.1 Solving an initial physical problem using the finite element method.

Use of the finite element method is oriented to the solving of a physical (technical) problem for which a mathematical model is selected, see Figure 11.1.

11.2 Linear Analysis of Structural Behavior

Assume that the displacements of the structure are infinitesimally small and material behavior is linearly elastic. Also, let the boundary conditions remain unchanged during the application of the load on the considered structure. The structural equilibrium equation derived under such conditions corresponds to linear static analysis of the structure. In other words, the vector of nodal displacements (**U**) of the structure is a linear function of the applied loads presented by load vector $\mathbf{F}_R$. The request that the displacements must be small enters into the evaluation of structural stiffness matrix, **K**, and load vector, $\mathbf{F}_R$, since the integrations have been performed over the original volume of the finite elements and the strain-displacement matrix, **B**, is assumed to be constant and independent of the displacements of elements.

The assumption of a linear elastic material is involved as a result of the use of constant stress–strain matrix, **C**.

11.2.1 Formulations of Equilibrium Equations

The main topics in this section refer to the:

- Formulation (creation) of the *finite element equilibrium equation*;
- Formulation of the *structural equilibrium equation*.

The finite element equilibrium equation is sometimes called the *element stiffness equation*, while the structural equilibrium equation is also called the *structural stiffness equation* (or the global structural equation). The finite element equilibrium equation facilitates the analysis of the equilibrium of a finite element, while the structural equilibrium equation facilitates the analysis of the equilibrium of a construction subjected to loads. In other words, this initiative yields the description of formulative procedures for the *finite element equilibrium equation* at the level of the finite element as well as at the level of the entire structure. There are several formulations of the finite element method (FEM) which relate to the equilibrium equation (stiffness equation) of a finite element. These formulations differ from each other in the method of determining the equilibrium equation. Some of these formulations can be listed at the level of the finite element as follows: the *direct method of finite element equilibrium equation formulation* (*element stiffness equation, element force–displacement equation*); the *variational method of element equation formulation*; the *method of weighted residuals of element equation formulation*; the *energy balance method of element equation formulation.*

The finite element method represents a modern numerical method that is, at the same time, a very popular technique for computer solution of engineering problems and, in this sense, a very efficient and powerful tool in the engineering analysis of structural responses. Of course, its application has now spread to all areas of human activity. As stated previously, structural analysis requires the idealization of the considered structure as well as the applied load and boundary conditions into a form that can be analyzed. The development of the finite element method, supported by the appearance of increasingly highly capacitive computers as well as matrix notation for the applied load, boundary conditions and entire finite element equilibrium equations, has provided a successful procedure for structural analysis. Although it is difficult to mention the exact date or time period when this numerical method was "born", it can be said that its origin dates back to the time of its first use by several research groups such as applied mathematicians, engineers and physicists.

The global equilibrium equation or, simply, the structural equilibrium equation, defines the state of the equilibrium of the considered structure and it refers to the global coordinate system. Similarly, as a structure represents an assembly of structural elements, the structural equation represents the appropriate assembly of the finite element equilibrium equations. We will elaborate on this idea in the remainder of this chapter. Global and local coordinate systems, loading and some types of finite elements and boundary conditions are presented in Figure 11.2.

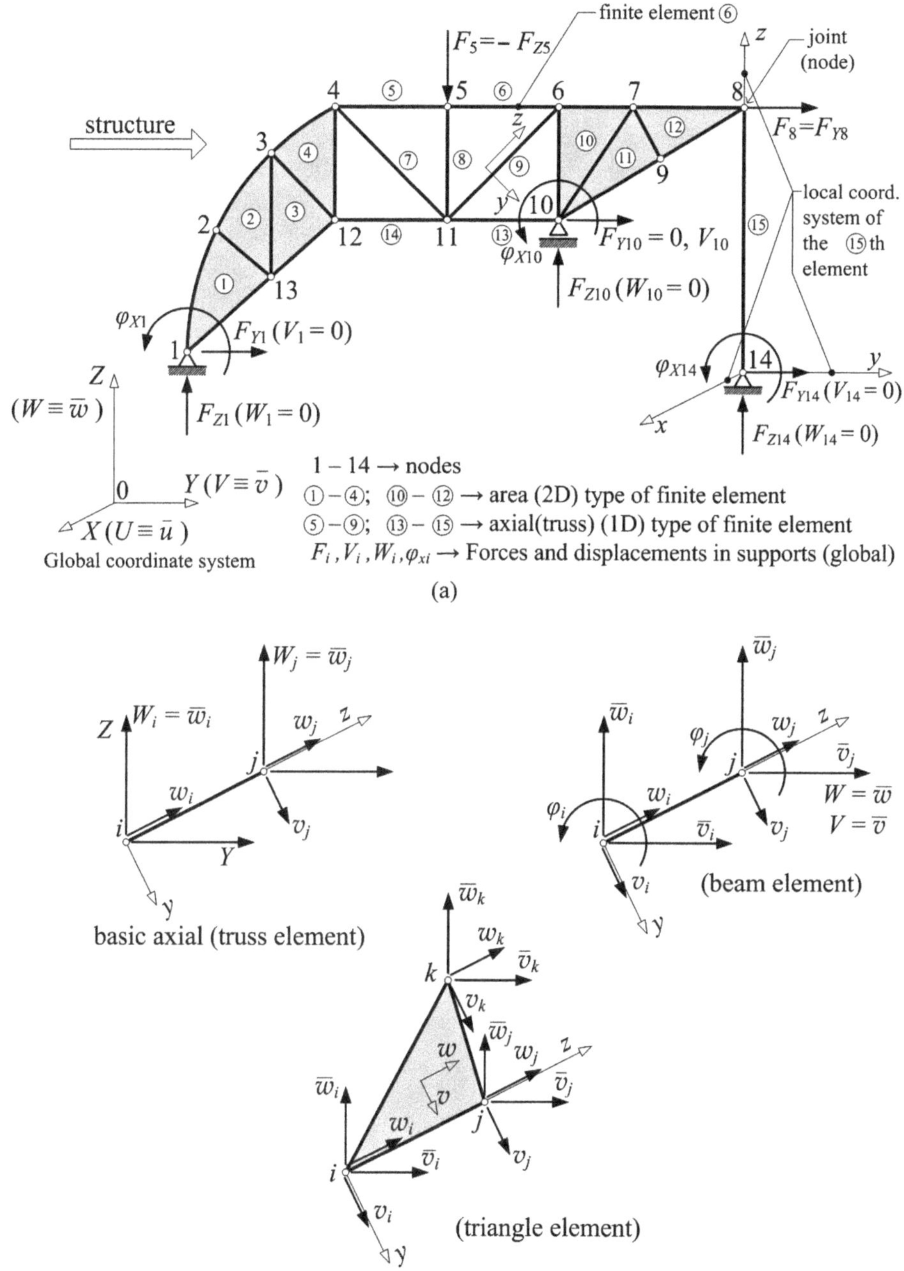

Figure 11.2 A simple planar structure and its discretization. (a) Structure and load; (b) finite elements: basic axial element and its nodal displacements; beam element and its nodal displacements (axial and rotation); basic triangular element and its axial nodal displacements.

11.2.1.1 Variational Formulation of the Finite Element (Equilibrium) Equation

For simplicity, instead of calling this section "The variational method for the formulation of the finite element equilibrium equation" we have chosen to write instead "Variational formulation of the finite element equation".

Depending on the problem at hand as well as the geometry of the structure that is being discretized, discretization of the structure can be made using the appropriate finite elements. A structure can be assembled from different structural elements (see Figure 11.2). The structure represents a continuum (domain) while a region within the domain is called a finite element (or subdomain), as shown in Figure 11.2(b). Each structural element (for example, a rod, beam and so on) can be discretized with a number of finite elements that are connected to nodes; that is, a structural element (engineering element) can be divided into a number of finite elements. The equilibrium equation that refers to a finite element is usually called the basic finite element equilibrium equation (or finite element stiffness equation) or, simply, the *finite element equation* and it is writen in the local coordinate system. Sometimes the local coordinate system can coincide with the global coordinate system. To formulate *the equation of the structure* (that is, the equilibrium equation of the structure or the global equilibrium equation), first it is necessary to formulate the finite element equilibrium equations for all the finite elements forming the structure. Subsequently, it is necessary to transform these finite element equations into the global coordinate system and then, according to the principle of belonging to the same node, these equations need to be assembled using an appropriate procedure into the equation of the structure. In Section 11.3, the variational formulation of the finite element equilibrium equation will be presented.

Polynomial Matrix

Consider the three-dimensional body D shown in Figure 11.3. Let S_q be the body surface upon which forces can act freely and S_u be the portion of the body surface where displacement components are prescribed. The following types of loads act on the considered body: joint (concentrated) forces, $\mathbf{F}_i$; surface tractions, $\mathbf{q}_{b_q}$, that is, forces per unit surface where loading $\mathbf{q}$ acts on the body b and on the portion S_q of its surface S; and body forces, q_V.These forces can have three components corresponding to the three coordinate axes.

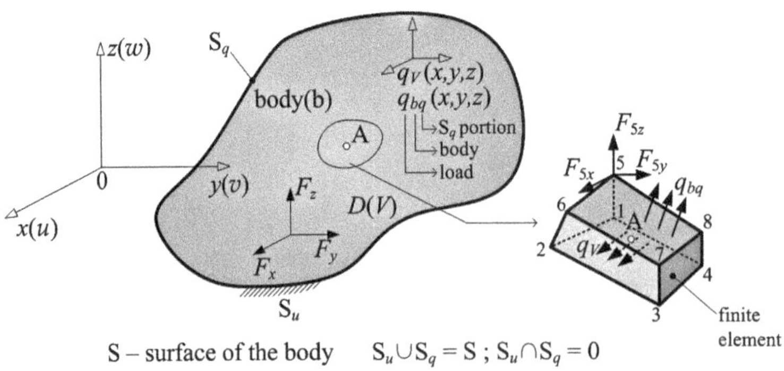

Figure 11.3 A domain subjected to forces.

As stated previously, the finite element is a subdomain consisting of a *finite element field* and *nodes* of the element (Figure 11.2(b)). The aforementioned nodes are usually called *node points*, but the term *joints* is also sometimes used. Although the terms *node* and *joint* may be used interchangeably in this context, the term joint may cause some confusion, as, in practical engineering applications, the term is usually used for the point where two or more engineering elements are connected to each other.

Let **u** denote a displacement vector in the field of a finite element. It is changing with the load in accordance with certain law in the function of the coordinates, but this variation is unknown in advance. The mentioned displacement vector in the field of the finite element can be written as follows:

$$\begin{aligned} [u]^{\mathrm{T}} &= [u \; v \; w] \\ u &= u(x, y, z) \\ v &= v(x, y, z) \\ w &= w(x, y, z) \end{aligned} \tag{11.1}$$

It is useful to note that, in this chapter, some quantities such as stress, strain, force, displacement, elastic constants and so on will appear. These quantities will appear in different equations or as stand-alone quantities, and some of them may appear using different notation, such as symbolic notation, matrix notation, vector notation or index notation.

To describe the variation in the displacement, **u**, between the end points (nodal points) of a finite element, that is, the variation in displacement in the field of a finite element, a polynomial can be used. The choice of polynomial should be such that it matches the shape of the considered element, and thus matches the boundary conditions of the considered finite element. The polynomial may be complete or reduced. The polynomial is complete to a given order (n) when it contains all terms of that order and below. For a one-dimensional problem, a complete polynomial is:

$$P_n(x) = \alpha_1 + \alpha_2 x + \alpha_3 x^2 + \ldots \alpha_{n+1} x^n = \sum_{i=0}^{n} \alpha_{i+1} x^i \tag{11.2}$$

It is clear that the number of terms in the polynomial is equal to $(n+1)$.

In two dimensions, a complete nth degree polynomial is:

$$P_n(x,y) = \sum_{k=0}^{n} \alpha_k x^i y^j, \; +j \leq k, \tag{11.2a}$$

where i and j are permuted accordingly and the number of terms is equal to $(n+1)(n+2)/2$. In two-dimensional or in three-dimensional problems, polynomials are chosen according to Pascal's triangular scheme or Pascal's tetrahedron (see Figure 11.4). For certain problems, full or reduced forms of the polynomials can be chosen. However, when polynomials are used under the assumption of a displacement distribution in the finite element field, the rate of convergence depends on the order of the polynomials used.

Displacements in the field of a finite element (*field displacements, element displacements field*), instead of depending on the coordinate of a considered point in the field

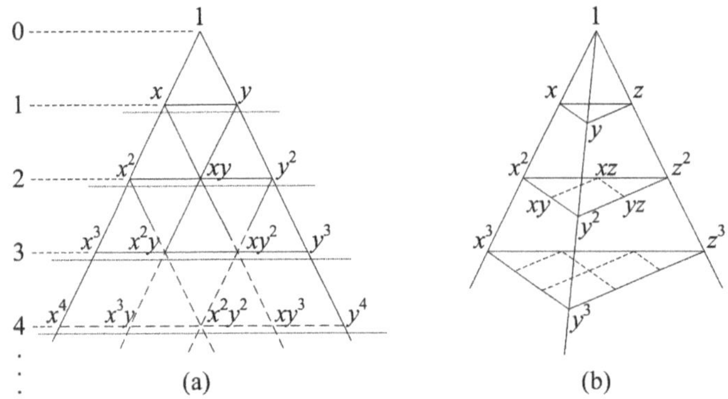

Figure 11.4 (a) Pascal's triangular scheme; (b) Pascal's tetrahedron.

of the finite element, can be displayed depending on the *finite element nodal* (*node*) *displacements* (*joint displacements*). This dependence can be established using so-called *shape functions* (*interpolation functions, approximation functions*). Approximation of the distribution of displacement in the field of the finite element, by applying a polynomial, can be written in the form (matrix and symbolic notation):

$$[u] = [a][\alpha], \rightarrow \mathbf{u} = \mathbf{a}\boldsymbol{\alpha} \tag{11.3}$$

where:

$[u]$ is the displacement vector in the field of the finite element,
$[a]$ is the polynomial matrix,
$[\alpha]$ the vector of constants (vector of generalized coordinates).

Since it is sometimes inconvenient to write the content of a vector in columnar fashion, this can be written in the form of a row vector in transpose form: $\lfloor u \rfloor^{\mathrm{T}}$. However, here, for the vector (single-column matrix) and for the matrix we use the notation shown above.

The constitution of the vector $[u]$, as the polynomial representation of the displacement in the field of a finite element, is of such form that it contains the above-mentioned constants (parameters) and the polynomial matrix. The vector of constants contains as many components as the finite element has degrees of freedom (d.o.f.); that is, the number of components in the vector of constants corresponds to the number of d.o.f. In this case, the nodal conditions of the finite element will be fulfilled. However, this means that the choice of the matrix **a** and the vector $\boldsymbol{\alpha}$ is not free but depends on the nodal displacements vector. In the case of boundary conditions, vector $[u]$ and matrix $[a]$ take the following forms:

$$[u] \rightarrow [u] = \left[u^{\mathrm{b}}\right] \equiv \left[u^{\mathrm{e}}\right] \tag{11.4a}$$

$$[a] \rightarrow [a] = \left[a^{\mathrm{b}}\right] \tag{11.4b}$$

In Equations (11.4), the designation "b" (boundary) means that the matrix refers to the node points of the element (nodes), that is, $[u^{\mathrm{e}}]$ and $[a^{\mathrm{b}}]$ are the nodal displacements vector and the polynomial matrix at the boundary, respectively. If Equation (11.3) is rewritten using Equation (11.4), it follows that [9]:

$$[u^{e}] = [a^{b}][\alpha] \tag{11.5}$$

Solving this equation with respect to $[\alpha]$, it follows that:

$$[\alpha] = [a^{b}]^{-1}[u^{e}] \rightarrow \boldsymbol{\alpha} = \mathbf{a}^{b^{-1}}\mathbf{u}^{e}, \tag{11.5a}$$

and the inclusion of this value in Equation (11.3) yields:

$$[u] = [a][a^{b}]^{-1}[u^{e}] = [N][u^{e}] \rightarrow \mathbf{u} = \mathbf{N}\mathbf{u}^{e} \tag{11.6}$$

Equation (11.6) represents the relationship between displacement in the field of a finite element and the finite element nodal displacements, where $[N]$ is the so-called interpolation matrix (shape functions matrix), that is, the matrix of interpolation (shape) functions:

$$[N] = [a][a^{b}]^{-1} \rightarrow \mathbf{N} = \mathbf{a}\mathbf{a}^{b^{-1}} \tag{11.6a}$$

Equations (11.3) and (11.6) are written in matrix notation, but similar equations are usually written in symbolic notation, as shown in the following section.

It is known that stress analysis is more accurate if a polynomial of higher order is used, although the computational work necessary to complete such an analysis is more complex and extensive.

Strain-displacement Matrix

A strain is defined as:

$$\boldsymbol{\varepsilon} = \mathbf{d}_{\varepsilon}\mathbf{u}, \tag{11.7}$$

where $\mathbf{d}_{\varepsilon}$ is a differential operator (it is a kinematic differential operator), while $\mathbf{u}$ is the vector of displacements in the finite element field and is defined by Equation (11.6). Using Equation (11.6), strain in the field of the finite element may be described depending on the nodal displacements as:

$$\boldsymbol{\varepsilon} = \mathbf{d}_{\varepsilon}\mathbf{N}\mathbf{u}^{e} = \mathrm{B}\mathbf{u}^{e}, \qquad \mathbf{B} = \mathbf{d}_{\varepsilon}\mathbf{N} \tag{11.8}$$

where $\mathbf{B}$ is the *strain-displacement matrix*. Now, stress can be calculated as:

$$\boldsymbol{\sigma} = \mathbf{C}\boldsymbol{\varepsilon} = \mathbf{C}\mathbf{B}\mathbf{u}^{e}. \tag{11.9}$$

Equations of the form presented in Equation (11.9) represent the so-called material constitutive relationships or material constitutive equations, sometimes simply known as constitutive equations (stress–strain relationships) and they refer to a material's mechanical properties. In short, such equations give the strains due to stresses. Material constitutive relationships were considered in Chapter 7. The task now is to show the full set of coefficients (**C**) that connect each stress component to all strain components. With this in mind, in Equation (11.9), C is the *matrix of elastic constants* (*matrix of elastic coefficients, elasticity matrix, stress–strain matrix, material matrix, material stiffness matrix*) that connects the components of stress with the strain components.

Finite Element Equation and Its Transformation into a Global Coordinate System

As mentioned earlier, the variational formulation will be used to derive the finite element equilibrium equation. The variational formulation is based on variational principles. Here, the principle of virtual work based on virtual displacements (the displacement-based method) will be used. The principle of virtual work is an independent, formulative procedure for obtaining the finite element (equilibrium) equation. Sometimes, instead of virtual displacements, virtual forces (the forces-based method) can be used [10]. In the principle of virtual work based on the displacement method, it is assumed that a structure which is subjected to load and is in a state of equilibrium is now subjected to virtual displacement defined by virtual displacement components at each point. These components must be kinematically admissible. The principle of virtual work is written in the form:

$$\delta U^{e} = \delta W^{e}, \tag{11.10}$$

and it signifies equality between the virtual work of internal and external forces acting on the observed body. If we consider the eight-node finite element that is subjected to external forces, shown in Figure 11.3, Equation (11.10) can be written as [10–15]:

$$\int_{V} \delta \boldsymbol{\varepsilon}^{T} \boldsymbol{\sigma} dV = \int_{V} \delta \mathbf{u}^{T} \mathbf{q}_{v} dV + \int_{S_q} \delta \mathbf{u}_{S_q}^{T} \mathbf{q}_{b_q} dS_q + \Sigma \delta \mathbf{u}_{i}^{T} \mathbf{F}_{i}. \tag{11.11}$$

Equation (11.11) refers to one finite element, and its left-hand side represents virtual strain energy, while its right-hand side represents the virtual work of external forces (body, surface and concentrated). Virtual strain energy may be written in the form:

$$\delta U^{e} = \int_{V} \delta \boldsymbol{\varepsilon}^{T} \boldsymbol{\sigma} dV = \int_{V} \delta \boldsymbol{\varepsilon}^{T} \mathbf{C} \boldsymbol{\varepsilon} dV \tag{11.12}$$

and, using Equation (11.8) and Equation (11.9), the following expression may be written:

$$\delta U^{e} = \delta \mathbf{u}^{e^{T}} \int_{V} \mathbf{B}^{T} \mathbf{C} \mathbf{B} \, dV \mathbf{u}^{e}. \tag{11.13}$$

In accordance with Equation (11.6), we have:

$$\delta \mathbf{u} = \mathbf{N} \delta \mathbf{u}^{e} \tag{11.14a}$$

and something similar applies for a portion of the surface S_q when boundary conditions are also prescribed, that is:

$$\delta \mathbf{u}_{S_q} = \mathbf{N}_{S_q} \delta \mathbf{u}^{e}. \tag{11.14b}$$

Using Equations (11.14) and (11.13), Equation (11.11) now becomes:

$$\delta \mathbf{u}^{e^{T}} \int_{V} \mathbf{B}^{T} \mathbf{C} \mathbf{B} dV \mathbf{u}^{e} = \delta \mathbf{u}^{e^{T}} \left[\int_{V} \mathbf{N}^{T} \mathbf{q}_{v} dV + \int_{S_q} \mathbf{N}_{S_q}^{T} \mathbf{q}_{b_q} dS_q + \mathbf{F} \right], \tag{11.15}$$

and from there we obtain:

$$\int_V \mathbf{B}^T\mathbf{C}\mathbf{B}dV\mathbf{u}^e = \underbrace{\int_V \mathbf{N}^T\mathbf{q}_v dV}_{\mathbf{F}_v} + \underbrace{\int_{S_q} \mathbf{N}_{S_q}^T\mathbf{q}_{b_q} dS_q}_{\mathbf{F}_{S_q}} + \mathbf{F}. \tag{11.16}$$

Equation (11.16) is the *finite element equilibrium equation* (*finite element equation, finite element stiffness equation*) and it can also be written in the following form:

$$\mathbf{k}^e\mathbf{u}^e = \mathbf{f}^e \tag{11.17}$$

where:

$$\mathbf{k}^e = \int_V \mathbf{B}^T\mathbf{C}\mathbf{B}\,dV \tag{11.18}$$

will be referred to as the *finite element stiffness matrix*, while $\mathbf{u}^e$ is the vector of finite element nodal displacements and $\mathbf{f}^e$, which represents the right-hand side of Equation (11.16), is the vector of finite element nodal forces consisting of joint forces (F), volume forces (F_v) and traction forces ($\mathbf{F}_{S_q}$). Since Equation (11.18) refers to a three-dimensional finite element [11], when a two-dimensional or one-dimensional finite element is considered, the stiffness matrix is of the form:

$$\mathbf{k}^e = \int_A \mathbf{B}^T\mathbf{C}\mathbf{B}h\,dA, \quad \mathbf{k}^e = \int_l \mathbf{B}^T\mathbf{C}\mathbf{B}\,dz, \tag{11.19}$$

where h (or t) represents the thickness of the element while A is the area of the finite element and l is the length of the one-dimensional element. The values relating to the matrix C will be presented when appropriate problems are being considered. Note that matrix C in Equation (11.18) and Equation $(11.19)_1$ contains only elastic properties of the material, and it is known as the matrix of elastic constants, while matrix C shown in Equation $(11.19)_2$, which refers to one-dimensional finite elements, includes geometrical properties as well, although we will also refer to it as the elasticity matrix.

The Transformation of the Vector of Degrees of Freedom (Nodal Displacements) To obtain the structural equation (that is, the global structural equilibrium equation), first it is necessary to transform all of the finite element equations from the local coordinate system to the global coordinate system. This means that all of the "elements" (members, items) of all finite element equations, such as nodal displacement vectors (degrees of freedom), vectors of nodal forces and finite element stiffness matrices, need to be transformed from the local to the global coordinate system.

A transformation of the vector of nodal displacements (degrees of freedom) for a considered planar one-dimensional finite element with two nodes (node points, joints) from the local to the global coordinate system can be made as shown in Figure 11.5. In this case, as can be seen, the longitudinal axis of the finite element is the x axis and it is rotated counterclockwise by an angle φ with respect to the global coordinate axis, X. The considered planar finite element is a rod (bar, truss, axial element) type element that is

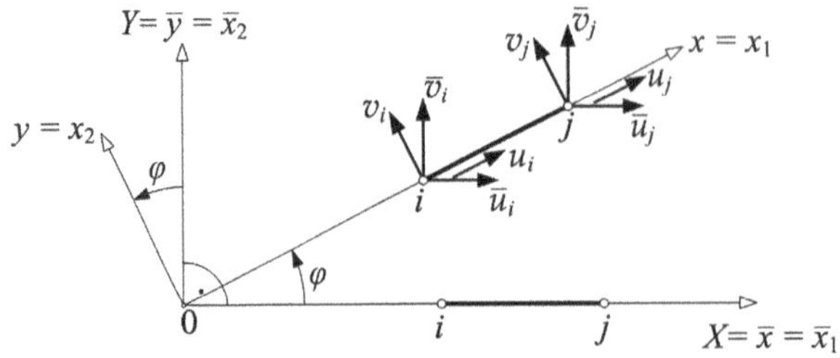

Figure 11.5 One-dimensional finite element subjected to planar translational displacements only: nodal displacements and their transformations.

subjected only to planar forces on its node points (nodes), so each of the nodal displacement vectors ($\mathbf{u}_i$, $\mathbf{u}_j$) consists of two translational displacements (u_i, v_i; u_j, v_j).

Let $X = \bar{x} = \bar{x}_1$ and $Y = \bar{y} = \bar{x}_2$ be the global Cartesian coordinate axes (global coordinate system), while $x = x_1$ and $y = x_2$ are the local Cartesian coordinate axes (local coordinate system). The relationship between the components of the vector displacement ($\mathbf{u}$) in the local coordinate system and the components of the vector displacement ($\bar{\mathbf{u}}$) in the global coordinate system of the considered planar finite element with two nodes, in accordance with Figure 11.5, for the node "i", is:

$$u_i = \bar{u}_i \cos\varphi + \bar{v}_i \sin\varphi$$
$$v_i = -\bar{u}_i \sin\varphi + \bar{v}_i \cos\varphi. \tag{11.20}$$

Equation (11.20) can be written in the form:

$$\begin{bmatrix} u_i \\ v_i \end{bmatrix} = \begin{bmatrix} \cos\varphi & \sin\varphi \\ -\sin\varphi & \cos\varphi \end{bmatrix} \begin{bmatrix} \bar{u}_i \\ \bar{v}_i \end{bmatrix} \tag{11.21}$$

In Equation (11.21), the basic transformation matrix, $\mathbf{t}_0$, is:

$$\mathbf{t}_0 = \begin{bmatrix} \cos\varphi & \sin\varphi \\ -\sin\varphi & \cos\varphi \end{bmatrix} \tag{11.21a}$$

Since the transformation matrix represents the relationships between the local and global coordinate systems, the result given by Equation (11.21a) is presented again based on Equation (11.23).

Equation (11.21) can now be written in the form:

$$[u_i] = [t_0][\bar{u}_i], (\mathbf{u} = \mathbf{t}_0\bar{\mathbf{u}}), \tag{11.22a}$$

and where it follows that:

$$[\bar{u}_i] = [t_0]^{\mathrm{T}}[u_i] \tag{11.22b}$$

In Equations (11.22), $[t_0]$ is, as previously stated, the so-called basic transformation matrix (or rotational matrix), and it establishes the relationship between displacements

$[u_i]$ and displacements $[\bar{u}_i]$. It represents the direction cosines between local and global coordinate axes and can be written in a general form as:

$$[t_0] = \begin{bmatrix} \cos(x,X) & \cos(x,Y) \\ \cos(y,X) & \cos(y,Y) \end{bmatrix} \tag{11.23}$$

If, in Equation (11.23), the appropriate values relating to the angle of rotation are inserted, it follows that:

$$\mathbf{t}_0 = \begin{bmatrix} \cos(x,X) = \cos\varphi & \cos(x,Y) = \cos(90-\varphi) = \sin\varphi \\ \cos(y,X) = \cos(90+\varphi) = -\sin\varphi & \cos(y,Y) = \cos\varphi \end{bmatrix},$$

which is the same value as that in Equation (11.21a).

The same is valid for the "*j*" node. The basic transformation matrix, $\mathbf{t}_0$ refers to the finite element, so it can be written $\mathbf{t}_0 = \mathbf{t}_0^{\mathrm{e}}$. If, instead of a one-dimensional planar finite element, we consider a one-dimensional spatial finite element, then the basic transformation matrix (relating local coordinate system (*x*, *y*, *z*) with the global coordinate system (*X*, *Y*, *Z*)), for one node is:

$$\mathbf{t}_0 = \begin{bmatrix} \cos(x,X) & \cos(x,Y) & \cos(x,Z) \\ \cos(y,X) & \cos(y,Y) & \cos(y,Z) \\ \cos(z,X) & \cos(z,Y) & \cos(z,Z) \end{bmatrix}. \tag{11.24}$$

In the case of coordinate transformation of the orthogonal coordinate axes, we have:

$$[t_0][t_0]^{\mathrm{T}} = [\mathrm{I}]. \tag{11.25}$$

In accordance with the definition of the inverse matrix:

$$[t_0][t_0]^{-1} = [\mathrm{I}]. \tag{11.26}$$

Comparing Equations (10.25) and (10.26), it follows that:

$$[t_0]^{-1} = [t_0]^{\mathrm{T}}. \tag{11.27}$$

A property defined by Equation (11.27) means that the inverse of the matrix is equal to its transpose and this matrix (here, the basic transformation matrix) is termed an *orthogonal matrix* [3]. For the considered planar one-dimensional finite element with two nodes, transformation of the vector of finite element nodal displacements is governed by:

$$\begin{bmatrix} \begin{bmatrix} u_i \\ v_i \end{bmatrix} \\ \begin{bmatrix} u_j \\ v_j \end{bmatrix} \end{bmatrix} = \begin{bmatrix} \begin{bmatrix} \cos(x,X) & \cos(x,Y) \\ \cos(y,X) & \cos(y,Y) \end{bmatrix} & [0] \\ [0] & \begin{bmatrix} \cos(x,X) & \cos(x,Y) \\ \cos(y,X) & \cos(y,Y) \end{bmatrix} \end{bmatrix} \begin{bmatrix} \begin{bmatrix} \bar{u}_i \\ \bar{v}_i \end{bmatrix} \\ \begin{bmatrix} \bar{u}_j \\ \bar{v}_j \end{bmatrix} \end{bmatrix}. \tag{11.28}$$

Equation (11.28) can be written as:

$$\begin{aligned} \mathbf{u}^{\mathrm{e}} &= \mathbf{t}^{\mathrm{e}}\bar{\mathbf{u}}^{\mathrm{e}} \\ \bar{\mathbf{u}}^{\mathrm{e}} &= \mathbf{t}^{\mathrm{e}^{\mathrm{T}}}\mathbf{u}^{\mathrm{e}} \end{aligned} \tag{11.29}$$

In Equation (11.29), $\mathbf{t}^e$ is the transformation matrix of the finite element:

$$\mathbf{t}^e \equiv [T] = \begin{bmatrix} [t_0^e] & \cdot \\ \cdot & [t_0^e] \end{bmatrix}, \quad [t_0] \equiv [t_0^e], \tag{11.30}$$

and it consists of two basic transformatiom matrices $[t_0]$ that correspond to two nodes. However, $\mathbf{t}^e$ corresponds to nodal vectors, $\mathbf{u}^e$; $\mathbf{f}^e$, that is, to the nodal displacements vector and the nodal forces vector, as given in Equation (11.17). Each of these two finite element nodal vectors in the finite element equation (Equation (11.17)) consists, for a one-dimensional finite element, actually of two vectors of nodal displacements and two vectors of nodal forces. Each of the vectors of nodal displacements corresponds to the node i and the node $j(i \rightarrow \mathbf{u}_i^e, j \rightarrow \mathbf{u}_j^e)$ and contains two translational displacements ($i \rightarrow u_i, v_i$; $j \rightarrow u_j, v_j$) while each of the nodal forces vectors ($\mathbf{f}_i^e; \mathbf{f}_j^e$) contains two forces ($F_{ix}$, F_{iy}; F_{jx}, F_{jy}), as previously explained. In Equation (11.29), $\mathbf{u}^e$ is a vector of the finite element nodal displacements in the local coordinate system, $\bar{\mathbf{u}}^e$ is a vector of the finite element nodal displacements in the global coordinate system and $\mathbf{t}^e$ is the finite element transformation matrix. For clarification, for the situation shown in Figure 11.5, one can write:

$$\mathbf{u}^{e^T} = \left[\mathbf{u}_i^{e^T} \ \mathbf{u}_j^{e^T}\right], \quad \mathbf{u}_i^{e^T} = [u_i \ v_i], \quad \mathbf{u}_j^{e^T} = [u_j \ v_j] \tag{11.30a}$$

A similar expression can be written for $\mathbf{f}^{eT}$.

The Transformation of the Vector of Nodal Forces For the transformation of the vector of finite element nodal forces, the same rule is applied as for nodal displacements, that is, from Equation (11.29):

$$\begin{aligned} \mathbf{f}^e &= \mathbf{t}^e \bar{\mathbf{f}}^e \\ \bar{\mathbf{f}}^e &= \mathbf{t}^{eT} \mathbf{f}^e \end{aligned} \tag{11.31}$$

The Transformation of the Finite Element Stiffness Matrix The procedure for the transformation of the finite element stiffness matrix is as follows.

Consider Equation (11.7), which defines the finite element stiffness matrix:

$$\mathbf{k}^e \mathbf{u}^e = \mathbf{f}^e. \tag{a}$$

By replacing $\mathbf{u}^e$ and $\mathbf{f}^e$ with the values defined by Equations (11.29_1) and (11.31_1), it follows that:

$$\mathbf{k}^e \mathbf{t}^e \bar{\mathbf{u}}^e = \mathbf{t}^e \bar{\mathbf{f}}^e. \tag{b}$$

If both sides of Equation (b) are multiplied by $\mathbf{t}^{e^{-1}}$, it follows that:

$$\mathbf{t}^{e^{-1}} \mathbf{k}^e \mathbf{t}^e \bar{\mathbf{u}}^e = \mathbf{t}^{e^{-1}} \mathbf{t}^e \bar{\mathbf{f}}^e \rightarrow \mathbf{t}^{e^T} \mathbf{k}^e \mathbf{t}^e \bar{\mathbf{u}}^e = \bar{\mathbf{f}}^e. \tag{c}$$

On the basis of Equation (c):

$$\bar{\mathbf{k}}^e \bar{\mathbf{u}}^e = \bar{\mathbf{f}}^e. \tag{11.32}$$

Equation (11.32) defines the finite element equation in the global coordinate system. Based on Equation (c) it is clear that transformation of the finite element stiffness matrix from the local to the global coordinate system is defined by:

$$\bar{\mathbf{k}}^{e} = \mathbf{t}^{e^{T}} \mathbf{k}^{e} \mathbf{t}^{e}. \tag{11.33}$$

Note: In further considerations, the longitudinal axis of the one-dimensional finite element (usually classified as a rod/bar/truss/axial element/bar shaft/beam element) will be designated the z axis. The reason for this is that the cross-sectional area, as the most important quantity of the structural element (and also of a finite element), will be defined by the x, y coordinates. This means that the finite element nodal displacements vector and the finite element nodal forces vector will be arranged using the following order of coordinate axes: z, x, y, when a spatial one-dimensional beam finite element is considered, then in the order: z, y, x, when planar bending in the z–y principal centroidal plane of inertia is considered, and in the order z, x, y, when planar bending in the z–x principal centroidal plane of inertia is considered.

In accordance with the above Note, the basic transformation matrix, Equation (11.24), for the case of a one-dimensional finite element will now be arranged as:

$$\mathbf{t}_0 = \begin{bmatrix} \cos(z,Z) & \cos(z,X) & \cos(z,Y) \\ \cos(x,Z) & \cos(x,X) & \cos(x,Y) \\ \cos(y,Z) & \cos(y,X) & \cos(y,Y) \end{bmatrix}. \tag{11.34}$$

Let the z–y plane of the finite element be chosen as one of the principal centroidal planes of inertia, as shown in Figure 11.6, and let the $(+z)$ axis of the finite element in its starting position coincide with the global $(+Z)$ axis of the coordinate system. Let's assume that the finite element may rotate for $\pm\varphi$ (that is, the finite element rotates around its x axis), as shown in Figure 11.6, and first that the element is subjected only to planar forces (F_y, F_z). In this case, only displacement components (w, v) exist in each node, (axial element/bar/rod).

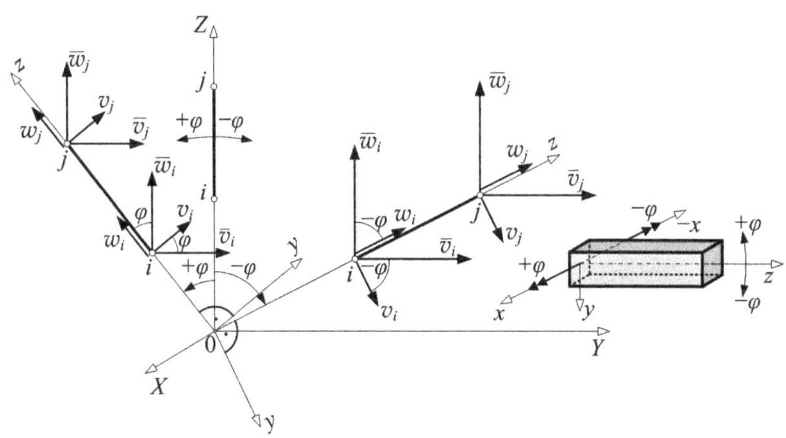

Figure 11.6 A finite element in the z–y plane subjected to planar forces.

As a second case, apart from planar forces, the influence of a bending moment around the x axis (beam element) will also be considered (the bending moment acts in the z–y principal centroidal plane of inertia). In accordance with the arranging of the nodal displacements (z, y, x), the basic transformation matrix in this case is of the form:

$$\mathbf{t}_0 = \begin{bmatrix} \cos(z,Z) & \cos(z,Y) & \cos(z,X) \\ \cos(y,Z) & \cos(y,Y) & \cos(y,X) \\ \cos(x,Z) & \cos(x,Y) & \cos(x,X) \end{bmatrix}. \tag{11.35}$$

In accordance with Figure 11.6 and Equation (11.35), when only translational displacements are considered, that is, translation ((Z, Y) to (z, y)) is considered, for both shown cases of rotation: $(-\varphi)$ and $(+\varphi)$, the basic transformation matrix can be obtained as follows:

- Case: $(-\varphi)$

$$w_i = \bar{w}_i \cos(-\varphi) + \bar{v}_i \sin(-\varphi)$$

$$v_i = \bar{v}_i \cos(-\varphi) - \bar{w}_i \sin(-\varphi)$$

$$\begin{bmatrix} w_i \\ v_i \end{bmatrix} = \begin{bmatrix} \cos\varphi & -\sin\varphi \\ \sin\varphi & \cos\varphi \end{bmatrix} \begin{bmatrix} \bar{w}_i \\ \bar{v}_i \end{bmatrix}, \tag{11.36a}$$

- Case: (φ)

$$w_i = \bar{w}_i \cos\varphi - \bar{v}_i \sin\varphi$$

$$v_i = \bar{v}_i \cos\varphi + \bar{w}_i \sin\varphi$$

$$\begin{bmatrix} w_i \\ v_i \end{bmatrix} = \begin{bmatrix} \cos\varphi & -\sin\varphi \\ \sin\varphi & \cos\varphi \end{bmatrix} \begin{bmatrix} \bar{w}_i \\ \bar{v}_i \end{bmatrix}, \tag{11.36b}$$

Thus, for the cases when only translations are considered (Figure 11.6, Equation (11.36)), the basic transformation matrix is of the form:

$$\mathbf{t}_0 = \begin{bmatrix} \cos\varphi & -\sin\varphi \\ \sin\varphi & \cos\varphi \end{bmatrix} \tag{11.36c}$$

However, when determining the basic transformation matrix, both the longitudinal axis of the element and the plane of rotation should be considered. Note that the longitudinal axis of the finite element in Figure 11.6 is z while in Figure 11.5 it is x.

If, in addition to the planar translations, rotation (caused by bending) around the x axis is also considered, then, in accordance with Equation (11.35), the basic transformation matrix is:

$$\mathrm{t}_0 = \begin{bmatrix} \cos\varphi & -\sin\varphi & 0 \\ \sin\varphi & \cos\varphi & 0 \\ 0 & 0 & 1 \end{bmatrix}. \tag{11.37}$$

Finally, the transformation matrix of a spatial one-dimensional beam finite element is of the form:

$$\mathbf{t}^{\mathrm{e}} \equiv \mathbf{T} = \begin{bmatrix} t_0^{\mathrm{e}} & \cdot & \cdot & \cdot \\ & t_0^{\mathrm{e}} & & \\ & & t_0^{\mathrm{e}} & \\ & & & t_0^{\mathrm{e}} \end{bmatrix} \tag{11.37a}$$

and for the planar bending in the z–y principal centroidal plane of inertia, the finite element transformation matrix of a beam finite element is:

$$\mathbf{t}^{e} \equiv \mathbf{T} = \begin{bmatrix} t_0^{\mathrm{e}} & \cdot \\ & t_0^{\mathrm{e}} \end{bmatrix} \tag{11.37b}$$

Structural Equation Consider the structure shown in Figure 11.7. It is a statically undetermined structure.

The structure is loaded as shown and is discretized using one-dimensional (truss, rod, axial, bar) finite elements and planar (triangular) finite elements. The global and local coordinate systems are also shown on the figure.

The structural equation (the global equation or equilibrium equation of the structure) has a similar form to the finite element equation given in Equation (11.32) and refers to the global coordinate system. The structural equation has the form:

$$\mathbf{KU} = \mathbf{F}_{\mathrm{R}} \tag{11.38}$$

This equation is obtained by bringing together all of Equations (11.32) for the finite elements in an appropriate way, as explaned below. In Equation (11.38), $\mathbf{K}$ is the global stiffness matrix (structural matrix), $\mathbf{U}$ is the vector of nodal displacements of the considered structure and $\mathbf{F}_{\mathrm{R}}$ is the vector of nodal forces of the considered structure,

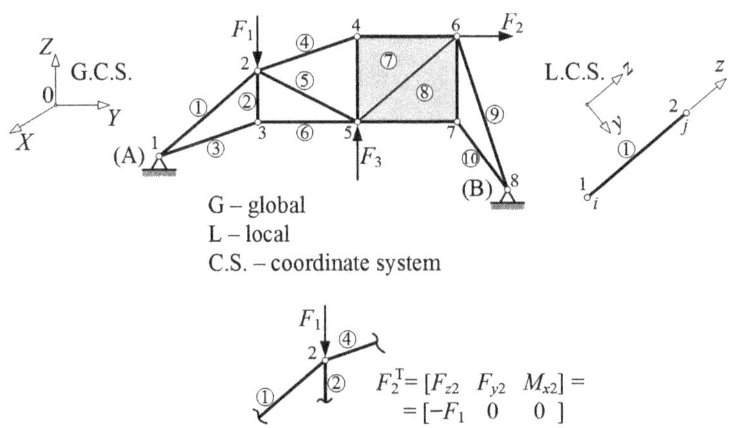

Figure 11.7 Planar structure.

all written in relation to the global coordinate system. The structural stiffness matrix and the displacement and force vectors are of the form:

$$\mathbf{K} = \begin{bmatrix} \mathbf{K}_{11} & \cdots & \mathbf{K}_{1n} \\ \vdots & \ddots & \vdots \\ \mathbf{K}_{n1} & \cdots & \mathbf{K}_{nn} \end{bmatrix}, \quad \mathbf{F}_R^T = [\mathbf{F}_1 \;\; .... \;\; \mathbf{F}_n], \quad \mathbf{U}^T = [\mathbf{U}_1 \;\; .... \;\; \mathbf{U}_n] \tag{11.39}$$

where all the members (sub-matrices) of the stiffness matrix as well as all the vectors refer to the global coordinate system. Since the matrix $\mathbf{K}$ is a symmetric matrix, for its sub-matrices:

$$\mathbf{K}_{rs} = \mathbf{K}_{sr}^T \tag{11.39a}$$

In a given case, for example, in the case of node 8, we have: $K_{nn} = K_{88}$; $F_n = F_8$; $U_n = U_8$. Also, for example, in the case of node 2, the vector of the nodal forces is: $\mathbf{F}_2^T = [F_{z2}\, F_{y2}\, M_{x2}] = [-F_1 \; 0 \; 0]$.

Static and dynamic finite element analysis, with a good enough level of finite element discretization, satisfies two conditions: equilibrium at the level of the node and equilibrium at the level of the finite element. The *structural equation* of the form given in Equation (11.38) is obtained by assembling all of the finite element equations that have previously been written into the global coordinate system; that is, the equations of the form in Equation (11.32) that belong to the considered node and refer to the same global number of degrees of freedom. Assembling the finite element equations into the structural equation is usually known as the *direct stiffness method*. To obtain each sub-matrix, $\mathbf{K}_{rs}$, of the matrix $\mathbf{K}$, it is necessary to partition the finite element equation (given in the global coordinate system) as is shown, for example, for one beam element and for one triangular element:

- **Beam element equation in the global coordinate system**:

$$\bar{\mathbf{k}}^e \bar{\mathbf{u}}^e = \bar{\mathbf{f}}^e$$

The equation of, for example, the "finite element 1" with nodes i–j (or 1–2), in Figure 11.7, after it has been transformed into the global coordinate system, is:

$$\begin{bmatrix} \bar{\mathbf{k}}_{ii}^1 & \bar{\mathbf{k}}_{ij}^1 \\ \bar{\mathbf{k}}_{ji}^1 & \bar{\mathbf{k}}_{jj}^1 \end{bmatrix} \begin{bmatrix} \bar{\mathbf{u}}_i^1 \\ \bar{\mathbf{u}}_j^1 \end{bmatrix} = \begin{bmatrix} \bar{\mathbf{f}}_i^1 \\ \bar{\mathbf{f}}_j^1 \end{bmatrix} \tag{11.40}$$

Note: The designation – above a letter means the global coordinate system; superscript ^{1} means that this quantity refers to the finite element "1" (that is, e = 1; subscript $_{11}$ or $_{ii}$ indicates that this quantity refers to the node "1" (or i).

- **Triangular finite element in the global coordinate system:**
The equation of, for example, the "finite element 8" with nodes $i-j-k$ (or, in general, 1–2–3), after it has been transformed into the global coordinate system (Figure 11.7), is:

$$\begin{bmatrix} \bar{\mathbf{k}}_{ii} & \bar{\mathbf{k}}_{ij} & \bar{\mathbf{k}}_{ik} \\ \bar{\mathbf{k}}_{ji} & \bar{\mathbf{k}}_{jj} & \bar{\mathbf{k}}_{jk} \\ \bar{\mathbf{k}}_{ki} & \bar{\mathbf{k}}_{kj} & \bar{\mathbf{k}}_{kk} \end{bmatrix} \begin{bmatrix} \bar{\mathbf{u}}_i \\ \bar{\mathbf{u}}_j \\ \bar{\mathbf{u}}_k \end{bmatrix} = \begin{bmatrix} \bar{\mathbf{f}}_i \\ \bar{\mathbf{f}}_j \\ \bar{\mathbf{f}}_k \end{bmatrix} \tag{11.41}$$

As stated previously, elements of the structural equation (the global equilibrium equation) given by Equation (11.38), that is, the global stiffness matrix, the vector of nodal forces and the vector of nodal displacements, take the forms shown in Equation (11.39). This means that all the element stiffness matrices have already been established in accordance with the global coordinate system. The global stiffness matrix (the structural stiffness matrix) represents the stiffness matrix of the complete assemblage of all finite elements and, in this sense, it is obtained on the basis of the stiffness matrices of individual finite elements using the so-called *direct stiffness method.* As such, the global stiffness matrix is calculated by direct addition of the element stiffness matrices:

$$\mathbf{K} = \sum_{\mathrm{e}} \bar{\mathbf{k}}^{\mathrm{e}} \rightarrow \mathbf{K} = \begin{bmatrix} \mathbf{K}_{11} & \cdots & \mathbf{K}_{1n} \\ \vdots & \ddots & \vdots \\ \mathbf{K}_{n1} & \cdots & \mathbf{K}_{nn} \end{bmatrix}; \ \mathbf{K}_{ij} = \sum_{e} \bar{\mathbf{k}}_{ij}^{\mathrm{e}} \tag{11.42}$$

In Equation $(11.42)_3$, the summation includes, as mentioned above, all of the finite elements which belong to the considered node and refer to the same global degrees of freedom. The application of the direct stiffness method is shown in example 4.11 in [1]. In addition, the procedure for analysis based on Equation (11.42) is evident from Examples 11.6 and 11.7 given below. The summation procedure is based on the following [1]. For the structure under consideration, one can display the global degrees of freedom. Further, each element of the structure can be shown separately and the possible global degrees of freedom as well as local ones, can also be displayed. When these two displayed systems are compared, the global stiffness matrix can be written based on the principle already outlined above. The other two elements of Equation (11.38), as shown in Equation (11.39), take the form:

$$\mathbf{F}_{\mathrm{R}}^{\mathrm{T}} = [\mathbf{F}_1 \ \ .... \ \ \mathbf{F}_n] \tag{11.43}$$

$$\mathbf{U}^{\mathrm{T}} = [\mathbf{U}_1 \ \ .... \ \ \mathbf{U}_n] \tag{11.44}$$

Boundary Conditions, Partition of the Structural Equation and Its Solution

A proposed structure that needs to be designed or a structure to be analyzed must be in stable equilibrium; that is, it must have supports. Because of this, some nodal displacements are known in advance. For example, if we consider the structure shown in Figure 11.8, the following displacements are known: $V_{\mathrm{A}} = 0$, $U_{\mathrm{B}} = V_{\mathrm{B}} = 0$, $V_{\mathrm{C}} = 0$.

The known displacements (or known forces) are boundary conditions.

Based on the conjugacy relations between displacements and forces, it is possible to determine the unknown values. In other words, where the values of displacements are known, the values of forces are unknown and vice versa. As such, the structural equation (Equation (11.38)) needs to be partitioned:

$$\begin{bmatrix} \mathbf{K}_{mm} & \mathbf{K}_{mr} \\ \mathbf{K}_{rm} & \mathbf{K}_{rr} \end{bmatrix} \begin{bmatrix} \mathbf{U}_m \\ \mathbf{U}_r \end{bmatrix} = \begin{bmatrix} \mathbf{F}_m \\ \mathbf{F}_r \end{bmatrix}. \tag{11.45}$$

In Equation (11.45):

$\mathbf{U}_m$ represents the unknown components of the vector **U**
$\mathbf{U}_r$ represents the known components of the vector **U**

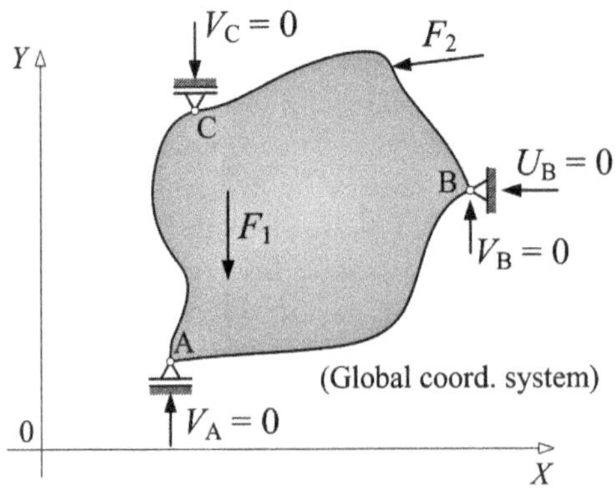

Figure 11.8 Boundary conditions.

$\mathbf{F}_m$ represents the known components of the vector $\mathbf{F}_\mathrm{R}$
$\mathbf{F}_r$ represents the unknown components of the vector $\mathbf{F}_\mathrm{R}$.

The known components of the vector **U**, that is, the vector $\mathbf{U}_r$, contains known displacements of the supports and they are usually equal to zero. Solving Equation (11.45), it follows that:

$$\mathbf{K}_{mm}\mathbf{U}_m + \mathbf{K}_{mr}\mathbf{U}_r = \mathbf{F}_m \tag{11.46a}$$

$$\mathbf{K}_{rm}\mathbf{U}_m + \mathbf{K}_{rr}\mathbf{U}_r = \mathbf{F}_r. \tag{11.46b}$$

From Equation (11.46a) follow the unknown components of the vector **U**, that is, vector $\mathbf{U}_m$:

$$\mathbf{U}_m = \mathbf{K}_{mm}^{-1}(\mathbf{F}_m - \mathbf{K}_{mr}\mathbf{U}_r). \tag{11.47}$$

Since $\mathbf{U}_r$ is usually equal to zero, Equation (11.47) takes the form:

$$\begin{aligned} &\mathbf{U}_m = \mathbf{K}_{mm}^{-1}\mathbf{F}_m \\ &(\mathbf{F}_m = \mathbf{K}_{mm}\mathbf{U}_m) \end{aligned} \tag{11.48}$$

And from there, $\mathbf{U}_m$ (the unknown components of the vector **U**) can be obtained. From Equation (11.46b), it finally follows that $(\mathbf{U}_r = 0)$:

$$\mathbf{F}_r = \mathbf{K}_{rm}\mathbf{U}_m. \tag{11.49}$$

Condensed Stiffness Matrix

The finite element stiffness matrix, $\bar{\mathbf{k}}^e$ in Equation (11.33) refers to all finite element nodal displacements and forces, while the structural stiffness matrix, **K** in Equation (11.38) refers to all structural nodal displacements and forces. In some engineering applications, only some nodal forces are of interest. Let these nodal parameters be designated "important" parameters, and those that are not of interest be designated "unimportant" parameters. In these cases it is more appropriate (or necessary) to calculate a new stiffness matrix that will contain only important parameters of nodal vectors, that is, certain degrees of freedom. This new matrix is known as a *condensed stiffness matrix* [12–14]. In the literature, under the term "condensed matrix" or the more general "condensation" may be found several explanations. One is the one we've just mentioned, that is, when only some parameters are of interest; a second explanation involves reducing the stiffness matrix in accordance with the nodal displacements that are equal to zero and, for example, in dynamic structural analysis, in cases of structural eigenvalue calculation. In such a case, a system with a large number of degrees of freedom may be reduced to a smaller number of degrees of freedom without a significant impact on the accuracy of the eigenvalue results. This approach to the determination of eigenvalues is called the approach of the condensation of the system or, simply, the method of reducing the number of degrees of freedom.

Let's return to the case where only some nodal forces are of interest. The term "condensation" (usually known as static condensation) refers to the contraction in size of a matrix by elimination of certain degrees of freedom. For example, imagine a planar structure that is assembled from some rods, some beams and some flat surfaces. The rods and beams are modeled by one-dimensional finite elements, while the flat surfaces are modeled by, for example, triangular finite elements. Let these flat surfaces, modeled by triangular finite elements, have only two degrees of freedom (two translations) per node, while all of the other elements have three degrees of freedom per node. In the analysis of such a structure, it is necessary to arrange all nodes that represent joints between the area elements and the line elements (rod elements) in such a way that the nodal vectors are of the same rank, and this implies editing the stiffness matrix of the considered elements. Let "a" represent the "important" degrees of freedom (parameters), and let "r" represent the "unimportant" parameters. If the element equation (Equation (11.32)) is now partitioned similar to Equation (11.45), then, with respect to the "important" and "unimportant" parameters, it follows that:

$$\begin{bmatrix} \bar{\mathbf{k}}^e_{aa} & \bar{\mathbf{k}}^e_{ar} \\ \bar{\mathbf{k}}^e_{ra} & \bar{\mathbf{k}}^e_{rr} \end{bmatrix} \begin{bmatrix} \bar{\mathbf{u}}^e_a \\ \bar{\mathbf{u}}^e_r \end{bmatrix} = \begin{bmatrix} \bar{\mathbf{f}}^e_a \\ \bar{\mathbf{f}}^e_r \end{bmatrix}. \tag{11.50}$$

Let forces $\bar{\mathbf{f}}^e_r = 0$, then, from the second equation of Equation (11.50), it follows that:

$$0 = \bar{\mathbf{k}}^e_{ra}\bar{\mathbf{u}}^e_a + \bar{\mathbf{k}}^e_{rr}\bar{\mathbf{u}}_r \tag{a}$$

and then:

$$\bar{\mathbf{u}}_r = -\bar{\mathbf{k}}^{e\,-1}_{rr}\,\bar{\mathbf{k}}^e_{ra}\bar{\mathbf{u}}^e_a r. \tag{b}$$

If the quantity in Equation (b) is introduced into the first of Equations (11.50), we obtain:

$$\left(\bar{\mathbf{k}}^e_{aa} - \bar{\mathbf{k}}^e_{ar}\bar{\mathbf{k}}^{e\,-1}_{rr}\,\bar{\mathbf{k}}^e_{ra}\right)\bar{\mathbf{u}}^e_a = \bar{\mathbf{f}}^e_a \tag{c}$$

Equation (c) can be written in the form:

$$\bar{\mathbf{k}}^e_c\bar{\mathbf{u}}^e_a = \bar{\mathbf{f}}^e_a, \tag{d}$$

where:

$$\bar{\mathbf{k}}^e_c = \bar{\mathbf{k}}^e_{aa} - \bar{\mathbf{k}}^e_{ar}\bar{\mathbf{k}}^{e\,-1}_{rr}\bar{\mathbf{k}}^e_{ra} \tag{11.51}$$

is the condensed stiffness matrix.

11.2.2 Structures

Chapter 1 gave only a brief overview of the possible types of structures. In general, structures can be classified with respect to the spatial arrangement of the structural elements that constitute them, and may be considered planar or spatial structures. On the other hand, if structures are classified with respect to the types of structural elements that make up the structure, then they can be treated as line structures, surface structures or (body) solid structures.

The type of analysis that may be performed on a considered structure, from the viewpoint of its dimensionality (1-D, 2-D or 3-D), depends on the applied finite elements selected for structural modeling. This selection needs to be in accordance with the issue discussed above. For example, one planar or spatial frame structure consisting of beam structural elements can be modeled, for example, with 1-D finite elements, or with 3-D finite elements, respectively. In Figure 11.9, some types of structures and their modeling are shown. In Figure 11.9(a), a one-dimensional beam is modeled using 1-D beam finite elements; in Figure 11.9(e), a cantilever solid beam is modeled using solid elements.

11.2.3 Finite Elements

The finite elements by which a structure is modeled (discretized) need to be chosen in such a way that the geometry of the structure, applied load and boundary conditions are modeled as the most credible, and that the analysis can give the required data. With respect to their shape, one-dimensional finite elements (trusses/rods/bars and beam elements) can be straight or curved; two-dimensional finite elements (triangular, rectangular and other area elements) and three-dimensional finite elements (solid elements) can have straight or curved sides (see Figure 11.9(f)). A one-dimensional truss (rod) element will be understood as the kind of element that can be subjected to axial force (that is, axial displacement) only, while a beam element will be understood to be element that can be subjected to an arbitrary load (that is, an arbitrary displacement).

Each finite element has nodes (node points) and a finite element field. A finite element may also have external and internal nodes; a finite element is connected to another by external (outside) nodes. Finite elements can be compatible or noncompatible, or, let's say, conformable or nonconformable. If the conditions of continuity (compatibility) at the boundaries between the elements are met, then the elements are compatible (conformable). There are different levels of continuity, such as continuity of the function: C^0, continuity of the function and its first derivative: C^1 and so on. The function mentioned is a function describing the field of variable parameters (that is, displacement) in the field of a finite element and it is called the shape (or interpolation, approximative) function.

11.2.4 Shape Functions – Cartesian and Natural (Dimensionless) Coordinate Systems

Under the analytical description of finite element field behavior is considered a description of the change in the finite element displacement field, and this is done by a polynomial series which contains the generalized parameters (coordinates), α_i. The polynomial form of an assumed displacement field in the finite element depends on the coordinates of the points in the element field. This polynomial form provides the means by which to

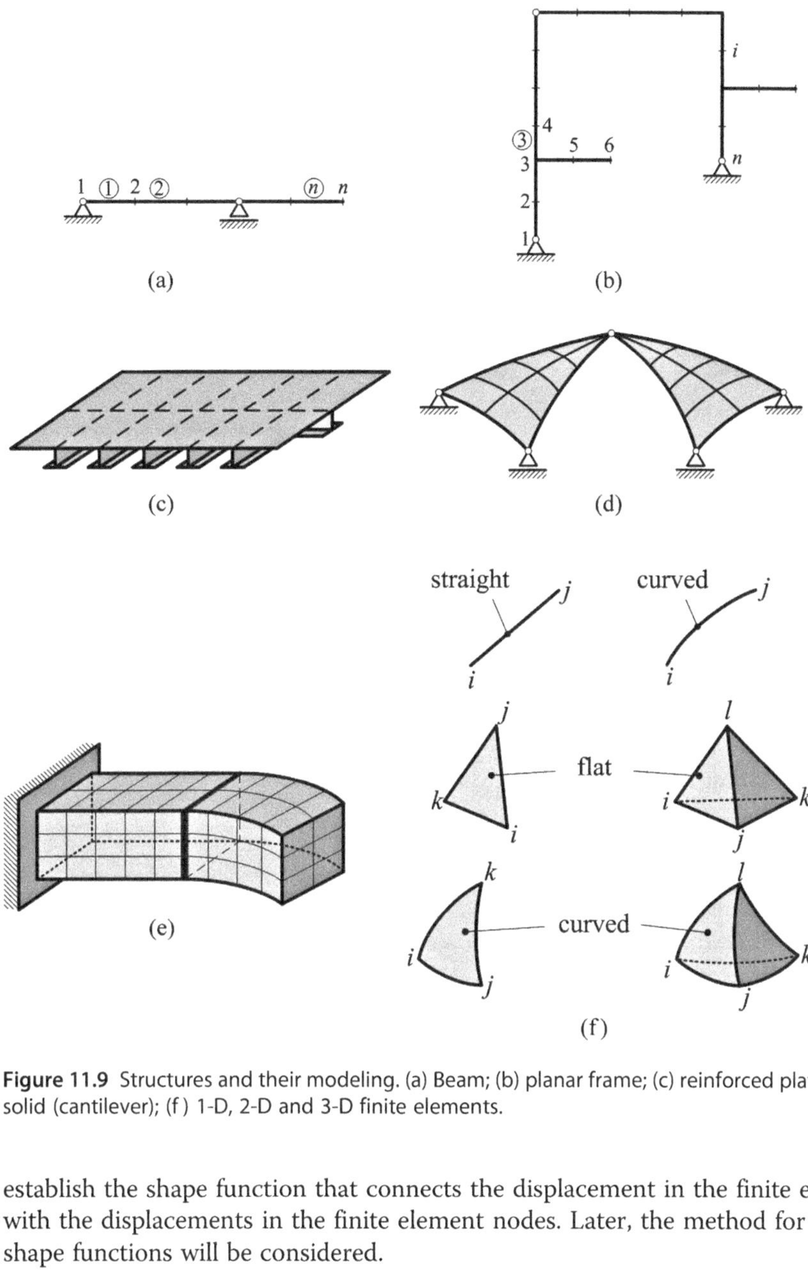

Figure 11.9 Structures and their modeling. (a) Beam; (b) planar frame; (c) reinforced plate; (d) shell; (e) solid (cantilever); (f) 1-D, 2-D and 3-D finite elements.

establish the shape function that connects the displacement in the finite element field with the displacements in the finite element nodes. Later, the method for establishing shape functions will be considered.

11.2.4.1 Cartesian Coordinate System

Using the shape function establishes a direct relationship between the variable parameter in the field of the finite element and the same parameter in the nodes of the element, that is, between the value of the function in the field of the finite element and its nodal values. The shape function matrix (interpolation matrix), **N**, contains the shape functions (Equation (11.6)). Continuity between the elements depends on the choice of interpolation functions. Definition of the interpolation function can be via the *generalized coordinates* (*method of generalized coordinates*) or via the *direct procedure* (*direct construction method through interpolation*). Generalized coordinates, α_i, are parameters independent of each other that define the intensity of the function whose form is determined by the order of the polynomial. The procedure for defining the interpolation function requiring the use of generalized coordinates is quite simple, but in cases involving more complex finite elements, it is no longer quite so simple due to the necessary inversion the matrix, $\mathbf{a}^{\mathrm{b}}$ (that is, $\mathbf{a}^{\mathrm{b}^{-1}}$). It is also important to know that the shape function, N_i has the property that its value is equal to one (1) in node "i" and is equal to zero (0) for all other nodes. The interpolation function (shape function) must ensure a good approximation of the considered parameter (for example, displacement) in the element and continuity between the elements; it must also be as simple as possible. These properties have polynomials, which are mainly used as interpolation functions. There are several functions in the form of a polynomial that can be used: Lagrange polynomials, Hermite polynomials and Serendipity functions [1–4, 9–16]. In the following two sections, both methods, that is, the method of generalized coordinates and the direct construction method in shape function determination will be shown for cases involving one-dimensional finite elements. More information about shape functions can be found in [3, 7, 10, 11] and many others.

Application of the Method of Generalized Coordinates in Shape Function Determination

Application of the method of generalized coordinates will be demonstrated briefly here, although we will return to it later. The method will first be considered on a basic, one-dimensional truss finite element.

Example 11.1.

Consider the *method of generalized coordinates* to determine the shape functions for the *basic truss (axial) finite element* shown in Figure 11.10. This finite element has two node points and one degree of freedom per node (one axial displacement per node).

Data:

A basic, one-dimensional axial finite element is shown in Figure 11.10.

Solution:

As can be seen, this finite element has two degrees of freedom, one per node. These two degrees of freedom are two axial displacements in its nodes, $\mathbf{u}^{\mathrm{e}^{\mathrm{T}}} = [w_1 \; w_2]$. Assuming a

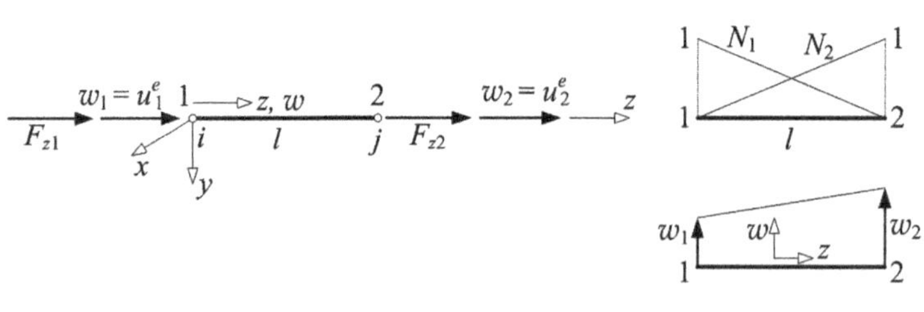

Figure 11.10 A basic one-dimensional truss (axial, rod) finite element subjected to axial nodal displacements (that is, axial nodal forces).

linear distribution of (axial) displacement in the field of the finite element, in accordance with Equation (11.2) and in accordance with the degrees of the freedom, the following polynomial may be used for its interpolation:

$$w = \alpha_1 + \alpha_2 z. \tag{11.52}$$

The polynomial matrix and the vector of generalized coordinates are of the following forms:

$$\mathbf{a} = \begin{bmatrix} 1 & z \end{bmatrix}, \quad \boldsymbol{\alpha} = \begin{bmatrix} \alpha_1 \\ \alpha_2 \end{bmatrix}. \tag{11.53}$$

In accordance with Equation (11.52) and Figure 11.10, the boundary conditions are:

$$z = 0 \rightarrow w = w_i = w_1 = \alpha_1; z = l \rightarrow w = w_j = w_2 = \alpha_1 + \alpha_2 l \tag{11.54}$$

Now, Equation (11.52), written for boundary conditions, that is, formulated in accordance with Equation (11.5), takes the form:

$$\begin{bmatrix} w_i \\ w_j \end{bmatrix} = \begin{bmatrix} 1 & 0 \\ 1 & l \end{bmatrix} \begin{bmatrix} \alpha_1 \\ \alpha_2 \end{bmatrix} \rightarrow \mathbf{u}^{\mathrm{e}} = \mathbf{a}^{\mathrm{b}} \boldsymbol{\alpha}. \tag{11.55}$$

Further:

$$\mathbf{a}^{\mathrm{b}^{-1}} = \begin{bmatrix} 1 & 0 \\ -\frac{1}{l} & \frac{1}{l} \end{bmatrix}; \quad \left(\mathbf{note}: \mathbf{a}^{\mathrm{b}^{-1}} \rightarrow \left(\mathbf{a}^{\mathrm{b}}\right)^{-1} \right) \tag{11.56}$$

In accordance with Equation (11.6a), the interpolation matrix has a value:

$$\mathbf{N} = \mathbf{N}_w = \mathbf{a}\mathbf{a}^{\mathrm{b}^{-1}} = \begin{bmatrix} 1 & z \end{bmatrix} \begin{bmatrix} 1 & 0 \\ -\frac{1}{l} & \frac{1}{l} \end{bmatrix} = \begin{bmatrix} 1 - \frac{z}{l} & \frac{z}{l} \end{bmatrix}, \tag{11.57}$$

where the shape functions, N_{w1}, N_{w2}, are:

$$\mathbf{N} = \mathbf{N}_w = \begin{bmatrix} N_{w1} & N_{w2} \end{bmatrix} = \begin{bmatrix} 1 - \frac{z}{l} & \frac{z}{l} \end{bmatrix} \tag{11.58}$$

and they are shown in Figure 11.10. The subscript "w" indicates that the shape functions refer to axial displacements.

Application of the Direct Construction Method in Shape Function Determination

Application of the *direct construction method* in determining the shape functions through interpolation will be demonstrated through the use of Lagrange and Hermite polynomials.

Lagrangian Interpolation (Lagrange Interpolating Polynomials) The distribution of a variable parameter (for example, displacement) in the field of a finite element can be assumed using a polynomial, as shown in Equation (11.2):

$$u(z) = \alpha_1 + \alpha_2 z + \alpha_3 z^2 + \ldots + \alpha_{n+1} z^n = \sum_{i=0}^{n} \alpha_{i+1} z^i \tag{11.59}$$

Since this equation must also be satisfied in the element node points, $u_i = u_i^e$, it follows, from Equation (11.6) that:

$$\mathbf{u} = \mathbf{N}\mathbf{u}^e \tag{a}$$

Function $u(z) = \mathbf{u}$, can be approximated by a polynomial and the values of the function at its discrete points, $u_i = u_i^e$, are:

$$u(z) = l_1(z)u_1 + l_2(z)u_2 + \ldots + l_n(z)u_n = \sum_{i=1}^{n} l_i(z)u_i, \tag{11.60}$$

where the coefficients $l_i(z)$ are of the form:

$$l_i(z) = \frac{\prod_{\substack{j=1 \\ j\neq i}}^{n} (z - z_j)}{\prod_{\substack{j=1 \\ j\neq i}}^{n} (z_i - z_j)} \tag{11.61}$$

In Equation (11.61), symbol "Π" means the product as usual, while n is the number of joints in the z direction. It is evident that the ith Lagrange interpolating polynomial, $(l_i(z))$, takes the value "one" (1) at the ith point, while for all other node points, the value of the polynomial is equal to "zero" (0):

$$l_i(z) = 1 \text{ for } z_i,\ l_i(z) = 0 \text{ for } z_j \text{ for } j \neq i \tag{11.62}$$

On the basis of Equations (11.6) and (11.60), we can write:

$$u(z) = \mathbf{u} = \sum_{i=1}^{n} l_i(z)u_i^e = \sum N_i u_i^e \;\rightarrow \mathbf{u} = \mathbf{N}\mathbf{u}^e \tag{11.63}$$

As stated, it is clear that the displacement, $u(z)$ (that is, the displacement in the field of the element) will take the value u_i (that is, the value of the nodal displacement) if $N_i = 1$. Thus, the shape function is equal to one (1) for the same considered point "i", and is equal to zero (0) for all other points. It can be noted that u is used as a general designation for the displacement.

Example 11.2.
Consider the *direct construction method* in the form of Lagrange interpolation to determine the shape functions for the previously considered one-dimensional, two-node-point finite element (a basic truss finite element) subjected to axial nodal displacements, as shown in Figure 11.10 and described in Example 11.1 above.

Data and Solution:
If the origin of the coordinate system is placed in the left node, that is, $z_1 = 0$, and the right node is defined by $z_2 = l$, then, using Equation (11.61), the Lagrangian polynomials are:

$$\begin{aligned} l_1(z) &= \frac{z - z_2}{z_1 - z_2} = \frac{z - l}{0 - l} = 1 - \frac{z}{l} \equiv N_1 \equiv N_{w_1} \\ l_2(z) &= \frac{z - z_1}{z_2 - z_1} = \frac{z - 0}{l - 0} = \frac{z}{l} \equiv N_2 \equiv N_{w_2}. \end{aligned} \tag{11.64}$$

As may be noted, the shape functions are of the same values as previously obtained in Equation (11.58), and thus they are presented in Figure 11.10. As such, Equation (11.60) or Equation (11.63) can be written as:

$$u(z) = l_1(z)u_1 + l_2(z)u_2 \rightarrow w(z) = l_1(z)w_1 + l_2(z)w_2$$
$$= N_1 w_1 + N_2 w_2 = \left(1 - \frac{z}{l}\right) w_1 + \frac{z}{l} w_2 \tag{11.65}$$

where w_1 and w_2 are the values of the nodal displacements.

Hermitian Interpolation (Hermite Interpolating Polynomials)

Lagrange interpolating polynomials generally provide continuity of the function (for example, displacement) at the boundaries between the finite elements, that is, in the finite element node points, and this is so-called C^0 continuity. However, a problem such as flexure requires continuity of the function and its first derivative, that is, so-called C^1 continuity. In this case, as basic unknowns at the finite element node points, along with the value of the function, its first derivative must also be introduced, and Hermite polynomials are adopted as shape functions [3]. In certain cases, even higher derivatives may be required.

Example 11.3.

Consider the two-node-point basic beam planar finite element (*basic flexural beam element*) subjected to transversal displacements and planar rotations (that is, nodal transversal forces and planar nodal bending moments), shown in Figure 11.11.

Using the direct construction method in the form of Hermite polynomials, determine the shape functions.

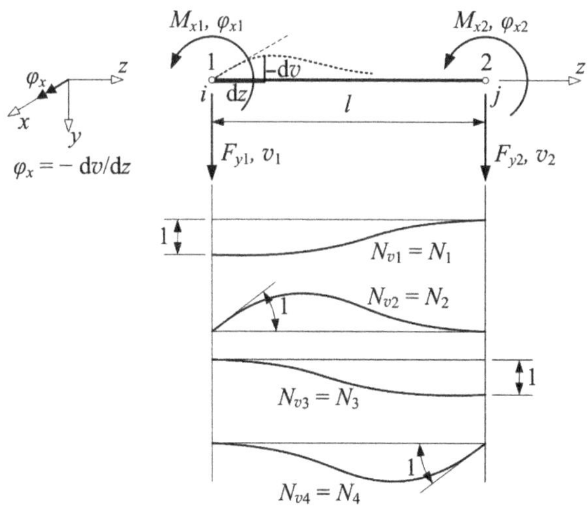

Figure 11.11 Basic, two-node-point beam element subjected to bending in the *z–y* principal centroidal plane of inertia (transversal and rotational displacements) and its shape functions.

Solution:
In the considered case, the vector of nodal displacements is defined as follows:

$$\mathbf{u}_{\mathrm{v}}^{\mathrm{e}^{\mathrm{T}}} = \left[\mathbf{u}_1^{\mathrm{e}^{\mathrm{T}}} \ \ \mathbf{u}_2^{\mathrm{e}^{\mathrm{T}}}\right], \ \mathbf{u}_i^{\mathrm{e}^{\mathrm{T}}} = \left[v_i \ \ \varphi_{xi}\right], i = 1,2 \tag{11.66}$$

The subscript "v" indicates that the shape functions refer to transversal displacements.

As can be seen, the degrees of freedom are two translational displacements and two rotational displacements. In accordance with Equation (11.2) or Equation (11.59), the distribution of the displacement in the field of the finite element can be assumed by the polynomial:

$$v(z) = \alpha_1 + \alpha_2\xi + \alpha_3\xi^2 + \alpha_4\xi^3, \tag{11.67a}$$

which has four generalized coordinates corresponding to four degrees of freedom, and where $\xi = z/l$. Translational displacement is represented by the value of the function (polynomial), while rotational displacement is represented by its derivative:

$$\varphi_x = -\mathrm{d}v/\mathrm{d}z, \tag{11.67b}$$

as can be seen from Figure 11.11.

A general record of the relationship between the displacement in the field of the finite element and the finite element nodal displacements is (from Equation (11.6)):

$$\mathbf{u} = \mathbf{N}\mathbf{u}^{\mathrm{e}}. \tag{a}$$

In the considered case, the shape function matrix contains four shape functions:

$$\mathbf{N} = [N_1 \ N_2 \ N_3 \ N_4], \tag{b}$$

where each of them is a Hermite polynomial. Each of the shape functions can be assumed to have the following form (from Equation (11.67a)):

$$N_i = \alpha_1 + \alpha_2\xi + \alpha_3\xi^2 + \alpha_4\xi^3; (= v(z) \equiv y(z)). \tag{c}$$

The derivative of the function in (c) is:

$$N_i' = \alpha_2 + 2\alpha_3\xi + 3\alpha_4\xi^2 = \varphi_x = -\mathrm{d}v/\mathrm{d}z. \tag{d}$$

Consider the determination of the first shape function, assuming that it is in the form (from Equation (c)):

$$N_1 = \alpha_1 + \alpha_2\xi + \alpha_3\xi^2 + \alpha_4\xi^3. \tag{11.68}$$

Its first derivative is (from Equation (d)):

$$N_1' = \alpha_2 + 2\alpha_3\xi + 3\alpha_4\xi^2 \tag{11.69}$$

Requirements that are placed on the shape function are:

$$N_1 = 1\,(z = 0); \ \ N_1 = 0\,(z = l); \ \ N_1' = 0\,(z = 0; z = l), \tag{e}$$

or, using substitution $\xi = z/l$, these requirements can be written in the form:

$$N_1(\xi = 0) = 1\ ; \ \ N_1(\xi = 1) = 0\ ; \ \ N_1'(\xi = 0) = 0\ ; \ \ N_1'(\xi = 1) = 0. \tag{f}$$

On the basis of these requirements and in accordance with Equation (11.68) and Equation (11.69), the following data (equations) can be obtained:

$$1 = \alpha_1;\ \ 0 = \alpha_2;\ \ 0 = \alpha_1 + \alpha_2 + \alpha_3 + \alpha_4;\ \ 0 = \alpha_2 + 2\alpha_3 + 3\alpha_4. \tag{g}$$

By solving the equations contained in Equation (g), the following values for the coefficients are obtained:

$$\alpha_1 = 1,\ \ \alpha_2 = 0,\ \ \alpha_3 = -3,\ \ \alpha_4 = 2, \tag{h}$$

and the first shape function (Equation (11.68)) takes the form:

$$N_1 = 1 - 3\xi^2 + 2\xi^3. \tag{i}$$

Using similar procedures, the other shape functions can also be determined. The shape functions are:

$$\begin{aligned} &N_1 = 1 - 3\xi^2 + 2\xi^3; N_2 = -l\xi(\xi - 1)^2; \\ &N_3 = 3\xi^2 - 2\xi^3; N_4 = -l\xi^2(\xi - 1) \end{aligned} \tag{11.70}$$

The shape functions for this element are shown in Figure 11.11.

11.2.4.2 Natural (Dimensionless) Coordinate System

Previously, we've shown an assumed displacement field within the finite element in correlation with the finite element nodal displacements (the joint d.o.f.). The correlation has been established through the shape function matrix, as shown in Equation (11.63). The shape functions relate also to the coordinates of the node points of the finite element; this means that the values of the coordinates depend on the size of the element. As such, the Cartesian coordinate system is sometimes inappropriate for this representation. As an alternative method of representation of the coordinates of a finite element's node points, one can use local, dimensionless coordinates known as *natural coordinates*; these are functions of Cartesian coordinates. The values of these coordinates are between zero and one (0, 1) [10]. There are *natural one-dimensional coordinates, natural area coordinates* and *natural volume coordinates*. For example, *triangular coordinates* belong to the category of natural area coordinates, while *tetrahedral coordinates* belong to the category of natural volume coordinates. In a natural coordinate system, the element sides are defined by the mentioned coordinate in a region regardless of the shape, physical size or element orientation in global coordinates [3, 8]. Apart from the above-mentioned natural coordinates, there are other natural coordinates whose values can be between (1, −1). The advantage of the use of natural coordinates is that the shape functions are much simpler.

Example 11.4.

Consider the triangular finite element of first order ($n = 1$) shown in Figure 11.12(a). Determine the shape functions using natural coordinates.

Figure 11.12(a) shows global (X, Y) and local (x, y) Cartesian coordinates as well as area natural coordinates (triangular coordinates), L_i, and natural coordinates (ξ, η). Element node points (1, 2) are placed on the x axis, that is, these joints coincide with the x axis. Figure 11.12 also shows some other types of triangular finite elements.

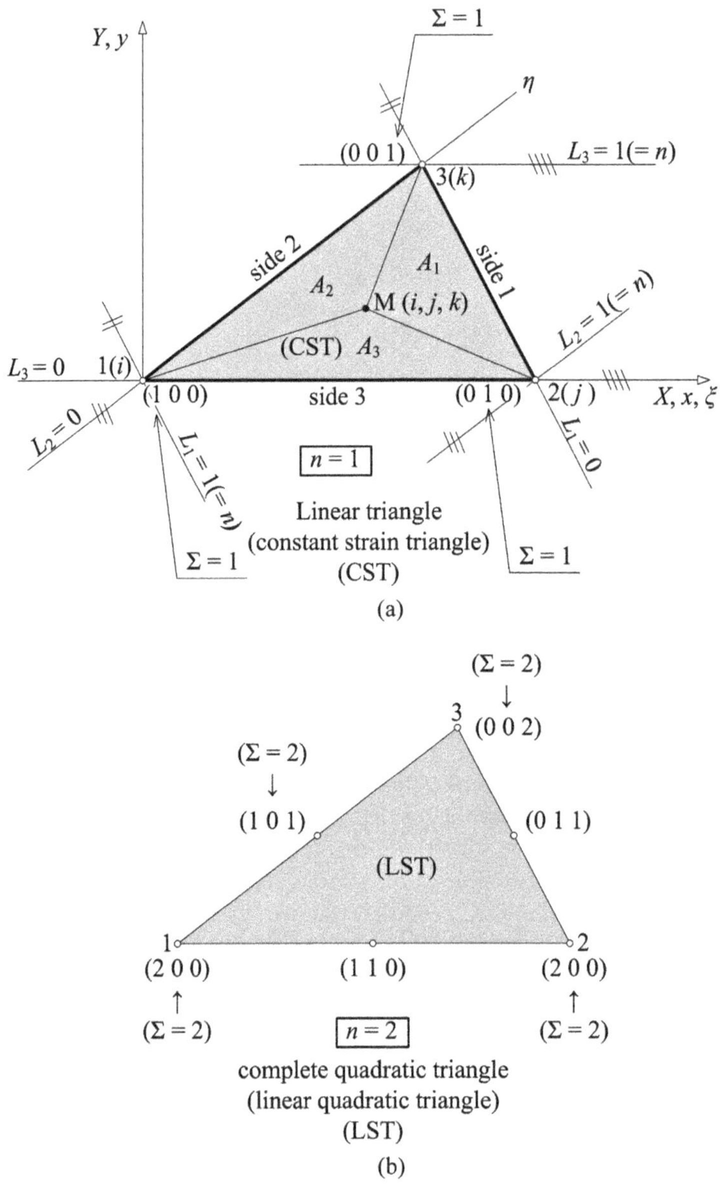

Figure 11.12 A triangular finite element: Cartesian and natural coordinates (Example 11.12). (a) Triangular element of first order (linear element, n = 1; 3 nodes; CST element/constant strain triangle); (b) triangular element of second order (quadratic element, n = 2; 6 nodes; LST element/linear strain triangle); (c) triangular element of third order (cubic element, n = 3; 10 nodes; QST element/quadratic strain triangle); (d) shape functions for the triangular finite element of first order shown in (a), that is, for the linear triangle.

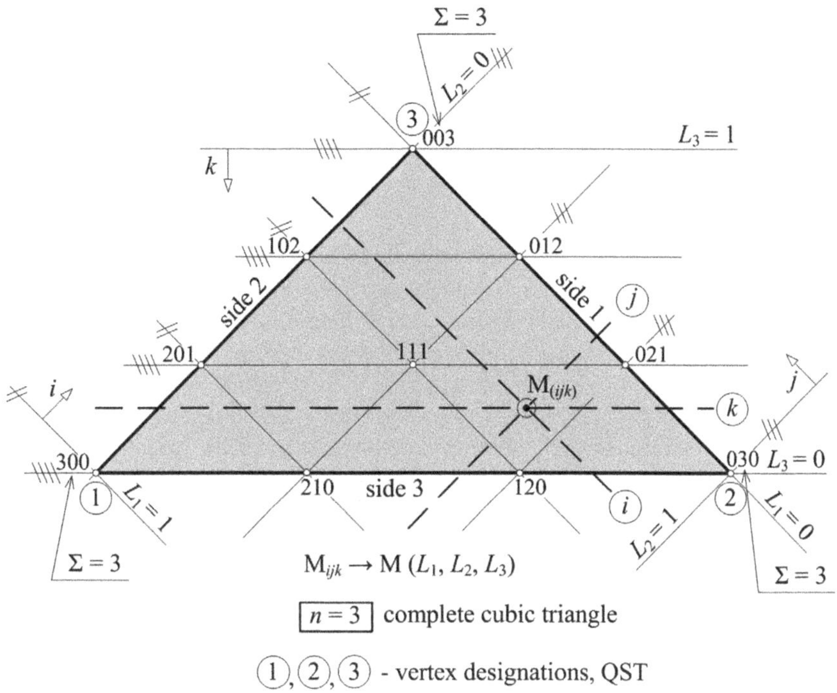

(c)

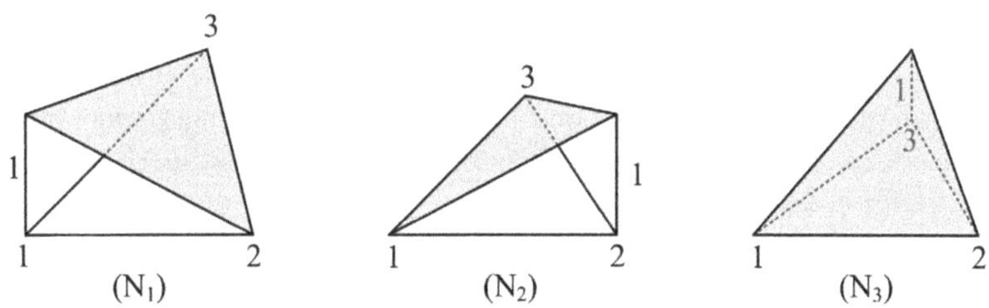

(d)

Figure 11.12 *(Continued)*

Solution:

Coordinate L_i has a value 1 at the node i, $(i = 1, 2, 3)$, while its value is equal to zero at the nodes on the opposite side to the considered node (the side that connects the other two nodes), see Figure 11.12(a). The triangular (natural) coordinates, L_i, are defined by the following ratios [3, 17]:

$$L_1 = \frac{A_1}{A}, \quad L_2 = \frac{A_2}{A}, \quad L_3 = \frac{A_3}{A}, \tag{11.71}$$

wherein:

$$A = \text{area of the triangle } (\Delta): 1,2,3; A_1 = \text{area of the } \Delta: M,2,3; A_2$$
$$= \text{area of the } \Delta: M,1,3; A_3 = \text{area of the } \Delta: M,1,2.$$

Also:

$$\sum_{i=1}^{3} L_i = L_1 + L_2 + L_3 = 1, \qquad \sum_{i=1}^{3} A_i = A_1 + A_2 + A_3 = A. \tag{11.72}$$

As can be seen from Figure 11.12(a), the sides of the triangle are identified by the opposite vertex, for example, side 2 and vertex 2 are on opposite sides, and so on. The vertices of the triangle are sometimes designated by numbers (1, 2, 3) and sometimes by letters (i, j, k). The point inside the triangle, for example, point M shown in Figure 11.12(a) and in Figure 11.12 (c), can also be identified by three digits (i, j, k). This point, M, is obtained as an intersection of three directions (i, j, k), where each of these directions starts from the vertex and is parallel to the opposite side of the triangle. The sum of the three digits at the point M is equal to "n", that is, to the order of the triangle (polynomial order). This means that, at any vertex: $i + j + k = n$. The considered point M may be any point, including a node point. The lines starting from the vertex "1" and parallel to side 1 (Figure 11.12(c)) are known as the "1" direction grid. Similar terminology is valid for the other lines and vertices.

The shape functions need to meet the conditions: $N_{ijk} = 1$ at point (ijk) and $N_{ijk} = 0$ at the other points. The applicable formula for the definition of the shape function can be obtained by performing a Lagrangian interpolation in each direction (Figure 11.12(a) and Figure 11.12(c)):

$$N_M = N_{ijk}(L_1, L_2, L_3) = l_i(L_1) l_j(L_2) l_k(L_3). \tag{11.73}$$

L_i are the area coordinates (natural area coordinates) defined by Equation (11.71) and polynomials l_i are the (one-dimensional) polynomials from Lagrangian interpolation in each direction. As such, for example, the first term on the right-hand side of Equation (11.73) is defined by:

$$l_i(L_1) = \begin{cases} \prod_{p=1}^{i} \left(\dfrac{nL_1 - p + 1}{p} \right), & \text{for } i \geq 1 \\ 1 & \textit{for } \; i = 0 \end{cases}, \tag{11.74}$$

where n is the order of the polynomial. The same is valid for $l_j(L_2)$ and for $l_k(L_3)$.

Consider the shape function for node 1 (Figure 11.12(a)), bearing in mind that the triangular element is of first order $(n = 1)$. In accordance with Equation (11.73), we have:

$$N_M = N_{ijk}(L_1, L_2, L_3) = N_1 = N_{100}(L_1, L_2, L_3) = l_1(L_1) l_0(L_2) l_0(L_3) \tag{a}$$

Further (from the second row of Equation (11.74)):

$$l_0(L_2) = l_0(L_3) = 1. \tag{b}$$

In Equation (a), N_{100} indicates that the first node is being considered and that the order of the polynomial is equal to 1 $(n = 1)$. Also, in accordance with Equation (11.74), it is clear that:

$$l_i(L_1) = l_1(L_1) = \frac{1L_1 - 1 + 1}{1} = L_1 \tag{c}$$

so, finally:

$$N_{100} = L_1 = N_1 \tag{d}$$

In this case, the shape function related to the node 1 is identical to the natural (area) coordinate for the first node for this triangle. The relationships between the Cartesian coordinates and the triangular coordinates for the considered point (M) can be written in the form:

$$\begin{aligned} x &= L_1 x_1 + L_2 x_2 + L_3 x_3 = \sum_{i=1}^{3} L_i x_i \\ y &= L_1 y_1 + L_2 y_2 + L_3 y_3 = \sum_{i=1}^{3} L_i y_i, . \end{aligned} \tag{11.75}$$

Also:

$$L_3 = 1 - L_1 - L_2 \tag{11.76}$$

These two equations (Equations (11.75)and (11.76)) can be collected as:

$$\begin{bmatrix} 1 \\ x \\ y \end{bmatrix} = \begin{bmatrix} 1 & 1 & 1 \\ x_1 & x_2 & x_3 \\ y_1 & y_2 & y_3 \end{bmatrix} \begin{bmatrix} L_1 \\ L_2 \\ L_3 \end{bmatrix} \tag{11.77}$$

As a result of the solutions of Equation (11.77) with respect to L_i, the following equations can be obtained:

$$\begin{aligned} L_1(x,y) &= \frac{1}{2A}(a_1 + b_1 x + c_1 y) \\ L_2(x,y) &= \frac{1}{2A}(a_2 + b_2 x + c_2 y). \\ L_3(x,y) &= \frac{1}{2A}(a_3 + b_3 x + c_3 y). \end{aligned} \tag{11.78}$$

In Equation (11.78), the coefficients are:

$$\begin{aligned} &a_1 = x_2 y_3 - x_3 y_2 \quad && a_2 = x_3 y_1 - x_1 y_3 \quad && a_3 = x_1 y_2 - x_2 y_1 \\ &b_1 = y_2 - y_3 && b_2 = y_3 - y_1 && b_3 = y_1 - y_2 \quad , \\ &c_1 = x_3 - x_2 && c_2 = x_1 - x_3 && c_3 = x_2 - x_1 . \end{aligned} \tag{11.79}$$

and the finite element area is ($A_e = A$):

$$A = \frac{1}{2}[(x_2 y_3 - x_3 y_2) - (x_1 y_3 - x_3 y_1) + (x_1 y_2 - x_2 y_1)]. \tag{11.80}$$

So, if, in the first of Equations (11.78), we insert the values of the coefficients from Equation (11.79), and given that points 1 and 2 (Figure 11.12(a)) are placed on the x axis ($x_1 = y_1 = y_2 = 0$), the value of the shape function determined earlier in Equation (d), in the Cartesian coordinate system, is:

$$N_{100} = L_1 = N_1 = \frac{1}{x_2 y_3}(x_2 y_3 - x y_3 - x_2 y + x_3 y) \tag{e}$$

However, as stated earlier, the shape function takes a value equal to one (1) at the point to which it refers, and is equal to zero at the other node points, as can be seen from Equation (e).

Analogously, the other shape functions can be obtained, so all of the shape functions for the considered case can be written as follows:

$$\begin{aligned} N_{100} &= L_1 = N_1 = \frac{1}{x_2 y_3}(x_2 y_3 - x y_3 - x_2 y + x_3 y) \\ N_{010} &= L_2 = N_2 = \frac{1}{x_2 y_3}(x y_3 - x_3 y) \\ N_{001} &= L_3 = N_3 = \frac{y}{y_3} \end{aligned} \tag{11.81}$$

The displacement field, u (Equation (11.63)), for this triangular finite element can be written in the form:

$$u = N_{100} u_{100} + N_{010} u_{010} + N_{001} u_{001}. \tag{11.81a}$$

The shape functions for the linear triangle formed by Equation (11.81) are shown in Figure 11.12(d), and they are the same regardless of whether one considers the translational displacement "u" or "v".

Example 11.5.
Using the natural coordinate system, determine the shape functions for the one-dimensional finite element presented earlier in Figure 11.10. This element has one d.o.f. per node. The shape functions were previously determined in the Cartesian coordinate system using the method of generalized coordinates (see Example 11.1) and also using the direct construction method with Lagrangian interpolation (see Example 11.2).

Let the natural coordinates also be depicted by L_1 and L_2, where the former refers to node point 1 and the latter refers to node point 2. The direction of the coordinate L_1 is from node 2 to node 1 while L_2 is going from 1 to 2.

Similar to Equation (11.75), coordinate z of a point in the field of the finite element is defined as:

$$z = L_1 z_1 + L_2 z_2 = \sum L_i z_i \tag{a}$$

where z_1 and z_2 are the nodal coordinates. Also, according to Equation (11.72):

$$\sum L_i = 1 = L_1 + L_2 \tag{b}$$

In accordance with Equation (11.77), combining Equations (a) and (b), it follows that:

$$\begin{bmatrix} 1 \\ z \end{bmatrix} = \begin{bmatrix} 1 & 1 \\ z_1 & z_2 \end{bmatrix} \begin{bmatrix} L_1 \\ L_2 \end{bmatrix} \tag{c}$$

Solving this equation with respect to L_i, we find:

$$\begin{bmatrix} L_1 \\ L_2 \end{bmatrix} = \frac{1}{z_2 - z_1} \begin{bmatrix} z_2 & -1 \\ -z_1 & 1 \end{bmatrix} \begin{bmatrix} 1 \\ z \end{bmatrix} \tag{d}$$

For example, if we determine L_1, it is:

$$L_1 = \frac{1}{z_2 - z_1}[z_2 \quad -1]\begin{bmatrix} 1 \\ z \end{bmatrix} = \frac{z_2 - z}{z_2 - z_1}, \tag{e}$$

And, since $z_2 = l$, $z_1 = 0$, this can be written as:

$$L_1 = \frac{l - z}{l} = 1 - \frac{z}{l}, \tag{f}$$

and thus, consequently, $L_1 = 1$ at point 1 ($z = 0$) and $L_1 = 0$ at point 2 ($z = l$). Similarly, L_2 can be determined as:

$$L_2 = \frac{z}{l}, \tag{g}$$

and $L_2 = 1$ at point 2 and $L_2 = 0$ at point 1. Further, in accordance with Equation (11.73), we have:

$$N_M = N_{ij}(L_1, L_2) = l_i(L_1)l_j(L_2), \tag{h}$$

And, using Equation (11.74) and taking:

$$l_0(L_1) = l_0(L_2) = 1,$$

it follows that:

$$N_1 = N_{10} = l_1(L_1)l_0(L_2) = \frac{1 \cdot L_1 - 1 + 1}{1} \cdot 1 = L_1 \tag{i}$$

$$N_2 = N_{01} = l_0(L_1)l_1(L_2) = 1 \cdot \frac{1 \cdot L_2 - 1 + 1}{1} = L_2 \tag{j}$$

Since $N_i = L_i$, the obtained results are the same as those in Equation (11.58) and Equation (11.64).

11.2.5 One-dimensional Finite Elements

11.2.5.1 Basic 1-D Finite Elements

Usually, finite elements are classified as either basic elements or finite elements of higher order. A basic finite element will be considered as an element that is described by the vertices (nodes) which define its shape and by basic degrees of freedom (translations and rotations) in its vertices (node points). A finite element that has additional degrees of freedom in its vertices (nodes) or has additional nodes along its sides or in its field will be treated as a finite element of higher order. This issue is discussed, for example, in [3, 7, 10, 18].

As previously described, a shape function can only approximately describe the distribution of the displacement in the field of a basic finite element if the considered element is continuously loaded in its field. In an effort to achieve better results (greater accuracy), it is necessary to use a polynomial of higher order for the distribution of displacement. This is associated with increasing degrees of freedom, that is, with the addition of new nodes or with the addition of new degrees of freedom in existing nodes.

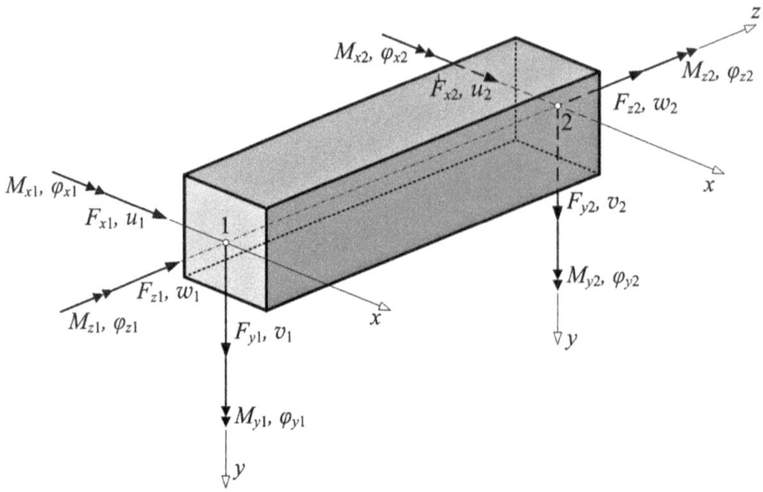

Figure 11.13 A spatial one-dimensional, two-node-point finite element.

One-dimensional finite elements belong to a simpler group of finite elements and are mainly intended for the analysis of one-dimensional engineering elements or structures; that is, engineering members whose geometry can be discretized. Using one-dimensional finite elements, the internal forces can be obtained as a function of the coordinate of a point along the element. In Figure 11.13, a basic, spatial, one-dimenisional finite element is presented. Its vector of nodal forces and its vector of nodal displacements are also shown. All of the relevant quantities, such as the modulus of elasticity (E), the shear modulus (G), displacement (u), torsional rigidity (GI_t), axial rigidity (AE), the axial moment of inertia (I_x), the elasticity matrix (C), the nodal displacements vector ($\mathbf{u}^e$), the nodal forces vector ($\mathbf{f}^e$), the strain–displacement matrix (**B**), the finite element stiffness matrix ($\mathbf{k}^e$) and so on have been explained in previous sections.

The vector of nodal displacements and the vector of nodal forces are:

$$\mathbf{u}^{e^T} = \left[\mathbf{u}_1^{e^T}\mathbf{u}_2^{e^T}\right], \quad \mathbf{u}_i^{e^T} = \left[w_i \;\; u_i \;\; v_i \;\; \varphi_{zi} \;\; \varphi_{xi} \;\; \varphi_{yi}\right], i = 1,2 \tag{11.82}$$

$$\mathbf{f}^{e^T} = \left[\mathbf{f}_1^{e^T}\mathbf{f}_2^{e^T}\right], \quad \mathbf{f}_i^{e^T} = \left[F_{zi} \;\; F_{xi} \;\; F_{yi} \;\; M_{zi} \;\; M_{xi} \;\; M_{yi}\right], i = 1,2 \tag{11.83}$$

As can be seen, the considered one-dimensional, spatial finite element has six degrees of freedom per node (three translational displacements and three rotational displacements) and six forces per node (three forces and three moments), which means that its stiffness matrix has 12 x 12 members. In the case of a planar finite element, the stiffness matrix has 6 x 6 members.

Truss (Rod) Finite Element

Axial Forces (Nodal Displacements Only) A one-dimensional, two-node-point finite element subjected only to axial nodal displacements (a *basic truss finite element*) is presented once again in Figure 11.14.

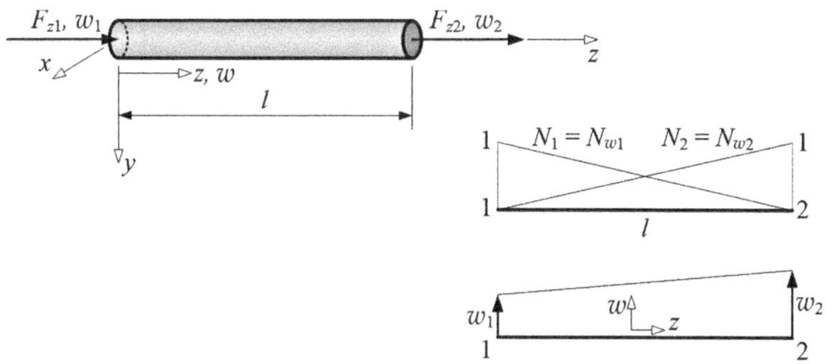

Figure 11.14 One-dimensional truss (axial) finite element subjected to axial nodal displacements (basic truss finite element).

The vector of nodal displacements and the vector of nodal forces are:

$$\mathbf{u}_{\mathrm{w}}^{\mathrm{e}^{\mathrm{T}}} = \left[\mathbf{u}_1^{\mathrm{e}^{\mathrm{T}}} \mathbf{u}_2^{\mathrm{e}^{\mathrm{T}}}\right] = [w_1 \quad w_2] \tag{11.84}$$

$$\mathbf{f}_{\mathrm{w}}^{\mathrm{e}^{\mathrm{T}}} = \left[\mathbf{f}_1^{\mathrm{e}^{\mathrm{T}}} \mathbf{f}_2^{\mathrm{e}^{\mathrm{T}}}\right] = [F_{z1} \quad F_{z2}] \tag{11.85}$$

Subscript "w" indicates that only axial displacements (forces) are being considered.

The shape functions for this finite element are defined by Equation (11.58):

$$\mathbf{N} = \mathbf{N}_{\mathrm{w}} = [N_{\mathrm{w}1} \; N_{\mathrm{w}2}] = \left[1 - \frac{z}{l} \; \frac{z}{l}\right] \tag{a}$$

and they are shown in Figure 11.10 (and also in Figure 11.14). The procedure for the determination of these shape functions is shown in Example 11.1 using the generalized coordinates method and in Example 11.2 using the direct construction method with Lagrangian interpolation. The strain–displacement matrix is (from Equation (11.8)):

$$\mathbf{B} = \mathbf{d}_{\varepsilon}\mathbf{N}, \tag{b}$$

where the differential operator for an axially loaded element is:

$$\mathbf{d}_{\varepsilon} = \frac{\mathrm{d}}{\mathrm{d}z}. \tag{c}$$

Based on Equations (b) and (c):

$$\begin{aligned} \mathbf{B}_{\mathrm{w}} &= \frac{\mathrm{d}}{\mathrm{d}z}\left[1 - \frac{z}{l} \; \frac{z}{l}\right] \\ \mathbf{B}_{\mathrm{w}} &= [-1 \quad 1]\frac{1}{l}. \end{aligned} \tag{11.86}$$

In accordance with the constitutive equation (Equation (11.9)) for a rod as a one-dimensional finite element, this equation is:

$$\boldsymbol{\sigma} = \mathbf{C} \cdot \boldsymbol{\varepsilon} \tag{d}$$

Since, in this case:

$$\sigma = \frac{N}{A} = E\varepsilon_z, \tag{e}$$

constitutive Equation (d) can be written in the form:

$$[N] = [EA][\varepsilon_z], \tag{f}$$

where N is the axial force, E is the modulus of elasticity, A is the cross-sectional area of the finite element, while ε_z is the axial strain (normal strain) in the z direction, that is, along the longitudinal axis of the considered structural (finite) element. The elasticity matrix, C, is simply obtained from the isotropic stress–strain relationship in Equation (f), and in this case is:

$$\mathbf{C}_\mathrm{w} = [EA], \tag{11.87}$$

where subscript w indicates that the stress–strain relationship has been obtained as a result of an action of axial force, that is, axial displacement, w, is caused by axial force. In accordance with the definition of the stiffness matrix (Equation (11.19)), its value is:

$$\mathbf{k}_\mathrm{w}^\mathrm{e} = \int_0^l \mathbf{B}_\mathrm{w}^\mathrm{T}\mathbf{C}_\mathrm{w}\mathbf{B}_\mathrm{w}\mathrm{d}z = \int_0^l \begin{bmatrix} -1 \\ 1 \end{bmatrix} \frac{1}{l} EA[-1 \quad 1]\frac{1}{l}\mathrm{d}z$$

$$\mathbf{k}_\mathrm{w}^\mathrm{e} = \frac{EA}{l}\begin{bmatrix} 1 & -1 \\ -1 & 1 \end{bmatrix}. \tag{11.88}$$

If this finite element is considered as a standalone element then its local and global coordinate axes coincide. If the finite element is a part of a planar structure, then some operations need to be taken in order to enable the transformation from local to global coordinate system. A planar structure may have two translational displacements and therefore, the stiffness matrix, before being transformed to the global coordinate system, must be extended to 4 x 4 terms (members):

$$\mathbf{k}_\mathrm{w}^\mathrm{e} = \frac{EA}{l}\begin{bmatrix} 1 & 0 & -1 & 0 \\ 0 & 0 & 0 & 0 \\ -1 & 0 & 1 & 0 \\ 0 & 0 & 0 & 0 \end{bmatrix}, \tag{11.89}$$

or, if a spatial problem is under consideration, this extension is:

$$\mathbf{k}_\mathrm{w}^\mathrm{e} = \frac{EA}{l}\begin{bmatrix} 1 & 0 & 0 & -1 & 0 & 0 \\ 0 & 0 & 0 & 0 & 0 & 0 \\ 0 & 0 & 0 & 0 & 0 & 0 \\ -1 & 0 & 0 & 1 & 0 & 0 \\ 0 & 0 & 0 & 0 & 0 & 0 \\ 0 & 0 & 0 & 0 & 0 & 0 \end{bmatrix}. \tag{11.90}$$

Beam Finite Element

Bending in the *z–y* Principal Centroidal Plane of Inertia Consider a planar, two-node-point beam finite element (a *basic beam finite element*) subjected to nodal transversal displacements and nodal rotations in the *z–y* principal centroidal plane of inertia, as shown in Figure 11.15(a) (and also considered earlier in Example 11.3).

The vector of nodal displacements is defined by Equation (11.66):

$$\mathbf{u}_{\mathrm{v}}^{\mathrm{e}^{\mathrm{T}}} = \left[\mathbf{u}_1^{\mathrm{e}^{\mathrm{T}}} \;\; \mathbf{u}_2^{\mathrm{e}^{\mathrm{T}}}\right], \;\; \mathbf{u}_i^{\mathrm{e}^{\mathrm{T}}} = [v_i \;\; \varphi_{xi}], \;\; i = 1,2 \tag{a}$$

while the vector of nodal forces is:

$$\mathbf{f}_{\mathrm{u}}^{\mathrm{e}^{\mathrm{T}}} = \left[\mathbf{f}_1^{\mathrm{e}^{\mathrm{T}}} \;\; \mathbf{f}_2^{\mathrm{e}^{\mathrm{T}}}\right], \;\; \mathbf{f}_i^{\mathrm{e}^{\mathrm{T}}} = \left[F_{yi} \;\; M_{xi}\right], \;\; i = 1,2. \tag{11.91}$$

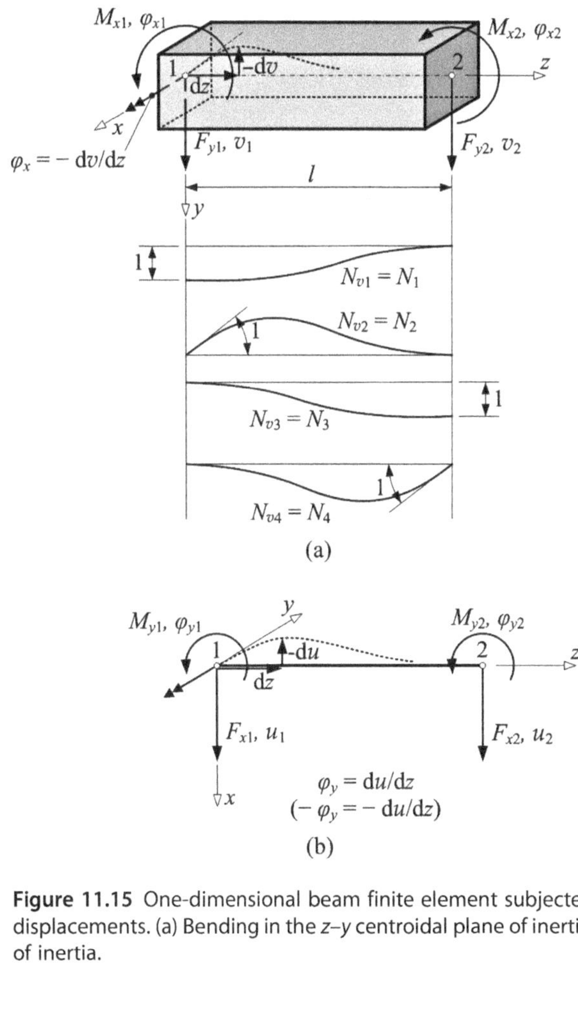

Figure 11.15 One-dimensional beam finite element subjected to nodal transversal and rotational displacements. (a) Bending in the *z–y* centroidal plane of inertia; (b) bending in the *z–x* centroidal plane of inertia.

Since this finite element has four degrees of freedom, the following one-dimensional polynomial of the third order may be used as an assumption of the distribution of displacement in the field of the finite element (from Equation (11.67a)):

$$v = \alpha_1 + \alpha_2 z + \alpha_3 z^2 + \alpha_4 z^3 = \mathbf{a}\boldsymbol{\alpha}. \tag{11.92}$$

The rotational displacement (see Figure 11.15(a)) is (from Equation (11.67b)):

$$\varphi_x = -\mathrm{d}v/\mathrm{d}z. \tag{b}$$

The polynomial matrix and the vector of constants are:

$$\mathbf{a} = \begin{bmatrix} 1 & z & z^2 & z^3 \end{bmatrix}, \quad \boldsymbol{\alpha}^{\mathrm{T}} = \begin{bmatrix} \alpha_1 & \alpha_2 & \alpha_3 & \alpha_4 \end{bmatrix}. \tag{11.93}$$

On the basis of Equation (11.92), its first derivative is:

$$v = \alpha_1 + \alpha_2 z + \alpha_3 z^2 + \alpha_4 z^3$$

$$\frac{\mathrm{d}v}{\mathrm{d}z} = \alpha_2 + 2\alpha_3 z + 3\alpha_4 z^2$$

Also, on the basis of the boundary conditions, we have:

$$\begin{aligned} &z = 0 \rightarrow \text{node 1} \rightarrow v = v_1 = \alpha_1; \varphi_{x1} = -\frac{\mathrm{d}v}{\mathrm{d}z} = -\alpha_2 \\ &z = l \rightarrow \text{node 2} \rightarrow v = v_2 = \alpha_1 + \alpha_2 l + \alpha_3 l^2 + \alpha_4 l^3; \\ &\varphi_{x2} = -\frac{\mathrm{d}v}{\mathrm{d}z} = -\alpha_2 - 2\alpha_3 l - 3\alpha_4 l^2. \end{aligned} \tag{11.94}$$

The vector of nodal displacements is now:

$$\mathbf{u}_{\mathrm{v}}^{\mathrm{e}} = \begin{bmatrix} \mathbf{u}_1^{\mathrm{e}} \\ \mathbf{u}_2^{\mathrm{e}} \end{bmatrix} = \begin{bmatrix} v_1 \\ \varphi_{x1} \\ v_2 \\ \varphi_{x2} \end{bmatrix} = \begin{bmatrix} 1 & 0 & 0 & 0 \\ 0 & -1 & 0 & 0 \\ 1 & l & l^2 & l^3 \\ 0 & -1 & -2l & -3l^2 \end{bmatrix} \begin{bmatrix} \alpha_1 \\ \alpha_2 \\ \alpha_3 \\ \alpha_4 \end{bmatrix}. \tag{11.95}$$

In accordance with Equation (11.6), it follows that:

$$v = \mathbf{a}\mathbf{a}^{\mathrm{b}^{-1}} \mathbf{u}_{\mathrm{v}}^{\mathrm{e}} = \mathbf{N}_{\mathrm{v}} \mathbf{u}_{\mathrm{v}}^{\mathrm{e}}. \tag{c}$$

The shape function matrix is:

$$\begin{aligned} \mathbf{N}_{\mathrm{v}} = \mathbf{a}\mathbf{a}^{\mathrm{b}^{-1}} &= \begin{bmatrix} 1 & z & z^2 & z^3 \end{bmatrix} \begin{bmatrix} 1 & 0 & 0 & 0 \\ 0 & -1 & 0 & 0 \\ 1 & l & l^2 & l^3 \\ 0 & -1 & -2l & -3l^2 \end{bmatrix}^{-1} \\ &= \begin{bmatrix} 1 & z & z^2 & z^3 \end{bmatrix} \begin{bmatrix} 1 & 0 & 0 & 0 \\ 0 & -1 & 0 & 0 \\ -3/l^2 & 2/l & 3/l^2 & 1/l \\ 2/l^3 & -1/l^2 & -2/l^3 & -1/l^2 \end{bmatrix} \\ &= \begin{bmatrix} 1 - 3\frac{z^2}{l^2} + 2\frac{z^3}{l^3} & -z + 2\frac{z^2}{l} - \frac{z^3}{l^2} & 3\frac{z^2}{l^2} - 2\frac{z^3}{l^3} & \frac{z^2}{l} - \frac{z^3}{l^2} \end{bmatrix}. \end{aligned} \tag{11.96}$$

Now, the shape functions are:

$$\mathbf{N}_{\mathrm{v}} = \left[N_{v1} \;\; N_{v2} \;\; N_{v3} \;\; N_{v4}\right]$$
$$= \left[1 - 3\frac{z^2}{l^2} + 2\frac{z^3}{l^3} \quad -z + 2\frac{z^2}{l} - \frac{z^3}{l^2} \quad 3\frac{z^2}{l^2} - 2\frac{z^3}{l^3} \quad \frac{z^2}{l} - \frac{z^3}{l^2}\right]. \tag{11.97}$$

These shape functions (N_{v1}, N_{v2}, N_{v3} and N_{v4}) were previously determined by the direct construction method using interpolation (Hermite polynomials) in Equation (11.70), and are graphically displayed in Figure 11.11 and Figure 11.15(a). Further, the strain–displacement matrix, $\mathbf{B}$, (Equation (11.8)) that is defined as:

$$\mathbf{B} = \mathbf{d}_{\varepsilon}\mathbf{N}, \tag{d}$$

and where the differential operator (in the case of beam bending in the z–y plane) is now:

$$\mathbf{d}_{\varepsilon} = \left[-\frac{\mathrm{d}^2}{\mathrm{d}z^2}\right], \tag{11.98}$$

can be written in the form:

$$\mathbf{B}_{\mathrm{v}} = \left[-\frac{\mathrm{d}^2 N_{v1}}{\mathrm{d}z} \quad -\frac{\mathrm{d}^2 N_{v2}}{\mathrm{d}z} \quad -\frac{\mathrm{d}^2 N_{v3}}{\mathrm{d}z} \quad -\frac{\mathrm{d}^2 N_{v4}}{\mathrm{d}z}\right]$$
$$= \left[\frac{6}{l^2} - \frac{12z}{l^3} \quad -\frac{4}{l} + \frac{6z}{l^2} \quad -\frac{6}{l^2} + \frac{12z}{l^3} \quad -\frac{2}{l} + \frac{6z}{l^2}\right]. \tag{11.99}$$

The matrix of elastic constants (the elasticity matrix), $\mathbf{C}$, is:

$$\mathbf{C}_{\mathrm{v}} = \left[EI_x\right]. \tag{11.100}$$

This quantity is obtained in a similar manner to that used for a case of axial loading in Equation (11.87). If the constitutive Equation (11.9) for the case of bending in the z–y principal centroidal plane of inertia is written in the form:

$$\boldsymbol{\sigma} = \mathbf{C} \cdot \boldsymbol{\varepsilon} \rightarrow \left[M_x\right] = \left[EI_x\right]\left[\kappa_x\right], \tag{e}$$

the value of matrix $\mathbf{C}_{\mathrm{v}}$ defined by Equation (11.100) can be seen. In Equation (11.100), I_x is the axial moment of inertia with respect to the x axis, E is the modulus of elaticity and κ_x in Equation (e) is the curvature. In Equation (e), all subscripts are writen as "x" and sometimes this can cause confusion. On the basis of Equation (6.115), Equation (6.116) was derived and this represents the so-called moment–curvature relationship. It is clear that the curvature is given by:

$$\kappa = -\frac{\mathrm{d}^2 v}{\mathrm{d}z^2} = -\frac{\mathrm{d}^2}{\mathrm{d}z^2}(v). \tag{f}$$

In accordance with the definition of curvature (curvature in the z direction), the curvature given by Equation (f) relates to the curvature of the elastic curve of the beam along the z axis, so it can be written as $\kappa = \kappa_z$. In addition, the bending (flexure) moment along the z axis may be designated $M_{\mathrm{f}} = M_z$, but it is designated M_x, since this bending moment causes flexure (rotation) of the beam around the x axis. To avoid confusion about subscripts, it is written as in Equation (e). When comparing Equation (f) with Equation (11.7):

$$\boldsymbol{\varepsilon} = \mathbf{d}_{\varepsilon}\mathbf{u}, \tag{g}$$

the differential operator (Equation (11.98)) can be obtained.

If the values defined by Equation (11.99) and Equation (11.100) are entered into Equation (11.19), it follows that the stiffness matrix of the one-dimensional beam finite element subjected to nodal transversal displacements and nodal rotational displacements (bending) in the z–y principal centroidal plane of inertia is:

$$\mathbf{k}_{\mathrm{v}}^{\mathrm{e}} = \int_l \mathbf{B}_{\mathrm{v}}^{\mathrm{T}} \mathbf{C}_{\mathrm{v}} \mathbf{B}_{\mathrm{v}} \mathrm{d}z, \tag{h}$$

$$\mathbf{k}_{\mathrm{v}}^{\mathrm{e}} = EI_x \begin{bmatrix} \frac{12}{l^3} & -\frac{6}{l^2} & -\frac{12}{l^3} & -\frac{6}{l^2} \\ -\frac{6}{l^2} & \frac{4}{l} & \frac{6}{l^2} & \frac{2}{l} \\ -\frac{12}{l^3} & \frac{6}{l^2} & \frac{12}{l^3} & \frac{6}{l^2} \\ -\frac{6}{l^2} & \frac{2}{l} & \frac{6}{l^2} & \frac{4}{l} \end{bmatrix} = \begin{bmatrix} b & -c & -b & -c \\ -c & d & c & e \\ -b & c & b & c \\ -c & e & c & d \end{bmatrix}, \tag{11.101}$$

where:

$$b = \frac{12EI_x}{l^3}, \quad c = \frac{6EI_x}{l^2}, \quad d = \frac{4EI_x}{l}, \quad e = \frac{2EI_x}{l}. \tag{11.102}$$

Bending in the z–x Principal Centroidal Plane of Inertia Consideration of the bending process in the z–x principal centroidal plane of inertia (Figure 11.15(b)) can be carried out in a similar way to planar bending in the z–y plane. As such, the nodal vectors are:

$$\mathbf{u}_{\mathrm{u}}^{\mathrm{e}^{\mathrm{T}}} = \left[\mathbf{u}_1^{\mathrm{e}^{\mathrm{T}}} \; \mathbf{u}_2^{\mathrm{e}^{\mathrm{T}}}\right] = \left[u_1 \; \varphi_{y1} \; u_2 \; \varphi_{y2}\right] \tag{11.103}$$

$$\mathbf{f}_{\mathrm{u}}^{\mathrm{e}^{\mathrm{T}}} = \left[\mathbf{f}_1^{\mathrm{e}^{\mathrm{T}}} \; \mathbf{f}_2^{\mathrm{e}^{\mathrm{T}}}\right] = \left[F_{x1} \; M_{y1} \; F_{x2} \; M_{y2}\right]. \tag{11.104}$$

The assumption regarding the distribution of displacement in the field of the finite element can be similar to that in Equation (11.92). Also, now:

$$\varphi_y = \frac{\mathrm{d}u}{\mathrm{d}z}. \tag{11.105}$$

After conducting a similar procedure to that used previously in the case of bending in the z–y principal centroidal plane of inertia, the matrix of shape functions and the strain–displacement matrix can be obtained.

The matrix of elastic constants (the elasticity matrix) can be written as:

$$\mathbf{C}_{\mathrm{u}} = \left[EI_y\right], \tag{11.106}$$

and it is obtained similarly to $\mathbf{C}_\mathrm{v}$. Finally, the finite element stiffness matrix is:

$$\mathbf{k}_\mathrm{u}^\mathrm{e} = \int_l \mathbf{B}_\mathrm{u}^\mathrm{T}\mathbf{C}_\mathrm{u}\mathbf{B}_\mathrm{u}\mathrm{d}z = EI_y \begin{bmatrix} \frac{12}{l^3} & \frac{6}{l^2} & -\frac{12}{l^3} & \frac{6}{l^2} \\ \frac{6}{l^2} & \frac{4}{l} & \frac{6}{l^2} & \frac{2}{l} \\ -\frac{12}{l^3} & \frac{6}{l^2} & \frac{12}{l^3} & -\frac{6}{l^2} \\ \frac{6}{l^2} & \frac{2}{l} & -\frac{6}{l^2} & \frac{4}{l} \end{bmatrix}. \tag{11.107}$$

Torsion (Twisting) The nodal vectors and shape functions for a finite element subjected to torsion (Saint Venant torsion) are shown in Figure 11.16.

The nodal vectors are:

$$\mathbf{u}_\varphi^{\mathrm{e}^\mathrm{T}} = \left[\mathbf{u}_1^{\mathrm{e}^\mathrm{T}} \ \ \mathbf{u}_2^{\mathrm{e}^\mathrm{T}}\right] = [\varphi_{z1} \ \ \varphi_{z2}] \tag{11.108}$$

$$\mathbf{f}_\varphi^{\mathrm{e}^\mathrm{T}} = \left[\mathbf{f}_1^{\mathrm{e}^\mathrm{T}} \ \ \mathbf{f}_2^{\mathrm{e}^\mathrm{T}}\right] = [M_{z1} \ \ M_{z2}] \tag{11.109}$$

Displacement (rotation) in the field of the finite element can be written in the form shown in Equation (11.52) since two degrees of freedom are considered.

Using the same procedure for determining the shape functions by the method of generalized coordinates as in the previous proceedings, the same values for the shape functions are obtained as those in Equation (11.58) [19]:

$$\mathbf{N}_\varphi = [N_{\varphi1} \ \ N_{\varphi2}] = \left[1-\frac{z}{l} \ \ \frac{z}{l}\right]. \tag{11.110}$$

The strain–displacement matrix takes an identical form to that for the basic truss finite element in Equation (11.86).

The matrix of elastic constants, $\mathbf{C}$, is:

$$\mathbf{C}_\varphi = [GI_\mathrm{t}]. \tag{11.111}$$

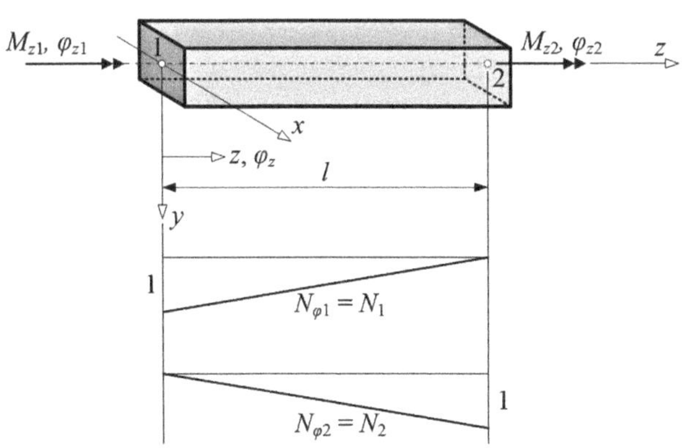

Figure 11.16 A finite element subjected to torsion.

The matrix shown in Equation (11.111) is obtained using a similar procedure as that used for bending.

The finite element stiffness matrix is:

$$\mathbf{k}_{\varphi}^{\mathrm{e}} = \int_{l} \mathbf{B}_{\varphi}^{\mathrm{T}} \mathbf{C}_{\varphi} \mathbf{B}_{\varphi} \mathrm{d}z = \frac{GI_{\mathrm{t}}}{l} \begin{bmatrix} 1 & -1 \\ -1 & 1 \end{bmatrix}. \tag{11.112}$$

Spatial Beam Finite Element – General Load Case The general case of the load for a spatial finite element includes: the axial load ($N = F_z$); bending moments in both principal centroidal planes of inertia ($Q_x = F_x$, M_y; $Q_y = F_y$, M_x) and torque (torsional moment) ($M_z = M_{\mathrm{t}}$). In accordance with this, the element stiffness matrix is:

$$\mathbf{k}^{\mathrm{e}} = \mathbf{k}_{\mathrm{w}}^{\mathrm{e}} + \mathbf{k}_{\mathrm{u}}^{\mathrm{e}} + \mathbf{k}_{\mathrm{v}}^{\mathrm{e}} + \mathbf{k}_{\varphi}^{\mathrm{e}} \tag{11.113}$$

where:

$$\mathbf{k}_{\mathrm{w}}^{\mathrm{e}} = \mathbf{k}_{N}^{\mathrm{e}};\ \mathbf{k}_{\mathrm{u}}^{\mathrm{e}} = \mathbf{k}_{M_y,\,Q_x}^{\mathrm{e}};\ \mathbf{k}_{\mathrm{v}}^{\mathrm{e}} = \mathbf{k}_{M_x,\,Q_y}^{\mathrm{e}};\ \mathbf{k}_{\varphi}^{\mathrm{e}} = \mathbf{k}_{M_{\mathrm{t}}}^{\mathrm{e}};\ M_{\mathrm{t}} \equiv M_z,$$

while N, Q_x, Q_y, M_x, M_y, M_{t} are forces: three forces in the z, x and y directions and three moments. Subscripts (w, u, v, φ) indicate displacement that is the result of a load effect, for example, in the case $\mathbf{k}_{\mathrm{w}}^{\mathrm{e}} = \mathbf{k}_{N}^{\mathrm{e}}$, subscript w means axial displacement that is the result of the action of axial force (N).

The stiffness matrix for the one-dimensinal finite element defined by Equation (11.118) is:

$$\mathbf{k}^{\mathrm{e}} = \left[\begin{array}{cccccc:cccccc}
a & \cdot & \cdot & \cdot & \cdot & \cdot & -a & \cdot & \cdot & \cdot & \cdot & \cdot \\
\cdot & b_1 & \cdot & \cdot & \cdot & c_1 & \cdot & -b_1 & \cdot & \cdot & \cdot & c_1 \\
\cdot & \cdot & b_2 & \cdot & -c_2 & \cdot & \cdot & \cdot & -b_2 & \cdot & -c_2 & \cdot \\
\cdot & \cdot & \cdot & t & \cdot & \cdot & \cdot & \cdot & \cdot & -t & \cdot & \cdot \\
\cdot & \cdot & -c_2 & \cdot & d_2 & \cdot & \cdot & \cdot & c_2 & \cdot & e_2 & \cdot \\
\cdot & c_1 & \cdot & \cdot & \cdot & d_1 & \cdot & -c_1 & \cdot & \cdot & \cdot & e_1 \\
\hdashline
-a & \cdot & \cdot & \cdot & \cdot & \cdot & a & \cdot & \cdot & \cdot & \cdot & \cdot \\
\cdot & -b_1 & \cdot & \cdot & \cdot & -c_1 & \cdot & b_1 & \cdot & \cdot & \cdot & -c_1 \\
\cdot & \cdot & -b_2 & \cdot & c_2 & \cdot & \cdot & \cdot & b_2 & \cdot & c_2 & \cdot \\
\cdot & \cdot & \cdot & -t & \cdot & \cdot & \cdot & \cdot & \cdot & t & \cdot & \cdot \\
\cdot & \cdot & -c_2 & \cdot & e_2 & \cdot & \cdot & \cdot & c_2 & \cdot & d_2 & \cdot \\
\cdot & c_1 & \cdot & \cdot & \cdot & e_1 & \cdot & -c_1 & \cdot & \cdot & \cdot & d_1
\end{array}\right] \tag{11.114}$$

The members of this matrix are:

$$\begin{aligned} a &= \frac{EA}{l},\ b_1 = \frac{12EI_y}{l^3},\ b_2 = \frac{12EI_x}{l^3},\ c_1 = \frac{6EI_y}{l^2},\ c_2 = \frac{6EI_x}{l^2} \\ d_1 &= \frac{4EI_y}{l},\ d_2 = \frac{4EI_x}{l},\ e_1 = \frac{2EI_y}{l},\ e_2 = \frac{2EI_x}{l},\ \mathrm{t} = \frac{GI_{\mathrm{t}}}{l}. \end{aligned} \tag{11.115}$$

Example 11.6.

Consider the cantilever beam element subjected to a bending moment at its free end shown in Figure 11.17(a). Determine the displacements of the free end of the considered beam.

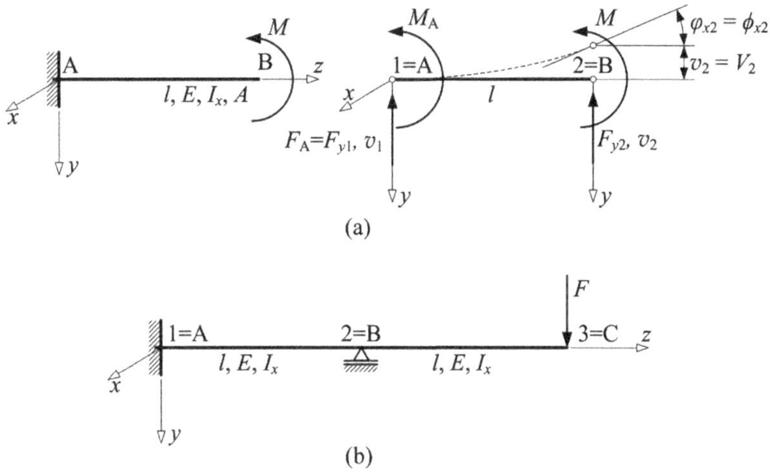

Figure 11.17 A beam structure subjected to load. (a) A cantilever beam element subjected to a bending moment at its free end; (b) a beam structural element subjected to force at its free end.

Data:
Known data are: M, l, A, E, I_x (Figure 11.17(a)).

Solution:
Since the beam is modeled using one finite element only, the structural equation, Equation (11.38), here is identical to the finite element equation in the global coordinate system, Equation (11.32):

$$\mathbf{KU} = \mathbf{F}_\mathrm{R} \qquad \leftrightarrow \qquad \bar{\mathbf{k}}^\mathrm{e}\bar{\mathbf{u}}^\mathrm{e} = \bar{\mathbf{f}}^\mathrm{e} \tag{a}$$

In accordance with Equations (11.42) and (11.101), since only one finite element is considered and bending is performed in the $z-y$ principal centroidal plane, and given that the global and local coordinate systems coincide, it follows that:

$$\mathbf{K} = \begin{bmatrix} \mathbf{K}_{11} & \mathbf{K}_{12} \\ \mathbf{K}_{21} & \mathbf{K}_{22} \end{bmatrix}, \quad \mathbf{K}_{ij} = \sum_e \bar{\mathbf{k}}_{ij}^\mathrm{e} \rightarrow \mathbf{K}_{11} = \bar{\mathbf{k}}_{ii}^1 = \mathbf{k}_{ii}^1 = \begin{bmatrix} b & -c \\ -c & d \end{bmatrix} \tag{b}$$

If other submatrices are calculated in a similar way, the global structural equation is:

$$\begin{bmatrix} F_{y1} = F_A \\ M_{x1} = M_A \\ \boldsymbol{F}_{y2} = \boldsymbol{F}_\mathbf{B} = 0 \\ \boldsymbol{M}_{x2} = \boldsymbol{M}_\mathbf{B} = M \end{bmatrix} = \begin{bmatrix} b & -c & -b & -c \\ -c & d & c & e \\ -b & c & \boldsymbol{b} & \mathbf{c} \\ -c & e & \mathbf{c} & \boldsymbol{d} \end{bmatrix} \begin{bmatrix} V_1 = V_A = 0 \\ \Phi_{x1} = \Phi_A = 0 \\ \boldsymbol{V}_2 = \boldsymbol{V}_\mathbf{B} \\ \boldsymbol{\Phi}_{x2} = \boldsymbol{\Phi}_\mathbf{B} \end{bmatrix}. \tag{c}$$

Note: The considered beam (that now represents the structure) is modeled as one finite element, and the nodal displacements are written using capital letters.

On the basis of Equation (c), taking the members of the matrices which are written in bold letters (that is, taking into account the boundary conditions), it follows that:

$$\begin{aligned} bV_2 + c\Phi_{x2} &= 0 \\ cV_2 + d\Phi_{x2} &= M. \end{aligned} \tag{d}$$

Since the quantities b, c and d from Equation (c) are defined by Equation (11.102), it follows that:

$$\begin{aligned} V_2 &= -\frac{Ml^2}{2EI_x} \\ \Phi_{x2} &= \frac{Ml}{EI_x} \end{aligned} \tag{e}$$

Also, from Equation (c), it follows that:

$$M_A = cV_2 + e\Phi_{x2} \tag{f}$$

$$M_A = -M \tag{g}$$

$$F_A = -bV_2 - c\Phi_{x2} = 0 \tag{h}$$

Example 11.7.
Consider the beam structure in Figure 11.17(b), subjected to a force acting at its free end. Show the procedure for determining the reactive forces and nodal displacements.

Data:
Known data are: F, l, A, E, I_x.

Solution:
In accordance with Figure 11.17(b) and Equations (11.38), (11.42$_2$) and (11.101), since the beam structure is modeled using two beam finite elements, the following structural equation can be written:

$$\begin{bmatrix} \mathbf{K}_{11} & \mathbf{K}_{12} & 0 \\ \mathbf{K}_{21} & \mathbf{K}_{22} & \mathbf{K}_{23} \\ 0 & \mathbf{K}_{32} & \mathbf{K}_{33} \end{bmatrix} \begin{bmatrix} \mathbf{U}_1 \\ \mathbf{U}_2 \\ \mathbf{U}_3 \end{bmatrix} = \begin{bmatrix} \mathbf{F}_1 \\ \mathbf{F}_2 \\ \mathbf{F}_3 \end{bmatrix} \tag{a}$$

Submatrix $[\mathbf{K}_{ij}]$ of the structural matrix $[\mathbf{K}]$ shown in Equation (a) can be determined on the basis of Equation (11.101). The global and local coordinate systems coincide. Equation (a) now takes the form:

$$\begin{bmatrix} F_{y1} = M_A \\ M_{x1} = M_A \\ F_{Y2} = F_B \\ \boldsymbol{M_{x2} = M_B = 0} \\ \boldsymbol{F_{y3} = F} \\ \boldsymbol{M_{x3} = 0} \end{bmatrix} = \begin{bmatrix} b & -c & -b & -c & 0 & 0 \\ -c & d & c & e & 0 & 0 \\ -b & c & 2b & 0 & -b & -c \\ -c & e & 0 & \mathbf{2d} & \mathbf{c} & \mathbf{e} \\ 0 & 0 & -b & \mathbf{c} & \mathbf{b} & \mathbf{c} \\ 0 & 0 & -c & \mathbf{e} & \mathbf{c} & \mathbf{d} \end{bmatrix} \begin{bmatrix} V_1 = V_A = 0 \\ \Phi_{x1} = \Phi_A = 0 \\ V_2 = V_B = 0 \\ \boldsymbol{\Phi_{x2} = \Phi_B} \\ \boldsymbol{V_3} \\ \boldsymbol{\Phi_{x3}} \end{bmatrix} \tag{b}$$

On the basis of Equation (b), it follows that:

$$\begin{aligned} 2d\Phi_{x2} + cV_3 + e\Phi_{x3} &= 0 \\ c\Phi_{x2} + bV_3 + c\Phi_{x3} &= F, \\ e\Phi_{x2} + cV_3 + d\Phi_{x3} &= 0 \end{aligned} \tag{c}$$

and, further, all unknowns (reactive forces and displacements) can be determined.

11.2.5.2 Finite Elements of Higher Order

Truss and Beam Finite Elements of Higher Order

This next section gives only some ideas with regard to the existence of two, one-dimensional finite elements of higher order without determining the values that define the stiffness matrices of the considered finite elements.. As such, one truss finite element of higher order and one beam element of higher order are shown in Figure 11.18.

Truss (Rod) Finite Element of Higher Order The truss (axial) finite element shown in Figure 11.18(a) is known as a *quadratic three-node finite element* [4, 11, 15]. The vector of nodal displacements is:

$$\mathbf{u}_{\mathrm{w}}^{\mathrm{e}^{\mathrm{T}}} = \left[\mathbf{u}_1^{\mathrm{e}^{\mathrm{T}}}\ \mathbf{u}_2^{\mathrm{e}^{\mathrm{T}}}\ \mathbf{u}_3^{\mathrm{e}^{\mathrm{T}}}\right] = [w_1\ w_2\ w_3] \tag{11.116}$$

The following polynomial may be used to describe the displacement in the finite element field:

$$w = \alpha_1 + \alpha_2 z + \alpha_3 z^2. \tag{11.117}$$

Using the standard procedure for shape function determination, the following values can be obtained:

$$\begin{aligned} \mathbf{N}_{\mathrm{w}} &= [N_{w1}\ N_{w2}\ N_{w3}] \\ &= \left[\frac{2}{l^2}z^2 - \frac{3}{l}z + 1 \quad -\frac{4}{l^2}z^2 + \frac{4z}{l} \quad \frac{2}{l^2}z^2 - \frac{z}{l}\right]. \end{aligned} \tag{11.118}$$

The shape functions are shown in Figure 11.18(a). The strain–displacement matrix (Equation (11.8)) is then calculated:

$$\mathbf{B}_{\mathrm{w}} = \frac{\mathrm{d}}{\mathrm{d}z}\mathbf{N}_{\mathrm{w}}, \tag{a}$$

$$\mathbf{B}_{\mathrm{w}} = \left[\frac{4}{l^2}z - \frac{3}{l} \quad -\frac{8}{l^2}z + \frac{4}{l} \quad \frac{4}{l^2}z - \frac{1}{l}\right], \tag{11.119}$$

and, since the elasticity matrix is (from Equation (11.87)):

$$\mathbf{C}_{\mathrm{w}} = [EA], \tag{b}$$

the finite element stiffness matrix, using Equation (11.19):

$$\mathbf{k}_{\mathrm{w}}^{\mathrm{e}} = \int_l \mathbf{B}_{\mathrm{w}}^{\mathrm{T}}\mathbf{C}_{\mathrm{w}}\mathbf{B}_{\mathrm{w}}\mathrm{d}z, \tag{c}$$

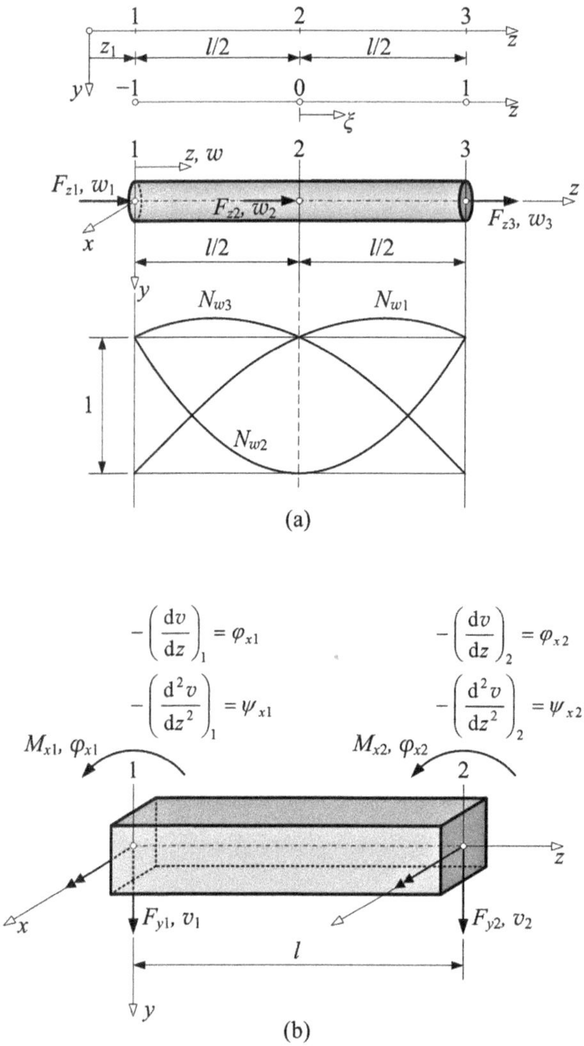

Figure 11.18 One-dimensional finite elements of higher order. (a) Truss (axial) finite element of higher order (quadratic, three-node finite element); (b) beam finite element of higher order.

can be determined in the form:

$$\mathbf{k}_{\mathrm{w}}^{\mathrm{e}} = \frac{EA}{3l}\begin{bmatrix} 7 & -8 & 1 \\ -8 & 16 & -8 \\ 1 & -8 & 7 \end{bmatrix} \tag{11.120}$$

If, instead of the Cartesian coordinate system, the node points are shown using natural (dimensionless) one-dimensional coordinate, ξ (Figure 11.18(a)), then, in accordance with Equation (11.61), the shape functions can be derived by Lagrangian interpolating polynomials in the form:

$$N_i(\xi) = l_i(\xi) = \frac{\prod_{\substack{j=1 \\ j \neq i}}^{n} (\xi - \xi_j)}{\prod_{\substack{j=1 \\ j \neq i}}^{n} (\xi_i - \xi_j)} \tag{d}$$

$$\begin{aligned} N_1 &= \frac{(\xi-\xi_2)(\xi-\xi_3)}{(\xi_1-\xi_2)(\xi_1-\xi_3)} = \frac{\xi(\xi-1)}{(-1-0)(-1-1)} = \frac{1}{2}\xi(\xi-1) \\ N_2 &= \frac{(\xi-\xi_1)(\xi-\xi_3)}{(\xi_2-\xi_1)(\xi_2-\xi_3)} = 1-\xi^2 \\ N_3 &= \frac{(\xi-\xi_1)(\xi-\xi_2)}{(\xi_3-\xi_1)(\xi_3-\xi_2)} = \frac{1}{2}\xi(\xi+1) \end{aligned} \tag{11.121}$$

The coordinate z can now be written as:

$$\begin{aligned} z &= \sum N_i z_i = N_1 z_1 + N_2 z_2 + N_3 z_3 \\ z &= \frac{1}{2}\xi(\xi-1)z_1 + \left(1-\xi^2\right)z_2 + \frac{1}{2}\xi(\xi+1)z_3, \end{aligned} \tag{11.122}$$

or expressed through z_1:

$$\begin{aligned} z &= \frac{1}{2}\xi(\xi-1)z_1 + \left(1-\xi^2\right)\left(z_1 + \frac{l}{2}\right) + \frac{1}{2}\xi(\xi+1)(z_1+l) \\ &= z_1 + \frac{l}{2}\xi + \frac{l}{2} \end{aligned} \tag{e}$$

The strain–displacement matrix is defined by Equation (11.8), repeated above in Equation (a), while the differential operator is:

$$d_\varepsilon = \frac{\partial}{\partial z} \tag{f}$$

Also:

$$\frac{\partial N_i}{\partial \xi} = \frac{\partial z}{\partial \xi}\frac{\partial N_i}{\partial z} \rightarrow \frac{\partial}{\partial \xi} = \mathbf{J}\frac{\partial}{\partial z} \tag{g}$$

where the Jacobian operator, $\mathbf{J}$, is:

$$\mathbf{J} = \frac{\partial z}{\partial \xi} \tag{h}$$

In accordance with Equations (e) and (h):

$$\mathbf{J} = \frac{l}{2} \rightarrow \mathbf{J}^{-1} = \frac{2}{l} \tag{i}$$

From Equations (g) and (i), it follows that:

$$\frac{\partial}{\partial z} = \mathbf{J}^{-1}\frac{\partial}{\partial \xi} = \frac{2}{l}\frac{\partial}{\partial \xi} \tag{j}$$

On the basis of Equation (11.8) or Equation (a):

$$\mathbf{B} = \mathbf{B}_{\mathrm{w}} = \frac{\partial}{\partial z}\mathbf{N} \tag{k}$$

$$\mathbf{B} = \frac{\partial}{\partial z}\mathbf{N} = \frac{2}{l}\frac{\partial}{\partial \xi}\left[\frac{1}{2}\xi(\xi-1) \quad 1-\xi^2 \quad \frac{1}{2}\xi(\xi+1)\right]$$

$$\mathbf{B} = \frac{2}{l}\left[\xi - \frac{1}{2} \quad -2\xi \quad \xi + \frac{1}{2}\right] \tag{l}$$

The finite element stiffness matrix from Equation (11.19) or Equation (a) is:

$$\mathbf{k}^{\mathrm{e}} = \int_0^l \mathbf{B}^{\mathrm{T}}\mathbf{C}\mathbf{B}\mathrm{d}z = \int_{-1}^{1} \mathbf{B}^{\mathrm{T}} EA\mathbf{B}\mathbf{J}\mathrm{d}\xi \tag{11.123}$$

$$\mathbf{k}^{\mathrm{e}} = \int_{-1}^{1} EA\frac{2}{l}\begin{bmatrix} \xi - \frac{1}{2} \\ -2\xi \\ \xi + \frac{1}{2} \end{bmatrix}\left[\xi - \frac{1}{2} \quad -2\xi \quad \xi + \frac{1}{2}\right]\frac{2}{l}\frac{l}{2}\mathrm{d}\xi$$

$$\mathbf{k}^{\mathrm{e}} = EA\frac{2}{l}\int_{-1}^{1}\begin{bmatrix} \left[\xi - \frac{1}{2}\right]^2 & -2\xi\left(\xi - \frac{1}{2}\right) & \left(\xi - \frac{1}{2}\right)\left(\xi + \frac{1}{2}\right) \\ -2\xi\left(\xi - \frac{1}{2}\right) & 4\xi^2 & -2\xi\left(\xi + \frac{1}{2}\right) \\ \left(\xi - \frac{1}{2}\right)\left(\xi + \frac{1}{2}\right) & -2\xi\left(\xi + \frac{1}{2}\right) & \left[\xi + \frac{1}{2}\right]^2 \end{bmatrix}\mathrm{d}\xi. \tag{11.124}$$

After mathematical calculations, the same value for the stiffness matrix is obtained as that presented in Equation (11.120):

$$\mathbf{k}^{\mathrm{e}} \equiv \mathbf{k}_{\mathrm{w}}^{\mathrm{e}} = \frac{EA}{3l}\begin{bmatrix} 7 & -8 & 1 \\ -8 & 16 & -8 \\ 1 & -8 & 7 \end{bmatrix} \tag{m}$$

Subscript "w" indicates that the element is subjected to axial displacements only.

Finally, instead of Cartesian coordinates or a natural, one-dimensional (dimensionless) coordinate, ξ, the natural, dimensionless coordinates, L_i, may also be used. Since the considered element has three nodes and three axial displacements (degrees of freedom) in total, for the distribution of displacement in the field of the element, a polynomial of third order is used (Equation (11.117)). From node 1 to the last node 3, the natural coordinates, L_i, have the values shown in brackets as follows: 1 (2 0), 2 (1 1), 3 (0 2). The sum in the bracket for each node is equal to the order of the polynomial. Using Equation (11.74), it follows that:

$$N_1 = N_{20} = 2L_1^2 - L_1; \quad N_2 = N_{11} = 4L_1L_2; \quad N_3 = N_{02} = 2L_2^2 - L_2 \tag{n}$$

For example, N_1 is obtained as follows:

$$N_1 = N_{20} = l_i(L_1)l_j(L_2) = l_2(L_1)l_0(L_2) = \frac{2L_1-1+1}{1} \cdot \frac{2L_1-2+1}{2} \cdot 1$$

$$N_1 = N_{20} = 2L_1^2 - L_1; \text{ (It is known that } l_0(L_2) = 1\text{)}$$

Beam Finite Element of Higher Order A beam finite element of higher order can be obtained, for example, if one degree of freedom per node is added to the basic beam finite element in Figure 11.15(a).

The vector of nodal displacements (Figure 11.18(b)) is:

$$\mathbf{u}_v^{e^T} = \left[\mathbf{u}_1^{e^T} \;\; \mathbf{u}_2^{e^T}\right] = \left[v_1 \;\; \varphi_{x1} \;\; \psi_{x1} \;\; v_2 \;\; \varphi_{x2} \;\; \psi_{x2}\right] \tag{11.125}$$

and the polynomial that can be used for an assessmnent of the displacement distribution in the field of the finite element is:

$$v = \alpha_1 + \alpha_2 z + \alpha_3 z^2 + \alpha_4 z^3 + \alpha_5 z^4 + \alpha_6 z^5 = \mathbf{a}\boldsymbol{\alpha}. \tag{11.126}$$

Using the already-known procedure, the stiffness matrix of the form presented in Equation $(11.19)_2$

$$\mathbf{k}_v^e = \int_l \mathbf{B}_v^T \mathbf{C}_v \mathbf{B}_v \mathrm{d}z \tag{o}$$

can be determined.

11.2.6 Two-dimensional Finite Elements

The plane stress state and the plane strain state are considered two-dimensional problems of the theory of elasticity (Figure 11.19). In the analysis of these problems, two-dimensional finite elements are usually used. Typically it is said that a structural element is in a state of plane stress (biaxial state of stress/biaxial stress state) when all of the stress components are in planes that are parallel to each other. In this case, each point of the element is in the plane stress state. A similar scenario occurs when the plane strain state is considered; that is, all of the strain components are in mutually parallel planes. This is illustrated in Figure 11.19.

Since the state of stress is defined with respect to a considered point in the solid body, it does not hurt to mention that, in each of the considered points in the solid body, a state of plane stress may exist, but the solid body as a whole does not have to be in a state of plane stress. For example, the stress components of the plate shown in Figure 11.19(a) lie in mutually parallel planes while this is not the case with a circular shaft subjected to torque, although at each point of the shaft, the plane stress state is present.

Similar to one-dimensional elements, here, also, basic elements and elements of higher order can be considered. As such, a finite element that has additional degrees of freedom in its vertices (nodes) or has additional nodes along its sides or in its field will be treated as a finite element of higher order. If we consider the plane stress state and the plane strain state in an isotropic material (Figure 11.19), neglecting initial stress and initial

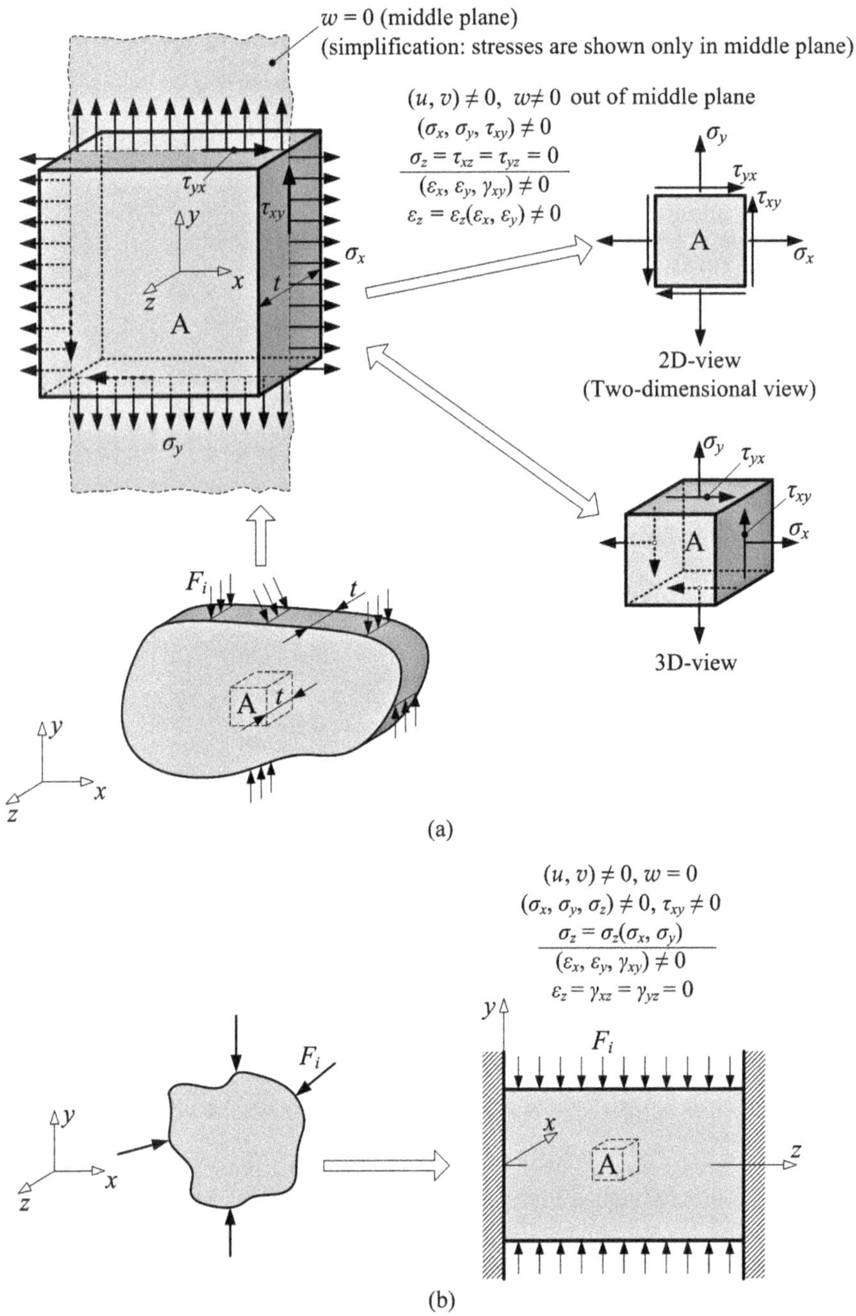

Figure 11.19 (a) Plane stress state; (b) plane strain state.

strain, the stress, strain, displacement and elasticity matrix can be written as outlined below [1, 6, 15].

- **Plane stress state**: Thin-walled planar structural element (plate) stressed in the plane of the structure (Figure 11.19(a))
  - Vectors of stresses, strains and displacements (general case):

$$\begin{aligned}
&\boldsymbol{\sigma}^{\mathrm{T}} = \left[\sigma_x \;\; \sigma_y \;\; \tau_{xy}\right]; \;\; \sigma_z = \tau_{xz} = \tau_{yz} = 0 \\
&\boldsymbol{\varepsilon}^{\mathrm{T}} = \left[\varepsilon_x \;\; \varepsilon_y \;\; \gamma_{xy}\right], \;\; \varepsilon_z = \frac{-\nu}{1-\nu}\left(\varepsilon_x + \varepsilon_y\right) \\
&\mathbf{u}^{\mathrm{T}} = [u \;\; v], w \neq 0 \rightarrow \text{out of middle plane}
\end{aligned} \tag{11.127}$$

In the case where stresses $\sigma_z = \tau_{xz} = \tau_{yz} = 0$ everywhere, the state of stress in each point of the structural element is said to be *plane stress.*

  - Hooke's Law in a componental form with respect to strain
    The method of superposition will be applied:

$$\begin{aligned}
&\varepsilon_{x^*} = \varepsilon_x^{(\sigma_x)} + \varepsilon_x^{(\sigma_y)} = \frac{1}{E}\left(\sigma_x - \nu\sigma_y\right) \\
&\varepsilon_y = \varepsilon_y^{(\sigma_y)} + \varepsilon_y^{(\sigma_x)} = \frac{1}{E}\left(\sigma_y - \nu\sigma_x\right)
\end{aligned} \tag{11.128}$$

$$\begin{aligned}
&\varepsilon_z = \varepsilon_z^{(\sigma_x)} + \varepsilon_z^{(\sigma_y)} = -\frac{\nu}{E}\left(\sigma_x + \nu\sigma_y\right) \\
&\gamma_{xy} = \gamma_{yx} = \frac{\tau_{xy}}{G} = \frac{2(1+\nu)}{E}\tau_{xy} \\
&\varepsilon^{(\tau)^{**}} = \gamma^{(\sigma)} = 0
\end{aligned} \tag{11.129}$$

* The meaning of Equation (11.128$_1$) is: the normal strain in the x direction is the sum of the normal strains in this direction due to normal stresses, σ_x and σ_y.

** It is known that normal stress does not cause shear strain and shear stress does not cause normal strain; that is, shear strain components due to normal stresses as well as normal strain components due to shear stress are equal to zero (Equation (7.22)).

  - Elasticity matrix:

$$\mathbf{C} = [C] = \frac{E}{1-\nu^2}\begin{bmatrix} 1 & \nu & 0 \\ \nu & 1 & 0 \\ 0 & 0 & \dfrac{1-\nu}{2} \end{bmatrix}, \tag{11.130}$$

is obtained solving the above equations for the stresses.

- **Plane strain state** (Figure11.19(b))
  - Vectors of stresses, strains and displacements

$$\begin{aligned}
&\boldsymbol{\sigma}^{\mathrm{T}} = \left[\sigma_x \;\; \sigma_y \;\; \tau_{xy}\right]; \;\; \sigma_z = \nu\left(\sigma_x + \sigma_y\right) \\
&\boldsymbol{\varepsilon}^{\mathrm{T}} = \left[\varepsilon_x \;\; \varepsilon_y \;\; \gamma_{xy}\right]; \;\; \varepsilon_z = \gamma_{xz} = \gamma_{yz} = 0 \\
&\mathbf{u}^{\mathrm{T}} = [u \;\; v]; \;\; w = 0
\end{aligned} \tag{11.131}$$

– Hooke's Law in a componental form with respect to strain

$$\varepsilon_x = \varepsilon_x^{(\sigma_x)} + \varepsilon_x^{(\sigma_y)} + \varepsilon_x^{(\sigma_z)} = \frac{1}{E}\left[\sigma_x - \nu\left(\sigma_y + \sigma_z\right)\right]$$

$$\varepsilon_y = \varepsilon_y^{(\sigma_y)} + \varepsilon_y^{(\sigma_x)} + \varepsilon_y^{(\sigma_z)} = \frac{1}{E}\left[\sigma_y - \nu\left(\sigma_x + \sigma_z\right)\right] \tag{11.132}$$

$$\varepsilon_z = 0 = \varepsilon_z^{(\sigma_z)} + \varepsilon_z^{(\sigma_x)} + \varepsilon_z^{(\sigma_y)} = \frac{1}{E}\left[\sigma_z - \nu\left(\sigma_x + \sigma_y\right)\right]$$

$$\rightarrow \sigma_z = \nu\left(\sigma_x + \sigma_y\right) \tag{a}$$

$$\gamma_{xy} = \gamma_{yx} = \frac{\tau_{xy}}{G} = \frac{2(1+\nu)}{E}\tau_{xy} \tag{11.133}$$

– Elasticity matrix:

$$\mathbf{C} = [C] = \frac{E}{(1+\nu)(1-2\nu)}\begin{bmatrix} 1-\nu & \nu & 0 \\ \nu & 1-\nu & 0 \\ 0 & 0 & \dfrac{1-2\nu}{2} \end{bmatrix}. \tag{11.134}$$

We can make the following remarks:

1) The plane stress state occurs in thin, plate-like members (for example, the web of a plate girder).
2) The plane strain state occurs in long, cylindrical structures (for example, dams, tunnels); that is, in bodies whose loading is independent of z and whose ends are fixed. In this case, displacement w is zero at the ends. Since the body is long and the load and geometry don't vary in the longitudinal direction, the displacement w is approximately zero everywhere and strain is also zero in the longitudinal direction.
3) The plane stress state is defined by: $\sigma_z = \tau_{xz} = \tau_{yz} = 0$, and the vectors of stresses, strains and displacements are given in Equation (11.127).
4) The plane strain state is defined by: $\varepsilon_z = \gamma_{xz} = \gamma_{yz} = 0$, and the vectors of stresses, strains and displacements are shown in Equation (11.131).
5) If $\sigma_z = 0$, as shown in Equations (11.127), for a linearly elastic material, it follows that $\varepsilon_z = \frac{-\nu}{1-\nu}\left(\varepsilon_x + \varepsilon_y\right)$, as shown in Equations (11.127), or $\varepsilon_z = -\nu/E\left(\sigma_x + \sigma_y\right)$, as shown in Equations (11.128). This means that normal strain, ε_z, in the z direction is not zero even though normal stress in the z direction is zero. This is the consequence of Poisson's effect.
6) Also, if $\varepsilon_z = 0$, as given in Equation (11.131), it follows that $\sigma_z = \nu\left(\sigma_x + \sigma_y\right)$, as given in Equation (11.131), or $\sigma_z = \frac{E\nu}{(1+\nu)(1-2\nu)}\left(\varepsilon_x + \varepsilon_y\right)$, as shown by Equation (2.30).

In conclusion, we can say that plane stress ($\sigma_z = 0$) does not generally lead to $\varepsilon_z = 0$, and plane strain ($\varepsilon_z = 0$) requires non-zero σ_z.

11.2.6.1 Basic 2-D Finite Elements

Triangular Finite Elements

The triangular finite element that was shown in Figure 11.12(a) and considered in Example 11.4 is presented again in Figure 11.20(a).

The nodal vectors can be written as follows:

$$\mathbf{u}^{e^T} = \left[\mathbf{u}_1^e \ \mathbf{u}_2^e \ \mathbf{u}_3^e\right] = \left[u_1 \ v_1 \ u_2 \ v_2 \ u_3 \ v_3\right] \tag{11.135}$$

$$\mathbf{f}^{e^T} = \left[\mathbf{f}_1^e \ \mathbf{f}_2^e \ \mathbf{f}_3^e\right] = \left[F_{x1} \ F_{y1} \ F_{x2} \ F_{y2} \ F_{x3} \ F_{y3}\right] \tag{11.136}$$

The displacement within the element is uniquely defined by six values (the element has six degrees of freedom), so it can be represented by two linear polynomials of the form:

$$\begin{aligned} u &= \alpha_1 + \alpha_2 x + \alpha_3 y \\ v &= \alpha_4 + \alpha_5 x + \alpha_6 y \end{aligned} \tag{11.137}$$

The polynomial matrix and vector of generalized coordinates can be written as:

$$\mathbf{a} = \begin{bmatrix} 1 & x & y & 0 & 0 & 0 \\ 0 & 0 & 0 & 1 & x & y \end{bmatrix} \tag{11.138}$$

$$\boldsymbol{\alpha}^T = \left[\alpha_1 \ \alpha_2 \ \alpha_3 \ \alpha_4 \ \alpha_5 \ \alpha_6\right] \tag{11.139}$$

The constants α_i can be determined on the basis of boundary conditions. For the boundary conditions (node points), Equation (11.137) gives:

$$\begin{aligned} u_1 &= \alpha_1 + \alpha_2 x_1 + \alpha_3 y_1 \\ u_2 &= \alpha_1 + \alpha_2 x_2 + \alpha_3 y_2 \\ u_3 &= \alpha_1 + \alpha_2 x_3 + \alpha_3 y_3 \end{aligned} \tag{11.140}$$

Similar equations can be written for the displacement, v.

For the considered triangular finite element shown in Figure 11.20(a), where nodes 1 and 2 are on the x axis and : $x_1 = y_1 = y_2 = 0$, Equation (11.140) can be written in the form:

$$\begin{bmatrix} u_1 \\ u_2 \\ u_3 \end{bmatrix} = \begin{bmatrix} 1 & 0 & 0 \\ 1 & x_2 & 0 \\ 1 & x_3 & y_3 \end{bmatrix} \begin{bmatrix} \alpha_1 \\ \alpha_2 \\ \alpha_3 \end{bmatrix} \rightarrow \mathbf{u}^e = \mathbf{a}^b \boldsymbol{\alpha} \tag{11.141}$$

On the basis of Equation (11.141), it follows that:

$$\boldsymbol{\alpha} = \mathbf{a}^{b^{-1}} \mathbf{u}^e \tag{a}$$

and, if the quantitiy in Equation (a) is inserted into Equation (11.3), we obtain (Equation (11.6)):

$$u = \mathbf{a}\boldsymbol{\alpha} = \mathbf{a}\mathbf{a}^{b^{-1}} \mathbf{u}^e = \mathbf{N}\mathbf{u}^e \tag{b}$$

The shape functions $[N_1 \ N_2 \ N_3] = \mathbf{N}$ can be obtained by inverting the polynomial matrix $\mathbf{a}^b$, defined by Equation (11.141), and inserting the obtained result into Equation (b). Finally, we have:

$$u = N_1 u_1 + N_2 u_2 + N_3 u_3 \tag{11.142a}$$

$$v = N_1 v_1 + N_2 v_2 + N_3 v_3. \tag{11.142b}$$

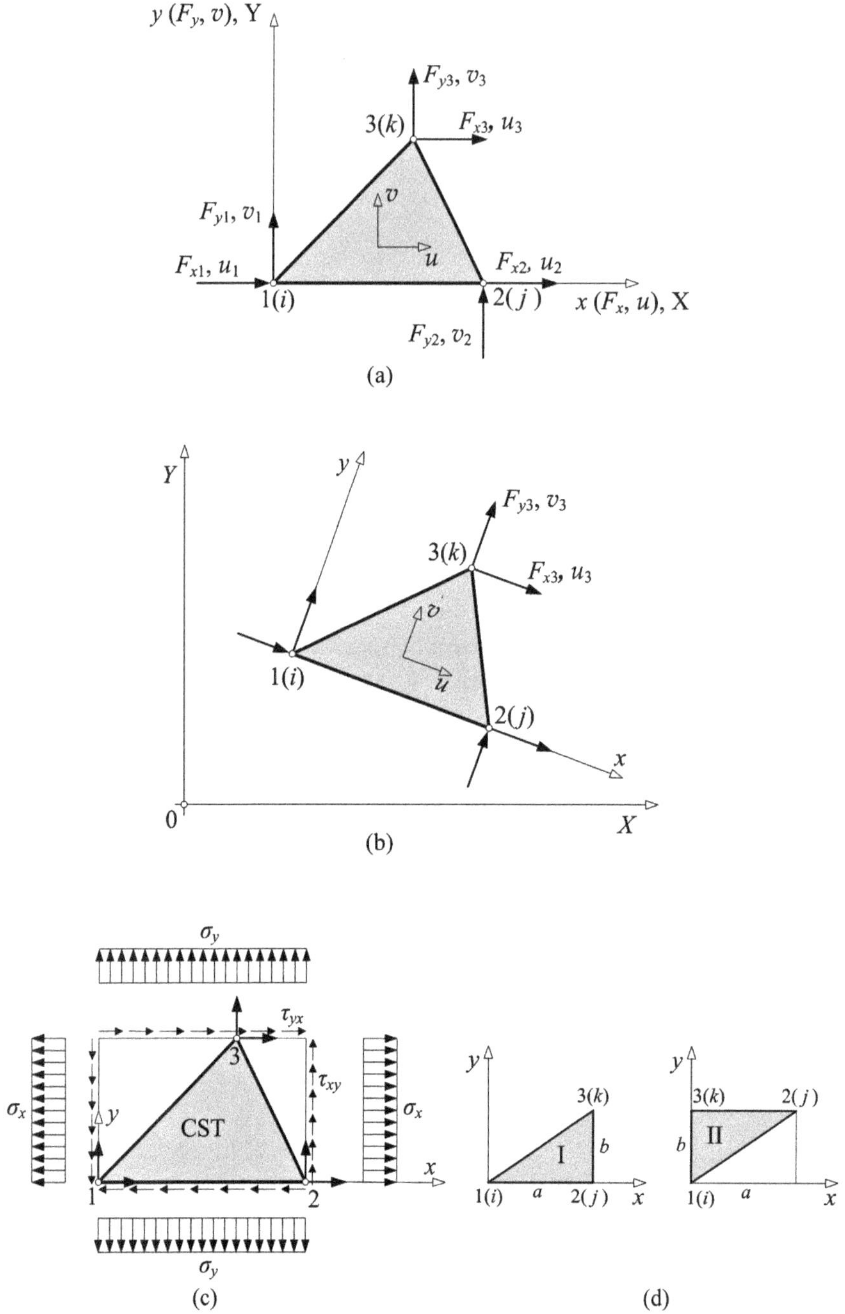

Figure 11.20 A basic triangular finite element. (a) Local and global coordinate systems coincide (nodes 1 and 2 are on the x axis); (b) the local coordinate system does not coincide with the global coordinate system; (c) constant strain triangular element (CST); (d) special position of the triangular finite element.

In Equation (11.142), the shape functions N_i are the same for both translational directions u and v and, using these equations, they may be obtained for any orientation of the triangular finite element in the coordinate system. However, if these functions refer to the finite element considered in Figure 11.20(a), that is, for the prescribed $x_1 = y_1 = y_2 = 0$, they have already been developed using natural (area) coordinates in Equation (11.81). In this case, the forms of these shape functions are:

$$\begin{aligned} N_1 &= L_1 = \frac{1}{x_2 y_3}(x_2 y_3 - x y_3 - x_2 y + x_3 y) \\ N_2 &= L_2 = \frac{1}{x_2 y_3}(x y_3 - x_3 y) \\ N_3 &= L_3 = \frac{y}{y_3} \end{aligned} \tag{11.143}$$

As already mentioned, regardless of the orientation of the triangular finite element in the coordinate system, the shape function can be written in a general form (from Equation (11.78)):

$$\begin{aligned} N_1 &= \frac{1}{2A_e}\left[x_2 y_3 - x_3 y_2 + x(y_2 - y_3) + y(x_3 - x_2)\right] \\ N_2 &= \frac{1}{2A_e}\left[x_3 y_1 - x_1 y_3 + x(y_3 - y_1) + y(x_1 - x_3)\right] \\ N_3 &= \frac{1}{2A_e}\left[x_1 y_2 - x_2 y_1 + x(y_1 - y_2) + y(x_2 - x_1)\right], \end{aligned} \tag{11.144}$$

where the finite element area is:

$$A_e = \frac{1}{2}\begin{vmatrix} 1 & x_1 & y_1 \\ 1 & x_2 & y_2 \\ 1 & x_3 & y_3 \end{vmatrix}. \tag{11.145}$$

Equations (11.144) are obtained using Equation (11.78) and Equation (11.79).

The shape functions of the linear triangular finite element defined by Equations (11.143) or by Equations (11.144) were shown in Figure 11.12(d).

In accordance with Equation (11.8), in this case the strain–displacement matrix is:

$$\mathbf{B} = \mathbf{d}_\varepsilon \mathbf{N} \rightarrow \mathbf{B} = \begin{bmatrix} \frac{\partial}{\partial x} & 0 \\ 0 & \frac{\partial}{\partial y} \\ \frac{\partial}{\partial y} & \frac{\partial}{\partial x} \end{bmatrix} [\mathbf{N}] \tag{c}$$

From Equation (c), the value of the differential operator is also cleare.
Since:

$$\mathbf{B} = [\mathbf{B}_1 \ \mathbf{B}_2 \ \mathbf{B}_3]; \ \mathbf{N} = [N_1 \ N_2 \ N_3], \ \mathbf{N}_i = \begin{bmatrix} N_i & 0 \\ 0 & N_i \end{bmatrix} \tag{d}$$

we obtain:

$$\mathbf{B}_i = \begin{bmatrix} \dfrac{\partial N_i}{\partial x} & 0 \\ 0 & \dfrac{\partial N_i}{\partial y} \\ \dfrac{\partial N_i}{\partial y} & \dfrac{\partial N_i}{\partial x} \end{bmatrix}; \ i = 1, 2, 3. \tag{11.146}$$

Applying Equation (11.146) to the values in Equation (11.144), for the strain–displacement matrix we obtain:

$$\mathbf{B} = \frac{1}{2A_\mathrm{e}} \begin{bmatrix} y_2 - y_3 & 0 & y_3 - y_1 & 0 & y_1 - y_2 & 0 \\ 0 & x_3 - x_2 & 0 & x_1 - x_3 & 0 & x_2 - x_1 \\ x_3 - x_2 & y_2 - y_3 & x_1 - x_3 & y_3 - y_1 & x_2 - x_1 & y_1 - y_2 \end{bmatrix}. \tag{11.147}$$

All the terms in this matrix are constants.

Since the strain in the field of the finite element (the strain distribution) is defined by Equation (11.8):

$$\boldsymbol{\varepsilon} = \mathbf{B}\mathbf{u}^\mathrm{e}, \tag{e}$$

it follows that the strain in the field of the finite element remains constant (that is, the terms in the strain–displacement matrix are constants). These parameters are nodal coordinates. Therefore, this triangular element (linear element) is called a finite element with constant strain: a *CST element* (constant strain triangular element). As is clear from Equation (11.137), formulation of this element is based upon linear polynomials to describe the u and v displacements. This phenomenon of constant strain is represented in Figure 11.20(c). The stress is defined by Equation (11.9):

$$\boldsymbol{\sigma} = \mathbf{C}\boldsymbol{\varepsilon} = \mathbf{C}\mathbf{B}\mathbf{u}^\mathrm{e}. \tag{f}$$

The stiffness matrix for a triangular finite element is defined by Equation (11.19):

$$\mathbf{k}^\mathrm{e} = \int_A \mathbf{B}^\mathrm{T}\mathbf{C}\mathbf{B}h\mathrm{d}A, \ (A = A_\mathrm{e}). \tag{g}$$

As already stated, "h" (or "t") is the designation for the thickness of the finite element. The stiffness matrix in Equation (g) can be determined by introducing into it the quantity from Equation (11.147) for the strain–displacement matrix and the quantity from

Equation (11.130) for the elasticity matrix. The stiffness matrix of the triangular finite element is of the form shown in Equation (11.41):

$$\mathbf{k}^{\mathrm{e}} = \begin{bmatrix} \mathbf{k}_{ii} & \mathbf{k}_{ij} & \mathbf{k}_{ik} \\ \mathbf{k}_{ji} & \mathbf{k}_{jj} & \mathbf{k}_{jk} \\ \mathbf{k}_{ki} & \mathbf{k}_{kj} & \mathbf{k}_{kk} \end{bmatrix}, \tag{h}$$

where submatrices $\mathbf{k}_{ij}$ are of the order 2 x 2 and are of the following form:

$$\mathbf{k}_{ij} = \frac{Eh}{4A_{\mathrm{e}}(1-\nu^2)} \begin{bmatrix} b_i b_j + \frac{1}{2}(1-\nu)c_i c_j & \nu b_i c_j + \frac{1}{2}(1-\nu)c_i b_j \\ \nu c_i b_j + \frac{1}{2}(1-\nu)b_i c_j & c_i c_j + \frac{1}{2}(1-\nu)b_i b_j \end{bmatrix}. \tag{11.148}$$

In Equation (11.148) the coefficients are:

$$\begin{aligned} &b_1 = y_2 - y_3 \quad b_2 = y_3 - y_1 \quad b_3 = y_1 - y_2 \\ &c_1 = x_3 - x_2 \quad c_2 = x_1 - x_3 \quad c_3 = x_2 - x_1 . \\ &b_i = b_1; \; b_j = b_2; \; b_k = b_3; \; c_i = c_1; \; c_j = c_2; \; c_k = c_3 \end{aligned} \tag{11.149}$$

In a case where the triangular finite element has a special shape and orientation (position), as shown in Figure 11.20(d), then, for a type I or type II triangular finite element, we have (see Equation (11.149)):

$$\begin{aligned} &\text{Type I}: && \text{Type II}: \\ &b_i = -b \quad c_i = 0 && b_i = 0 \quad c_i = -a \\ &b_j = b \quad c_j = -a && b_j = b \quad c_j = 0 \\ &b_k = 0 \quad c_k = a && b_k = -b \quad c_k = a . \end{aligned} \tag{11.150}$$

For example, on the basis of Equation (11.148), the submatrix, $\mathbf{k}_{ii}$, of the stiffness matrix for a type I triangular finite element is:

$$\begin{aligned} \mathbf{k}_{ii} &= \frac{Eh}{4A_e(1-\nu^2)} \begin{bmatrix} b_i b_i + \frac{1}{2}(1-\nu)c_i c_i & \nu b_i c_i + \frac{1}{2}(1-\nu)c_i b_i \\ \nu c_i b_i + \frac{1}{2}(1-\nu)b_i c_i & c_i c_i + \frac{1}{2}(1-\nu)b_i b_i \end{bmatrix} \\ &= \frac{Eh}{4A_e(1-\nu^2)} \begin{bmatrix} b^2 & 0 \\ 0 & \frac{1}{2}(1-\nu)b^2 \end{bmatrix} \end{aligned} \tag{i}$$

Finally, for the triangular finite element of type I, shown in Figure 11.20(d), the finite element equation defined by Equation (f), that is, $\mathbf{k}^{\mathrm{e}} = \mathbf{k}^{\mathrm{I}}$, is:

$$\mathbf{k}^{\mathrm{I}} = \frac{Eh}{4A_{\mathrm{e}}(1-\nu^2)}\begin{bmatrix} b^2 & 0 & -b^2 & \nu ab & 0 & -\nu a\, b \\ & \frac{1}{2}(1-\nu)b^2 & \frac{1}{2}(1-\nu)a\, b & -\frac{1}{2}(1-\nu)b^2 & -\frac{1}{2}(1-\nu)a\, b & 0 \\ & & b^2+\frac{1}{2}(1-\nu)a^2 & -a\, b\frac{1}{2}(1+\nu) & -\frac{1}{2}(1-\nu)a^2 & \nu ab \\ & & & a^2+\frac{1}{2}(1-\nu)b^2 & \frac{1}{2}(1-\nu)a\, b & -a^2 \\ & & & & \frac{1}{2}(1-\nu)a^2 & 0 \\ & & & & & a^2 \end{bmatrix} \tag{11.151a}$$

while for the triangular finite element of type II, also shown in Figure 11.20(d), the finite element equation defined by Equation (f), that is, $\mathbf{k}^{\mathrm{e}} = \mathbf{k}^{\mathrm{II}}$, is:

$$\mathbf{k}^{\mathrm{II}} = \frac{Eh}{4A_{\mathrm{e}}(1-\nu^2)}\begin{bmatrix} \frac{1}{2}(1-\nu)a^2 & 0 & 0 & -\frac{1}{2}(1-\nu)ab & -\frac{1}{2}(1-\nu)a^2 & \frac{1}{2}(1-\nu)ab \\ & a^2 & -\nu ab & 0 & \nu ab & -a^2 \\ & & b^2 & 0 & -b^2 & \nu ab \\ & & & \frac{1}{2}(1-\nu)b^2 & \frac{1}{2}(1-\nu)ab & -\frac{1}{2}(1-\nu)b^2 \\ & & & & b^2+\frac{1}{2}(1-\nu)a^2 & -\frac{1}{2}(1+\nu)ab \\ & & & & & a^2+\frac{1}{2}(1-\nu)b^2 \end{bmatrix} \tag{11.151b}$$

Rectangular Finite Elements

A bilinear rectangular finite element is presented in Figure 11.21.

For this finite element, the nodal vectors are:

$$\begin{aligned} \mathbf{u}^{\mathrm{e}^{\mathrm{T}}} &= \left[\mathbf{u}_1^{\mathrm{e}}\ \mathbf{u}_2^{\mathrm{e}}\ \mathbf{u}_3^{\mathrm{e}}\ \mathbf{u}_4^{\mathrm{e}}\right],\ \mathbf{u}_i = \left[u_i\ v_i\right] \\ \mathbf{f}^{\mathrm{e}^{\mathrm{T}}} &= \left[\mathbf{f}_1^{\mathrm{e}}\ \mathbf{f}_2^{\mathrm{e}}\ \mathbf{f}_3^{\mathrm{e}}\ \mathbf{f}_4^{\mathrm{e}}\right],\ \mathbf{f}_i^{\mathrm{e}} = \left[F_{xi}\ F_{yi}\right],\ i = 1,2,3,4. \end{aligned} \tag{11.152}$$

Since the considered element has eight degrees of freedom (that is, two translational displacements per node), a displacement in the field of the finite element can be represented by a bilinear polynomial:

$$\begin{aligned} u &= \alpha_1 + \alpha_2 x + \alpha_3 y + \alpha_4 xy \\ v &= \alpha_5 + \alpha_6 x + \alpha_7 y + \alpha_8 xy \end{aligned}. \tag{11.153}$$

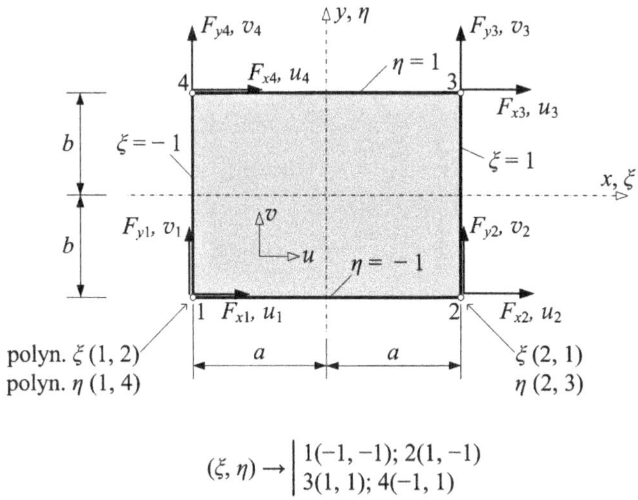

Figure 11.21 A bilinear rectangular finite element (Lagrange).

The matrix of shape functions is defined by Equation (11.6a):

$$\mathbf{N} = \mathbf{a}\mathbf{a}^{b^{-1}} = [\mathbf{N}_1\ \mathbf{N}_2\ \mathbf{N}_3\ \mathbf{N}_4]. \tag{a}$$

Taking:

$$\xi = \frac{x}{a}, \quad \eta = \frac{y}{b}$$

as dimensionless coordinates, and taking ξ_i and η_i as the unit coordinates of the nodes, each shape function can be written in the form:

$$\mathbf{N}_i = \begin{bmatrix} N_i & 0 \\ 0 & N_i \end{bmatrix}, \quad N_i = \frac{1}{4}(1 + \xi\xi_i)(1 + \eta\eta_i), i = 1,2,3,4. \tag{11.154}$$

For example, the unit coordinates ξ_i and η_i for node 1 are $(-1, -1)$.

Using standard procedures for determining the strain–displacement matrix, **B**, and the element stiffness matrix, $\mathbf{k}^e$, the following values are obtained:

Strain–displacement matrix (Equation (11.8)):

$$\mathbf{B} = \frac{1}{4}\begin{bmatrix} -\frac{1}{a}(1-\eta) & 0 & \frac{1}{a}(1-\eta) & 0 & \frac{1}{a}(1+\eta) & 0 & -\frac{1}{a}(1+\eta) & 0 \\ 0 & -\frac{1}{b}(1-\xi) & 0 & -\frac{1}{b}(1+\xi) & 0 & \frac{1}{b}(1+\xi) & 0 & \frac{1}{b}(1-\xi) \\ -\frac{1}{b}(1-\xi) & -\frac{1}{a}(1-\eta) & -\frac{1}{b}(1+\xi) & \frac{1}{a}(1-\eta) & \frac{1}{b}(1+\xi) & \frac{1}{a}(1+\eta) & \frac{1}{b}(1-\xi) & -\frac{1}{a}(1+\eta) \end{bmatrix} \tag{11.155}$$

Element stiffness matrix (Equation (11.9)):

$$\mathbf{k}^{\mathrm{e}} = \int_A \mathbf{B}^{\mathrm{T}}\mathbf{C}\mathbf{B}h\mathrm{d}A, \ (A = A_{\mathrm{e}}).$$

$$\left(\mathbf{k}^{\mathrm{e}} = \int_A \mathbf{B}^{\mathrm{T}}\mathbf{C}\mathbf{B}h\mathrm{d}A = \int_{-a}^{a}\int_{-b}^{b} \mathbf{B}^{\mathrm{T}}\mathbf{C}\mathbf{B}h\mathrm{d}x\mathrm{d}y\right) \tag{b}$$

$$\mathbf{k}^{\mathrm{e}} = \begin{bmatrix} \mathbf{k}_{11}^{\mathrm{e}} & \mathbf{k}_{12}^{\mathrm{e}} & \mathbf{k}_{13}^{\mathrm{e}} & \mathbf{k}_{14}^{\mathrm{e}} \\ \mathbf{k}_{21}^{\mathrm{e}} & \mathbf{k}_{22}^{\mathrm{e}} & \mathbf{k}_{23}^{\mathrm{e}} & \mathbf{k}_{24}^{\mathrm{e}} \\ \mathbf{k}_{31}^{\mathrm{e}} & \mathbf{k}_{32}^{\mathrm{e}} & \mathbf{k}_{33}^{\mathrm{e}} & \mathbf{k}_{34}^{\mathrm{e}} \\ \mathbf{k}_{41}^{\mathrm{e}} & \mathbf{k}_{42}^{\mathrm{e}} & \mathbf{k}_{43}^{\mathrm{e}} & \mathbf{k}_{44}^{\mathrm{e}} \end{bmatrix}. \tag{c}$$

After entering the required quantities in Equation (b) – strain–displacement matrix, **B** (from Equation (11.155)), elasticity matrix, **C** (from Equation (11.130)) – and after the necessary mathematical calculations, we obtain the finite element stiffness matrix for the plane stress state, which is of the form shown in Equation (11.156) [10, 18, 19]. We will show only half of the matrix, since it is symmetric. The designations used in Equation (11.156) are: $\gamma = a/b; \delta = b/a; \nu =$ Poisson's ratio.

$$\mathbf{k}^{\mathrm{e}} = \frac{Eh}{12(1-\nu^2)}\begin{bmatrix} 4\delta+2(1-\nu)\gamma & (3/2)(1+\nu) & -4\delta+(1-\nu)\gamma & -(3/2)(1-3\nu) & -2\delta-(1-\nu)\gamma & -(3/2)(1+\nu) & 2\delta-2(1-\nu)\gamma & (3/2)(1-3\nu) \\ & 4\gamma+2(1-\nu)\delta & (3/2)(1-3\nu) & 2\gamma-2(1-\nu)\delta & -(3/2)(1+\nu) & -2\gamma-(1-\nu)\delta & -(3/2)(1-3\nu) & -4\gamma+(1-\nu)\delta \\ & & 4\delta+2(1-\nu)\gamma & -(3/2)(1+\nu) & 2\delta-2(1-\nu)\gamma & -(3/2)(1-3\nu) & -2\delta-(1-\nu)\gamma & (3/2)(1+\nu) \\ & & & 4\gamma+2(1-\nu)\delta & (3/2)(1-3\nu) & -4\gamma+(1-\nu)\delta & (3/2)(1+\nu) & -2\gamma-(1-\nu)\delta \\ & & & & 4\delta+2(1-\nu)\gamma & (3/2)(1+\nu) & -4\delta+(1-\nu)\gamma & -(3/2)(1-3\nu) \\ & & & & & 4\gamma+2(1-\nu)\delta & (3/2)(1-3\nu) & 2\gamma-2(1-\nu)\delta \\ & & & & & & 4\delta+2(1-\nu)\gamma & -(3/2)(1+\nu) \\ & & & & & & & 4\gamma+2(1-\nu)\delta \end{bmatrix}. \tag{11.156}$$

The process used to determine the shape function matrix as defined by Equation (a), that is, Equation (11.6a), is not always effective and has some disadvantages. Occasionally, for example, the inverse of matrix $\mathbf{a}^{\mathrm{b}}$ may not exist, and this process then runs into some difficulties. It is therefore worthwile considering another method of determination – writing it directly. However, each shape function, N_i, as we know, has a value equal to "1" in the ith node point and the value "0" in other node points. The rectangular finite element presented in Figure 11.21, for which the finite element stiffness matrix is derived in the form presented in Equation (11.156), may also be recognized as a *Lagrange and serendipity* finite element. A method of generating shape functions can be achieved by multiplying the polynomials in the two coordinate directions, that is:

$$N_M \equiv N_{ij} = l_i^m(\xi)l_j^n(\eta) \tag{11.157}$$

In Equation (11.157), m and n represent the number of subdivisions in each direction (ξ, η), that is, m and n joints, as shown in Figure 11.23. As such, in an arbitrary node, M, the coordinates (ξ_i, η_j) are associated and the shape function in this node is defined by Equation (11.157): $N_M \equiv N_{ij} = l_i(\xi)l_j(\eta)$. The polynomials involved in this expression are known as *Lagrange polynomials* (see, for example, Equation (11.61)); that is:

$$l_i(\xi) = \frac{\prod_{\substack{k=1 \\ k \neq i}}^{n} (\xi - \xi_k)}{\prod_{\substack{k=1 \\ k \neq i}}^{n} (\xi_i - \xi_k)} \quad \text{(valid also for } l_j(\eta)) \tag{d}$$

In general, n is the number of joints in the considered direction.

Using the above-mentioned equations (Equations (11.157) and (d)), and in accordance with Figure 11.21, the shape functions N_i take the form*:

$$\begin{aligned}
N_1 &= \frac{\xi - \xi_2}{\xi_1 - \xi_2} \cdot \frac{\eta - \eta_4}{\eta_1 - \eta_4} = \frac{\xi - 1}{-1-1} \cdot \frac{\eta - 1}{-1-1} = \frac{1}{4}(1-\xi)(1-\eta) \\
N_2 &= \frac{\xi - \xi_1}{\xi_2 - \xi_1} \cdot \frac{\eta - \eta_3}{\eta_2 - \eta_3} = \frac{\xi + 1}{1+1} \cdot \frac{\eta - 1}{-1-1} = \frac{1}{4}(1+\xi)(1-\eta) \\
N_3 &= \frac{\xi - \xi_4}{\xi_3 - \xi_4} \cdot \frac{\eta - \eta_2}{\eta_3 - \eta_2} = \frac{\xi + 1}{1+1} \cdot \frac{\eta + 1}{1+1} = \frac{1}{4}(1+\xi)(1+\eta) \\
N_4 &= \frac{\xi - \xi_3}{\xi_4 - \xi_3} \cdot \frac{\eta - \eta_1}{\eta_4 - \eta_1} = \frac{\xi - 1}{-1-1} \cdot \frac{\eta + 1}{1+1} = \frac{1}{4}(1-\xi)(1+\eta)
\end{aligned} \tag{e}$$

*Equation (d) requires the same procedure for N_i (that is, l_i) determination as that defined by Equation (11.61). Since the starting node in each coordinate direction (ξ, η) is the same, that is, the starting node for ξ and η is the same, but the next node is different for each direction, in Equations (e), different subscripts arise relating to the η ccordinate. For example, the ξ direction is 1–2 for node 1, while the η direction for this node is 1–4.

It is clear that the meaning of l_i^m in Equation (11.157) and the meaning of $l_i(z)$ in Equation (11.61) is the same. The polynomial shown in Equation (11.157) is written that way for comparison of two finite elements of the same order (Figure 11.23(b) and Figure 11.23(e)). An explanation for the polynomials used in these elements of the same order is given below Figure 11.23.

In Equation (d), the appropriate values for (ξ_i, η_i) are inserted in accordance with Figure (11.21). Thus, the following shape functions for the (basic) rectangular finite element, known also as a bilinear rectangular finite element or a (basic) Lagrange rectangular finite element, presented in Figure 11.21 are obtained:

$$\begin{aligned}
N_1 &= l_1(\xi)l_1(\eta) = \frac{1}{4}(1-\xi)(1-\eta) \\
N_2 &= l_2(\xi)l_1(\eta) = \frac{1}{4}(1+\xi)(1-\eta) \\
N_3 &= l_2(\xi)l_2(\eta) = \frac{1}{4}(1+\xi)(1+\eta) \\
N_4 &= l_1(\xi)l_2(\eta) = \frac{1}{4}(1-\xi)(1+\eta)
\end{aligned} \tag{11.158}$$

11.2.6.2 Finite Elements of Higher Order

Triangular Finite Elements

Triangular finite elements of higher order have already been shown in Figure 11.12(b) and Figure 11.12(c). The triangular finite element shown in Figure 11.12(b) (a quadratic element) is called a finite element with linear strain – an *LST element* (linear strain triangular element) – while the element shown in Figure 11.12(c) (a cubic element) is called a *QST element* (quadratic strain triangular element) and it is shown again in Figure 11.22.

Example 11.8.

Consider the triangular finite element (QST element) shown in Figure 11.22. For this finite element of third order it is necessary to show the procedure for determining the shape functions relating to the vertices of the triangle using natural L_i coordinates.

Data:

A sketch of the element with the nodes and natural coordinates is shown in Figure 11.22. This finite element has 10 nodes, that is, 20 degrees of freedom, see Pascal's triangle. To describe the displacement in the field of the finite element, it is necessary to use a polynomial with 20 generalized coordinates (coefficients), that is, a polynomial of third order.

Solution:

Referring to Figure 11.22 and Equations (11.73) and (11.74), the shape functions will be derived as the product of functions of the respective coordinates. As such, the following procedure can be presented (M is considered a node; it may also be a vertex/corner):

$$N_M = N_{ijk}(L_1, L_2, L_3) = l_i(L_1)l_j(L_2)l_k(L_3)$$

Since $n = 3$, the shape function related to node 1 is:

$$N_1(L_1, L_2, L3) = N_{300}(L_1, L_2, L3) = l_3(L_1)l_0(L_2)l_0(L_3)$$

$$l_i(L_1) = \begin{cases} \Pi^i_{p=1}\left(\dfrac{nL_1 - p + 1}{p}\right) & \ldots\ldots\ i \geq 1 \\ 1 & \ldots\ldots\ i = 0 \end{cases}$$

$$l_0(L_2) = l_0(L_3) = 1$$

$$N_1(L_1, L_2, L_3) = l_3(L_1)\cdot 1\cdot 1$$

$$l_3(L_1) = \Pi^i_{p=1}\left(\frac{nL_1 - p + 1}{p}\right) = \frac{3L_1 - 1 + 1}{1}\cdot\frac{3L_1 - 2 + 1}{2}\cdot\frac{3L_1 - 3 + 1}{3}$$

$$= \frac{9L_1^3 - 9L_1^2 + 2L_1}{2}$$

$$N_1(L_1, L_2, L_3) = l_3(L_1)\cdot 1\cdot 1 = \frac{9L_1^3 - 9L_1^2 + 2L_1}{2}$$

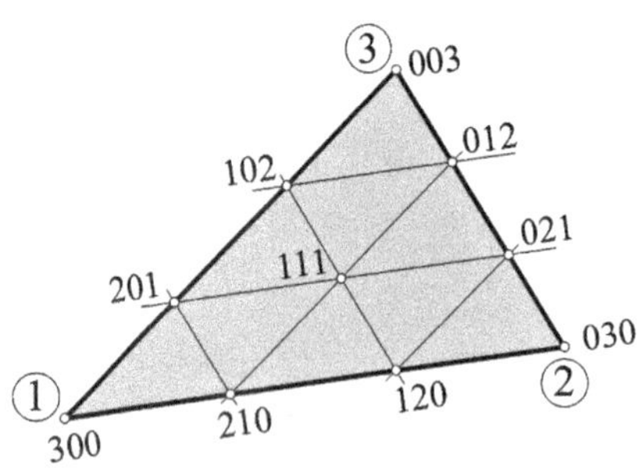

Figure 11.22 A cubic triangular finite element (QST element).

The shape functions for the other vertices of the triangle are:

$$N_2(L_1,L_2,L_3) = N_{030}(L_1,L_2,L_3) = l_0(L_1)l_3(L_2)l_0(L_3)$$
$$= \frac{9L_2^3 - 9L_2^2 + 2L_2}{2}$$
$$N_3(L_1,L_2,L_3) = N_{003}(L_1,L_2,L_3) = l_0(L_1)l_0(L_2)l_3(L_3)$$
$$= \frac{9L_3^3 - 9L_3^2 + 2L_3}{2}$$

In the procedures for determining N_2 and N_3, also:

$$l_0(L_1) = l_0(L_2) = l_0(L_3) = 0$$

Rectangular Finite Elements: Lagrange Family and Serendipity Family

In addition to the bilinear rectangular finite element shown in Figure 11.21, some rectangular finite elements of higher order within the Lagrange family and some within the serendipity family are presented in Figure 11.23.

In Figure 11.23(a) and Figure 11.23(b), Lagrange quadratic and Lagrange cubic rectangular finite elements for $(m{\cdot}n)$ polynomial terms are shown. In this case, m and n represent the number of subdivisions in each direction, as shown in Figure 11.23(a).

Consider Lagrange finite elements where $m = n$. In accordance with Pascal's triangle [6], it is clear that, for these elements, a larger number of polynomial terms is used than

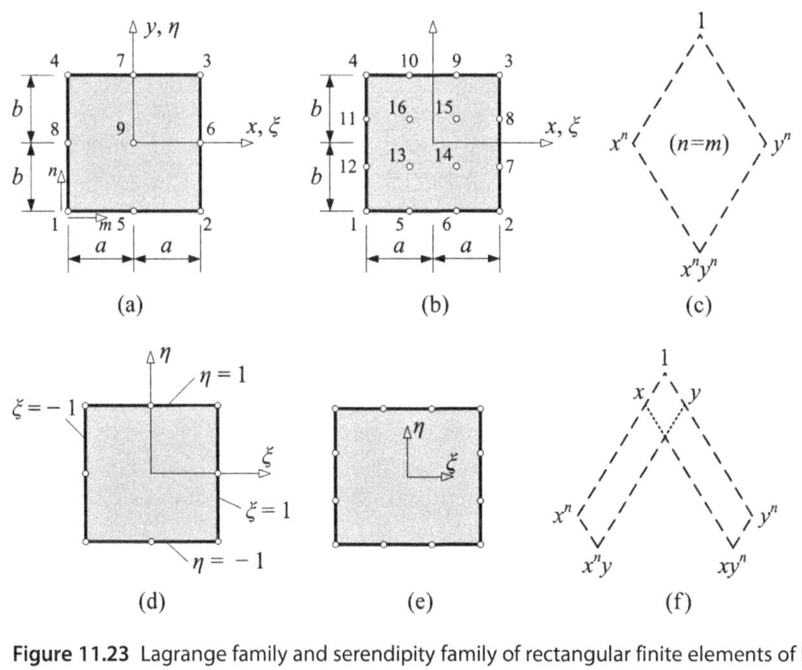

Figure 11.23 Lagrange family and serendipity family of rectangular finite elements of higher order. (a) Lagrange, quadratic; (b) Lagrange, cubic; (c) appropriate polynomial scheme (terms), Lagrange (complete polynomial, $m = n$); (d) serendipity, quadratic; (e) serendipity, cubic; (f) appropriate polynomial scheme (terms), serendipity.

is necessary for complete polynomial expansion. In other words, we can see how many terms belong to the appropriate order (n) of the polynomial and how many terms are used to describe all of the degrees of freedom for the considerd finite element. For example, the cubic finite element (Lagrange family) shown in Figure 11.23(b) uses 16 terms (bicubic interpolation) to describe all of the degrees of freedom, and that is six surplus terms with respect to the complete polynomial expansion of third order. Figure 11.23(c) shows the necessary number of polynomial terms to describe the degrees of freedom of the considered cubic finite element in the Lagrange family.

If we compare this with a cubic finite element in the serendipity family (Figure 11.23 (e)), it is clear that only two surplus terms arise with respect to the complete polynomial expansion of third order.

Figure 11.23(d) shows a quadratic serendipity finite element.

Terms that are necessary to describe the edge shape functions in serendipity finite elements are shown in Figure 11.23(f). Since it is more efficient to use nodes on the element boundary, the serendipity family of finite elements was developed. Take Equation $(11.158)_3$:

$$\frac{1}{4}(\xi+1)(\eta+1) \tag{11.159}$$

as a starting equation for the determination of the shape functions of the bilinear Lagrange finite element in Figure 11.21. If, instead of ξ, η, we insert into this equation the products $\xi\xi_i$ and $\eta\eta_i$, then the shape functions written in Equation (11.58) that refer to the bilinear element (Lagrange, serendipity) may be written in a general form (Equation (11.154)):

$$N_i = \frac{1}{4}(1+\xi\xi_i)(1+\eta\eta_i), \tag{a}$$

where ξ_i and η_i are unit coordinates of the node point. Equation (a), that is, Equation (11.154), can be used to determine the shape functions of corner nodes (nodes at the vertices of the element), for example, for the quadratic serendipity element [7] in Figure 11.23(d). In this case, it takes the form:

$$N_i = \frac{1}{4}(1+\xi\xi_i)(1+\eta\eta_i)(\xi\xi_i+\eta\eta_i-1). \tag{11.160}$$

For the element's mid-side nodes, or for a cubic element, this formula takes a different form.

11.2.6.3 Transformation Procedure for the Finite Element Equation

Similar to the one-dimensional finite elements, here it is necessary to transform the equation of the two-dimensional finite element into the global coordinate system. Let the two-dimensional structure be in the x–y plane (as shown in Figure 11.24), similar to the one-dimensional finite element in the z–y plane shown earlier in Figure 11.6.

Rotation can be performed in a clockwise or counterclockwise direction. The basic side (of the triangle or rectangle) under consideration in rotation can be placed, for example, on the y axis or the x axis. The transformation matrix can be adopted in a similar way as for rotation in a one-dimensional element.

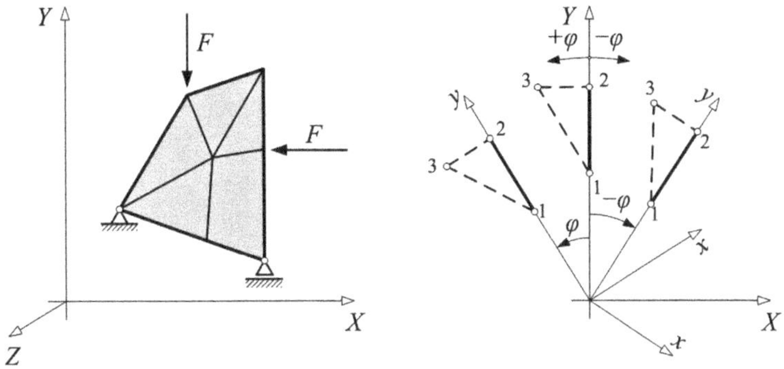

Figure 11.24 A planar structure and displacement transformation.

Example 11.9.
Consider the 2-D structural element, let's call it a bulkhead (or a high, thin beam element/structure), shown in Figure 11.25(a). Determine the displacement in the middle of the structural element if modeling of this structure is made by basic triangular finite elements (CST) and basic rectangular finite elements, and compare these results with the theoretically obtained results. Use several finite element meshes.

Data:
Data relating to the geometry, load and material properties are known (Figure 11.25): $l = 400$ mm, $h = 225$ mm, $t = 20$ mm, $E = 210\ 000$ MPa, $\nu = 0.3$, $q = 4$ kN/mm and the moment of inertia is:

$$I_x = \frac{450^3 \cdot 20}{12} = 151875000\,\text{mm}^4$$

Solution:
For the given data, in accordance with the known solution presented in [20], the displacement in the middle of the structural element is:

$$v_C = \frac{5}{24}\frac{ql^4}{EI_x}\left[1 + \frac{12}{5}\frac{h^2}{l^2}\left(\frac{4}{5} + \frac{\nu}{2}\right)\right],$$

$$v_C = 1.1514\text{mm}.$$

To obtain the result using the proposed triangular CST finite elements and basic rectangular elements, the system is modeled as follows: on the left side and on the right side of the considered structural element are placed supports perpendicular to the mentioned sides. In this way, the reactive forces (F_A, F_B) are obtained as the sum of the reactive forces of the vertically placed supports.

11.2.7 Three-dimensional Finite Elements

So far, the structural elements we've considered were analyzed as one-dimensional or two-dimensional structures. They were created by taking one or two dimensions of the considered structure as very small or negligible in the analysis. However,

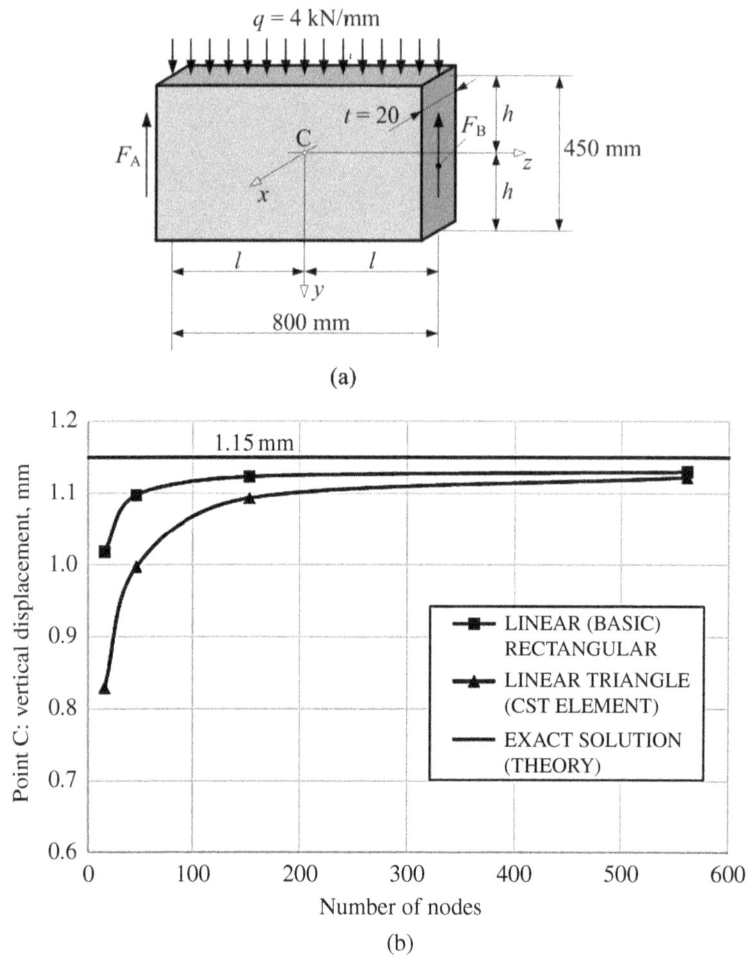

Figure 11.25 A planar structural 2-D element – high, thin-beam structure, modeling and analysis. (a) Structural element and load; (b) determination of the displacement in the middle of the element using triangular and rectangular finite elements. (Numerical calculation made by Professor Marko Canadija, DSc, Faculty of Engineering, University of Rijeka.)

one-dimensional and two-dimensional problems are considered to be economical from the point of view of their solution and they do not always meet real conditions. Many structures or machines can be described as three-dimensional structures and they need to be analyzed using three-dimensional (solid) finite elements. In this case, under consideration are fully three-dimensional problems of stress analysis. Basic and higher orders of one-dimensional, two-dimensional and three-dimensional finite elements can be considered, and they may be straight or curved types. In one-dimensional problems, the simplest finite elements used in analysis are *truss* and *beam* finite elements; in two-dimensional problems, we looked at *triangular* and *rectangular* finite elements (planar finite elements); in three-dimensional problems, there are *tetrahedral* and *hexahedral* (rectangular prism) finite elements (that is, based on a tetrahedron and a hexahedron). There also exist triangular prism elements. One-dimensional finite elements were first applied in the shipbuilding and aircraft industries. Just as the triangular finite element was first applied in the analysis of a planar structure, so the tetrahedral finite element was first introduced in the analysis of three-dimensional structures. It can be said that solid elements are three-dimensional generalizations of planar elements and the basic relationships among these finite elements belong to the linear theory of elasticity.

Bearing in mind the relationship between stress and strain, the following value for the elasticity matrix related to three-dimensional problems can be written:

$$\mathbf{C} = \frac{E}{(1+\nu)(1-2\nu)} \begin{bmatrix} 1-\nu & \nu & \nu & 0 & 0 & 0 \\ \nu & 1-\nu & \nu & 0 & 0 & 0 \\ \nu & \nu & 1-\nu & 0 & 0 & 0 \\ 0 & 0 & 0 & \frac{1-2\nu}{2} & 0 & 0 \\ 0 & 0 & 0 & 0 & \frac{1-2\nu}{2} & 0 \\ 0 & 0 & 0 & 0 & 0 & \frac{1-2\nu}{2} \end{bmatrix} \tag{11.161}$$

In this analysis, as was shown previously, the finite element stiffness matrix is defined as (from Equation (11.18)):

$$\mathbf{k}^{\mathrm{e}} = \int_V \mathbf{B}^{\mathrm{T}}\mathbf{C}\mathbf{B}\,\mathrm{d}V. \tag{a}$$

11.2.7.1 Basic 3-D Finite Elements

Tetrahedral Elements

A linear tetrahedral finite element (tetrahedron) is shown in Figure 11.26.

In the case of a linear tetrahedral finite element, the nodal vectors are:

$$\mathbf{u}^{\mathrm{e}^{\mathrm{T}}} = [\mathbf{u}_1^{\mathrm{e}}\ \mathbf{u}_2^{\mathrm{e}}\ \mathbf{u}_3^{\mathrm{e}}\ \mathbf{u}_4^{\mathrm{e}}],\ \mathbf{f}^{\mathrm{e}^{\mathrm{T}}} = [\mathbf{f}_1^{\mathrm{e}}\ \mathbf{f}_2^{\mathrm{e}}\ \mathbf{f}_3^{\mathrm{e}}\ \mathbf{f}_4^{\mathrm{e}}]. \tag{11.162}$$

For each of the subvectors within the nodal vectors, we have:

$$\mathbf{u}_i^{\mathrm{e}^{\mathrm{T}}} = [u_i\ v_i\ w_i],\quad \mathbf{f}_i^{\mathrm{e}^{\mathrm{T}}} = [F_{xi}\ F_{yi}\ F_{zi}],\quad i = 1\ldots4. \tag{11.163}$$

This tetrahedral finite element has 12 d.o.f. per element, or 3 d.o.f. per joint (node); that is, three translational displacements per node. Consequently, a linear polynomial can be used to describe the change in displacement in the field of the element:

$$\begin{aligned} u &= \alpha_1 + \alpha_2 x + \alpha_3 y + \alpha_4 z \\ v &= \beta_1 + \beta_2 x + \beta_3 y + \beta_4 z \rightarrow \mathbf{u} = \mathbf{a}\boldsymbol{\alpha} \\ w &= \gamma_1 + \gamma_2 x + \gamma_3 y + \gamma_4 z \end{aligned} \tag{11.164}$$

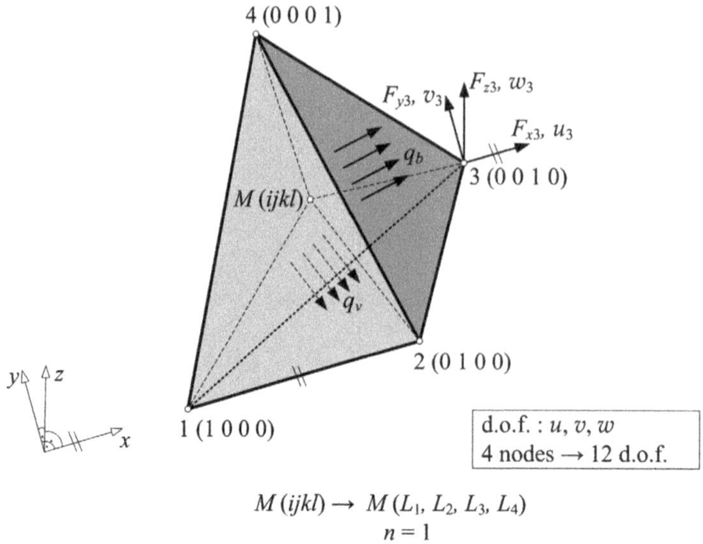

Figure 11.26 A linear tetrahedral finite element.

In Equation (11.164):

$$\mathbf{u} = \begin{bmatrix} u \\ v \\ w \end{bmatrix} \rightarrow \mathbf{u} = \mathbf{a}\boldsymbol{\alpha}, \tag{11.165}$$

where:

$$\mathbf{a} = \begin{bmatrix} \bar{\mathbf{a}} & 0 & 0 \\ 0 & \bar{\mathbf{a}} & 0 \\ 0 & 0 & \bar{\mathbf{a}} \end{bmatrix}, \qquad \bar{\mathbf{a}} = [1 \;\; x \;\; y \;\; z] \tag{11.166}$$

On the boundary, Equation $(11.166)_1$ takes the boundary value, $\mathbf{a}^{\mathrm{b}}$.

$$\boldsymbol{\alpha}^{\mathrm{T}} = \left[\boldsymbol{\alpha}_x \;\; \boldsymbol{\alpha}_y \;\; \boldsymbol{\alpha}_z\right], \boldsymbol{\alpha}_x^{\mathrm{T}} = [\alpha_1 \ldots\ldots \alpha_4], \boldsymbol{\alpha}_y^{\mathrm{T}} = [\beta_1 \ldots\ldots \beta_4] \\ \boldsymbol{\alpha}_z^{\mathrm{T}} = [\gamma_1 \ldots\ldots \gamma_4] \tag{11.167}$$

For the node points:

$$\mathbf{u}^{\mathrm{e}} = \mathbf{a}^{\mathrm{b}}\boldsymbol{\alpha} \rightarrow \boldsymbol{\alpha} = \mathbf{a}^{\mathrm{b}^{-1}}\mathbf{u}^{\mathrm{e}} \rightarrow \mathbf{u} = \mathbf{a}\cdot\boldsymbol{\alpha} = \mathbf{a}\mathbf{a}^{\mathbf{b}^{-1}}\mathbf{u}^{\mathbf{e}} = \mathbf{N}\mathbf{u}^{\mathbf{e}} \tag{a}$$

The shape function matrix is of the form:

$$\mathbf{N} = [\mathbf{N}_1 \;\; \mathbf{N}_2 \;\; \mathbf{N}_3 \;\; \mathbf{N}_4], \tag{11.168}$$

where:

$$\mathbf{N}_j = \begin{bmatrix} N_i & 0 & 0 \\ 0 & N_i & 0 \\ 0 & 0 & N_i \end{bmatrix} \quad j = 1\ldots4, \tag{b}$$

and finally:

$$\mathbf{N} = \begin{bmatrix} N_1 & 0 & 0\ldots & N_4 & 0 & 0 \\ 0 & N_1 & 0\ldots & 0 & N_4 & 0 \\ 0 & 0 & N_1\ldots & 0 & 0 & N_4 \end{bmatrix} \tag{11.169}$$

The volume of the considered tetrahedron is:

$$\sum_{i=1}^{i=4} V_i = V = \frac{1}{6} \begin{vmatrix} 1 & x_1 & y_1 & z_1 \\ 1 & x_2 & y_2 & z_2 \\ 1 & x_3 & y_3 & z_3 \\ 1 & x_4 & y_4 & z_4 \end{vmatrix} \tag{11.170}$$

If the tetrahedron with node points (1, 2, 3, 4) in Figure 11.26 is divided into four sub-tetrahedrons where M is their common node, then we can write:

$$V = V_1 + V_2 + V_3 + V_4. \tag{11.171}$$

If the following ratios are introduced:

$$L_1 = \frac{V_1}{V}, \ L_2 = \frac{V_2}{V}, \ L_3 = \frac{V_3}{V}, \ L_4 = \frac{V_4}{V}, \tag{11.172}$$

where L_i are volume coordinates, then:

$$L_1 + L_2 + L_3 + L_4 = 1 \tag{11.173}$$

The volume coordinates can now be used to describe the rectangular coordinates:

$$x = L_1 x_1 + L_2 x_2 + L_3 x_3 + L_4 x_4, x = (x, y, z) \tag{11.174}$$

In Equation (11.174), x, y and z are coordinates of the considered point, M, within the element, while x_i, y_i, z_i are coordinates of the vertices of the finite element. Equation (11.174) and Equation (11.173) can be combined as:

$$\begin{bmatrix} 1 \\ x \\ y \\ z \end{bmatrix} = \begin{bmatrix} 1 & 1 & 1 & 1 \\ x_1 & x_2 & x_3 & x_4 \\ y_1 & y_2 & y_3 & y_4 \\ z_1 & z_2 & z_3 & z_4 \end{bmatrix} \begin{bmatrix} L_1 \\ L_2 \\ L_3 \\ L_4 \end{bmatrix}. \tag{11.175}$$

As mentioned earlier for Equation (11.77), now, as a result of the solutions of Equation (11.175) with respect to L_i, we can write:

$$\begin{bmatrix} L_1 \\ L_2 \\ L_3 \\ L_4 \end{bmatrix} = \frac{1}{6V} \begin{bmatrix} a_1 & b_1 & c_1 & d_1 \\ a_2 & b_2 & c_2 & d_2 \\ a_3 & b_3 & c_3 & d_3 \\ a_4 & b_4 & c_4 & d_4 \end{bmatrix} \begin{bmatrix} 1 \\ x \\ y \\ z \end{bmatrix}, (V = V_e). \tag{11.176}$$

The coefficients are:

$$a_i = \begin{vmatrix} x_j & y_j & z_j \\ x_k & y_k & z_k \\ x_l & y_l & z_l \end{vmatrix}, \quad b_i = -\begin{vmatrix} 1 & y_j & z_j \\ 1 & y_k & z_k \\ 1 & y_l & z_l \end{vmatrix},$$
$$c_i = -\begin{vmatrix} x_j & 1 & z_j \\ x_k & 1 & z_k \\ x_l & 1 & z_l \end{vmatrix}, \quad d_i = -\begin{vmatrix} x_j & y_j & 1 \\ x_k & y_k & 1 \\ x_l & y_l & 1 \end{vmatrix}. \tag{11.177}$$

A tetrahedral shape function for the point $M\ (i\, j\, k\, l)$ is designated $N_M = N_{ijkl}$ and it can be defined similarly to Equation (11.73), that is, as the product of functions of the respective volume coordinates:

$$N_M = N_{ijkl}(L_1, L_2, L_3, L_4) = l_i(L_1)l_j(L_2)l_k(L_3)l_l(L_4). \tag{11.178}$$

At point M: $i + j + k + l = n$; $n \rightarrow$ polynomial order.

An appropriate formula for l_i $(i = i, j, k, l)$ is given by Equation (11.74). Each shape function can be defined as:

$$N_i = \frac{1}{6V_e}(a_i + b_i x + c_i y + d_i z), \quad i = 1, .., 4 \quad N_i = f(L_i), \tag{11.179}$$

Each submatrix, $\mathbf{N}_j$, in interpolation matrix $\mathbf{N}$ (Equations (11.168) and (11.169)), contains three interpolation functions (N_i). In addition, displacements of the considered point in the field of the element are defined as:

$$u = \sum_{i=1}^{4} N_i u_i; \quad v = \sum_{i=1}^{4} N_i v_i; \quad w = \sum_{i=1}^{4} N_i w_i \tag{11.180}$$

After the shape functions have been determined, on the basis of Equation (11.8), matrix $\mathbf{B}$ can also be determined:

$$\mathbf{B} = \mathbf{d}_\varepsilon \mathbf{N}. \tag{c}$$

In this case, the differential operator is:

$$\mathbf{d}_\varepsilon = \begin{bmatrix} \partial_x & 0 & 0 \\ 0 & \partial_y & 0 \\ 0 & 0 & \partial_z \\ \partial_y & \partial_x & 0 \\ 0 & \partial_z & \partial_y \\ \partial_z & 0 & \partial_x \end{bmatrix}. \tag{11.181}$$

The strain–displacement matrix is:

$$\mathbf{B} = [\mathbf{B}_1\ \mathbf{B}_2\ \mathbf{B}_3\ \mathbf{B}_4], \tag{11.182}$$

and:

$$\mathbf{B}_i = \begin{bmatrix} \frac{\partial N_i}{\partial x} & 0 & 0 \\ 0 & \frac{\partial N_i}{\partial y} & 0 \\ 0 & 0 & \frac{\partial N_i}{\partial z} \\ \frac{\partial N_i}{\partial y} & \frac{\partial N_i}{\partial x} & 0 \\ 0 & \frac{\partial N_i}{\partial z} & \frac{\partial N_i}{\partial y} \\ \frac{\partial N_i}{\partial z} & 0 & \frac{\partial N_i}{\partial x} \end{bmatrix}, \quad \mathbf{B}_i = \frac{1}{6V_e}\begin{bmatrix} b_i & 0 & 0 \\ 0 & c_i & 0 \\ 0 & 0 & d_i \\ c_i & b_i & 0 \\ 0 & d_i & c_i \\ d_i & 0 & b_i \end{bmatrix}, \quad i = 1,2,3,4 \tag{11.183}$$

Since the change in displacement in the field of the element is linear, the strain state in the finite element is constant, similar to the case of the linear (basic) triangular element. As such, this element is called a *CSTh element* (constant strain tetrahedral element).

Example 11.10.
Consider the basic tetrahedron (CSTh element) in Figure 11.26. Determine the shape functions (that is, the shape functions relating to the vertices of the element) using natural L_i coordinates.

Data:
The geometry of the considered tetrahedral finite element is shown in Figure 11.26.

Solution:
Since the considered element has 12 d.o.f. (that is, three translational displacements per node), it is sufficient to use a linear polynomial ($n = 1$) to describe the displacement. In accordance with Equations (11.178) and (11.74):

$$N_M = N_{ijkl}(L_1, L_2, L_3, L_4) = l_i(L_1)l_j(L_2)l_k(L_3)l_l(L_4)$$

$$l_i(L_1) = \begin{cases} \prod_{p=1}^{i}\left(\frac{nL_1 - p + 1}{p}\right), & \text{for } i \geq 1 \\ 1 & \textit{for } i = 0 \end{cases}$$

Consider node 1 (Figure 11.26), where:

$$\begin{aligned} N_M = N_1 = N_{ijkl}(L_1, L_2, L_3, L_4) &= N_{1000}(L_1, L_2, L_3, L_4) \\ &= l_1(L_1)l_0(L_2)l_0(L_3)l_0(L_4) \\ l_0(L_2) &= l_0(L_3) = l_0(L_4) = 1 \end{aligned}$$

$$N_1 = \frac{1 \cdot L_1 - 1 + 1}{1} \cdot 1 \cdot 1 \cdot 1 = L_1$$

In this way, for node 1 we can write:

$$\mathbf{N}_1 = \begin{bmatrix} N_1 & 0 & 0 \\ 0 & N_1 & 0 \\ 0 & 0 & N_1 \end{bmatrix} = \begin{bmatrix} L_1 & 0 & 0 \\ 0 & L_1 & 0 \\ 0 & 0 & L_1 \end{bmatrix}$$

In the same way:

$$N_2 = L_2, N_3 = L_3, N_4 = L_4$$

Finally, the shape function matrix is:

$$\mathbf{N} = \begin{bmatrix} L_1 & 0 & 0 \ldots & L_4 & 0 & 0 \\ 0 & L_1 & 0 \ldots & 0 & L_4 & 0 \\ 0 & 0 & L_1 \ldots & 0 & 0 & L_4 \end{bmatrix}$$

Rectangular Hexahedral Elements

A basic rectangular hexahedral element (that is, a rectangular/right prism or rectangular hexahedron) is a *linear rectangular hexahedron*, also known as a *linear Lagrange/serendipity hexahedral element* (Figure 11.27).

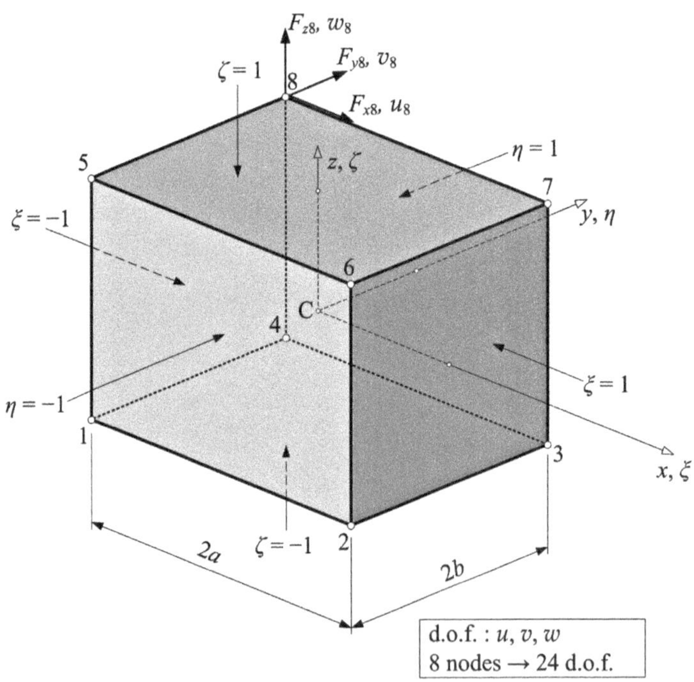

Figure 11.27 A linear (basic) rectangular hexahedral element (linear hexahedron – Lagrange linear element/serendipity element).

For this linear rectangular hexahedron, non-dimensional coordinates (ξ, η, ζ) are placed at the point C (the center of gravity), as shown in Figure 11.27. The geometry of the finite element is defined by 2a x 2b x 2c. The considered element has 24 d.o.f.; that is, three translational displacements per joint (node). Each node is placed at a vertex of the hexahedron. This element belongs to the Lagrange family and is known as a linear Lagrange element. The displacement field is constructed using the Lagrangian interpolation concept (see Section 11.2.4.2). For this finite element, approximation of the displacement field can be made using the following polynomial:

$$\begin{aligned} u &= \alpha_1 + \alpha_2 x + \alpha_3 y + \alpha_4 z + \alpha_5 xy + \alpha_6 yz + \alpha_7 zx + \alpha_8 xyz \\ v &= \beta_1 + \beta_2 x + \beta_3 y + \beta_4 z + \beta_5 xy + \beta_6 yz + \beta_7 zx + \beta_8 xyz \\ w &= \gamma_1 + \gamma_2 x + \gamma_3 y + \gamma_4 z + \gamma_5 xy + \gamma_6 yz + \gamma_7 zx + \gamma_8 xyz \end{aligned} \tag{11.184}$$

Using the conventional procedure (Equation (11.3) to Equation (11.6)), we know that:

$$\mathbf{u}^{\mathrm{e}} = \mathbf{a}^{\mathrm{b}}\boldsymbol{\alpha} \rightarrow \boldsymbol{\alpha} = \mathbf{a}^{\mathrm{b}^{-1}}\mathbf{u}^{\mathrm{e}} \rightarrow u = \mathbf{a}\cdot\boldsymbol{\alpha} = \mathbf{a}\mathbf{a}^{\mathrm{b}^{-1}}\mathbf{u}^{\mathrm{e}} = \mathbf{N}\mathbf{u}^{\mathrm{e}}. \tag{a}$$

If non-dimensional coordinates are applied:

$$\xi = \frac{x}{a}, \quad \eta = \frac{y}{b}, \quad \zeta = \frac{z}{c} \tag{b}$$

then, the shape (interpolation) functions take the following form:

$$N_i = \frac{1}{8}(1 + \xi\xi_i)(1 + \eta\eta_i)(1 + \zeta\zeta_i) \tag{11.185}$$

where ξ_i, η_i, ζ_i are unit coordinates. When the shape functions are determined by non-dimensional coordinates, ξ, η, ζ, then, similarly to Equation (11.157), we have:

$$N_M \equiv N_{ijk} = l_i^m(\xi)l_j^n(\eta)l_k^o(\zeta), \tag{11.186}$$

where $N_M \equiv N_{ijk}$ is a typical node; n, m, o are subdivisions along each side (direction) and l_i^m and so on are given by Equation (11.61).

Example 11.11.

Consider the basic hexahedron (linear prism element, linear Lagrange hexahedral rectangular element) in Figure 11.27. Determine the shape functions for node points 1 and 8.

Data:

The geometry of the considered hexahedron is shown in Figure 11.27.

Solution:

Since the considered element has 24 d.o.f. (that is, three translational displacements per node), it is sufficient to use a linear polynomial ($n = 1$) to describe the displacement. In accordance with Figure 11.27 and Equation (11.186):

$$N_1 = l_i(\xi)l_j(\eta)l_k(\zeta) = l_1(\xi)l_1(\eta)l_1(\zeta),$$

and, using the appropriate Lagrangian polynomials (Equation (11.61)), we have:

$$N_1 = \frac{\xi-\xi_2}{\xi_1-\xi_2}\cdot\frac{\eta-\eta_4}{\eta_1-\eta_4}\cdot\frac{\zeta-\zeta_5}{\zeta_1-\zeta_5} = \frac{\xi-1}{-1-1}\cdot\frac{\eta-1}{-1-1}\cdot\frac{\zeta-1}{-1-1}$$
$$= \frac{1}{8}(1-\xi)(1-\eta)(1-\zeta),$$
$$N_8 = \frac{\xi-\xi_7}{\xi_8-\xi_7}\cdot\frac{\eta-\eta_5}{\eta_8-\eta_5}\cdot\frac{\zeta-\zeta_4}{\zeta_8-\zeta_4} = \frac{\xi-1}{-1-1}\cdot\frac{\eta+1}{1-(-1)}\cdot\frac{\zeta+1}{1+1}$$
$$= \frac{1}{8}(1-\xi)(1+\eta)(1+\zeta).$$

This result corresponds to Equation (11.185). Using the same procedure for other shape functions, it follows that:

$$\mathbf{N} = \begin{bmatrix} N_1 & 0 & 0\ldots & N_8 & 0 & 0 \\ 0 & N_1 & 0\ldots & 0 & N_8 & 0 \\ 0 & 0 & N_1\ldots & 0 & 0 & N_8 \end{bmatrix}$$

11.2.7.2 Finite Elements of Higher Order

Tetrahedral Elements

Some tetrahedral elements of higher order are presented in Figure 11.28.

For the quadratic tetrahedron (with 10 node points, 30 d.o.f.) in Figure 11.28(a), the relationships between the displacement in the field of the element and displacement in the nodes of the finite element can be written in a general form:

$$u = \sum_{i=1}^{n} N_i u_i; \quad v = \sum_{i=1}^{n} N_i v_i; \quad w = \sum_{i=1}^{n} N_i w_i, \tag{11.187}$$

where displacement in the field of the element can be described by a (complete) polynomial of second order (10 generalized coordinates). Here, we are referring to the displacement, u, v and w.

This finite element is known as a tetrahedral finite element with complete quadratic displacement field (the field of strain in the element is approximated by linear functions). As such, this element is called an *LSTh element* (linear strain tetrahedral element). The most convenient way to describe this element is to use the system of natural (volume) coordinates.

Example 11.12.

Consider the quadratic tetrahedron (LSTh element) in Figure 11.28(a). Determine the shape functions.

Data:

The geometry of the considered tetrahedron is shown in Figure 11.28(a).

Solution:

The considered element has 10 nodes (30 d.o.f.), that is, three translational displacements per node. To describe the displacement, the quadratic polynomial ($n = 2$) can be used.

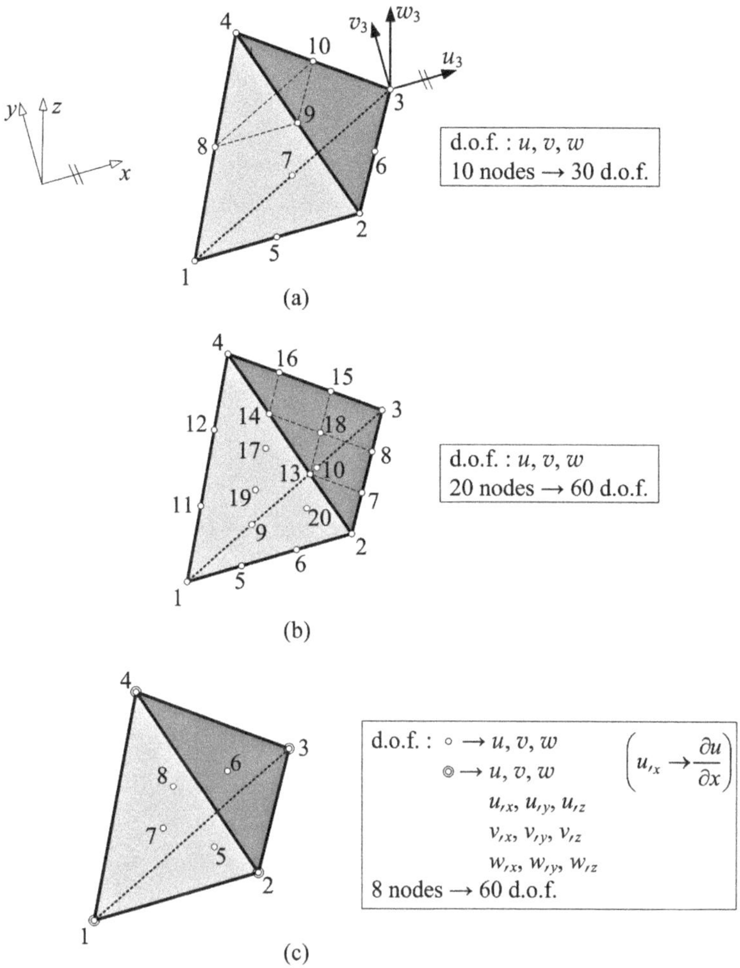

Figure 11.28 Tetrahedral finite elements of higher order. (a) Quadratic (10 node points, 30 d.o.f.); (b) cubic (20 node points, 60 d.o.f.); (c) element with 8 node points (4 node points x 12 d.o.f. plus 4 node points x 3 d.o.f., total 60 d.o.f.).

The most suitable coordinate system in this analysis is the natural (volume) coordinate system. In accordance with Equations (11.178) and (11.74), we have:

$$N_M = N_{ijkl}(L_1, L_2, L_3, L_4) = l_i(L_1)l_j(L_2)l_k(L_3)l_l(L_4) \quad \text{(a)}$$

$$l_i(L_1) = \begin{cases} \prod_{p=1}^{i}\left(\dfrac{nL_1 - p + 1}{p}\right), & \text{for } i \geq 1 \\ 1 & \textit{for } i = 0 \end{cases} \quad \text{(b)}$$

Since $n = 2$, the shape function for the first node (node point 1) is:

$$N_M = N_{ijkl} = N_{2000} = N_1 = N_1(L_1, L_2, L_3, L_4) = l_2(L_1)l_0(L_2)l_0(l_3)l_0(L_4)$$

$$l_0(L_2) = l_0(l_3) = l_0(L_4) = 1$$

$$l_2(L_1) = \prod_{p=1}^{2} \frac{2L_1 - 1 + 1}{1} \cdot \frac{2L_1 - 2 + 1}{2} = L_1(2L_1 - 1)$$

$$N_1 = L_1(2L_1 - 1) \cdot 1 \cdot 1 \cdot 1 = L_1(2L_1 - 1)$$

Further, for example, the shape function for the node point 10 is:

$$N_M = N_{ijkl} = N_{0011} = N_{10} = N_{10}(L_1, L_2, L_3, \mathrm{L}_4)$$
$$= l_0(L_1)l_0(L_2)l_1(l_3)l_1(L_4)$$

$$l_0(L_1) = l_0(l_2) = 1$$

$$l_1(L_3) = \prod_{p=1}^{1} \frac{2L_3 - 1 + 1}{1} = 2L_3; \quad l_1(L_4) = \prod_{p=1}^{1} \frac{2L_4 - 1 + 1}{1} = 2L_4$$

$$N_{10} = 1 \cdot 1 \cdot 2L_3 \cdot 2L_4 = 4L_3L_4$$

Using similar procedures, the following shape functions can be obtained:

$$\begin{aligned}
&N_1(2000) = L_1(2L_1 - 1), && N_2(0200) = L_2(2L_2 - 1),\\
&N_3(0020) = L_3(2L_3 - 1), && N_4(0002) = L_4(2L_4 - 1),\\
&N_5(1100) = 4L_1L_2, && N_6(0110) = 4L_2L_3,\\
&N_7(1010) = 4L_3L_1, && N_8(1001) = 4L_1L_4,\\
&N_9(0101) = 4L_2L_4, && N_{10}(0011) = 4L_3L_4.
\end{aligned} \tag{11.188}$$

The shape function matrix, $\mathbf{N}$, contains submatrices, $\mathbf{N}_i$, related to the node points (1 to 10). Each submatrix, $\mathbf{N}_i$, for example, submatrix $\mathbf{N}_1$, contains three shape functions, N_1, similar to those shown for the basic hexahedron in Figure 11.27 (Example 11.11). If the values defined by Equation (11.188) are entered into Equation (11.8), the relationship between the strain vector and the displacement vector is clear, confirming the linear distribution of deformation. In other words, this is an LSTh element (a linear strain tetrahedron).

Rectangular Hexahedral Elements: Lagrange Family and Serendipity Family

Some hexahedral elements of higher order are presented in Figure 11.29.

Shape functions for some of the mentioned finite elements can be obtained by assessing changes in the displacement as outlined below.

For the element shown in Figure 11.29(a) [18]:

$$\begin{aligned}
u = {}&\alpha_1 + \alpha_2 x + \alpha_3 y + \alpha_4 z + \alpha_5 xy + \alpha_6 yz + \alpha_7 zx + \alpha_8 xyz + \alpha_9 x^2 + \alpha_{10} y^2\\
&+ \alpha_{11} z^2 + \alpha_{12} x^2 y + \alpha_{13} x^2 z + \alpha_{14} y^2 x + \alpha_{15} y^2 z + \alpha_{16} z^2 x + \alpha_{17} z^2 y +\\
&\alpha_{18} x^2 yz + \alpha_{19} xy^2 z + \alpha_{20} xyz^2.
\end{aligned} \tag{11.189}$$

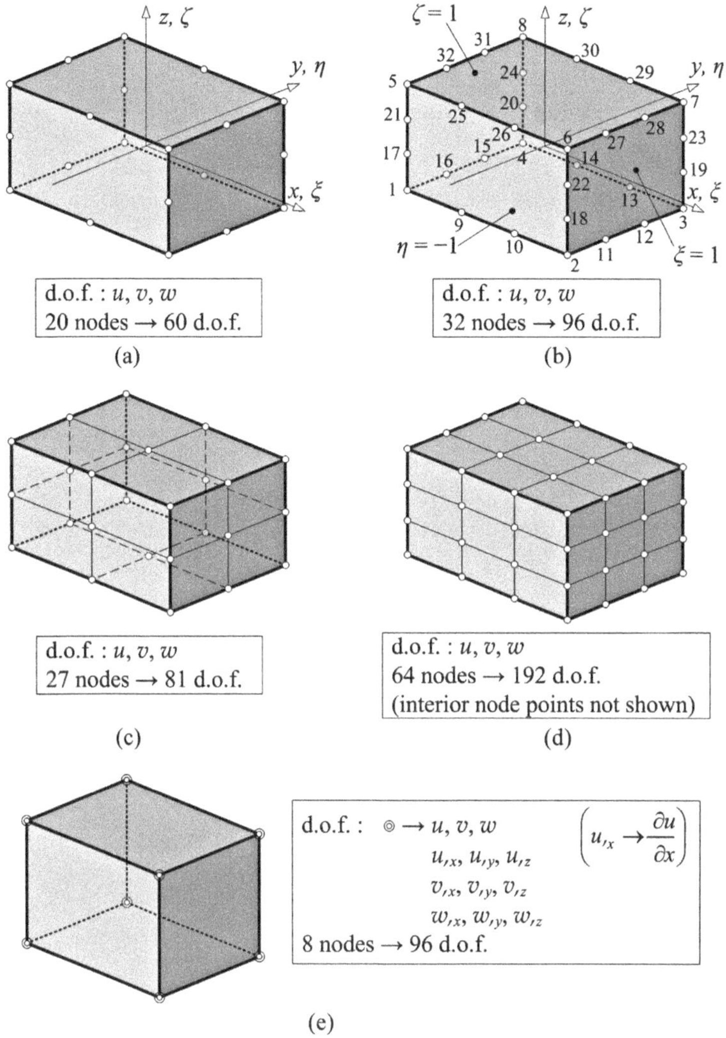

Figure 11.29 Rectangular hexahedral (right prism) finite elements of higher order.

The same is valid for v and w.

The shape function can be written in the following form [7, 10, 18]:

$$N_i(\xi,\eta,\zeta) = \frac{1}{8}(1+\xi\xi_i)(1+\eta\eta_i)(1+\zeta\zeta_i)(\xi\xi_i+\eta\eta_i+\zeta\zeta_i-2) \tag{11.190}$$

for the node points at the vertices (the corner nodes) of the hexahedron, and for a typical, mid-side node:

$$\xi_i = 0,\ \eta_i = \pm 1,\ \zeta_i = \pm 1$$
$$N_i(\xi,\eta,\zeta) = \frac{1}{4}\left(1-\xi^2\right)(1+\eta\eta_i)(1+\zeta\zeta_i) \tag{11.191}$$

For the element shown in Figure 11.29(b):

$$\begin{aligned} u = &\alpha_1 + \alpha_2 x + \alpha_3 y + \alpha_4 z + \alpha_5 xy + \alpha_6 yz + \alpha_7 zx + \alpha_8 xyz + \alpha_9 x^2 \\ &+ \alpha_{10} y^2 + + \alpha_{11} z^2 + \alpha_{12} x^2 y + \alpha_{13} x^2 z + \alpha_{14} y^2 x + \alpha_{15} y^2 z + \alpha_{16} z^2 x \\ &+ \alpha_{17} z^2 y + \alpha_{18} x^2 yz + \alpha_{19} xy^2 z + \alpha_{20} xyz^2 + \ldots\ldots.\ \alpha_{32} z^3 yx. \end{aligned} \tag{11.192}$$

The same is valid for v and w.

Applying Equations (11.186) and (11.61), for the shape function of node point 1, Figure 29b, we have:

$$N_1 = l_i(\xi) l_j(\eta) l_k(\zeta) = l_1(\xi) l_1(\eta) l_1(\zeta),\ l_i(\xi) = \frac{\prod_{\substack{k=1 \\ k \neq i}}^{n} (\xi - \xi_k)}{\prod_{\substack{k=1 \\ k \neq i}}^{n} (\xi_i - \xi_k)} \tag{a}$$

$$l_1(\xi) = \frac{(\xi - \xi_9)}{(\xi_1 - \xi_9)} \cdot \frac{(\xi - \xi_{10})}{(\xi_1 - \xi_{10})} \cdot \frac{(\xi - \xi_2)}{(\xi_1 - \xi_2)} = \frac{\left(\xi + \frac{1}{3}\right)}{\left(-1 + \frac{1}{3}\right)} \cdot \frac{\left(\xi - \frac{1}{3}\right)}{\left(-1 - \frac{1}{3}\right)} \cdot \frac{(\xi - 1)}{(-1 - 1)}$$

$$= -\frac{9}{16}\left(\xi^3 - \xi^2 - \frac{1}{9}\xi + \frac{1}{9}\right)$$

$$l_1(\eta) = \frac{(\eta - \eta_{16})}{(\eta_1 - \eta_{16})} \cdot \frac{(\eta - \eta_{15})}{(\eta_1 - \xi_{15})} \cdot \frac{(\eta - \eta_4)}{(\eta_1 - \eta_4)}$$

$$= \frac{\left(\eta + \frac{1}{3}\right)}{\left(-1 + \frac{1}{3}\right)} \cdot \frac{\left(\eta - \frac{1}{3}\right)}{\left(-1 - \frac{1}{3}\right)} \cdot \frac{(\eta - 1)}{(-1 - 1)} = -\frac{9}{16}\left(\eta^3 - \eta^2 - \frac{1}{9}\eta + \frac{1}{9}\right)$$

In the same way:

$$l_1(\zeta) = -\frac{9}{16}\left(\zeta^3 - \zeta^2 - \frac{1}{9}\zeta + \frac{1}{9}\right)$$

Finally, for the node point 1:

$$\begin{aligned} N_1 &= l_1(\xi) l_1(\eta) l_1(\zeta) \\ &= \left[-\frac{9}{16}\left(\xi^3 - \xi^2 - \frac{1}{9}\xi + \frac{1}{9}\right)\right]\left[-\frac{9}{16}\left(\eta^3 - \eta^2 - \frac{1}{9}\eta + \frac{1}{9}\right)\right]\left[-\frac{9}{16}\left(\zeta^3 - \zeta^2 - \frac{1}{9}\zeta + \frac{1}{9}\right)\right] \\ &= \left[-\frac{1}{16}(9\xi^3 - 9\xi^2 - \xi + 1)\right]\left[-\frac{1}{16}(9\eta^3 - 9\eta^2 - \eta + 1)\right]\left[-\frac{1}{16}(9\zeta^3 - 9\zeta^2 - \zeta + 1)\right]. \end{aligned}$$

If the values $\xi = -1, \eta = -1, \zeta = -1$ (coordinates related to node 1) are inserted into the above equation, it follows that:

$$N_1 = 1 \cdot 1 \cdot 1 = 1.$$

For all other nodes lying in directions ξ, η, ζ from node 1, the value of the shape function, N_1, is equal to zero.

Otherwise, for this element, the shape functions can be written in the form [7, 10, 18]:

$$N_i(\xi,\eta,\zeta) = \frac{1}{64}(1+\xi\xi_i)(1+\eta\eta_i)(1+\zeta\zeta_i)\left[9\left(\xi^2+\eta^2+\zeta^2\right)-19\right], \tag{11.193}$$

for node points at the vertices (corner nodes) of the hexahedron, and for a typical, mid-side node:

$$\begin{aligned} &\xi_i = \pm\frac{1}{3}, \quad \eta_i = \pm 1, \quad \zeta_i = \pm 1, \\ &N_i(\xi,\eta,\zeta) = \frac{9}{64}\left(1-\xi^2\right)(1+9\xi\xi_i)(1+\eta\eta_i)(1+\zeta\zeta_i) \end{aligned} \tag{11.194}$$

As is clear, if Equation (11.193) is used, and if the values for the coordinate (ξ, η, ζ) of node point 1 are inserted, we also obtain $N_1 = 1$.

Some of the elements shown are serendipity elements, where incomplete polynomials are used to describe/approximate the field of the displacement.

11.2.8 Isoparametric Finite Elements

11.2.8.1 Introduction

With the development of constructions has come the need for a better approximation of structural geometry with a relatively small number of finite elements. This means that there is a need for elements that can describe, as realistically as possible, complex structural geometry [2, 3, 6]. The geometry of a real structure that is modeled using finite elements is always somewhat approximated and this can be a source of error in the analysis. It has been found that modeling of displacements and modeling of geometry may be of equal importance. As such, our previously presented simple rectangular, triangular, tetrahedral and other similar forms of elements are no longer sufficient in the modeling of complex structures. Thus, this section is concerned with distorting such simple forms of finite elements into other forms that are more appropriate in structural modeling. Basic one-dimensional, two-dimensional and three-dimensional finite elements can be mapped into distorted forms and such forms can better describe real structural forms. As shown in Figure 11.30, the ξ, η, ζ (or L_1, L_2, L_3, L_4,) local non-dimensional coordinates may be transformed (mapped) into a distorted set that may be of straight or curvilinear form when plotted using the Cartesian system, x, y, z. When the same (iso-) parameters are employed for both geometry and displacement approximation, the concept of representation is known as *isoparametric representation.* Isoparametric finite elements are those for which the representation of their deformational behavior and the representation of their geometries is the same (that is, if the displacement field in the finite element is described by the same shape functions as the finite element geometry, the finite element is an isoparametric element).

Figure 11.30 shows some finite elements in their basic, non-dimensional coordinates and mapped forms.

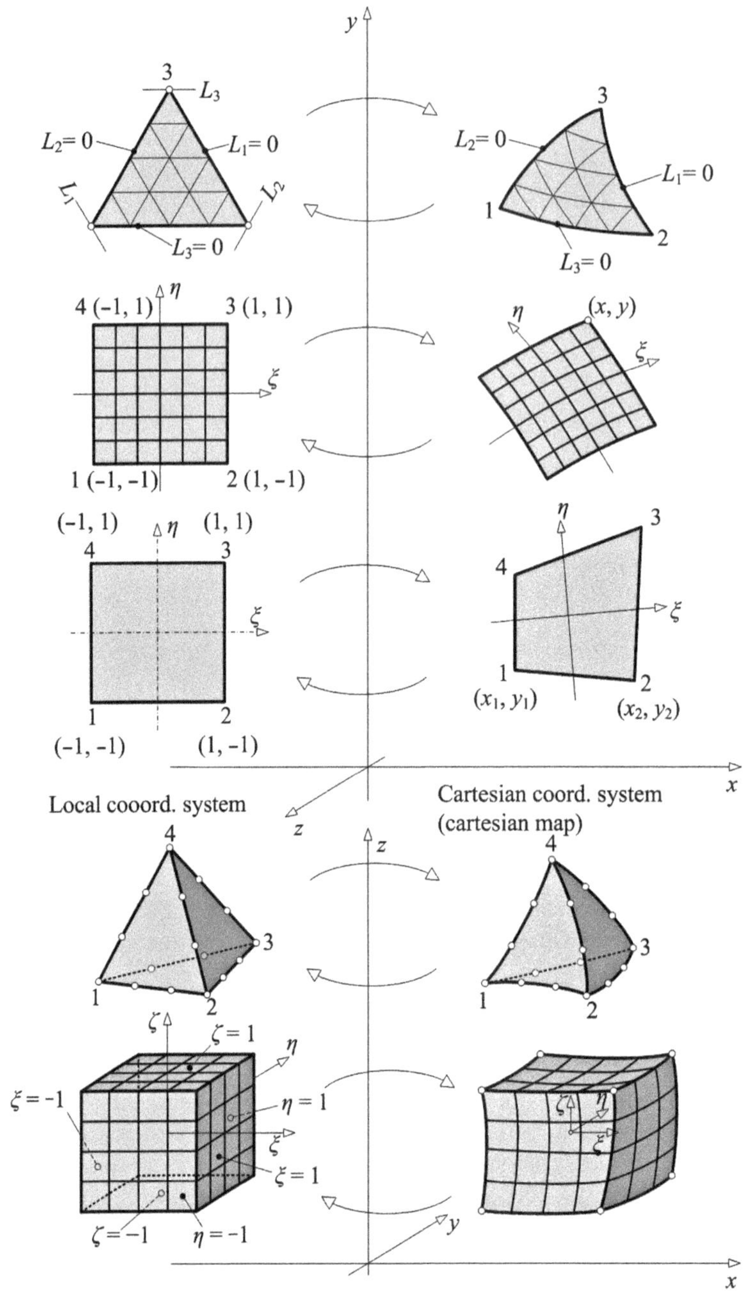

Figure 11.30 Elements of the basic forms in local non-dimensional coordinates, [ξ, η, ζ (or L_1, L_2, L_3, L_4)], are mapped into distorted forms when plotted in the x, y, z coordinate system.

The relationship between Cartesian and curvilinear coordinates can be established:

$$\begin{bmatrix} x \\ y \\ z \end{bmatrix} = \begin{bmatrix} f_x(\xi, \eta, \zeta) \\ f_y(\xi, \eta, \zeta) \\ f_z(\xi, \eta, \zeta) \end{bmatrix} \quad \text{or} \quad x_i = f_i(L_i) \tag{11.195}$$

It is of interest to discuss the so-called isoparametric form of Equation (11.195). As mentioned above, in an isoparametric representation, the element's geometry and the displacement field in the element are described by the same shape functions. The construction of an isoparametric finite element means mapping the 1-D, 2-D or 3-D non-dimensional finite element (with a defined number of node points) into distorted forms with the same number of node points. Thus, the non-dimensional coordinates, ξ, η, ζ or L_i, will be mapped (that is, distorted) to a new coordinate set in the Cartesian coordinate system. After the mapping, the local coordinate axes are no longer perpendicular to each other and may pass into curvilinear coordinates. If the geometric functional representation is of a higher order than that for displacement, the element formulation is known as *superparametric*, and if the geometric representation is of a lower order than that for displacement, we are looking at a *subparametric* representation.

After the brief discussion conducted so far, we can see that the main idea of isoparametric element formulation and its use in structural analysis is better modeling of a complex structure and the achievement of the relationship between the element displacements at any point and the element nodal point displacements directly using shape functions. In this case, the inverse polynomial matrix at the boundary, that is, at the node points ($\mathbf{a}^{\mathrm{b}^{-1}}$) is not evaluated.

11.2.8.2 Isoparametric Representation

In accordance with the above, an element's geometry (coordinates) and displacements can be described by the same shape functions. If we consider, for example, a two-dimensional, four-sided element, we have:

$$x = \sum_{i=1}^{N} N_i(\xi, \eta) x_i \qquad y = \sum_{i=1}^{N} N_i(\xi, \eta) y_i \tag{11.196}$$

$$u = \sum_{i=1}^{N} N_i(\xi, \eta) x_i \qquad v = \sum_{i=1}^{N} N_i(\xi, \eta) x_i \tag{11.197}$$

In Equations (11.196) and (11.197), N is the number of element node points, N_i are shape functions, while (x_i, y_i) are the coordinates of nodes. As stated previously, the shape function N_i is equal to one (1) at node "i" and is equal to zero (0) at all other node points.

A very important part of the structural analysis is the calculation of the finite element stiffness matrix. To construct an element stiffness matrix, first it is necessary to determine the matrix B. In addition, to obtain the strain tensor at a point (ξ, η) (Equation $(11.8)_1$), it is also necessary to determine matrix B. As such, it is necessary to find a relationship between the derivatives of the shape functions with respect to x and y and with respect to ξ and η. Therefore (from Equation $(11.8)_2$):

$$\mathbf{B} = \mathbf{d}_\varepsilon \mathbf{N}. \tag{a}$$

The matrix B, for a two-dimensional problem, is similar to that given in Equation (11.146):

$$B = \begin{bmatrix} \frac{\partial N_1}{\partial x} & 0 & \dots & \frac{\partial N_N}{\partial x} & 0 \\ 0 & \frac{\partial N_1}{\partial y} & \dots & 0 & \frac{\partial N_N}{\partial y} \\ \frac{\partial N_1}{\partial y} & \frac{\partial N_1}{\partial x} & \dots & \frac{\partial N_N}{\partial y} & \frac{\partial N_N}{\partial x} \end{bmatrix} \tag{11.198}$$

The procedure of differentiation is:

$$\begin{aligned} \frac{\partial N_i}{\partial \xi} &= \frac{\partial N_i}{\partial x}\frac{\partial x}{\partial \xi} + \frac{\partial N_i}{\partial y}\frac{\partial y}{\partial \xi} \\ \frac{\partial N_i}{\partial \eta} &= \frac{\partial N_i}{\partial x}\frac{\partial x}{\partial \eta} + \frac{\partial N_i}{\partial y}\frac{\partial y}{\partial \eta} \end{aligned} \tag{11.199}$$

and this can be written in the form:

$$\begin{bmatrix} \frac{\partial N_i}{\partial \xi} \\ \frac{\partial N_i}{\partial \eta} \end{bmatrix} = \underbrace{\begin{bmatrix} \frac{\partial x}{\partial \xi} & \frac{\partial y}{\partial \xi} \\ \frac{\partial x}{\partial \eta} & \frac{\partial y}{\partial \eta} \end{bmatrix}}_{\mathbf{J}} \begin{bmatrix} \frac{\partial N_i}{\partial x} \\ \frac{\partial N_i}{\partial y} \end{bmatrix} \Rightarrow \begin{bmatrix} \frac{\partial}{\partial \xi} \\ \frac{\partial}{\partial \eta} \end{bmatrix} N_i = \mathbf{J} \begin{bmatrix} \frac{\partial}{\partial x} \\ \frac{\partial}{\partial y} \end{bmatrix} N_i, \tag{11.200}$$

where $\mathbf{J}$ is the Jacobian matrix. Further:

$$\begin{bmatrix} \frac{\partial N_i}{\partial x} \\ \frac{\partial N_i}{\partial y} \end{bmatrix} = \mathbf{J}^{-1} \begin{bmatrix} \frac{\partial N_i}{\partial \xi} \\ \frac{\partial N_i}{\partial \eta} \end{bmatrix} \tag{11.201}$$

$$\mathbf{J}^{-1} = \begin{bmatrix} \frac{\partial x}{\partial \xi} & \frac{\partial y}{\partial \xi} \\ \frac{\partial x}{\partial \eta} & \frac{\partial y}{\partial \eta} \end{bmatrix}^{-1} = \frac{1}{\det \mathbf{J}} \begin{bmatrix} \frac{\partial y}{\partial \eta} & -\frac{\partial y}{\partial \xi} \\ -\frac{\partial x}{\partial \eta} & \frac{\partial x}{\partial \xi} \end{bmatrix} \tag{11.202}$$

The necessary derivatives in Equation (11.202) can be found as:

$$\frac{\partial x}{\partial \xi} = \sum_{i=1}^{N} \frac{\partial N_i}{\partial \xi} x_i. \tag{11.203}$$

Similar expressions can be written for other derivatives.

We can write:

$$\mathbf{J}^{-1}(\xi,\eta) = \frac{1}{\det\mathbf{J}}\begin{bmatrix} \sum_{i=1}^{N}\frac{\partial N_i}{\partial \eta}y_i & -\sum_{i=1}^{N}\frac{\partial N_i}{\partial \xi}y_i \\ -\sum_{i=1}^{N}\frac{\partial N_i}{\partial \eta}x_i & \sum_{i=1}^{N}\frac{\partial N_i}{\partial \xi}x_i \end{bmatrix} \tag{11.204}$$

If, for example, a bilinear isoparametric finite element is being considered (such as the last one shown in the first coordinate system in Figure 11.30), then its shape function matrix is already defined (from Equation (11.158)) as:

$$\mathbf{N} = [N_1\, N_2\, N_3\, N_4], \tag{b}$$

where the shape functions are:

$$\begin{aligned} N_1 &= l_1(\xi)l_1(\eta) = \frac{1}{4}(1-\xi)(1-\eta) \\ N_2 &= l_2(\xi)l_1(\eta) = \frac{1}{4}(1+\xi)(1-\eta) \\ N_3 &= l_2(\xi)l_2(\eta) = \frac{1}{4}(1+\xi)(1+\eta) \\ N_4 &= l_1(\xi)l_2(\eta) = \frac{1}{4}(1-\xi)(1+\eta). \end{aligned} \tag{c}$$

Taking into account Equations (b), (c) and (11.199), we now have:

$$\begin{aligned} \frac{\partial \mathbf{N}}{\partial \xi} &= \frac{1}{4}[-(1-\eta)\ (1-\eta)\ (1+\eta)\ -(1+\eta)] \\ \frac{\partial \mathbf{N}}{\partial \eta} &= \frac{1}{4}[-(1-\xi)\ -(1-\xi)\ (1+\xi)\ (1+\xi)] \end{aligned}$$

and, for example, the upper-left coefficient of $\mathbf{J}$ in Equation (11.200), in accordance with Equation (11.203), is:

$$\mathbf{J}_{11} = \frac{1}{4}[(1-\eta)(x_2-x_1) + (1+\eta)(x_3-x_4)].$$

Other $\mathbf{J}_{ij}$ coefficients can also be found using similar procedures.

The element stiffness matrix is (from Equation (11.19)):

$$\mathbf{k}^{\mathrm{e}} = \int_{A} \mathbf{B}^{\mathrm{T}}\mathbf{C}\mathbf{B}h\,\mathrm{d}A, \tag{d}$$

and in the case of plane strain conditions, $h = 1$, while $\mathrm{d}A = \mathrm{d}x\mathrm{d}y$ in a rectangular coordinate system. Since a finite element is defined in the natural coordinate system (ξ,η), it is necessary to change the variables of integration. Thus:

$$\mathrm{d}x\mathrm{d}y = \det\mathbf{J}\,\mathrm{d}\xi\,\mathrm{d}\eta, \tag{11.205}$$

and, finally, the element stiffness matrix is:

$$\mathbf{k}^{\mathrm{e}} = \int_{-1}^{1}\int_{-1}^{1} \mathbf{B}^{\mathrm{T}}\mathbf{C}\mathbf{B}h \det \mathbf{J}\, \mathrm{d}\xi\, \mathrm{d}\eta. \tag{11.206}$$

Given the complexity of the integrand, the integration is carried out numerically. The term $\mathbf{B}^{\mathrm{T}}\mathbf{C}\mathbf{B}\,h \det \mathbf{J}$ is multiplied by a weigting factor depending on the element's Gauss point, and this will be used in the element stiffness matrix determination (Equation (11.206)).

11.2.9 Bending of Elastic Flat Plates

11.2.9.1 Deformation Theories for Elastic Plates

In the literature, it is possible to find a number of references that discuss deformation in elastic plates. Here, we present only a brief overview that provides some insight into the possibility of using elastic flat plates and solving the problem of bending in plates.

A plate may be defined as an elastic solid body supporting a transverse load that causes it to bend. The subject of the deformation (bending/flexure) of plates is one of the most interesting problems for designers of structures. Usually, it is thought that there is an analogy between a beam-type structure as a one-dimensional structure and a plate as a two-dimensional structure. However, a beam may be straight or curved, while a plate is considered to be a flat structural element. The occurrence of curvature in a plate could make it a shell [8].

To solve the problems with plates, certain assumptions with respect to their behavior under deformation were introduced.

A plate is a structure that can be encountered in all areas of engineering: mechanical engineering (platforms, depositories and so on), civil engineering (buildings, platforms, bridges and so on), shipbuilding and ocean engineering (platforms, decks and so on) as well as in other types of structural engineering. A plate is usually of constant thickness, and its lateral dimensions are large compared to its thickness. The central part (plane) of the plate is bounded by two parallel planes. In this sense, the plate is an extension of a beam in two dimensions. As regards the shape of a plate, it may be rectangular, circular or another shape. In addition, a plate may be made of steel, concrete or another type of material. In engineering practice, so-called laminated plates may also be mentioned [21].

The deformation of a plate is usually described by *classical plate theory* or by *first-order shear deformation plate theory* [1, 4, 13]. Classical plate theory is treated as an extension of *Euler–Bernoulli beam theory* and is also known as the *Kirchoff plate theory* (or thin plate theory). On the other hand, first-order shear deformation plate theory is an extension of *Timoshenko beam theory* and is also known as the *Reissner–Mindlin plate theory*. Plates (and shells) are essentially special cases of three-dimensional structures. The Kirchoff plate theory is associated with thin plates, and a relaxation of the assumptions made by Reissner–Mindlin is associated with the application of the theory to tick plates. In the Kirchoff plate theory, that is, in classical plate theory, the assumptions are as follows:

1) A straight line that is initially perpendicular to the middle surface of the unstrained plate remains straight and perpendicular to the middle surface after deformation of the plate and is inextensible.

2) The middle surface is unstrained.
3) The normal stress in the normal direction (z) to the plate is negligible, that is, $\sigma_z \approx 0$. Thus, the plate is in a state of plane stress.

The first assumption is similar to the Euler–Bernoulli hypothesis in simple beam theory. This also implies a linear stress distribution throughout the plate's thickness. Assumption "1" may be expressed as:

$$\varepsilon_z = 0, \quad \gamma_{xz} = \gamma_{yz} = 0 \tag{11.207}$$

Further, the basic assumption of thin plate bending theory, mentioned in assumption "1", that a straight line that is initially perpendicular to the midsurface remains straight during deformation, implies the equality of the angular displacement and the middle surface slope, that is, $\alpha_y = -\partial w/\partial x$ and $\alpha_x = \partial w/\partial y$, where derivatives are interpreted as the rotation of the line initially normal to the midsurface ($z = 0$) of the plate (Figure 11.31).

In the first-order theory, the normality condition of the straight line after its rotation is removed. In this sense, a straight line remains straight but no longer perpendicular to the tangent of the middle surface in the considered point. So, a zero value for the lateral (normal) strain remains, and a constant state of tranverse shear strain through the plate's thickness is added:

$$\varepsilon_z = 0, \quad \gamma_{xz} = f(x, y), \quad \gamma_{yz} = f(x, y) \tag{11.208}$$

A rectangular plate of uniform thickness, transversally loaded, is shown in Figure 11.31.

Governing Equations

Consider a displacement field and stress field in accordance with Figures 11.31(a) and (b). In accordance with the previously mentioned assumptions, in the Kirchoff theory (for thin plates), shear deformations are neglected (Equation (11.207)) and a straight line remains perpendicular to the midsurface, while in the Reissner–Mindlin theory, shear deformations are included.

Classical Plate Theory Consider displacements $u\ (x, y)$ and $v\ (x, y)$ of the point C. Point A is a point on the middle surface (Figures 11.31(a) and (b)).

- Displacements

$$\begin{aligned}
&\alpha_y(+);\ \alpha_y = -\frac{\partial w}{\partial x}\\
&\alpha_x(-);\ \alpha_x = \frac{\partial w}{\partial y};\ \left(-\alpha_x = -\frac{\partial w}{\partial y}\right)\\
&u(x,y) = z\alpha_y = -z\frac{\partial w}{\partial x}\\
&v(x,y) = -z\alpha_x = -z\frac{\partial w}{\partial y};
\end{aligned} \tag{11.209}$$

$w(x, y)$ = vertical displacement of the middle surface.

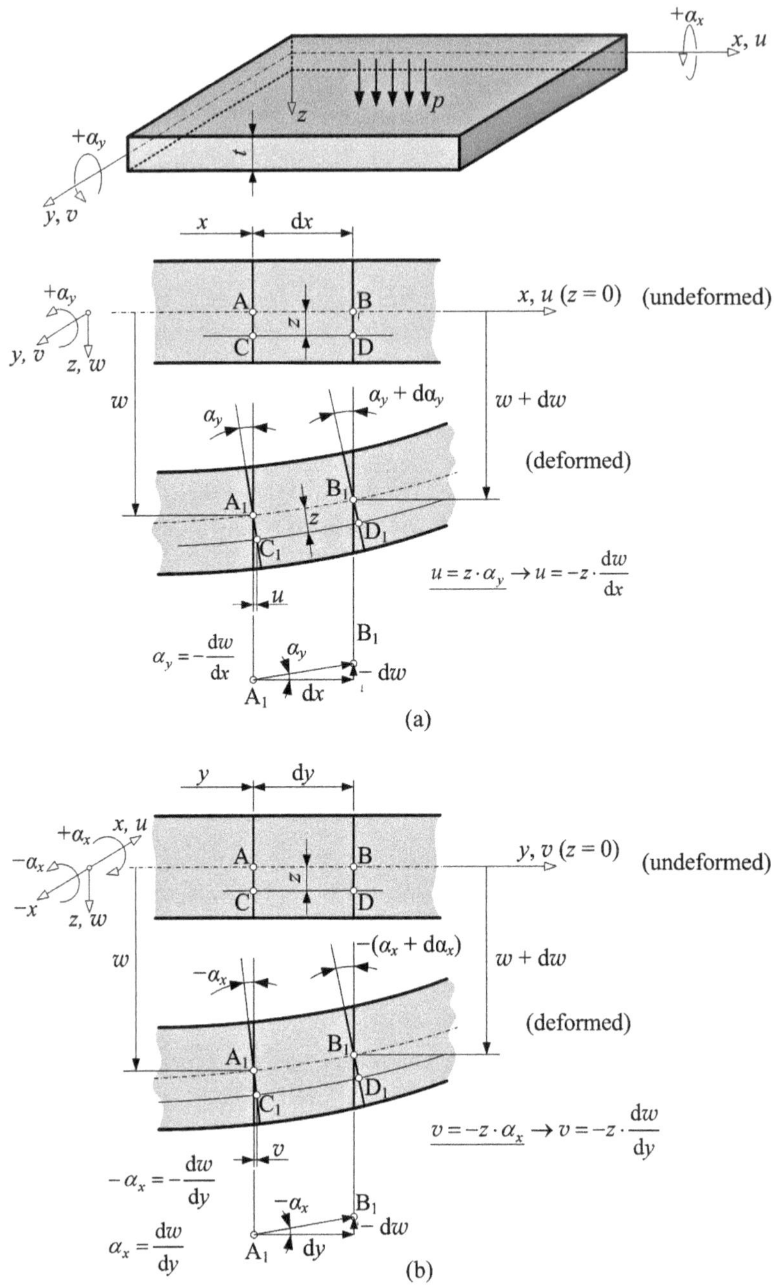

Figure 11.31 A plate and its geometry. (a) Classical plate theory (CPT) – bending around the y axis; (b) classical plate theory (CPT) – bending around the x axis; (c) first-order shear deformation plate theory (FSDT) – bending about the y axis.

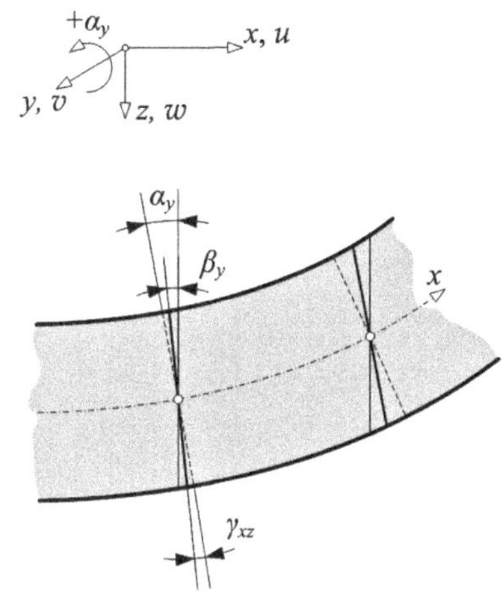

Figure 11.31 *(Continued)*

However, $\alpha_y = -\dfrac{\partial w}{\partial x}$ represents flexure in the x direction (that is, rotation about the y axis) and $\alpha_x = \dfrac{\partial w}{\partial y}$, represents flexure in the y direction (that is, rotation about the x axis). It is understood than angle α is very small.

The change in angular displacements (α_y, α_x) of the lines perpendicular to the middle surface is described by the so-called direct (κ_x, κ_y) and twisting (κ_{xy}, κ_{yx}) curvatures written depending on w.

The curvatures of the midsurface $w(x, y)$ in the x and y directions are defined as:

$$\kappa_x = \frac{\partial \alpha_y}{\partial x} = -\frac{\partial^2 w}{\partial x^2} \quad \kappa_y = \frac{\partial(-\alpha_x)}{\partial y} = -\frac{\partial^2 w}{\partial y^2}; (A_1B_1 \sim AB = \mathrm{d}x) \tag{11.209a}$$

The twisting of the element (dx dy) of the midsurface $w(x, y)$, is:

$$\kappa_{xy} = \frac{\partial \alpha_y}{\partial y} = -2\frac{\partial^2 w}{\partial x \partial y}; \quad \kappa_{yx} = \frac{\partial(-\alpha_x)}{\partial x} = -2\frac{\partial^2 w}{\partial x \partial y} \to \kappa_{xy} = \kappa_{yx} \tag{11.209b}$$

The matrix of change in curvature (the vector of curvature) [3, 13] is:

$$\boldsymbol{\kappa} = \begin{bmatrix} \kappa_x & \kappa_y & \kappa_{xy} \end{bmatrix}^{\mathrm{T}} = -\begin{bmatrix} \dfrac{\partial^2 w}{\partial x^2} & \dfrac{\partial^2 w}{\partial y^2} & 2\dfrac{\partial^2 w}{\partial x \partial y} \end{bmatrix}^{\mathrm{T}} \tag{11.209c}$$

In Figures 11.31(a) and (b), the rotation of the transverse (perpendicular) line $\overline{AC}$ in the x–z plane about the y axis, to a new position, $\overline{A_1C_1}$, is shown as $(-\partial w/\partial x) \to (\alpha_y)$. The in-plane displacements (u, v) are usually the subject of discussion in the field of stress and strain. Here, considering linear theory based on infinitesimal strains and plane stress conditions, we will discuss only the equations governing the deflection w (affected by bending).

- **Strain field**
  On the basis of Equations (2.24a), (3.16), (11.127), (11.131) and (11.209), we have:

$$\begin{bmatrix} \varepsilon_x \\ \varepsilon_y \\ \gamma_{xy} \end{bmatrix} = \begin{bmatrix} \dfrac{\partial u}{\partial x} \\ \dfrac{\partial v}{\partial y} \\ \dfrac{\partial u}{\partial y} + \dfrac{\partial v}{\partial x} \end{bmatrix} = -z \begin{bmatrix} \dfrac{\partial^2 w}{\partial x^2} \\ \dfrac{\partial^2 w}{\partial y^2} \\ 2\dfrac{\partial^2 w}{\partial x \partial y} \end{bmatrix} = z \begin{bmatrix} \dfrac{\partial \alpha_y}{\partial x} \\ -\dfrac{\partial \alpha_x}{\partial y} \\ \dfrac{\partial \alpha_y}{\partial y} - \dfrac{\partial \alpha_x}{\partial x} \end{bmatrix} \tag{11.210}$$

- **Stresses**

$$\begin{bmatrix} \sigma_x \\ \sigma_y \\ \tau_{xy} \end{bmatrix} = \frac{E}{1-\nu^2} \begin{bmatrix} 1 & \nu & 0 \\ \nu & 1 & 0 \\ 0 & 0 & \dfrac{1-\nu}{2} \end{bmatrix} \begin{bmatrix} \varepsilon_x \\ \varepsilon_y \\ \gamma_{xy} \end{bmatrix} \tag{11.211}$$

First-order Shear Deformation Plate Theory In accordance with Equations (11.209) and Figures 11.31(a), (b) and (c), it follows that:

- **Displacement field**

$$\begin{aligned} u(x,y) &= z\beta_y \\ v(x,y) &= -z\beta_x \end{aligned} \tag{11.212}$$

$w(x, y)$ = vertical displacement of the middle surface,

where β_y and β_x are the rotations of the perpendicular straight line in the x–z (about the y axis) and y–z (about the x axis) planes, respectively (Figure 11.31(c)).

- **Strain field**

$$\begin{bmatrix} \varepsilon_x \\ \varepsilon_y \\ \gamma_{xy} \\ \gamma_{yz} \\ \gamma_{zx} \end{bmatrix} = \begin{bmatrix} \dfrac{\partial u}{\partial x} \\ \dfrac{\partial v}{\partial y} \\ \dfrac{\partial v}{\partial x} + \dfrac{\partial u}{\partial y} \\ \dfrac{\partial w}{\partial y} + \dfrac{\partial v}{\partial z} \\ \dfrac{\partial u}{\partial z} + \dfrac{\partial w}{\partial x} \end{bmatrix} = \begin{bmatrix} z\partial\beta_y/\partial x \\ -z\partial\beta_x/\partial y \\ z\left(\partial\beta_y/\partial y - \partial\beta_x/\partial x\right) \\ \partial w/\partial y - \beta_x \\ \partial w/\partial x + \beta_y \end{bmatrix} \tag{11.213}$$

- **Stresses**

$$\begin{bmatrix} \sigma_x \\ \sigma_y \\ \tau_{xy} \end{bmatrix} = z\frac{E}{1-\nu^2}\begin{bmatrix} 1 & \nu & 0 \\ \nu & 1 & 0 \\ 0 & 0 & \frac{1-\nu}{2} \end{bmatrix}\begin{bmatrix} \partial\beta_y/\partial x \\ -\partial\beta_x/\partial y \\ \left(\partial\beta_y/\partial y - \partial\beta_x/\partial x\right) \end{bmatrix} \tag{11.214a}$$

$$\begin{bmatrix} \tau_{yz} \\ \tau_{zx} \end{bmatrix} = \frac{E}{2(1+\nu)}\begin{bmatrix} \partial w/\partial y - \beta_x \\ \partial w/\partial x + \beta_y \end{bmatrix} \tag{11.214b}$$

Generalized Elasticity Matrix (Bending) The components of stress and the components of strain are interconnected by the elasticity matrix, C, also known as the stress–strain matrix or the material matrix. The elasticity matrix for three-dimensional problems is represented by Equation (11.161), for a plane stress state by Equation (11.130) and for a plane strain state by Equation (11.134).

A generalized elasticity matrix for plate bending analysis can be obtained as follows. Note, the term "generalized" is included in our consideration because curvature and moments are included in the analysis. In the analysis of the plate, it is assumed that each layer of the plate acts under plane stress conditions. By integrating the normal stresses in the considered plate to obtain moments per unit length, the *generalized elasticity matrix* is obtained as:

$$\mathbf{C}_b = \int_{-t/2}^{t/2} z^2\frac{E}{1-\nu^2}\begin{bmatrix} 1 & \nu & 0 \\ \nu & 1 & 0 \\ 0 & 0 & \frac{1-\nu}{2} \end{bmatrix} dz = \frac{Et^3}{12(1-\nu^2)}\begin{bmatrix} 1 & \nu & 0 \\ \nu & 1 & 0 \\ 0 & 0 & \frac{1-\nu}{2} \end{bmatrix} = D\begin{bmatrix} 1 & \nu & 0 \\ \nu & 1 & 0 \\ 0 & 0 & \frac{1-\nu}{2} \end{bmatrix}, \tag{11.215}$$

where

$$D = \frac{Et^3}{12(1-\nu^2)} \tag{11.215a}$$

is called the plate flexural rigidity (the bending stiffness). The elasticity matrix in Equation (11.215) refers to isotropic thin-plate flexure. The elasticity matrix for shear stresses is:

$$\mathbf{C}_s = \frac{Etk}{2(1+\nu)}\begin{bmatrix} 1 & 0 \\ 0 & 1 \end{bmatrix} \tag{11.216}$$

where k is a constant that takes into account the non-uniformity of the shear stresses [2].

From *Strength of Materials*, we know that, for example, when a rectangular cross-sectional shape of area A is subjected to shear force Q, the maximum shear stress is greater than an average value, and is $1.5Q/A$. The average shear stress is $\tau_{av} = Q/A_s$, where A_s is the shear area and here it is taken as the area (A) of the rectangle. In reality, the shear area (A_s) is smaller than the area of the rectangle, A. The total rotation (α_y) of the perpendicular line/plane (Figure 11.31(c)) consists of the rotation β_y and the shear deformation $\gamma_{xz} = \gamma$, $(\beta_y = \alpha_y - \gamma; \alpha_y = \partial w/\partial x)$. This shearing strain ($\gamma$) is a constant across the section. Shearing stress τ and sharing strain γ vary over the cross-section (there is a parabolic distribution of τ across the rectangular section). The shearing strain γ is a constant strain on a shear stress area A_s and Q is the shear force in the considered section. The constant k (the shear correction factor) is known from *Strength of Materials*, and it is usually 5/6 for a rectangular cross-section. This factor (k) is evaluated on the basis of equality of shear strain energy due to constant shear stress on the shear area and actual shear stress on the actual area.

Internal Forces, Equilibrium Equations and Stresses

Here, we will consider a thin, flat, rectangular plate in a plane stress state ($\sigma_z \approx 0$). Figure 11.32 shows the internal forces and stresses [22–24].

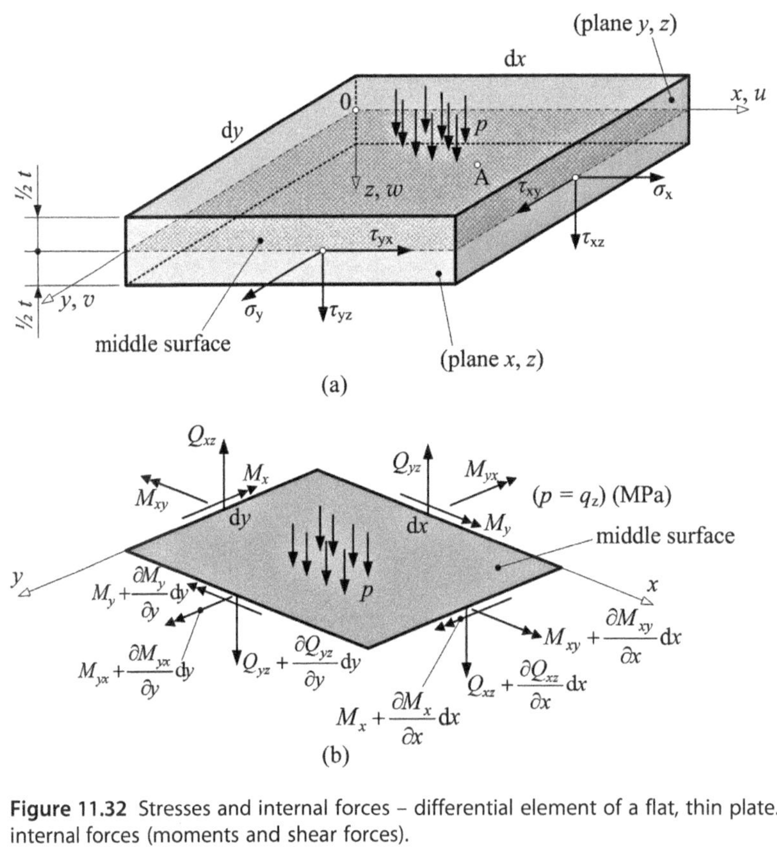

Figure 11.32 Stresses and internal forces – differential element of a flat, thin plate. (a) Stresses; (b) internal forces (moments and shear forces).

Based on Hooke's Law, the stresses in the thin plate, defined by Equation (11.211), are:

$$\sigma_x = \frac{E}{1-\nu^2}\left(\varepsilon_x + \nu\varepsilon_y\right), \quad \sigma_y = \frac{E}{1-\nu^2}\left(\varepsilon_y + \nu\varepsilon_x\right)$$
$$\tau_{xy} = \frac{E}{2(1+\nu)}\gamma_{xy}. \tag{11.217}$$

Also, if the quantities defined by Equation (11.210) are inserted into Equation (11.217), it follows that:

$$\sigma_x = -\frac{E}{1-\nu^2}\left(\frac{\partial^2 w}{\partial x^2} + \nu\frac{\partial^2 w}{\partial y^2}\right)z, \quad \sigma_y = -\frac{E}{1-\nu^2}\left(\frac{\partial^2 w}{\partial y^2} + \nu\frac{\partial^2 w}{\partial x^2}\right)z$$
$$\tau_{xy} = -\frac{E}{1+\nu}\left(\frac{\partial^2 w}{\partial x \partial y}\right)z. \tag{11.218}$$

Stresses σ_x, σ_y and τ_{xy} are distributed linearly across the thickness of the plate. Since a thin, flat plate in a state of plane stress is being considered (Kirchoff plate theory), in accordance with Equation (11.207), shear strains γ_{xz} and γ_{yz} are considered to be negligible. Thus, Hooke's Law cannot be used to obtain shear stresses τ_{xz} and τ_{yz}. If, instead, we use the first two equilibrium equations from Equations (2.4) and integrate them with respect to z (assuming that these shear stresses are equal to zero for $z = \pm t/2$), the results can be obtained. Transverse shear stresses τ_{yz} and τ_{xz} are also present and their distributions through the thickness of the plate are parabolic. Internal forces (moments) in the plate are usually written in such a way that they relate to the unit of length of the plate. For example, bending moment M_x is defined as:

$$M_x = \int_{-t/2}^{t/2} \sigma_x z\,\mathrm{d}y\,\mathrm{d}z(\mathrm{Nm}) \rightarrow /\mathrm{d}y \rightarrow \int_{-t/2}^{t/2} \sigma_x z\,\mathrm{d}z,$$
$$\uparrow$$
$$\text{(divided by dy)}$$

and index "x" refers to the bending moment and to the stress that causes the moment, while the vector of the considered moment is perpendicular to the x axis (along the y axis). The stresses defined by Equation (11.210) (or Equation (11.218)) give rise to bending moments M_x and M_y as well as to twisting moment M_{xy}, which are known as line loadings. In accordance with Figure 11.32, the internal forces are as follows:

$$M_x = \int_{-t/2}^{t/2} \sigma_x z\,\mathrm{d}z \quad M_y = \int_{-t/2}^{t/2} \sigma_y z\,\mathrm{d}z \quad M_{xy} = \int_{-t/2}^{t/2} \tau_{xy} z\,\mathrm{d}z$$
$$Q_{xz} = \int_{-t/2}^{t/2} \tau_{xz} z\,\mathrm{d}z \quad Q_{yz} = \int_{-t/2}^{t/2} \tau_{yz} z\,\mathrm{d}z \tag{11.219}$$

If the quantities defined by Equation (11.218) are inserted into Equations (11.219), the following values for the moments and shear forces can be obtained. For example, moment M_x is:

$$M_x = -\frac{E}{1-\nu^2}\left(\frac{\partial^2 w}{\partial x^2} + \nu\frac{\partial^2 w}{\partial y^2}\right)\int_{-t/2}^{t/2} z^2 \mathrm{d}z$$

$$= -\frac{Et^3}{12(1-\nu^2)}\left(\frac{\partial^2 w}{\partial x^2} + \nu\frac{\partial^2 w}{\partial y^2}\right) = -D\left(\frac{\partial^2 w}{\partial x^2} + \nu\frac{\partial^2 w}{\partial y^2}\right).$$

After all the quantities have been calculated, the final results are:

$$M_x = -D\left(\frac{\partial^2 w}{\partial x^2} + \nu\frac{\partial^2 w}{\partial y^2}\right), \quad M_y = -D\left(\frac{\partial^2 w}{\partial y^2} + \nu\frac{\partial^2 w}{\partial x^2}\right)$$
$$M_{xy} = -D(1-\nu)\frac{\partial^2 w}{\partial x \partial y}. \tag{11.220}$$

In Equations (11.220), D is the plate flexural rigidity from Equation (11.215a).

Setting the equilibrium equations for the differential element of the plate in accordance with Figure 11.32(b) and neglecting small quantities of higher order, it follows that:

$$\frac{\partial Q_{xz}}{\partial x} + \frac{\partial Q_{yz}}{\partial y} + p = 0$$
$$\frac{\partial M_x}{\partial x} + \frac{\partial M_{yx}}{\partial y} - Q_{xz} = 0 \tag{11.221}$$
$$\frac{\partial M_y}{\partial y} + \frac{\partial M_{xy}}{\partial x} - Q_{yz} = 0$$

In Equation (11.221), p is the external load. On the basis of Equations (11.221) and (11.220), it follows that:

$$Q_{xz} = -D\frac{\partial}{\partial x}\left(\frac{\partial^2 w}{\partial x^2} + \frac{\partial^2 w}{\partial y^2}\right)$$
$$Q_{yz} = -D\frac{\partial}{\partial y}\left(\frac{\partial^2 w}{\partial x^2} + \frac{\partial^2 w}{\partial y^2}\right) \tag{11.222}$$

If the quantities defined by Equation (11.222) are inserted in the first of Equations (11.221), one can obtain the differential equation of the bending of a thin rectangular plate:

$$\frac{\partial^4 w}{\partial x^4} + 2\frac{\partial^4 w}{\partial x^2 \partial y^2} + \frac{\partial^4 w}{\partial y^4} = \frac{p}{D} \tag{11.223}$$

Finally, comparing Equation (11.218) and Equation (11.220), and using Equation (11.215), it follows that:

$$\sigma_x = \frac{12M_x}{t^3}z \quad \sigma_y = \frac{12M_y}{t^3}z \quad \tau_{xy} = \frac{12M_{xy}}{t^3}z \tag{11.224a}$$

Since the maximum stresses occur at: $z = z_{\max} = \pm t/2$, from Equation (11.224a), it follows that:

$$\sigma_{x_{\max}} = \pm \frac{6M_x}{t^2} \quad \sigma_{y_{\max}} = \pm \frac{6M_y}{t^2} \quad \tau_{xy_{\max}} = \pm \frac{6M_{xy}}{t^2} \tag{11.224b}$$

Knowing several factors, such as the curvature as the basic measure of deformation in thin plate flexure, the internal forces (moments per unit of length) and the matrix of elasticity, it is possible, in analogy with plane elasticity, to write the constitutive relationships in isotropic, thin plate flexure:

$$\mathbf{M} = \mathbf{C}_\mathrm{b}\boldsymbol{\kappa} \tag{11.225}$$

In Equation (11.225):

$$\mathbf{M} = \left[M_x \;\; M_y \;\; M_{xy}\right] \rightarrow \text{Equation (11.219)}; \;\; \mathbf{C}_\mathrm{b} \rightarrow \text{Equation (11.215)};$$
$$\boldsymbol{\kappa} \rightarrow \text{Equation (11.209c)}$$

11.2.9.2 Finite Elements Based on Kirchhoff Plate Theory

When dealing with the analysis of plate bending, bearing in mind that the plate is a very widely used and important structural element, it is necessary to describe the solution of the problem as thoroughly and accurately as possible. As stated previously, the state of deformation may be represented by one quantity – displacement w – that is, the lateral displacement of the plate's middle surface. This formulation is irreducible. This introduces second derivatives of w into the definition of strain. Continuity conditions between the elements have to be imposed on this quantity and on its derivatives (so-called C^1 continuity) [1, 3, 10, 13, 15] to ensure that the plate remains continuous. In other words, at the nodes, we need to use the displacement w and its first derivatives (slopes). Satisfying C^1 continuity makes the element conform. For plate bending analysis, a certain number of elements is developed that ensure only displacement continuity, that is, they belong to a non-conforming element, but by using them it is possible to achieve convergence of the solution. In the analysis of plates, conforming as well as non-conforming elements are used. In our short review, only two elements will be shown, since many types of finite element used in plate-bending analysis can be found in [1, 4, 8, 10, 13, 18].

Rectangular, Thin, Flat Plate Finite Element → 12 d.o.f.

This non-conforming finite element and its degrees of freedom are shown in Figure 11.33.

This type of finite element is the simplest element used in plate-bending analysis. Since the element has three degrees of freedom per node, this is an element with 12 degrees of freedom. The vector of nodal displacements and the vector of nodal forces are:

$$\mathbf{u}^{\mathrm{e}^\mathrm{T}} = \left[\mathbf{u}_1^{\mathrm{e}^\mathrm{T}} \; \mathbf{u}_2^{\mathrm{e}^\mathrm{T}} \; \mathbf{u}_3^{\mathrm{e}^\mathrm{T}} \; \mathbf{u}_4^{\mathrm{e}^\mathrm{T}}\right]; \qquad \mathbf{u}_i^{\mathrm{e}^\mathrm{T}} = \left[w_i \;\; \varphi_{xi} \;\; \varphi_{yi}\right] \tag{11.226}$$

$$\mathbf{f}^{\mathrm{e}^\mathrm{T}} = \left[\mathbf{f}_1^{\mathrm{e}^\mathrm{T}} \; \mathbf{f}_2^{\mathrm{e}^\mathrm{T}} \; \mathbf{f}_3^{\mathrm{e}^\mathrm{T}} \; \mathbf{f}_4^{\mathrm{e}^\mathrm{T}}\right]; \qquad \mathbf{f}_i^{\mathrm{e}^\mathrm{T}} = \left[F_{zi} \;\; M_{xi} \;\; M_{yi}\right] \tag{11.227}$$

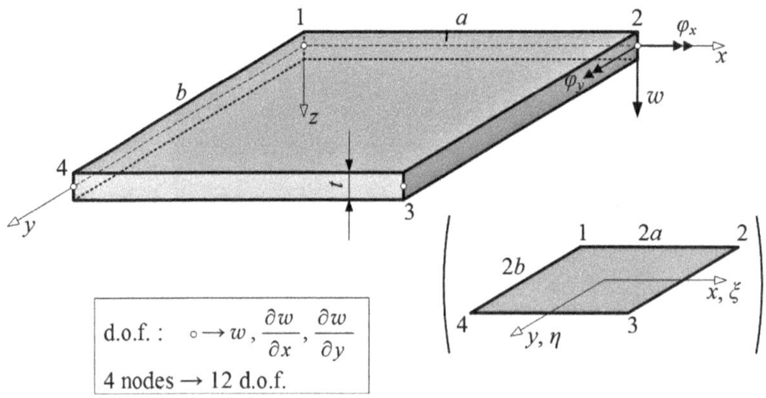

Figure 11.33 A rectangular, thin, flat plate – a non-conforming finite element with 12.d.o.f.

Here, as for rotation, the meaning is (from Equation (11.209)):

$$\varphi_x = \frac{\partial w}{\partial y}; \ \varphi_y = -\frac{\partial w}{\partial x} \tag{11.228}$$

The quantities defined by Equation (11.228) have been previously designated α_x and α_y and are shown in Figure 11.31, where the general theory of plate deformation is presented.

Since the element has 12 d.o.f., the polynomial that can be applied may have 12 generalized coefficients. However, a polynomial of third order has ten terms; that is, two more terms need to be chosen. Some terms from a complete fourth-order polynomial must be omitted. Usually, members x^3y and xy^3 are chosen. It can be seen that, along any x constant direction and y constant direction, the displacement (w) will vary as a cubic function and this is the case for element boundaries. It can therefore be seen that two displacements and two slopes at the ends will be defined uniquely. As such, the displacements and the rotations are:

$$\begin{aligned} w(x, y) = {} & \alpha_1 + \alpha_2 x + \alpha_3 y + \alpha_4 x^2 + \alpha_5 xy + \alpha_6 y^2 + \alpha_7 x^3 + \alpha_8 x^2 y \\ & + \alpha_9 xy^2 + \alpha_{10} y^3 + \alpha_{11} x^3 y + \alpha_{12} xy^3 \end{aligned} \tag{11.229a}$$

$$\begin{aligned} \varphi_x = \frac{\partial w}{\partial y} = {} & \alpha_3 + \alpha_5 x + 2\alpha_6 y + \alpha_8 x^2 + 2\alpha_9 xy + 3\alpha_{10} y^2 + \alpha_{11} x^3 \\ & + 3\alpha_{12} xy^2 \end{aligned} \tag{11.229b}$$

$$\varphi_y = -\frac{\partial w}{\partial x} = -\left(\alpha_2 + 2\alpha_4 x + \alpha_5 y + 3\alpha_7 x^2 + 2\alpha_8 xy + \alpha_9 y^2 + 3\alpha_{11} x^2 y + \alpha_{12} y^3\right) \tag{11.229c}$$

Along the side 1–2 (that is, along the x axis ($y = 0$)), we have:

$$w = \alpha_1 + \alpha_2 x + \alpha_4 x^2 + \alpha_7 x^3 \tag{11.230a}$$

$$\varphi_x = \frac{\partial w}{\partial y} = \alpha_3 + \alpha_5 x + \alpha_8 x^2 + \alpha_{11} x^3 \tag{11.230b}$$

$$\varphi_y = -\frac{\partial w}{\partial x} = -\left(\alpha_2 + 2\alpha_4 x + 3\alpha_7 x^2\right) \tag{11.230c}$$

By inclusion of the values $x = 0$ and $x = a$ in Equation (11.230), the unknowns at nodes 1 and 2 (w_1, w_2, φ_{x1}, φ_{x2}, φ_{y1}, φ_{y2}) should be determined. However, from six equations it is not possible to determine eight unknowns. Further, if only w and φ_y (displacement in the z direction and flexure in the x direction) are considered, then the unknowns that occur can be determined. This means that only continuity (compatibility) with respect to displacement w and to rotation φ_y (flexure in the x direction) is ensured. The similar is valid when the y axis is considered. Therefore, this finite element is a non-conforming element.

The strain–displacement matrix can be calculated as (from Equation (11.8)):

$$\mathbf{B} = \mathbf{d}_\varepsilon \mathbf{N}, \tag{a}$$

where the matrix of shape functions is [3, 10, 13]:

$$\mathbf{N} = [\mathbf{N}_1\ \mathbf{N}_2\ \mathbf{N}_3\ \mathbf{N}_4] \tag{11.231}$$

The element stiffness matrix relating the nodal forces (lateral force and two moments) is:

$$\mathbf{k}^e = \int_{-a}^{a}\int_{-b}^{b} \mathbf{B}^T\mathbf{C}\mathbf{B}\,dx\,dy; \quad \mathbf{C} = \mathbf{C}_b\ (\text{Equation (11.215)}) \tag{11.232}$$

The external forces, in accordance with the degrees of freedom at the node, can be calculated from distributed load $p \equiv q_z(x, y)$. As such, the vector of these forces is (from Equation (11.227)):

$$\mathbf{f}_i^e = \begin{bmatrix} F_{zi} \\ M_{xi} \\ M_{yi} \end{bmatrix} \tag{11.233}$$

The values of these forces when the load is constant $(p \equiv q_z(x, y) = q_0)$ can be found, for example, in [10, 13]. In addition, to achieve C^1 continuity along the element boundaries (element edges), it is necessary to increase the number of degrees of freedom. This can be realized by the addition of the second derivative, $\partial^2 w/\partial x \partial y$ to each node, and now the element has 16 d.o.f. and it is a rectangular conforming element. The displacement is described by an incomplete, two-dimensional polynomial of fourth order with 16 terms, which corresponds to bicubic interpolation.

Triangular, Thin, Flat Plate Finite Elements

Some triangular finite elements are shown in Figure 11.34.

The issue of triangular finite elements used in the analysis of plate bending is covered in more detail in many references, such as [2, 8, 10, 13], so, in this part, we will provide just a brief summary.

The simplest triangular element, with the displacement w at the element vertices and with the normal slopes on the middle side of the element, is shown in Figure 11.34(a). A non-conforming element with nine degrees of freedom is presented in Figure 11.34(b). The displacement here can be described by nine coefficients, that is, using a polynomial of third order. The problem that arises is that a complete polynomial

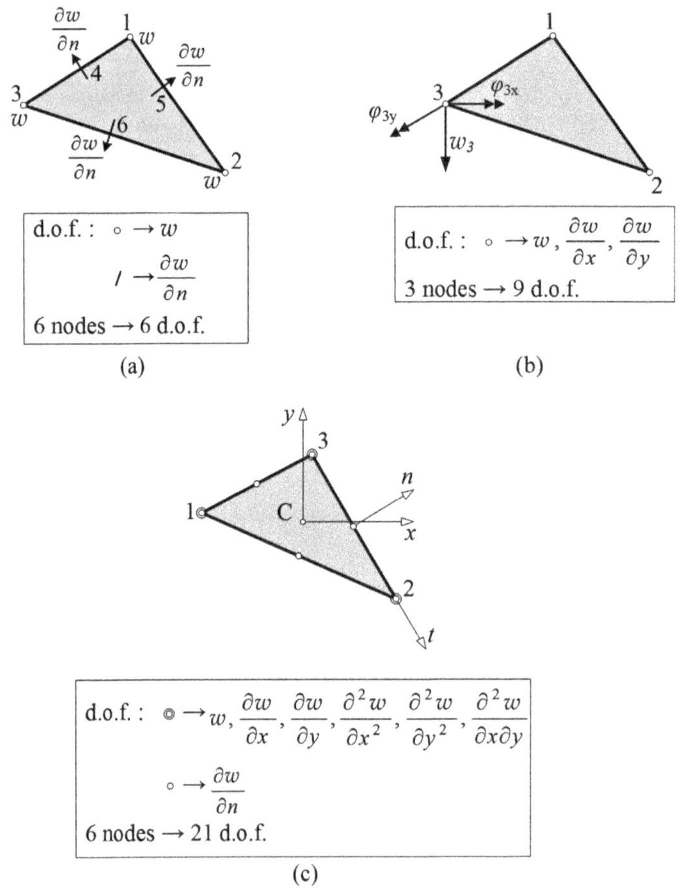

Figure 11.34 Triangular thin, flat plate elements. (a) w at the node points and $\partial w/\partial n$ at the sides of the element; (b) w and $\partial w/\partial x, \partial w/\partial y$ at the node points; (c) w, $\partial w/\partial x, \partial w/\partial y, \partial^2 w/\partial x^2, \partial^2 w/\partial y^2$ at the vertices and $\partial w/\partial n$ at the node points of the sides of the triangle.

of third order has ten terms. The first suggestion may be that one term (xy) can be omitted. The disadvantage of this is that the condition of constant strain in the element is not provided. That is, constant torsion is no longer preserved when the element becomes finally small in size. A second suggestion may be that one member of the polynomial could be a combination, perhaps of $(x^2y + xy^2)$. The problem now is that, while satisfying the continuity of displacement, the slope continuity along the edges of the element is not fulfilled. Element is non-conforming. To solve the problems, a new (internal) node with only one degree of freedom (w), usually placed at the center of the element, is added.

11.2.10 Basics of Dynamic Behavior of Elastic Structures

In this short review of the dynamic behavior of structures, only structures made up of one-dimensional finite elements will be considered. In terms of the dynamic response of structures made of different structural elements as well as in terms of determining the mass matrix for other types of finite elements, readers are referred to known references [2–4, 6, 9, 10, 13].

In previous discussions time as a variable was not included; that is, the problems were considered as static problems. When the displacement of a considered structure varies with time, there are two sets of additional forces. When the load is applied rapidly, inertia forces need to be considered; the second additional set of forces is that due to resistance (friction) opposing the motion. The inertia forces are characterized by ($\ddot{u}$), while the second set of forces may be due to air resistance, microstructural movements or similar, and they are velocity-dependent damping forces.

The static equivalent of the mentioned forces are $(-\rho\ddot{u})$ and $(-\mu\dot{u})$. Here, the inertia forces and the velocity-dependent damping forces are considered as part of the body forces. In this consideration, ($\ddot{u}$) is the acceleration and ρ is the mass density of the element while μ is the damping property parameter of the element. Equation (11.17) defines the finite element equilibrium equation in the case of a static analysis:

$$\mathbf{k}^{\mathrm{e}}\mathbf{u}^{\mathrm{e}} = \mathbf{f}^{\mathrm{e}}, \tag{a}$$

where the vector of finite element nodal forces, according to Equation (11.15) or (11.16), includes the effect of the element body forces ($\mathbf{F}_{\mathrm{V}}$), the effect of the element surface forces ($\mathbf{F}_{\mathrm{S_q}}$) and the concentrated loads ($\mathbf{F}$).

In dynamic analysis, in addition to the above mentioned forces inertia forces and velocity-dependent damping forces need to be added. The effect of distributed body forces is replaced by its equivalent ($\mathbf{F}_{\mathrm{V}} - \rho\ddot{u} - \mu\dot{u}$). As such, Equation (11.11) or Equation (11.15) or Equation (11.16) may now be constructed as:

$$\delta\mathbf{u}^{\mathrm{e^T}}\left[\int_{\mathrm{V}}\mathbf{B}^{\mathrm{T}}\mathbf{C}\mathbf{B}\,\mathrm{d}V\mathbf{u}^{\mathrm{e}} + \int_{\mathrm{V}}\mathbf{N}^{\mathrm{T}}\rho\mathbf{N}\,\mathrm{d}V\frac{\partial^2}{\partial t^2}\mathbf{u}^{\mathrm{e}} + \int_{\mathrm{V}}\mathbf{N}^{\mathrm{T}}\mu\mathbf{N}\,\mathrm{d}V\frac{\partial}{\partial t}\mathbf{u}^{\mathrm{e}}\right]$$
$$= \delta\mathbf{u}^{\mathrm{e^T}}\left[\int_{\mathrm{V}}\mathbf{N}^{\mathrm{T}}\mathbf{q}_{\mathrm{v}}\,\mathrm{d}V + \int_{\mathrm{V}}\mathbf{N}_{\mathrm{S_q}}^{\mathrm{T}}\mathbf{q}_{\mathrm{S_q}}\,\mathrm{d}S_q + \mathbf{F}\right]. \tag{11.234}$$

Equation (11.234) can be written in the form:

$$\mathbf{f}^{\mathrm{e}} = \mathbf{m}^{\mathrm{e}}\ddot{\mathbf{u}}^{\mathrm{e}} + \mathbf{c}^{\mathrm{e}}\dot{\mathbf{u}}^{\mathrm{e}} + \mathbf{k}^{\mathrm{e}}\mathbf{u}^{\mathrm{e}}, \tag{11.235}$$

where:

- $$\mathbf{m}^{\mathrm{e}} = \int_{\mathrm{V}}\mathbf{N}^{\mathrm{T}}\rho\mathbf{N}\,\mathrm{d}V, \tag{11.236}$$

is the mass matrix, while

- $$\mathbf{c}^{\mathrm{e}} = \int_{\mathrm{V}}\mathbf{N}^{\mathrm{T}}\mu\mathbf{N}\,\mathrm{d}V \tag{11.237}$$

is the damping matrix, and

- $$\mathbf{k}^{e} = \int_{V} \mathbf{B}^{T}\mathbf{C}\mathbf{B}\,dV, \tag{b}$$

is the well-known elastic element stiffness matrix (from Equation (11.18)). In a similar way, the structural equation can be written as:

$$\mathbf{M}\ddot{\mathbf{U}} + \mathbf{C}\dot{\mathbf{U}} + \mathbf{K}\mathbf{U} = \mathbf{F}_{R}. \tag{11.238}$$

In Equation (11.238):

$\mathbf{M}$ is the structural mass matrix
$\mathbf{C}$ is the structural damping matrix
$\mathbf{K}$ is the structural stiffness matrix
$\mathbf{F}_R$ is the structural external nodal load (forces) vector
$\mathbf{U}$ is the structural nodal displacements vector
$\dot{\mathbf{U}}$ is the structural velocity vector
$\ddot{\mathbf{U}}$ is the structural acceleration vector

In this case, a "structure" means a finite element assemblage.

Equation (11.238) represents a system of linear differential equations of second order. In Equation (11.238), in general, all the matrices are symmetric (for more information, see [6]) and this second-order equation reduces to the first-order when $\mathbf{M}$ is zero, for example, in the consideration of transient heat-conducting problems. Equation (11.238) may be nonlinear when the stiffness matrices are considered to be dependent on nonlinear material properties or when large deformations are involved. In this very short consideration of dynamic structural behavior, only linear cases, and primarily undamped, free vibrations, will be considered.

The transformation procedure from local to global coordinate system for each of term in Equation (11.235) has already been defined. The element's mass matrix will be transformed in the same way as the stiffness matrix. Equation (11.238) can also be written in the form:

$$\mathbf{F}_{\text{iner}}(t) + \mathbf{F}_{\text{damp}}(t) + \mathbf{F}_{\text{elast}}(t) = \mathbf{F}_{R}(t), \tag{11.239}$$

that is, it can be treated as being derived from a consideration of statics at time t. Thus, in fact, when a static equilibrium exists at time t, including the effect of acceleration-dependent inertia forces and velocity-dependent damping forces is treated as an application of dynamic analysis. In engineering practice, the choice of problem consideration as static (neglecting velocity-dependent forces and acceleration-dependent inertia forces) or dynamic is decided by judgement.

To construct a structural equation of the type shown in Equation (11.238), it is necessary to have previously transformed all of the element matrices into the global coordinate system. These transformations are performed according to the procedures defined by Equations (11.29), (11.31), (11.32) and (11.33). The element's mass matrix is transformed in the same way as the element's stiffness matrix. The procedure for constructing the structural mass matrix is the same as that for the structural stiffness matrix.

11.2.10.1 Mass Matrix of the Finite Element

Mass Matrix of a One-dimensional Finite Element

The finite element mass matrix is defined in accordance with Equation (11.236). With regard to the structure of the mass matrix, $\mathbf{m}^e$, it consists of the same shape functions as the finite element stiffness matrix, $\mathbf{k}^e$. When the same interpolation functions are used to produce the mass matrix as those used to produce the element stiffness matrix, the obtained mass matrix is called a *consistent mass matrix.*

Rod Element

- **Axial displacement**

  This type of finite element is shown in Figure 11.10/11.14.

  Let's assume: the mass density is constant, ρ = const.; A = the cross-sectional area of the element; l = the length of the finite element. The mass matrix is (from Equation (11.236)):

$$\mathbf{m}_{\mathrm{w}}^{\mathrm{e}} = \rho A \int_0^l \mathbf{N}_{\mathrm{w}}^{\mathrm{T}} \mathbf{N}_{\mathrm{w}} \, \mathrm{d}z \tag{a}$$

  Since the shape function matrix is (from Equation (11.58)):

$$\mathbf{N}_{\mathrm{w}} = \left[\mathbf{N}_{\mathrm{w}1} \;\; \mathbf{N}_{\mathrm{w}2}\right] = \left[1 - \frac{z}{l} \;\; \frac{z}{l}\right], \tag{b}$$

  it follows that:

$$\mathbf{m}_{\mathrm{w}}^{\mathrm{e}} = \rho A \int_0^l \begin{bmatrix} 1 - \dfrac{z}{l} \\ \dfrac{z}{l} \end{bmatrix} \left[1 - \frac{z}{l} \;\; \frac{z}{l}\right] \mathrm{d}z \rightarrow \mathbf{m}_{\mathrm{w}}^{\mathrm{e}} = \frac{\rho A l}{6} \begin{bmatrix} 2 & 1 \\ 1 & 2 \end{bmatrix} \tag{11.240}$$

Beam Element

- **Bending in the z–y plane**

  A beam element is shown in Figure 11.15. The mass matrix is (from Equation (11.236)):

$$\mathbf{m}_{\mathrm{v}}^{\mathrm{e}} = \rho A \int_0^l \mathbf{N}_{\mathrm{v}}^{\mathrm{T}} \mathbf{N}_{\mathrm{v}} \, \mathrm{d}z \tag{c}$$

  The shape function matrix is (from Equation (11.97)):

$$\begin{aligned} \mathbf{N}_{\mathrm{v}} &= \left[N_{v1} \;\; N_{v2} \;\; N_{v3} \;\; N_{v4}\right] \\ &= \left[1 - 3\frac{z^2}{l^2} + 2\frac{z^3}{l^3} \;\; -z + 2\frac{z^2}{l} - \frac{z^3}{l^2} \;\; 3\frac{z^2}{l^2} - 2\frac{z^3}{l^3} \;\; \frac{z^2}{l} - \frac{z^3}{l^2}\right]. \end{aligned} \tag{d}$$

$$\mathbf{m}_{v}^{e} = \rho A \int_{0}^{l} \begin{bmatrix} 1 - \frac{3z^2}{l^2} + \frac{2z^3}{l^3} \\ -z + \frac{2z^2}{l} - \frac{z^3}{l^2} \\ \frac{3z^2}{l^2} - \frac{2z^3}{l^3} \\ \frac{z^2}{l} - \frac{z^3}{l^2} \end{bmatrix} \begin{bmatrix} 1 - \frac{3z^2}{l^2} + \frac{2z^3}{l^3} & -z + \frac{2z^2}{l} - \frac{z^3}{l^2} & \frac{3z^2}{l^2} - \frac{2z^3}{l^3} & \frac{z^2}{l} - \frac{z^3}{l^2} \end{bmatrix} \mathrm{d}z.$$

After mathematical calculations, the consistent mass matrix of a basic beam finite element for bending in the y–z principal centroidal plane of inertia is:

$$\mathbf{m}_{v}^{e} = \frac{\rho A l}{420} \begin{bmatrix} 156 & -22l & 54 & 13l \\ -22l & 4l^2 & -13l & -3l^2 \\ 54 & -13l & 156 & 22l \\ 13l & -3l^2 & 22l & 4l^2 \end{bmatrix} \tag{11.241}$$

- **Bending in the z–y plane plus axial deformation**
  In the case of the superposition of bending in the z–y plane (Equation (11.241)) and axial deformation (Equation (11.240)), the mass matrix is:

$$\mathbf{m}_{w,v}^{e} = \frac{\rho A l}{420} \left[\begin{array}{ccc|ccc} 140 & . & . & 70 & . & . \\ . & 156 & -22l & . & 54 & 13l \\ . & -22l & 4l^2 & . & -13l & -3l^2 \\ \hline 70 & . & . & 140 & . & . \\ . & 54 & -13l & . & 156 & 22l \\ . & 13l & -3l^2 & . & 22l & 4l^2 \end{array} \right] = \tag{11.242}$$

$$= \begin{bmatrix} \mathbf{m}_{11}^{e} & \mathbf{m}_{12}^{e} \\ \mathbf{m}_{21}^{e} & \mathbf{m}_{22}^{e} \end{bmatrix}.$$

- **Saint Venant torsion**
  When Saint Venant torsion is considered, the mass matrix is:

$$\mathbf{m}_{\varphi}^{e} = \frac{\rho I_t l}{6} \begin{bmatrix} 2 & 1 \\ 1 & 2 \end{bmatrix} \tag{11.243}$$

11.2.10.2 Free, Undamped Vibrations of Constructions – Eigenvalues

Considerations of structural vibrations are very important problems in the study of structural behavior. However, here, only the determination of free, undamped vibrations of the structure will be considered, in order to simplify the determination of the solutions of a homogeneous system of differential equations of motion (Equation (11.238)). We will also only consider some very simple cases.

In engineering practice, it is commonly said that a structure has free vibration when it is moved from an equilibrium state (steady state) due to a single excitation force and then it is left to move without any external or internal excitation forces. If no forces or damping terms (members) exist in Equation (11.238), that is, if the problem of damping is neglected and in the absence of external forces, this equation reduces to the form:

$$\mathbf{M}\ddot{\mathbf{U}} + \mathbf{K}\mathbf{U} = 0 \tag{11.244}$$

Since this form of vibration represents harmonic vibrations, the particular integral of this equation (a general solution of such an equation) can be:

$$\mathbf{U} = \mathbf{q}e^{i\omega t} \tag{11.244a}$$

The harmonic response can be written in the form:

$$e^{i\omega t} = (\cos\omega t + i\ \sin\omega t) \tag{a}$$

The derivatives are as follows:

$$\begin{aligned} &\frac{\partial}{\partial t}\left(e^{i\omega t}\right) = (-\omega\ \sin\omega t + i\ \cos\omega t) = i\omega(\cos\omega t + i\ \sin\omega t);\ \ i^2 = -1 \\ &\frac{\partial}{\partial t}[i\omega(\cos\omega t + i\ \sin\omega t)] = i\omega(-\omega\ \sin\omega t + i\omega\ \cos\omega t) = -\omega^2 e^{i\omega t} \end{aligned} \tag{b}$$

If the quantity in Equation (b) is inserted into Equation (11.244), it follows that:

$$\begin{aligned} &-\omega^2\mathbf{M}\mathbf{q} + \mathbf{K}\mathbf{q} = 0, \\ &(-\omega^2\mathbf{M} + \mathbf{K})\mathbf{q} = 0. \end{aligned} \tag{11.245}$$

Equation (11.245) defines a general linear eigenvalue problem (or characteristic value problem), where **q** (the vector of amplitudes) and ω (the angular velocity) are unknowns. A non-zero solution for **q** (that is, $\mathbf{q} \neq 0$) can be obtained if:

$$\det\left(-\omega^2\mathbf{M} + \mathbf{K}\right) = 0 \tag{11.246}$$

The determinant in Equation (11.246) has, in general, n values of ω_i^2 and this corresponds to the size, n x n, of the matrices **K** and **M**. In this case, n corresponds to the number of degrees of freedom of the considered structure, ω_i are free vibration frequencies and $\boldsymbol{q_i}$ are the corresponding mode vectors, usually known as eigenvalues and eigenvectors. The solution to a problem such as Equation (11.246) is known as the solution of an eigenproblem. Our intention is to show the procedure for the determination of eigenvalues and eigenvectors. Since the determination of the mentioned quantities is very difficult without using a computer, two very simple examples using a coarse finite element mesh are presented.

Example 11.13.
A) **Simply supported beam – free, undamped transversal vibrations**

Consider the transversal, free, undamped vibration of the simply supported beam in Figure 11.35(a), which is modeled by a single finite element. Determine the eigenvalues, ω (frequencies) using finite element analysis.

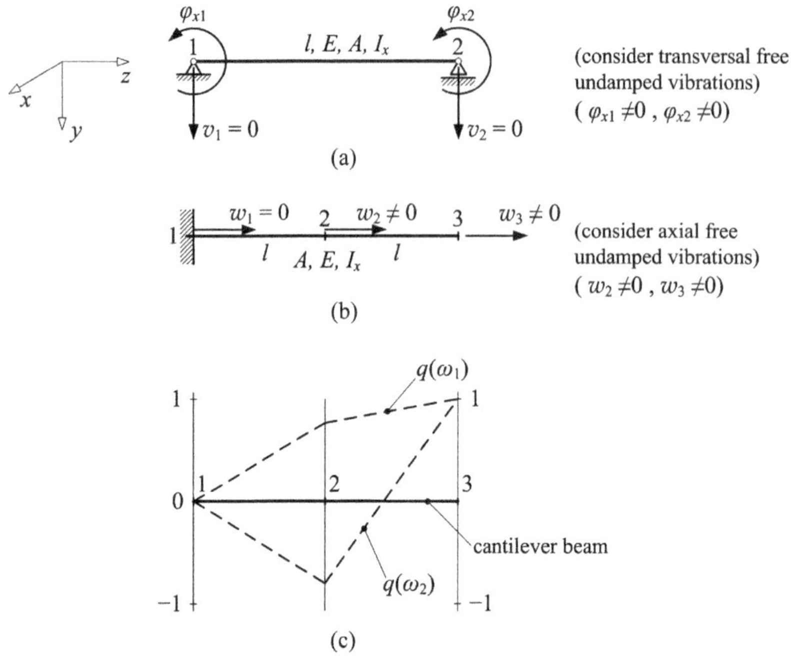

Figure 11.35 Transversal and axial, free, undamped vibration analysis for a beam. (a) Simply supported beam – considered transversal vibrations; (b) cantilever beam – considered axial vibrations; (c) vibration modes for the cantilever beam.

Data:

The geometry and properties of the considered beam are known: A, E, l, I_x.

Solution:

The considered structure (beam) is modeled using only one beam finite element. The matrices **K** and **M** of the system are shown. Using standard procedure, the frequencies are determined. Since transversal vibrations are being considered, *only transversal and rotational degrees of freedom are considered.* As such, as possible displacements we have: v_1, φ_{x1}; v_2, φ_{x2}. In accordance with the boundary conditions, it is clear that the appropriate structural matrices (**K** and **M**) in their final form contain only those terms that relate to the rotations about the x axis since the displacements (v) in the y direction related to the nodes 1 and 2 are constrained.

In accordance with Figure 11.35(a), the structural matrices and vector of nodal displacements are:

$$\mathbf{K} = \begin{bmatrix} \mathbf{K}_{11} & \mathbf{K}_{12} \\ \mathbf{K}_{21} & \mathbf{K}_{22} \end{bmatrix} \quad \mathbf{M} = \begin{bmatrix} \mathbf{M}_{11} & \mathbf{M}_{12} \\ \mathbf{M}_{21} & \mathbf{M}_{22} \end{bmatrix} \tag{a}$$

$$\mathbf{U}=\begin{bmatrix}\mathbf{U}_1\\\mathbf{U}_2\end{bmatrix}=\begin{bmatrix}v_1=0\\\varphi_{x1}\\v_2=0\\\varphi_{x2}\end{bmatrix} \tag{b}$$

Since the structure is discretized using only one finite element and since the local and global coordinate systems coincide, the structural matrices are therefore equal to the element matrices. In this sense, matrices **K** and **M** are (from Equations (11.101) and (11.241)):

$$\mathbf{k}_{\mathrm{v}}^{\mathrm{e}}=EI_x\begin{bmatrix}\frac{12}{l^3}&-\frac{6}{l^2}&-\frac{12}{l^3}&-\frac{6}{l^2}\\-\frac{6}{l^2}&\frac{4}{l}&\frac{6}{l^2}&\frac{2}{l}\\-\frac{12}{l^3}&\frac{6}{l^2}&\frac{12}{l^3}&\frac{6}{l^2}\\-\frac{6}{l^2}&\frac{2}{l}&\frac{6}{l^2}&\frac{4}{l}\end{bmatrix}=\begin{bmatrix}b&-c&-b&-c\\-c&d&c&e\\-b&c&b&c\\-c&e&c&d\end{bmatrix}=\mathbf{K} \tag{c}$$

$$\mathbf{m}_{\mathrm{v}}^{\mathrm{e}}=\frac{\rho Al}{420}\begin{bmatrix}156&-22l&54&13l\\-22l&4l^2&-13l&-3l^2\\54&-13l&156&22l\\13l&-3l^2&22l&4l^2\end{bmatrix}=\mathbf{M} \tag{d}$$

In accordance with the boundary conditions and taking into consideration that transversal vibrations are being considered, the final form of the structural matrices, in accordance with Equations (b), (c) and (d) is:

$$\mathbf{K}=\frac{2EI_x}{l}\begin{bmatrix}2&1\\1&2\end{bmatrix}\quad \mathbf{M}=\frac{\rho Al^3}{420}\begin{bmatrix}4&-3\\-3&4\end{bmatrix} \tag{e}$$

In accordance with Equation (11.245), it follows that:

$$\left(-\omega^2\mathbf{M}+\mathbf{K}\right)\mathbf{q}=0 \tag{f}$$

$$\left(-\omega^2\frac{\rho Al^3}{420}\begin{bmatrix}4&-3\\-3&4\end{bmatrix}+\frac{2EI_x}{l}\begin{bmatrix}2&1\\1&2\end{bmatrix}\right)[\mathbf{q}]=0 \tag{g}$$

Taking:

$$\lambda^2=\frac{\omega^2\rho Al^4}{840EI_x}, \tag{h}$$

in accordance with Equation (11.246), it follows that:

$$\det\left(-\omega^2\mathbf{M}+\mathbf{K}\right)=0 \tag{i}$$

$$\det\begin{bmatrix} 2(1-2\lambda^2) & 1+3\lambda^2 \\ 1+3\lambda^2 & 2(1-2\lambda^2) \end{bmatrix} = 0. \tag{j}$$

By solving the determinant, we obtain:

$$7\lambda^4 - 22\lambda^2 + 3 = 0 \tag{k}$$

and the following solutions for λ_i^2 and ω_i:

$$\lambda_1^2 = 0.142857 \rightarrow \omega_1 = 10.9544\sqrt{\frac{EI_x}{\rho A l^4}};\ \text{from (h)} \tag{l}$$

$$\lambda_2^2 = 3.0 \rightarrow \omega_2 = 50.2\sqrt{\frac{EI_x}{\rho A l^4}};\ \text{from (h)} \tag{m}$$

If one considers eigenvectors, then we find:

$$\begin{aligned} \mathbf{q}_{(\omega_1)} &= [q_1\ q_2]^{\mathrm{T}} \rightarrow q_1 = -q_2, \text{ and} \\ \mathbf{q}_{(\omega_2)} &= [q_1\ q_2]^{\mathrm{T}} \rightarrow q_1 = q_2 \end{aligned} \tag{n}$$

Note:

$$\mathbf{q} = [q_1\ q_2]^{\mathrm{T}} = [\varphi_{x1}\ \varphi_{x2}]^{\mathrm{T}}$$

B) **Cantilever beam – free, undamped axial vibrations**

Consider the axial, free, undamped vibration of the cantilever beam in Figure 11.35(b) which is modeled by two finite elements. Determine the eigenvalues ω_i (frequencies) using finite element analysis.

Data:

The geometry and properties of the considered cantilever beam are known: A, E, l, I_x.

Solution:

The considered cantilever beam structure (that is, there is beam rigidly supported at one end) is modeled using two beam finite elements. The matrices **K** and **M** of the system are shown. Since axial vibrations are being considered, only axial degrees of freedom are considered. In accordance with the boundary conditions, it is clear that the appropriate structural matrices (**K** and **M**) in their final form contain only those terms that are related to the axial displacements (z axis, nodes 2 and 3), since axial displacement in node 1 is constrained.

In accordance with Figure 11.35(b), the structural matrices **M** and **K** and the vector of nodal displacements, **U**, are:

$$\mathbf{K}=\begin{bmatrix} K_{11} & K_{12} & 0 \\ K_{21} & K_{22} & K_{23} \\ 0 & K_{32} & K_{33} \end{bmatrix};\ \mathbf{M}=\begin{bmatrix} M_{11} & M_{12} & 0 \\ M_{21} & M_{22} & M_{23} \\ 0 & M_{32} & M_{33} \end{bmatrix} \tag{a}$$

$$\mathbf{U}=\begin{bmatrix} \mathbf{U}_1 \\ \mathbf{U}_2 \\ \mathbf{U}_3 \end{bmatrix}=\begin{bmatrix} w_1=0 \\ w_2 \\ w_3 \end{bmatrix} \tag{b}$$

Matrices **K** and **M**, defined in Equation (a), taking into account the vector of nodal displacements including the boundary conditions in Equation (b) as well as the quantity in Equation (11.89), are:

$$\mathbf{K}=\begin{bmatrix} 2 & -1 \\ -1 & 1 \end{bmatrix}\frac{EA}{l},\ \mathbf{M}=\frac{\rho Al}{6}\begin{bmatrix} 4 & 1 \\ 1 & 2 \end{bmatrix} \tag{c}$$

Using the quantities defined in Equation (c), Equation (11.245) becomes:

$$\left(-\omega^2\frac{\rho Al}{6}\begin{bmatrix} 4 & 1 \\ 1 & 2 \end{bmatrix}+\frac{EA}{l}\begin{bmatrix} 2 & -1 \\ -1 & 1 \end{bmatrix}\right)[\mathbf{q}]=0 \tag{d}$$

and taking:

$$\lambda^2=\frac{\rho\omega^2 l^2}{6E}, \tag{e}$$

it follows that:

$$\left(-\lambda^2\begin{bmatrix} 4 & 1 \\ 1 & 2 \end{bmatrix}+\begin{bmatrix} 2 & -1 \\ -1 & 1 \end{bmatrix}\right)\begin{bmatrix} q_2 \\ q_3 \end{bmatrix}=0 \tag{f}$$

On the basis of Equation (11.246):

$$\det\begin{bmatrix} 2\left(1-2\lambda^2\right) & -\left(1+\lambda^2\right) \\ -\left(1+\lambda^2\right) & 1-2\lambda^2 \end{bmatrix}=0, \tag{g}$$

and

$$7\lambda^4-10\lambda^2+1=0 \tag{h}$$

Finally, the solutions of Equation (h) are:

$$\begin{aligned} &\lambda_1^2=0,108 \quad \omega_1=0,805\sqrt{\frac{E}{\rho l^2}} \\ &\lambda_2^2=1,32 \quad \omega_2=2,814\sqrt{\frac{E}{\rho l^2}}. \end{aligned} \tag{i}$$

When the solutions from Equations (i) are inserted into Equation (f), the following solutions are obtained. First, for λ_1^2:

$$\left(-0.108\begin{bmatrix}4 & 1\\1 & 2\end{bmatrix}+\begin{bmatrix}2 & -1\\-1 & 1\end{bmatrix}\right)\begin{bmatrix}q_2\\q_3\end{bmatrix}=0 \tag{j}$$

By solving Equation (j), we have:

$$\begin{bmatrix}1.568 & -1.108\\-1.108 & 0.784\end{bmatrix}\begin{bmatrix}q_2\\q_3\end{bmatrix}=0 \tag{k}$$

and, finally, on the basis of Equation (k), we have:

$$q_2=0.707q_3 \tag{l}$$

Setting $q_3 = 1$, it follows that:

$$q_{(\omega_1)}=\begin{bmatrix}q_2\\q_3\end{bmatrix}=\begin{bmatrix}\frac{\sqrt{2}}{2}\\1\end{bmatrix} \tag{m}$$

Similarly, for $\lambda_2^2 = 1.32$, inserted into Equation (f), it follows that:

$$\left(-1.32\begin{bmatrix}4 & 1\\1 & 2\end{bmatrix}+\begin{bmatrix}2 & -1\\-1 & 1\end{bmatrix}\right)\begin{bmatrix}q_2\\q_3\end{bmatrix}=0 \tag{n}$$

By solving Equation (n), we have:

$$\begin{bmatrix}-3.28 & -2.32\\-2.32 & -1.64\end{bmatrix}\begin{bmatrix}q_2\\q_3\end{bmatrix}=0, \tag{o}$$

and finally:

$$q_2=-0.707q_3. \tag{p}$$

Setting $q_3 = 1$, it follows that:

$$q_{(\omega_2)}=\begin{bmatrix}q_2\\q_3\end{bmatrix}=\begin{bmatrix}-\frac{\sqrt{2}}{2}\\1\end{bmatrix}. \tag{r}$$

Note:

$$\mathbf{q}=[q_2\ q_3]^{\mathrm{T}}=[w_2\ w_3]^{\mathrm{T}}. \tag{s}$$

On the basis of the eigenvectors in Equations (l) and (m), for frequencies ω_1 and ω_2, vibration modes (forms) are presented in Figure 11.35(c).

11.3 A Brief Introduction to Nonlinear Analysis of Structural Behavior

Generally speaking, nonlinearity can be categorized as either material nonlinearity or geometric nonlinearity. More about these problems and their solutions can be found in the literature, for example, in [2, 5, 13] and here we present only a few details.

A problem under consideration may include both types of nonlinearity. Since analysis that, besides material nonlinearity, includes geometric nonlinearity may refer to large displacements, large rotations and, for example, large or small strains, there are some typical formulations (total or updated Lagrangian, for example) that can be used, and also some stress and strain measures (engineering stress–strain or other types) involved. Figure 11.36 shows some cases of nonlinearities.

When considering a particular problem, it is necessary to decide to which category the problem belongs, because this will dictate which formulation will be used to describe the actual situation. The fundamental problem in nonlinear analysis is to establish the equilibrium equation for the considered structure that corresponds to the applied load. Assuming that an externally applied load is time-dependent, the equilibrium condition requires that:

$$ {}^{t}\mathbf{F}_R - {}^{t}\mathbf{F}_\sigma = 0 \qquad (11.247) $$

In Equation (11.247), ${}^{t}\mathbf{F}_R$ is a vector of externally applied nodal forces in the considered structure at time t, while ${}^{t}\mathbf{F}_\sigma$ is a vector of nodal forces that corresponds to the element stresses of the considered structure at the same time. This equation expresses the equilibrium of the structure in the current state, that is, in the current deformed geometry and considering the inclusion of all nonlinearities. If the considered problem is a dynamic one, inertia and damping forces need to be included. The equilibrium condition (state) defined by Equation (11.247) must be sustainable through the considered history of load application, that is, through the period of interest. In analysis, usually only stresses and strains at a particular load level or at specific times are required. When analysis refers to the path-dependent nonlinear conditions (geometric, material or phenomenon), the equilibrium relations in Equation (11.247) need to be solved for the time range of interest. It needs

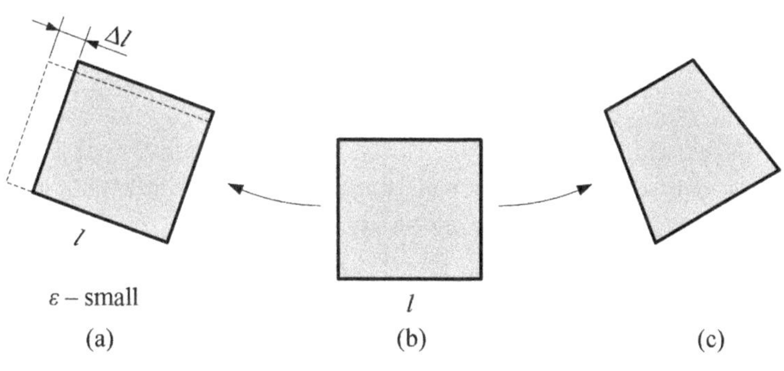

Figure 11.36 Symbolic representation of nonlinearity. (a) The original structure; (b) large displacements, large rotations, small strains, linear or nonlinear material behaviour; (c) large displacements, large rotations, large strains, linear or nonlinear material behavior.

to be said that, in a nonlinear analysis, the equilibrium in Equation (11.247) must be established in the current configuration. In an incremental formulation (step-by-step solution), it is assumed that the solution for the discrete time, t, is known while that for the discrete time $t + \Delta t$ is required, where Δt is the time increment. In this sense, for this discrete time $(t + \Delta t)$, Equation (11.247) can be written as:

$$^{t+\Delta t}\mathbf{F}_R - {}^{t+\Delta t}\mathbf{F}_\sigma = 0 \tag{11.248}$$

An approximation of the displacement $^{t+\Delta}t\mathbf{U}$ may be calculated as outlined below. This procedure is well explained in [2]. On the basis of the known solution at time t, we can write:

$$^{t+\Delta t}\mathbf{F}_\sigma = {}^{t}\mathbf{F}_\sigma + \mathbf{F}_{\sigma},\tag{11.249}$$

In Equation (11.249), $\mathbf{F}_\sigma$ is an increment in the vector of nodal forces that corresponds to the increment in element displacements and stresses from t to $t + \Delta t$. It may be approximated using tangential stiffness matrix, $^{t}\mathbf{K}$, at time t, that is:

$$\mathbf{F}_\sigma = {}^{t}\mathbf{K}\mathbf{U} \tag{11.250}$$

where U is an increment in the vector of nodal displacements which can be calculated on the basis of comparison among Equations (11.248) to (11.250):

$$^{t}\mathbf{K}\mathbf{U} = {}^{t+\Delta t}\mathbf{F}_{\mathrm{R}} - {}^{t}\mathbf{F}_\sigma \tag{11.251}$$

When U is calculated, then:

$$^{t+\Delta t}\mathbf{U} = {}^{t}\mathbf{U} + \mathbf{U} \tag{11.252}$$

11.4 Metal-forming Processes – Brief Overview

11.4.1 Introduction

In addition to its wide applications in the analysis of different types of structures, the finite element method has also found applications in metal-forming processes [17]. We have already discussed finite element modeling of a considered structure and the selection of the type of finite elements in the analysis of different types of structures. When talking about the FEM application to metal-forming processes, there are two types of formulation of the finite element method: the flow formulation and the solid formulation. In the flow formulation, it is assumed that the material that is deformed has a negligible elastic response, while in the solid formulation, elasticity is included, but this formulation in metal-forming problems remains limited.

In metal-forming processes, plastic strains are significant in relation to elastic strains. In accordance with this, it follows that rigid plastic (or rigid viscoplastic) material behavior idealization in metal forming is acceptable, and analysis based on this assumption is known as the flow formulation.

In the solid formulation of the finite element method for metal forming, the material is idealized as elastic–plastic (viscoplastic). This formulation has been based on the use of Prandtl–Reuss equations [17, 25], which are given in the rate form and under the assumption of infinitesimal deformation theory. Such an elastic–plastic formulation has some shortcomings: the large amount of rotation involved in forming rules out infinitesimal analysis and large-deformation analysis also has some difficulties. It has been shown that this problem may be solved when elastic deformation is neglected, that is, when all deformation is taken as plastic deformation in a flow problem.

11.4.2 Classification, Variables and Characteristics of Metal-forming Processes

Metal-forming processes are, in general, plastic deformation processes involving certain methods of forming the metal into the desired shape. The processes may be classified as: *massive metal-forming processes* or *sheet metal-forming processes* [17, 26]. In the literature, the metal-forming processes are also sometimes classified in terms of their dimensionality. In this case, the processes are recognized as *two-dimensional* or *three-dimensional processes* [27]. In three-dimensional processes (massive metal-forming) or bulk deformation processes, the material is deformed in three principal directions and the limiting factor is fracture of the material. In two-dimensional processes (sheet metal-forming processes), the deformation limiting factor is fracture, buckling or thinning.

Within the category of massive metal-forming processes some subcategories can be listed [17]:

- *Rolling* – ring rolling, shape rolling, sheet rolling, cross rolling, gear rolling, surface rolling and so on;
- *Forging* – closed-die forging, open-die forging, isothermal forging, radial forging, forward extrusion forging, backward extrusion forging, coining and so on);
- *Extrusion* – non-lubricated hot extrusion, lubricated direct hot extrusion, hydrostatic extrusion;
- *Drawing* – drawing with rolls, ironing and so on.

Within the category of sheet metal-forming processes can be listed:

- Roll bending;
- Creep forming;
- Vacuum forming;
- Roll forming;
- Deep drawing;
- Stretch forming;
- Bending and straight flanging.

In addition to those mentioned above, the literature also mentions other classifications of metal-forming processes, such as [28]: *hot working processes*, which involve extrusion, rolling, forging, drawing and so on, and *cold working processes*, which involve *bending* (angle/roll/drawing/flanging and so on), *shearing* (blanking, notching, shaving, trimming and so on), *drawing* (wire drawing, stretch forming, bar and tube drawing and so on) and *squeezing* (rolling, sizing, extrusion, peening, cold forging and so on). A hot working process is described as a plastic deformation process in metals that takes place above their recrystallization temperature, while a cold working process is described as a plastic deformation process that takes place below the recrystallization temperature, and usually happens at room temperature.

Processing (hot rolling, cold rolling, quenching and so on) is a means of developing and controlling a material's microstructure. For a particular material (for example, for a defined steel alloy) with a defined chemical composition, some material properties depend on the microstructure and these are known as structure-sensitive properties. These properties include, for example, yield strength and hardness. The yield stress and the workability are very important material variables in the analysis of metal-working processes. When the microstructure is known, yield stress is expressed as a function of strain, temperature and the strain rate and this relationship is obtained by tests.

Formability means the deformation of a material without failure; that is, the ability of a sheet metal to maintain its integrity while being plastically deformed. Similarly, workability may be specified as a technological concept that depends on a material's fracture resistance. Workability can be limited by crack formation, while a stress state that causes high compressive components that impede crack initiation enhances workability.

The selection of a machine for the considered process depends on some characteristics of the machine. The dimensions and surface finish of the product are influenced by process variables (temperature, strain and so on). In massive metal-forming processes, a workpiece undergoes large plastic deformation where significant cross-sectional changes are made and elastic recovery may be negligible; in sheet metal-forming processes, there are also significant changes in shape but not in the cross-section (sheet) and, in some cases, plastic and recoverable elastic deformations may be comparable.

Several basic types of metal-forming processes are shown in Figure 11.37.

Engineering metals (metallic materials), such as steel, aluminum and copper alloys, in their first phase of "shaping" are usually cast into ingots and after that they are again shaped, for example, by hot rolling, into other forms such as billets or similar. After their first shaping phase, they are known as semifinished products. The second stage comprises transformation of the semifinished products by rolling into different structural shapes such as bars (different types), rods, plates, sheets, tubes and so on. These structural shapes may be used in engineering practice as final products or in the manufacture of different engineering elements/members (screws, nuts, pressure vessels and so on).

From Figure 11.37, we can see how certain processes are conducted. So, an open-die forging process (Figure 11.37(a)) does not confine the flow of metal. The required force for this operation can be calculated on the basis of some parameters such as flow stress (yield stress) at specific strain, the friction coefficient and appropriate geometric parameters of the workpiece [28]. In a closed-die forging process (Figure 11.37(b)), shaping of the product is achieved in such a way that a heated (or cold) metal (workpiece, billet) with a controlled volume, is deformed by a punch. This is done in order to fill a die cavity. A similar case is presented in Figure 11.37(c).

Further, Figure 11.37(d), represents (schematically) the possible procedure for a hot or cold rolling process of a strip, sheet or plate. When a hot rolling process is considered, the thickness of the heated metal is reduced by its passing between the two rolls and, in this case, the initial thickness of the plate is higher than in the case of a cold rolling process.

In the extrusion process (Figure 11.37(e)), the cross-section of the workpiece is reduced and the product is formed when the metal (workpiece) is compressively forced to flow through a shaped die. This process can be hot or cold depending on the required force.

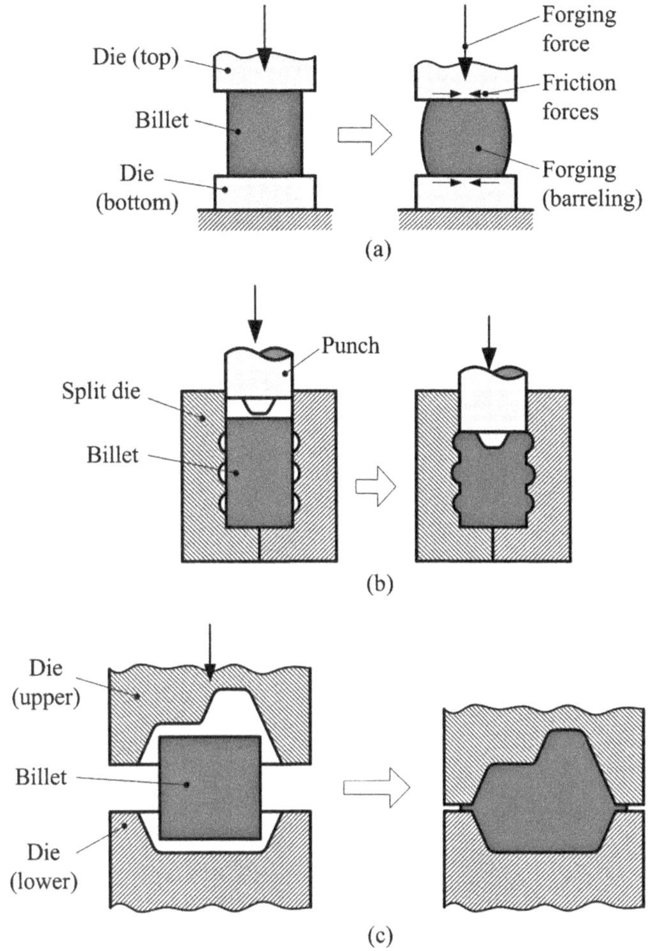

Figure 11.37 Representation of several metal-forming processes. (a) Open-die (hammer) forging – hot metal-forming process, shaping by hammering or pressing; (b) closed-die forging (without flash); (c) closed-die forging (with flash); (d) plate or sheet rolling, hot or cold forming process depending on the thickness of the workpiece – uses rotating rolls; (e) extrusion process; (f) drawing process of a rod or wire.

By pulling through a die (Figure 11.37(f)), using a hot or cold process, the cross-sectional area of a rod (bar) is reduced. This is a drawing process.

Let's just consider now the case shown in Figure 11.37(d), which represents a cold rolling process for a thin sheet. Our consideration is only made from a geometrical standpoint. In accordance with the mentioned figure, and taking the distance between the rolls as "t", the following equation can be written:

$$\overline{O_1O_2} = 2R + t = t_0 + 2R\cos\alpha \tag{a}$$

$$\cos\alpha = \frac{2R + t - t_0}{2R} \tag{b}$$

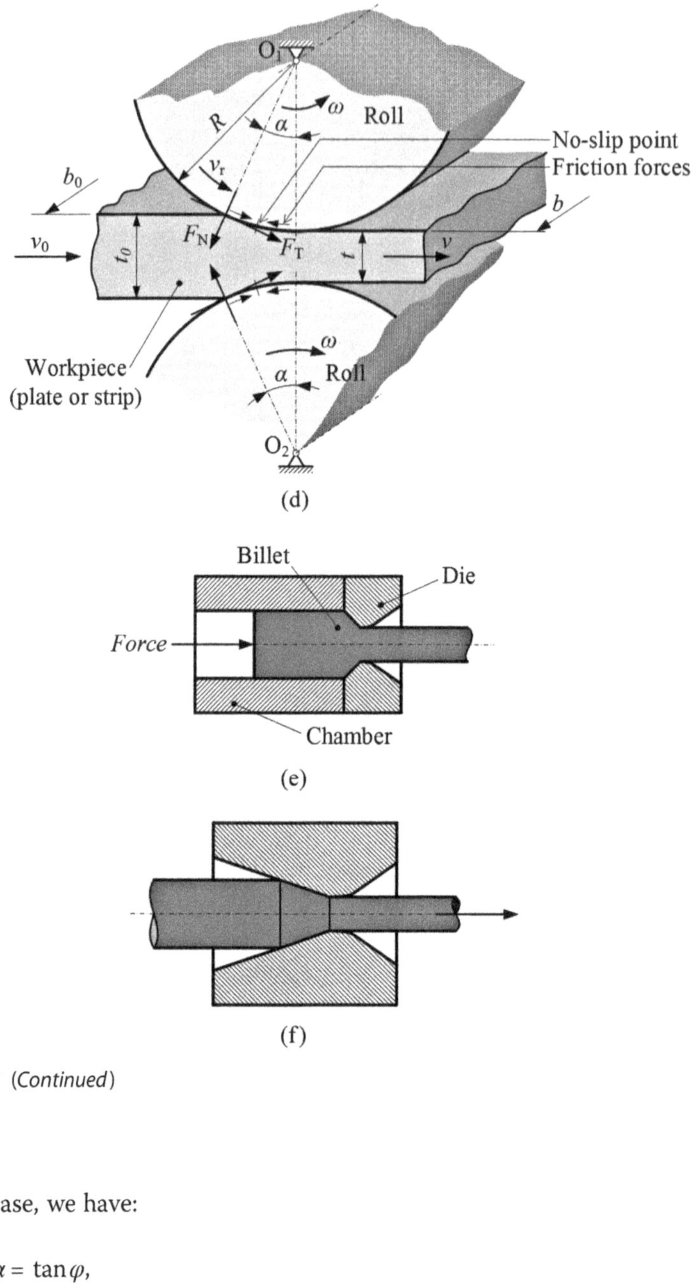

Figure 11.37 *(Continued)*

In a limit case, we have:

$$\tan\alpha = \tan\varphi, \tag{c}$$

where φ is the friction angle. From a mathematical point of view, the following relationship is valid:

$$\cos\alpha = \frac{1}{\sqrt{1+\tan^2\alpha}} \rightarrow \frac{1}{\sqrt{1+\mu^2}} \leq \frac{2R+t-t_0}{2R} \tag{d}$$

Finally,

$$t_0 \leq 2R + t - \frac{2R}{\sqrt{1+\mu^2}}, \tag{e}$$

where μ is the coefficient of friction.

The primary objective in a rolling process (hot or cold) is to reduce the cross-section of the incoming workpiece and obtain the desired section while improving the material properties. In addition to the distinction between hot and cold types, rolling processes may also be classified on the basis of metal flow during the process and the geometry of the rolled product. In this sense, we can mention: (a) *processes with a uniform reduction in thickness and with no change in width* and (b) *processes with a uniform reduction in thickness and an increase in width.* Processes with a non-uniform reduction in cross-section also exist. In any case, the rolling process is a metal-forming process for reducing the cross-sectional area of the workpiece.

Processes of type (a) mentioned above involve rolling of a strip, sheet or thin plate where the deformation is only in the directions of rolling and the thickness of the workpiece but there is no deformation in the width direction. This is so-called *plane strain deformation*. Usually, we say that this is realized when the width of the deformation zone is at least ten times the length of the deformation zone.

Processes of type (b) above involve thick plates. In this case, metal flow occurs in three directions: the rolled material is elongated in the rolled direction, spread in the lateral (width) direction and becomes thinner in the direction of thickness. The difference in deformation between the two mentioned cases may be recognized on the basis of given explanation. In engineering practice, the lateral spread is quite small (seldom greater than 1 to 2 percent) since the radii of the rolls are of the order of 100 times the thickness of the strip and the strip width to thickness ratio is greater than 10.

As stated, rolling processes can be divided into two main categories: hot rolling and cold rolling. When it is possible, the reasons for choosing cold rolling are to obtain better dimensional accuracy and surface finish.

For a successful analysis of a metal-forming process, it is necessary to know the yield stress of the considered metal and this has to be determined experimentally as a function of strain, the strain rate and temperature conditions that occur in the metal-forming process. At hot working temperatures (temperatures above the recrystallization temperature), the yield stress for metallic materials is very much strain-rate-dependent, while at room temperature it is only slightly strain-rate-dependent for most metallic materials.

Let's consider the cold rolling process (Figure 11.37(d)) that refers to a strip of material of initial thickness, t_0. The strip enters the rolls and then passes between the rolls. First it is compressed elastically until it yields, after that it is plastically deformed (work-hardened with an increase in strain) and finally it leaves the area between the rolls and undergoes elastic recovery while a reduced thickness is achieved. We assume plane strain deformation in this process and, in this sense, the material does not spread in the lateral (width) direction; that is, $b_0 = b$. Since the volume of material passing between the rolls remains constant, the velocity of the deformed material in the rolling direction must increase. At the point of no-slip, the velocity of the roll is equal to the velocity of the strip.

The rolling process is widely used in the metal-working industry. The finite element method is also used in modeling and solving rolling problems. Material properties and

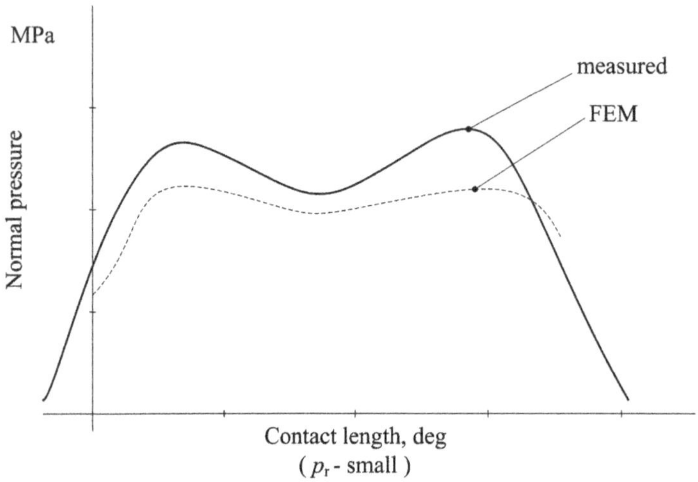

Figure 11.38 Schematic representation of cold sheet-rolling process.

roll geometry as well as workpiece thickness are the process variables involved in this operation. Usually, for a rolling process, comparison is made between the numerically obtained result and experimental data on contact pressure, torque and roll-separating force.

Many authors have dealt with rolling problems [29–31]. In cold sheet rolling processes, it has been observed that the pressure distribution curves have double peaks and there is a pressure drop in the middle of the contact length. This has been observed in processes where a parameter defined as: $p_r = R(t_0 - t)/t_0^2$ is of small value, and this phenomenon is shown, as a sketch, in Figure 11.38. At other values of the parameter p_r, the pressure distribution curves have different shapes.

In a cold rolling process, as represented by Figure 11.37(d), the yield stress characteristic of the metal is essentially independent of the deformation rate, and the frictional force between the rolls and the material is low. In contrast to this, in a hot rolling process, the metal yield stress characteristic is dependent on the stress rate and the frictional force is high.

Usually, the assumptions for a cold rolling process can be summarized as follows:

- The deformation is in the direction of rolling and sheet thickness (that is, there is no deformation in the width direction);
- The coefficient of friction is constant during contact;
- Plane sections perpendicular to the rolling direction remain plane;
- The material is rigid plastic work hardening.

11.4.2.1 Comparison of Hot and Cold Working Processes in Terms of Working Temperature, Shaping Force and Achieved Material Properties

The main characteristic of a metal-forming process is that the metal being processed is plastically deformed to shape it into a desired geometry. For plastic deformation of the material (metal), a force that exceeds the yield strength of the material is required.

In addition to hot and cold working processes, we should also mention the warm working process (or warm forming), which is a forming process carried out above the temperature range of cold working but below the recrystallization temperature of the metal. A warm working process may sometimes be preferred over cold forming because of the lower force required to perform the forming operation.

The processes of hot and cold metal forming differ from each other mainly in the temperature used in the process of forming and, as result of this, in the applied forces needed to perform the considered process. In a hot forming process that is performed above the recrystallization temperature, a considered metal has reduced yield strength, but also no strain hardening will occur as the material is plastically deformed. We need to add here "hot" does not necessarily imply a high temperature for plastic deformation of the material. For example, steels require temperatures of about 1100 °C, while some other materials require lower temperatures. As a consequence of the process, the produced structure has stronger properties in one direction than in another [28]. In most cases, hot working processes form massive products.

When cold forming processes are considered, as already stated, they are defined as processes of plastic deformation of metals at a temperature below the recrystallization temperature. Usually, such metal-forming processes are performed at room temperature. However, sometimes the working can be performed at elevated temperatures and this provides an increase in ductility and a reduction in strength.

In summary, we list below the relative advantages and disadvantages of hot and cold forming processes. The advantages of a cold working process compared to a hot working process are:

- Better product dimension control;
- Better surface finish;
- Better strength properties achieved;
- Heating is not required.

The disadvantages of a cold working process are:

- More power equipment is required;
- The achieved ductility is lower;
- The appearance of undesirable residual stresses;
- Deformation requires higher forces;
- The equipment used must be more powerful;
- Strain hardening occurs.

11.4.3 Basic Settings Related to the Theory of Metal-forming Processes

The considerations that have been made so far to deal with stress, strain and metal-forming processes can be supplemented with some further details. Two forms of describing the deformation of a continuum can be mentioned: the Lagrangian and the Eulerian forms. In the first (Lagrangian) form, the description employs the coordinates, x_i, of a considered particle in the *undeformed* (reference) state as the independent variables. In Eulerian description, as the independent variables one uses the coordinates, x_i, of a considered particle in the *deformed* state. In a case of infinitesimal deformation, products of the derivatives of the displacements may be neglected and, in accordance with this, no distinction needs to be made between these systems. Equation (6.4):

$$\sigma = F/A \tag{a}$$

which represents stress, is usually called true stress or Cauchy stress (the Cauchy stress tensor) and it describes the force intensity (F) acting in the deformed state on the area (A) of the cross-section; that is, A is the cross-sectional area of the element in its deformed state. The action of force intensity (F) in the deformed state on the area (A) measured in the undeformed state, that is, A is now the cross-sectional area in the undeformed state, is known as the first Piola–Kirchoff stress tensor. Further, the quantity defined by Equation (3.1) (or Equation (6.1)):

$$\varepsilon = \varepsilon_z = \frac{l_1 - l_0}{l_0} \tag{b}$$

is called engineering strain.

The strain rate is defined as:

$$\dot{\varepsilon} = \frac{\dot{l}_1}{l_0}. \tag{11.253}$$

11.4.3.1 Strain-rate Tensor and Data Relating to Yield Criteria

Strain-rate Tensor

Similarly to Equation (3.16), here, the strain-rate tensor $[\dot{\varepsilon}_{ij}]$ has the following strain-rate components:

$$\begin{aligned}
&\dot{\varepsilon}_x = \frac{\partial \dot{u}}{\partial x}; \quad \dot{\varepsilon}_y = \frac{\partial \dot{v}}{\partial y}; \quad \dot{\varepsilon}_z = \frac{\partial \dot{w}}{\partial z}; \left(\dot{u} = \frac{\partial u}{\partial t}\right), \ldots \\
&\dot{\varepsilon}_{xy} = \frac{1}{2}\left(\frac{\partial \dot{u}}{\partial y} + \frac{\partial \dot{v}}{\partial x}\right) = \frac{\dot{\gamma}_{xy}}{2}; \\
&\dot{\varepsilon}_{yz} = \frac{1}{2}\left(\frac{\partial \dot{v}}{\partial z} + \frac{\partial \dot{w}}{\partial y}\right) = \frac{\dot{\gamma}_{yz}}{2} \\
&\dot{\varepsilon}_{zx} = \frac{1}{2}\left(\frac{\partial \dot{w}}{\partial x} + \frac{\partial \dot{u}}{\partial z}\right) = \frac{\dot{\gamma}_{zx}}{2}
\end{aligned} \tag{11.254}$$

In Equation (11.254), $\dot{u}, \dot{v}$ and $\dot{w}$ (derivatives of the displacements with respect to time) are velocity components in considered points relating to the x, y and z axes. As the symbol v is usually used as a designation for velocity, instead of the symbols $\dot{u}, \dot{v}, \dot{w}$ (derivatives of displacements), the symbols v_i ($i = x, y, z$) may also be used.

Yield Criteria

The problem of yield criteria was considered in Chapter 5. A particular yield criterion defines the limit of elastic behavior of a body subjected to multiaxial stress, that is, the limit of elasticity of the material from which the body (structure) is made. The yield criterion may be written in the form presented in Chapter 5. Let the yield function be expressed as (from Equation (5.1a)):

$$f(\sigma_e, YS) = 0 \tag{11.255}$$

Since the yield of the material is assumed to be unaffected by hydrostatic pressure, it can be said that yielding depends only on the principal deviatoric components (S_1, S_2, S_3) of the deviatoric stress tensor $S_{ij} = \sigma_{ij} - \sigma_{ij}^0$ from Equations (2.7) and (2.9). The yield criterion is reduced to the form:

$$f(J_2, J_3) = \text{const.} \tag{11.256}$$

where J_i is the invariant of the deviatoric stress tensor.

A critical state of an engineering member is considered to be yielding or fracture. This means that a similar concept is used when yielding or failure by fracture is being considered. Tresca's criterion (the maximum shear stress criterion) and the von Mises criterion (the distortional energy density criterion) are the most used in metal-forming processes. The yield function in the Tresca criterion is given by Equation (5.9):

$$f = \sigma_e - \frac{YS}{2} \tag{a}$$

while effective stress is defined by Equation (5.10a):

$$\begin{aligned} &\sigma_2 - \sigma_3 = \pm YS \\ \sigma_e :\rightarrow\ &\sigma_1 - \sigma_3 = \pm YS \\ &\sigma_1 - \sigma_2 = \pm YS \end{aligned} \tag{b}$$

On the other hand, the yield function in the distortional energy criterion (the von Mises yield criterion) is given by Equation (5.16):

$$f = \sigma_e^2 - YS^2, \tag{c}$$

while effective stress is defined as (from Equation (5.15)):

$$\sigma_e = \sqrt{\frac{1}{2}(\sigma_1 - \sigma_2)^2 + (\sigma_2 - \sigma_3)^2 + (\sigma_3 - \sigma_1)^2} \tag{d}$$

In accordance with Equation (2.40a), the octahedral stress is:

$$\tau^{oct} = \frac{1}{3}\sqrt{(\sigma_1 - \sigma_2)^2 + (\sigma_2 - \sigma_3)^2 + (\sigma_3 - \sigma_1)^2}, \tag{e}$$

or, in accordance with Equation (5.18):

$$\tau^{oct} = \frac{\sqrt{2}}{3} YS. \tag{f}$$

In addition, on the basis of Equation (2.39):

$$\sigma^{oct} = \frac{\sigma_1 + \sigma_2 + \sigma_3}{3} = \frac{\sigma_x + \sigma_y + \sigma_z}{3} = \frac{1}{3} I_1, \tag{g}$$

while on the basis of Equation (2.40a), we have:

$$\tau^{\text{oct}} = \frac{1}{3}\sqrt{(\sigma_1-\sigma_2)^2+(\sigma_2-\sigma_3)^2+(\sigma_3-\sigma_1)^2}$$
$$= \frac{1}{3}\sqrt{(\sigma_x-\sigma_y)^2+(\sigma_y-\sigma_z)^2+(\sigma_z-\sigma_x)^2+6\left(\tau_{xy}^2+\tau_{yz}^2+\tau_{zx}^2\right.}$$
$$= \frac{\sqrt{2}}{3}\left(3I_2-I_1^2\right)^{1/2} \tag{h}$$

The second invariant of the deviatoric stress tensor is (from Equations (2.50b) and (5.17)):

$$J_2 = I_2 - \frac{1}{3}I_1^2$$
$$= -\frac{1}{6}\left[(\sigma_x-\sigma_y)^2+(\sigma_y-\sigma_z)^2+(\sigma_z-\sigma_x)^2+6\left(\tau_{xy}{}^2+\tau_{yz}{}^2+\tau_{zx}{}^2\right)\right] \tag{11.257}$$

On the basis of Equation (5.17b):

$$J_2 = -\frac{3}{2}\left(\tau^{\text{oct}}\right)^2 \tag{11.258}$$

The relationship between the deviatoric stress tensor (S_{ij}), the octahedral shear stress (τ^{oct}) and the second deviatoric invariant (J_2), can be written in the form:

$$J_2 = -\frac{3}{2}\left(\tau^{\text{oct}}\right)^2 = -\frac{1}{2}S_{ij}S_{ij} \tag{11.259}$$

Therefore, it can be shown that:

$$\frac{1}{2}S_{ij}S_{ij} = \frac{1}{2}\left[S_{1j}\cdot S_{1j}+S_{2j}\cdot S_{2j}+S_{3j}\cdot S_{3j}\right]$$
$$= \frac{1}{2}\left[S_{11}^2+S_{21}^2+S_{31}^2+S_{12}^2+S_{22}^2+S_{32}^2+S_{13}^2+S_{23}^2+S_{33}^2\right] \tag{11.260a}$$

In accordance withEquation (2.9), and taking S_{ij} in such a way that the stress components are presented:

$$S_{ij} = \begin{bmatrix} \dfrac{2\sigma_x-\sigma_y-\sigma_z}{3} & \tau_{xy} & \tau_{xz} \\ \tau_{yx} & \dfrac{2\sigma_y-\sigma_z-\sigma_x}{3} & \tau_{yz} \\ \tau_{zx} & \tau_{zy} & \dfrac{2\sigma_z-\sigma_x-\sigma_y}{3} \end{bmatrix}, \tag{11.260b}$$

it follows that:

$$\frac{1}{2}S_{ij}S_{ij} = \frac{1}{2}\left[\left(\frac{2\sigma_x-\sigma_y-\sigma_z}{3}\right)^2+\left(\frac{2\sigma_y-\sigma_z-\sigma_x}{3}\right)^2+\left(\frac{2\sigma_z-\sigma_x-\sigma_y}{3}\right)^2+2\tau_{xy}^2+2\tau_{xy}^2+2\tau_{xy}^2\right]$$
$$= \frac{1}{6}\left[(\sigma_x-\sigma_y)^2+(\sigma_y-\sigma_z)^2+(\sigma_z-\sigma_x)^2+6\left(\tau_{xy}{}^2+\tau_{yz}{}^2+\tau_{zx}{}^2\right)\right] \tag{11.260c}$$

On the basis of Equations (5.16), (5.18) and (11.259), it follows that:

$$S_{ij}S_{ij} = \frac{2}{3}YS^2 = 3(\tau^{oct})^2 \tag{11.261}$$

11.4.3.2 Virtual Work-rate Principle

Earlier in this chapter (within Section 11.2.1.1), the principle of virtual work was described (see Equations (11.10) – (11.16)). With regard to the virtual work-rate principle relating to metal-forming processes, one may state that: in a state of equilibrium of a body, which implies equilibrium of the internal forces (stresses) in the body and the applied surface tractions, the work rate of the stress field (σ_{ij}) inside the deforming body (the virtual work rate of the internal forces) equals the work rate of the surface tractions (F_j) for all velocity fields (virtual work rate of the external forces) that are continuous and continuously differentiable. This principle can be expressed as follows:

$$\int_V \sigma_{ij}\dot{\varepsilon}_{ij} \mathrm{d}V = \int_V F_j v_j \mathrm{d}S \tag{11.262}$$

where: V is the volume; S is the surface; v are velocity components.

If, for example, a two-dimensional problem is being considered, then:

$$\int_V \left(\sigma_x\dot{\varepsilon}_x + \sigma_y\dot{\varepsilon}_y + \tau_{xy}\dot{\gamma}_{xy}\right) dV = \int_S \left(F_x v_x + F_y v_y\right) dS \tag{11.263}$$

The strain rate tensor can be written as follows:

$$\dot{\varepsilon}_{ij} = \frac{1}{2}\left(v_{i,j} + v_{j,i}\right) \tag{11.264}$$

where:

$$\varepsilon_x = \frac{\partial u}{\partial x} \rightarrow \dot{\varepsilon}_x = \frac{\partial \dot{u}}{\partial x} = \frac{\partial v_x}{\partial x}, \ldots..$$

$$\dot{u} = v_x = \frac{\partial u}{\partial t},$$

while the equilibrium conditions were presented in Equation (2.4), if body forces are neglected, it may be written in the following form:

$$\frac{\partial \sigma_{ij}}{\partial x_j} = 0 \rightarrow \sigma_{ij,j} = 0 \tag{11.265}$$

11.4.3.3 The Prandtl–Reuss Equations

In Chapter 2, we showed that the stress tensor can be decomposed into two parts: the *mean (spherical) stress tensor* and the *deviatoric stress tensor*. From Chapter 3, the strain tensor can also be divided into two parts: the spherical tensor and the deviatoric tensor. As stated in those chapters, the spherical tensor describes the hydrostatic pressure and, consequently, the change in volume, while the deviatoric tensor describes the change in shape [9]. Engineering experiments performed in the plastic region, in accordance with the above, show that processes of plastic deformation are characterized by two main properties. The first is the hydrostatic pressure (represented by the hydrostatic part of the stress tensor – the hydrostatic stress tensor), which has negligible influence on plastic strains of isotropic materials, and the second is that the change in the plastic deformed volume of the considered material is equal to zero ($\Delta V^{pl} = 0$). This means that in plastic deformation, that is, in the development of plastic strain, only the deviatoric part of the stress state (the deviatoric stress tensor) participates [32]. As we know, the change in the volume is (from Equations (3.3) and (3.5b)):

$$\varepsilon_V = \frac{\Delta V}{V} = \varepsilon_x + \varepsilon_y + \varepsilon_z \tag{a}$$

When material behavior considerations are carried out in the area behind the yield point, then the assumption about small strains is no longer valid. In such a case, the strain is calculated in relation to the previous length, that is, as logarithmic strain. In analogy with the definition of logarithmic deformation, for logarithmic volumetric strain (logarithmic cubic dilatation) we can write:

$$\mathrm{d}\epsilon_V = \frac{\mathrm{d}V}{V} \rightarrow \epsilon_V = \ln\frac{V}{V_0} \tag{11.266}$$

The criterion of incompressibility, $\Delta V^{pl} = 0$, taking into consideration Equations (11.266) and (a) is:

$$\mathrm{d}\epsilon_V^{pl} = \mathrm{d}\epsilon_x^{pl} + \mathrm{d}\epsilon_y^{pl} + \mathrm{d}\epsilon_z^{pl} = 0. \tag{11.267}$$

Generally, strain can be divided into an elastic and a plastic part (also described as logarithmic):

$$\mathrm{d}\epsilon = \mathrm{d}\epsilon^{el} + \mathrm{d}\epsilon^{pl} \tag{11.268}$$

For the elastic and plastic parts, based on a uniaxial test, the following is valid (z – longitudinal axis):

$$\mathrm{d}\epsilon_x = \mathrm{d}\epsilon_y = -\nu \mathrm{d}\epsilon_z \tag{11.269}$$

Now we can write:

$$\mathrm{d}\epsilon_V^{pl} = \left(1 - 2\nu^{pl}\right)\mathrm{d}\epsilon_z^{pl} = 0 \rightarrow \nu^{pl} = 0.5 \,(\text{in the plastic region}) \tag{11.270}$$

while in the elastic region: $0 < \nu^{el} < 0.5$.

In the theory of structural design, generally, when the design criterion with respect to structural safety and functionality specifies that plastic deformation (strain) is undesirable, it is accepted that (if, for strain, for example, an engineering record is used):

$$\varepsilon_{ij} = \varepsilon_{ij}^{el}; \quad \varepsilon_{ij}^{pl} = 0 \tag{11.271}$$

In contrast to this, in metal-forming processes (rolling, extrusion, forging and so on), where the occurrence of plastic deformation is dominant, it is accepted that:

$$\epsilon_{ij} = \epsilon_{ij}^{el} + \epsilon_{ij}^{pl} \cong \epsilon_{ij}^{pl}; \quad \epsilon_{ij}^{pl} \gg \epsilon_{ij}^{el} \tag{11.272}$$

In other words, so-called logarithmic strain is defined as:

$$\epsilon = \ln\left(\frac{l + \Delta l}{l}\right) = \ln(1 + \varepsilon) \tag{11.273}$$

If this equation is developed in Taylor series $(\varepsilon = \varepsilon - (1/2)\varepsilon^2 + \ldots)$, taking

$$|\varepsilon| \gg \varepsilon^2 \rightarrow \epsilon \cong \varepsilon \tag{11.274}$$

the equality (11.274) can be used in the field of small elastic–plastic strains. When elastic and plastic strains need to be included in the considered process, then the Saint Venant equations discussed earlier are not sufficient. The Saint Venant equations are not complete – that is, they do not respect the fact that the total strain consists of elastic and plastic strains – and can only be used if elastic deformations are negligible. So, in processes where elastic and plastic strains need to be included, the so-called *Prandtl–Reuss equations* can be used and these complement the Saint Venant equations.

For plane strain deformation in an elastic–perfectly plastic material, the stress–strain relations were proposed by Prandtl (1924), while the general form of the equations was given by Reuss (1930) [33]. Since, for Prandtl–Reuss equations in the case of small elastic–plastic strains, elastic deformations are not negligible, the tensor of logarithmic deformation, $d\epsilon_{ij}$, can be replaced by tensor $d\varepsilon_{ij}$. As such, based on Equations (11.268) and (11.274), the tensor of small elastic–plastic strains can be written as:

$$d\varepsilon_{ij} = d\varepsilon_{ij}^{el} + d\varepsilon_{ij}^{pl} \tag{11.275}$$

The elastic part of this tensor follows Hooke's Law written in rheological form (see Equation (8.1)):

$$de_{ij}^{el} = \frac{1}{2\mu} dS_{ij}, \tag{11.276}$$

where μ is the Lame constant, while the plastic part follows the Saint Venant equations:

$$d\varepsilon_{ij}^{pl} = d\lambda \cdot S_{ij}. \tag{11.277}$$

In Equation (11.277), $d\lambda$ is a positive coefficient which changes depending on both load history and the yield point. The Saint Venant equations can be used in metal-forming processes of ideal plastic materials. Taking into consideration that Equation (3.47d) is for plastic strain, and since:

$$d\varepsilon_V^{pl} \equiv d\varepsilon_{kk}^{pl} = 0 \rightarrow d\varepsilon_{ij}^{pl} = de_{ij}^{pl}, \tag{11.278}$$

Equation (11.275) becomes:

$$de_{ij} = de_{ij}^{el} + de_{ij}^{pl}, \tag{11.279}$$

and finally, the stress–strain relationship is:

$$de_{ij} = \frac{1}{2\mu} dS_{ij} + d\lambda S_{ij} \tag{11.280}$$

Equations (11.280) represent the Prandtl–Reuss equations and these need to be solved together (complete deformation state) with Equation (6.43d):

$$d\varepsilon_{kk} = \frac{1-2\nu}{E} d\sigma_{kk} \tag{11.281}$$

For a case involving a uniaxial stress state, parameter $d\lambda$, mentioned in Equation (11.280), can be determined on the basis of the Saint Venant equations (Equations (11.277)) [9], and amounts to:

$$d\lambda = \frac{3}{2}\frac{d\,\epsilon_{e}^{pl}}{\sigma_{e}}, \tag{11.282}$$

where subscript "e" means equivalent. For a uniaxial stress state:

$$\sigma_1 = \sigma,\ \sigma_2 = \sigma_3 = 0. \tag{b}$$

As stated previously, in the development of plastic strain, only the deviatoric part of the stress state is significant. In this case, in accordance with Equation (2.9), we have:

$$S_1 = (2/3)\sigma,\ S_2 = S_3 = -(1/3)\sigma \tag{c}$$

In the plastic area, in accordance with Equations (11.279) and (11.270), the increments in the plastic strains are:

$$d\epsilon_1^{pl} = d\epsilon^{pl},\ d\epsilon_2^{pl} = d\epsilon_3^{pl} = -\frac{1}{2}d\epsilon_1^{pl} \tag{d}$$

A property that follows on the basis of Equations (c) and (d) is:

$$\frac{d\,\epsilon_1^{pl}}{S_1} = \frac{d\,\epsilon_2^{pl}}{S_2} = \frac{d\,\epsilon_3^{pl}}{S_3} = d\lambda \tag{11.283}$$

Comparing Equations (b), (c) and (d) for a case involving a uniaxial stress state, we obtain (from Equation (11.282)):

$$d\lambda = \frac{3}{2}\frac{d\epsilon^{pl}}{\sigma_e}, \tag{e}$$

and this is identical to our previous result from Equation (11.282).

In accordance with the considered stress state, equivalent stress is $\sigma_e = f(\sigma_{ij}) \geq 0$, and it is $\sigma_e = \sigma_1 \geq \sigma_{YS}$ for uniaxial stress. Equivalent stress is a measure which gives an answer as to whether the considered stress state meets the conditions for the appearance of plastic strain, and it must not be dependent on the hydrostatic (mean) part of the stress tensor. As such, parameter $d\lambda$ for a general stress state can be written in the form:

$$d\lambda = \frac{3}{2}\frac{d\,e_e^{pl}}{\sigma_e} \tag{f}$$

which is similar to the form in Equation (11.282), and $\varepsilon = e$, while subscript "e" means equivalent. There are some shortcomings of this formulation (small deformation theory), for example, the presence of large rotations during the process. Early applications in metal-forming problems included those related to applications of the rigid plastic finite element method in the analysis of compression processes. Improvements were made by the inclusion of the effects of the strain rate and temperature on material properties, and this has extended the applications to include analysis in the field of hot working process problems. The FEM has found applications in both two- and three-dimensional metal-forming problems.

11.4.3.4 The Governing Equations of Plastic Deformation

The governing equations for metal-forming problem solution in the field of plastic deformation of rigid plastic (and rigid viscoplastic) materials may be summarized as follows:

- **Equilibrium equations** (Equation (2.4)):

$$\frac{\partial \sigma_{ij}}{\partial x_j} = 0 \tag{11.284a}$$

- **Yield criterion** (based on Equations (5.16) and (11.309)):

$$\sigma_e = \sqrt{\frac{3}{2}}\left(S_{ij}S_{ij}\right)^{1/2} \tag{11.284b}$$

- **Constitutive equations** (in accordance with some previous equations, for example, Equations (11.277) and (11.282):

$$\dot{\varepsilon}_{ij} = \frac{3}{2}\frac{\dot{\varepsilon}_e}{\sigma_e} S_{ij} \tag{11.284c}$$

- **Compatibility conditions** (Equation (11.299)):

$$\dot{\varepsilon}_{ij} = \frac{1}{2}\left(v_{i,j} + v_{j,i}\right) \tag{11.284d}$$

$$\dot{\varepsilon}_x = \frac{\partial \dot{u}}{\partial x} = \frac{\partial v_x}{\partial x}, \ \ldots\ldots$$

11.4.3.5 Shape Functions

In accordance with Equation (11.6), the parameter function, such as displacement (u) in the field of the finite element, is written in the form:

$$\mathbf{u} = \mathbf{N}\,\mathbf{u}^{\mathrm{e}}$$

Here, the velocity field within the finite element can be written as:

$$\dot{\mathbf{u}} = \mathbf{N}\,\dot{\mathbf{u}}^{\mathrm{e}};\ (\rightarrow \mathbf{v} = \mathbf{N}\mathbf{v}^{\mathrm{e}}) \tag{11.285}$$

where the velocity components within the element may be: $\dot{\mathbf{u}} \rightarrow v_x, v_y..$; and the velocity components of the nodes are: $\dot{\mathbf{u}}^{\mathrm{e}} \rightarrow v_x^1, v_y^1 ..; v_x^2, v_y^2....$ ($i = 1, 2, \ldots$ denotes the number of the node). In Equation (11.285), $\mathrm{N} \rightarrow \mathbf{N}_1, \mathbf{N}_2, ..$ is the matrix of shape functions. If, for example, a linear triangular finite element is considered (Figure 11.12(a)), then Equation (11.285) can be written as:

$$\begin{aligned} v_x(L_1, L_2, L_3) &= \sum_i N_i(L_1, L_2, L_3) v_x^i; \quad (i = 1, 2, 3) \\ v_y(L_1, L_2, L_3) &= \sum_i N_i(L_1, L_2, L_3) v_y^i \end{aligned} \tag{11.286}$$

11.4.3.6 Strain-rate Matrix

Similar to the strain–displacement matrix, **B**, which connects strain with element nodal displacements and is shown in Equation (11.8):

$$\boldsymbol{\varepsilon} = \mathbf{d}_\varepsilon \mathbf{N} \mathbf{u}^e = \mathbf{B} \mathbf{u}^e,$$

here, we will first define the strain-rate matrix. Using Equation (11.254), the matrix **B**, which connects the strain rate with element nodal displacement velocities can be written in the form:

$$\dot{\boldsymbol{\varepsilon}} = \mathbf{B} \dot{\mathbf{u}}^e \rightarrow \dot{\boldsymbol{\varepsilon}} = \mathbf{B} \mathbf{v}^e, \tag{11.287}$$

and its number of columns is determined by the number of the element's degrees of freedom. This equation is the differential form of the standard strain–displacement relationship from Equation (11.8). In Section 11.2.1, several approaches for the derivation of the basic finite element equilibrium equation (or finite element stiffness equation) were described. The same is valid for the equilibruim equation of a finite element assemblage (structure). However, in metal-forming processes, the stiffness equation is nonlinear and the solution is obtained iteratively. The final solution is obtained by linearization and using convergence criteria. As for the stiffness equation of the type given in Equation (11.17), force and displacement depend on time.

11.5 The Application of the Finite Element Method in Structural Analysis

The finite element method is a very powerful tool in engineering analysis. It is widely used in mechanical engineering, civil engineering and many other disciplines. Static and dynamic analyses of structures, earthquake analysis and the analysis of metal-forming processes are just some of the possible applications. Discretization of a considered continuum or process as well as appropriate finite element selection, the establishment of the finite element equation and solution of the structural global equation are all parts of a finite element solution. Some of the possible formulations of finite element equations have already been mentiond in Section 11.2.1.

Some of the main advantages of the use of the finite element method are: the ability to model the structure or process appropriately, and the capture of data relating to stresses, strains, velocities, shapes, changes in temperature and so on.

11.5.1 One-dimensional Finite Elements: Finite Element Analysis of Truss Structure Deformation

Example 11.14.
Consider the truss structure in Figure 11.39, made of steel bars (tubes). It is necessary to calculate the diameter (D) of the tubes [34]. The considered structure is subjected to the force F at the point C. The vertical displacement at the node C has not to exceed the allowable displacement.

Data:
The material of the structure is steel S355JO (ASTM A709 Gr50) with the following properties: Ultimate tensile strength: $\sigma_{\mathrm{m}} = 550$ MPa; yield strength: $\sigma_{0.2} = 385$ MPa, modulus of elasticity: $E = 210$ GPa.

Load: Force, $F = 20$ kN.

The geometry is shown in Figure 11.39; the lengths of the bars are:
$l_1 = 4$ m, $l_2 = 4.472$ m, $l_3 = 2$ m, $l_4 = 2.236$ m, $l_5 = 2$ m, $l_6 = 2.236$ m, $l_7 = 2.236$ m.

Wall thickness of the bars, $t = 3$ mm.

Allowable stress and vertical displacement: $\sigma_{\mathrm{all}} = 200$ MPa, $V_{C,\mathrm{all}} = 5$ mm.

Note: The longitudinal axis of the bars is designated "x". In Figure 11.39, the global and local coordinate systems are shown.

Solution:
The structural equation is of the form:

$$\mathbf{F}_R = \mathbf{K}\mathbf{U} \tag{a}$$

$$\begin{bmatrix} \mathbf{F}_1 = \mathbf{F}_A \\ \mathbf{F}_2 \\ \mathbf{F}_3 = \mathbf{F}_C = \mathbf{F} \\ \mathbf{F}_4 \\ \mathbf{F}_5 = \mathbf{F}_B \end{bmatrix} = \begin{bmatrix} \mathbf{K}_{11} & \mathbf{K}_{12} & 0 & 0 & \mathbf{K}_{15} \\ \mathbf{K}_{21} & \mathbf{K}_{22} & \mathbf{K}_{23} & \mathbf{K}_{24} & \mathbf{K}_{25} \\ 0 & \mathbf{K}_{32} & \mathbf{K}_{33} & \mathbf{K}_{34} & \mathbf{K}_{35} \\ 0 & \mathbf{K}_{42} & \mathbf{K}_{43} & \mathbf{K}_{44} & \mathbf{K}_{45} \\ \mathbf{K}_{51} & \mathbf{K}_{52} & 0 & \mathbf{K}_{54} & \mathbf{K}_{55} \end{bmatrix} \begin{bmatrix} \mathbf{U}_1 \\ \mathbf{U}_2 \\ \mathbf{U}_3 \\ \mathbf{U}_4 \\ \mathbf{U}_5 \end{bmatrix} \tag{b}$$

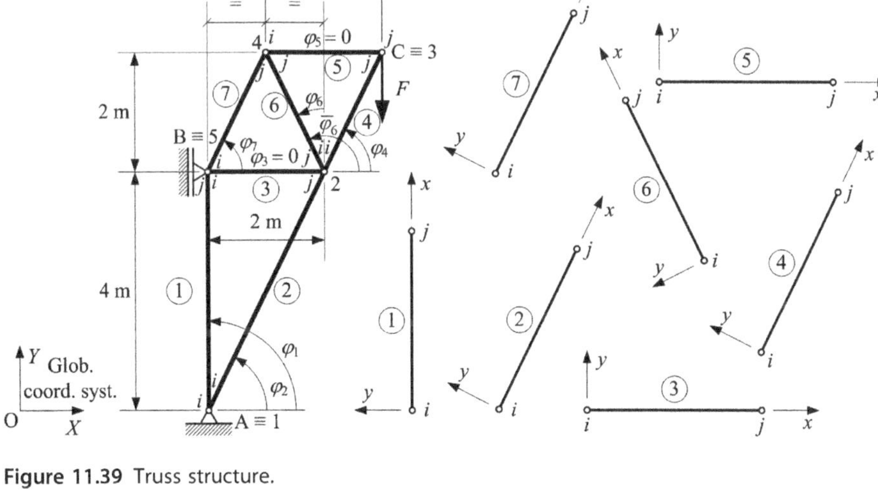

Figure 11.39 Truss structure.

Submatrices of the global equation are as follows. Since the considered structure is a planar structure, the stiffness matrix is of the known form.

The angles shown in Figure 11.39 are: $\varphi_2 = 63.43° = \varphi_4 = \varphi_7$; $\varphi_6 = 26.56°$; $\bar{\varphi}_6 = 90° + 26.56°$.

$$\mathbf{K}_{11} = \bar{\mathbf{k}}_{ii}^{1} + \bar{\mathbf{k}}_{ii}^{2} = \mathbf{t}_0^{1^{\mathrm{T}}} \mathbf{k}_{ii}^{1} \mathbf{t}_0^{1} + \mathbf{t}_0^{2^{\mathrm{T}}} \mathbf{k}_{ii}^{2} \mathbf{t}_0^{2}$$

$$= \begin{bmatrix} \cos\varphi_1 & -\sin\varphi_1 \\ \sin\varphi_1 & \cos\varphi_1 \end{bmatrix} \frac{EA}{l_1} \begin{bmatrix} 1 & 0 \\ 0 & 0 \end{bmatrix} \begin{bmatrix} \cos\varphi_1 & \sin\varphi_1 \\ -\sin\varphi_1 & \cos\varphi_1 \end{bmatrix}$$

$$+ \begin{bmatrix} \cos\varphi_2 & -\sin\varphi_2 \\ \sin\varphi_2 & \cos\varphi_2 \end{bmatrix} \frac{EA}{l_2} \begin{bmatrix} 1 & 0 \\ 0 & 0 \end{bmatrix} \begin{bmatrix} \cos\varphi_2 & \sin\varphi_2 \\ -\sin\varphi_2 & \cos\varphi_2 \end{bmatrix}$$

$$= \begin{bmatrix} 0 & -1 \\ 1 & 0 \end{bmatrix} \frac{EA}{l_1} \begin{bmatrix} 1 & 0 \\ 0 & 0 \end{bmatrix} \begin{bmatrix} 0 & 1 \\ -1 & 0 \end{bmatrix} + \begin{bmatrix} 0.447 & -0.894 \\ 0.894 & 0.447 \end{bmatrix} \frac{EA}{l_2} \begin{bmatrix} 1 & 0 \\ 0 & 0 \end{bmatrix} \begin{bmatrix} 0.447 & 0.894 \\ -0.894 & 0.447 \end{bmatrix}$$

$$= EA \begin{bmatrix} 0.0447 & 0.0894 \\ 0.0894 & 0.4289 \end{bmatrix} \rightarrow \left(\frac{N}{m}\right)$$

$$\mathbf{K}_{22} = \bar{\mathbf{k}}_{jj}^{2} + \bar{\mathbf{k}}_{ii}^{4} + \bar{\mathbf{k}}_{ii}^{6} + \bar{\mathbf{k}}_{jj}^{3} = \mathbf{t}_0^{2^{\mathrm{T}}} \mathbf{k}_{jj}^{2} \mathbf{t}_0^{2} + \mathbf{t}_0^{4^{\mathrm{T}}} \mathbf{k}_{ii}^{4} \mathbf{t}_0^{4} + \mathbf{t}_0^{6^{\mathrm{T}}} \mathbf{k}_{ii}^{6} \mathbf{t}_0^{6} + \bar{\mathbf{k}}_{jj}^{3}$$

$$= \begin{bmatrix} \cos\varphi_2 & -\sin\varphi_2 \\ \sin\varphi_2 & \cos\varphi_2 \end{bmatrix} \frac{EA}{l_2} \begin{bmatrix} 1 & 0 \\ 0 & 0 \end{bmatrix} \begin{bmatrix} \cos\varphi_2 & \sin\varphi_2 \\ -\sin\varphi_2 & \cos\varphi_2 \end{bmatrix}$$

$$+ \begin{bmatrix} \cos\varphi_4 & -\sin\varphi_4 \\ \sin\varphi_4 & \cos\varphi_4 \end{bmatrix} \frac{EA}{l_4} \begin{bmatrix} 1 & 0 \\ 0 & 0 \end{bmatrix} \begin{bmatrix} \cos\varphi_4 & \sin\varphi_4 \\ -\sin\varphi_4 & \cos\varphi_4 \end{bmatrix}$$

$$+ \begin{bmatrix} -\sin\varphi_6 & -\cos\varphi_6 \\ \cos\varphi_6 & -\sin\varphi_6 \end{bmatrix} \frac{EA}{l_6} \begin{bmatrix} 1 & 0 \\ 0 & 0 \end{bmatrix} \begin{bmatrix} -\sin\varphi_6 & \cos\varphi_6 \\ -\cos\varphi_6 & -\sin\varphi_6 \end{bmatrix} + \frac{EA}{l_3} \begin{bmatrix} 1 & 0 \\ 0 & 0 \end{bmatrix}$$

$$= \frac{EA}{l_2} \begin{bmatrix} 0.2 & 0.4 \\ 0.4 & 0.8 \end{bmatrix} + \frac{EA}{l_4} \begin{bmatrix} 0.2 & 0.4 \\ 0.4 & 0.8 \end{bmatrix} + \frac{EA}{l_6} \begin{bmatrix} 0.2 & -0.4 \\ -0.4 & 0.8 \end{bmatrix} + \frac{EA}{l_3} \begin{bmatrix} 1 & 0 \\ 0 & 0 \end{bmatrix} =$$

$$= EA \begin{bmatrix} 0.724 & 0.0894 \\ 0.0894 & 0.895 \end{bmatrix} \rightarrow \left(\frac{N}{m}\right)$$

$$\mathbf{K}_{33} = \bar{\mathbf{k}}_{jj}^{4} + \bar{\mathbf{k}}_{jj}^{5} = \mathbf{t}_0^{4^{\mathrm{T}}} \mathbf{k}_{jj}^{4} \mathbf{t}_0^{4} + \mathbf{k}_{jj}^{5}$$

$$= \frac{EA}{l_4} \begin{bmatrix} 0.2 & 0.4 \\ 0.4 & 0.8 \end{bmatrix} + \frac{EA}{l_5} \begin{bmatrix} 1 & 0 \\ 0 & 0 \end{bmatrix} = EA \begin{bmatrix} 0.589 & 0.179 \\ 0.179 & 0.358 \end{bmatrix} \rightarrow \left(\frac{N}{m}\right)$$

$$\mathbf{K}_{44}=\bar{\mathbf{k}}_{jj}^{7}+\bar{\mathbf{k}}_{jj}^{6}+\bar{\mathbf{k}}_{ii}^{5}=\mathbf{t}_0^{7^{\mathrm{T}}}\mathbf{k}_{jj}^{7}\mathbf{t}_0^{7}+\mathbf{t}_0^{6^{\mathrm{T}}}\mathbf{k}_{jj}^{6}\mathbf{t}_0^{6}+\mathbf{k}_{jj}^{5}$$

$$=EA\begin{bmatrix}0.679 & 0\\ 0 & 0.716\end{bmatrix}\rightarrow\left(\frac{N}{m}\right)$$

$$\mathbf{K}_{55}=\bar{\mathbf{k}}_{jj}^{1}+\bar{\mathbf{k}}_{ii}^{3}+\bar{\mathbf{k}}_{ii}^{7}=\mathbf{t}_0^{1^{\mathrm{T}}}\mathbf{k}_{jj}^{1}\mathbf{t}_0^{1}+\mathbf{t}_0^{6^{\mathrm{T}}}\mathbf{k}_{jj}^{6}\mathbf{t}_0^{6}+\mathbf{k}_{jj}^{5}$$

$$=EA\begin{bmatrix}0.5984 & 0.179\\ -0.179 & 0.608\end{bmatrix}\rightarrow\left(\frac{N}{m}\right)$$

$$\mathbf{K}_{12}=\bar{\mathbf{k}}_{ij}^{2}=\mathbf{t}_0^{2^{\mathrm{T}}}\mathbf{k}_{ij}^{2}\mathbf{t}_0^{2}=EA\begin{bmatrix}-0.0447 & -0.089\\ -0.089 & -0.179\end{bmatrix}\rightarrow\left(\frac{N}{m}\right)$$

$$\mathbf{K}_{21}=\mathbf{K}_{12}^{\mathrm{T}}$$

$$\mathbf{K}_{23}=\bar{\mathbf{k}}_{ij}^{4}=\mathbf{t}_0^{4^{\mathrm{T}}}\mathbf{k}_{ij}^{4}\mathbf{t}_0^{4}=EA\begin{bmatrix}-0.0894 & -0.179\\ -0.179 & -0.357\end{bmatrix}\rightarrow\left(\frac{N}{m}\right)$$

$$\mathbf{K}_{32}=\mathbf{K}_{23}^{\mathrm{T}}$$

$$\mathbf{K}_{24}=\bar{\mathbf{k}}_{ij}^{6}=\mathbf{t}_0^{6^{\mathrm{T}}}\mathbf{k}_{ij}^{6}\mathbf{t}_0^{6}=EA\begin{bmatrix}-0.0894 & 0.179\\ 0.179 & -0.357\end{bmatrix}\rightarrow\left(\frac{N}{m}\right)$$

$$\mathbf{K}_{42}=\mathbf{K}_{24}^{\mathrm{T}}$$

$$\mathbf{K}_{52}=\bar{\mathbf{k}}_{ij}^{3}=\frac{EA}{l_3}\begin{bmatrix}-1 & 0\\ 0 & 0\end{bmatrix}=EA\begin{bmatrix}-0.5 & 0\\ 0 & 0\end{bmatrix}\rightarrow\left(\frac{N}{m}\right)$$

$$\mathbf{K}_{25}=\mathbf{K}_{52}^{\mathrm{T}}$$

$$\mathbf{K}_{43}=\bar{\mathbf{k}}_{ij}^{5}=\frac{EA}{l_5}\begin{bmatrix}-1 & 0\\ 0 & 0\end{bmatrix}=EA\begin{bmatrix}-0.5 & 0\\ 0 & 0\end{bmatrix}\rightarrow\left(\frac{N}{m}\right)$$

$$\mathbf{K}_{34}=\mathbf{K}_{43}^{\mathrm{T}}$$

$$\mathbf{K}_{15}=\bar{\mathbf{k}}_{ij}^{1}=\mathbf{t}_0^{1^{\mathrm{T}}}\mathbf{k}_{ij}^{1}\mathbf{t}_0^{1}=\begin{bmatrix}0 & -1\\ 1 & 0\end{bmatrix}\frac{EA}{l_1}\begin{bmatrix}-1 & 0\\ 0 & 0\end{bmatrix}\begin{bmatrix}0 & 1\\ -1 & 0\end{bmatrix}$$

$$=EA\begin{bmatrix}0 & 0\\ 0 & -0.25\end{bmatrix}\rightarrow\left(\frac{N}{m}\right)$$

$$\mathbf{K}_{51}=\mathbf{K}_{15}^{\mathrm{T}}$$

$$\mathbf{K}_{54}=\bar{\mathbf{k}}_{ij}^{7}=\mathbf{t}_0^{7^{\mathrm{T}}}\mathbf{k}_{ij}^{1}\mathbf{t}_0^{7}=EA\begin{bmatrix}-0.0894 & -0.179\\ -0.179 & -0.357\end{bmatrix}\rightarrow\left(\frac{N}{m}\right)$$

$$\mathbf{K}_{45}=\mathbf{K}_{54}^{\mathrm{T}}$$

The structural (global) equation (b) is:

$$\begin{bmatrix} F_{A_{x(1)}} \\ F_{A_{y(1)}} \\ F_{2_x}=0 \\ F_{2_y}=0 \\ F_{3_x}=0 \\ F_{3_y}=-F \\ F_{4_x}=0 \\ F_{4_y}=0 \\ F_{B_{x(5)}} \\ F_{By_{(5)}}=0 \end{bmatrix} = EA \begin{bmatrix} 0.0447 & 0.0894 & -0.0447 & -0.089 & 0 & 0 & 0 & 0 & 0 & 0 \\ 0.0894 & 0.4289 & -0.089 & -0.179 & 0 & 0 & 0 & 0 & 0 & -0.25 \\ -0.0447 & -0.089 & 0.724 & 0.0894 & -0.0894 & -0.179 & -0.089 & 0.179 & -0.5 & 0 \\ -0.089 & -0.179 & 0.0894 & 0.895 & -0.179 & -0.357 & 0.179 & -0.357 & 0 & 0 \\ 0 & 0 & -0.0894 & -0.179 & 0.580 & 0.179 & -0.5 & 0 & 0 & 0 \\ 0 & 0 & -0.179 & -0.357 & 0.179 & 0.358 & 0 & 0 & 0 & 0 \\ 0 & 0 & -0.089 & 0.179 & -0.5 & 0 & 0.679 & 0 & -0.0894 & -0.179 \\ 0 & 0 & 0.179 & -0.357 & 0 & 0 & 0 & 0.716 & -0.179 & -0.357 \\ 0 & 0 & -0.5 & 0 & 0 & 0 & -0.0894 & -0.179 & 0.5894 & 0.179 \\ 0 & -0.25 & 0 & 0 & 0 & 0 & -0.179 & -0.357 & 0.179 & 0.608 \end{bmatrix} \begin{bmatrix} U_{1x}=0,(A) \\ U_{1y}=0,(A) \\ U_{2x} \\ U_{2y} \\ U_{3x} \\ U_{3y} \\ U_{4x} \\ U_{4y} \\ U_{5x}=0,\ (B) \\ U_{5y},\ (B) \end{bmatrix} \tag{c}$$

From Equation (c), the known components (data) of the vector $\mathbf{U}_\mathrm{r}$ can be written as:

$$\mathbf{U}_\mathrm{r} = \begin{bmatrix} U_{1_x}=0,(A) \\ U_{1_y}=0,(A) \\ U_{5_x}=0,(B) \end{bmatrix} \tag{d}$$

In accordance with Equation (11.48):

$$\mathbf{F}_\mathrm{m} = \mathbf{K}_\mathrm{mm}\mathbf{U}_\mathrm{m} \tag{e}$$

where $\mathbf{F}_\mathrm{m}$ are the known forces (known components of vector $\mathbf{F}_\mathbf{R}$) while $\mathbf{U}_\mathrm{m}$ are unknown displacement components (unknown components of vector $\mathbf{U}$). On the basis of Equation (c), taking into account the boundary conditions, Equation (e) now is:

$$\begin{bmatrix} F_{2_x}=0 \\ F_{2_y}=0 \\ F_{3_x}=0 \\ F_{3_y}=-F \\ F_{4_x}=0 \\ F_{4_y}=0 \\ F_{5_y,(B)}=0 \end{bmatrix} = EA \begin{bmatrix} 0{,}724 & 0{,}0894 & -0{,}0894 & -0{,}179 & -0{,}089 & 0{,}179 & 0 \\ 0{,}0894 & 0{,}895 & -0{,}179 & -0{,}357 & 0{,}179 & -0{,}357 & 0 \\ -0{,}0894 & -0{,}179 & 0{,}589 & 0{,}179 & -0{,}5 & 0 & 0 \\ -0{,}179 & -0{,}357 & 0{,}179 & 0{,}358 & 0 & 0 & 0 \\ -0{,}089 & 0{,}179 & -0{,}5 & 0 & 0{,}679 & 0 & -0{,}179 \\ 0{,}179 & -0{,}357 & 0 & 0 & 0 & 0{,}716 & -0{,}357 \\ 0 & 0 & 0 & 0 & -0{,}179 & -0{,}357 & 0{,}608 \end{bmatrix} \begin{bmatrix} U_{2_x} \\ U_{2_y} \\ U_{3_x} \\ U_{3_y} \\ U_{4_x} \\ U_{4_y} \\ U_{5_y,(B)} \end{bmatrix} \tag{f}$$

Taking:

$$E\left(\frac{\mathrm{N}}{\mathrm{m}^2}\right),\ F(\mathrm{N}),\ A(\mathrm{m}^2),\ U(\mathrm{m})$$

$$F = 20000\,(\mathrm{N}),\ E = 210\cdot 10^9\left(\frac{\mathrm{N}}{\mathrm{m}^2}\right),$$

and solving system (f) (by computer), it follows that:

$$\mathbf{U}_{\mathrm{m}} = \begin{bmatrix} U_{2_x} \\ U_{2_y} \\ U_{3_x} \\ U_{3_y} \\ U_{4_x} \\ U_{4_y} \\ U_{5_y,(B)} \end{bmatrix} = \begin{bmatrix} 9,116\cdot10^{-8} \\ -8,2688\cdot10^{-7} \\ 1,4297\cdot10^{-6} \\ -1,7599\cdot10^{-6} \\ 1,334\cdot10^{-6} \\ -3,383\cdot10^{-7} \\ 1,94\cdot10^{-7} \end{bmatrix} \frac{\mathrm{m}^3}{A} \tag{g}$$

On the basis of Equation (g) and taking the constraint related to point C as: $V_C = V_3 = 5\text{ mm} = V_{C,\text{ all}}$, it follows that:

$$\left|U_{3y}\right| = \frac{1.76\cdot10^{-6}\cdot10^{9}\mathrm{mm}^3}{D\pi t} = 5\,\mathrm{mm}$$

$$D = \frac{1.76\cdot10^{3}}{\pi\cdot3\cdot5} = 37.4\,\mathrm{mm}$$

Adopt: $D = 40$ mm.

The cross-sectional area is now $A = 376.8\text{ mm}^2$.

The reactive forces can be obtained from Equation (11.49) and Equation (c):

$$\mathbf{F}_{\mathrm{r}} = \mathbf{K}_{\mathrm{rm}}\cdot\mathbf{U}_{\mathrm{m}} \tag{h}$$

In Equation (h), $\mathbf{F}_{\mathrm{r}}$ are unknown components of the vector $\mathbf{F}_{\mathrm{R}}$, that is, reactive forces. Equations (h) are:

$$\begin{aligned} F_{\mathrm{A_x}} &= \left(-0,0447U_{2_x} - 0,0894U_{2_y}\right)EA \\ F_{\mathrm{A_y}} &= \left(-0,0894U_{2_x} - 0,1789U_{2_y} - 0,25U_{5_y}\right)EA \\ F_{\mathrm{B_x}} &= \left(-0,5U_{2_x} - 0,0894U_{4_x} - 0,179U_{4_y} + 0,179U_{5_y}\right)EA \end{aligned} \tag{i}$$

Taking into account the values of the members of the stiffness matrix being adapted to a certain number of decimal places, the calculated values of the reactive forces are somewhat different from the exact values. On the basis of Equations (i) and obtained values of displacements (g), the reactive forces are:

$$F_{A_x} \approx 15000\text{ N}$$

$$F_{A_y} \approx 20000\text{ N}$$

$$F_{B_x} \approx -15000\text{ N}$$

11.5.2 Two-dimensional Finite Elements: *J* Integral Calculation

Example 11.15.

(This example is courtesy of Associate Professor Goran Vukelic, DSc, Faculty of Maritime Studies, University of Rijeka.)

When dealing with growing cracks, J integral values can be determined for a range of crack extensions (Δa) and plotted in the form of J-resistance curves that are usually obtained experimentally following standardized procedures [35]. Extensive experimental procedures can be complemented successfully by numerical methods, for example, the finite element (FE) method, as in this example.

Here, in order to determine J values for a cracked disc compact tension (DCT) specimen, an experimental single-specimen test method following the elastic unloading compliance technique was numerically simulated. As we know, this test method uses measured crack mouth opening displacement to estimate the growing crack size. Collected J values are then plotted against crack extension to obtain J-resistance curves.

The DCT specimen usually used in the experimental procedure can be modeled using the FEM. Two-dimensional FE models of specimens with initial a/b ($b = 50$ mm) ratios of 0.25, 0.5 and 0.75 are developed (Figure 11.40). Material behavior is considered to be of the multilinear, isotropic hardening type. FE specimen models are meshed with 8-node isoparametric quadrilateral elements. In nonlinear fracture problems, the mesh must be refined in the regions where yielding occurs in order to capture the existing high deformation gradients. A quasi-static load is imposed on specimens to simulate the compliance procedure of the single-specimen test method. The node-releasing technique is used to simulate crack propagation, so it is necessary to ensure that the size of the finite elements (that is, the distance between the nodes that are to be released) corresponds to the desired measure of crack extension, Δa.

Using the advantage of symmetry through the crack, only half of the specimen has to be modeled. The stress analysis is run and the resulting sets of stresses and strains in integration points are recorded along with nodal displacements, as shown in Figure 11.41.

The stress analysis results in the integration points of selected elements are used to calculate the contribution of each element to the J integral path that encloses the crack tip.

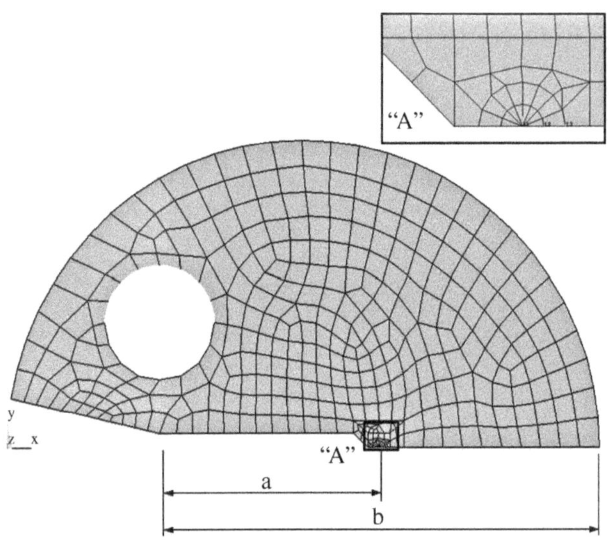

Figure 11.40 Finite element mesh of a DCT specimen with detail of the crack tip.

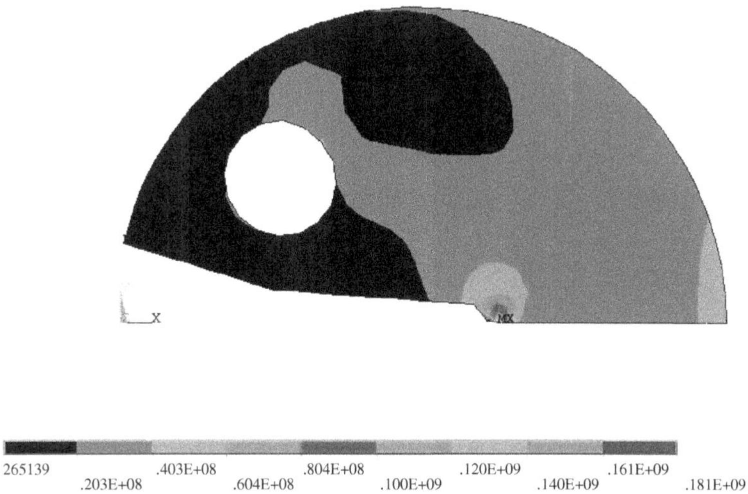

Figure 11.41 Von Mises stress distribution for a cracked DCT specimen.

By summing the values of dJ, the contribution of each element to J (the total value of the J integral) is calculated. For this purpose, the following equation is used [36]:

$$J = \sum_{p=1}^{np} W_{Gp} P_{Ip}\left(\xi_p, \eta_p\right), \tag{11.288}$$

where W_{Gp} is the Gauss weighting factor, np is the number of integration points and P_{Ip} is the integrand evaluated at each integration point, p:

$$P_{Ip} = \left\{ \frac{1}{2}\left[\sigma_{xx}\frac{\partial u_x}{\partial x} + \sigma_{xy}\left(\frac{\partial u_x}{\partial y} + \frac{\partial u_y}{\partial x}\right)\frac{\partial u_x}{\partial x} + \sigma_{yy}\frac{\partial u_y}{\partial y}\right]\frac{\partial y}{\partial \eta} - \left[\left(\sigma_{xx}n_1 + \sigma_{xy}n_2\right)\frac{\partial u_x}{\partial x} + \left(\sigma_{xy}n_1 + \sigma_{yy}n_2\right)\frac{\partial u_y}{\partial x}\right]\sqrt{\left(\frac{\partial x}{\partial \eta}\right)^2 + \left(\frac{\partial y}{\partial \eta}\right)^2} \right\} \tag{11.289}$$

Material properties, including yield points for considered materials, were experimentally determined [37, 38]. Since J integral values may vary in the vicinity and away from the crack tip because of high stress gradients, three different paths around the crack tip are defined and their average value is taken as final (Figure 11.42).

Numerically obtained J values are presented versus crack growth size (Δa) for two different materials (steels 1.4122 and 1.0553) on a DCT specimen with varying crack size in Figure 11.43. From the resulting values of the J integral, it can be observed that higher a/b ratios correspond with lower critical toughness values of materials and vice versa, which matches previous results of other authors [39].

Although this method cannot be a substitute for experimental procedures, it can act as a starting point in the investigation of J values for new materials, thereby reducing the time needed to perform experiments. Such results can be of great help in the process of material selection during the design of structures.

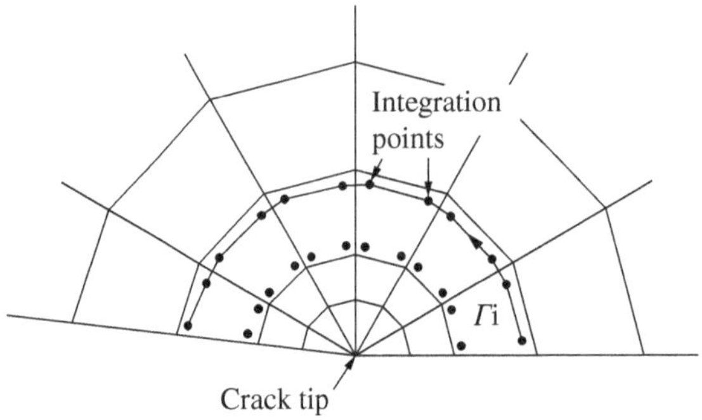

Figure 11.42 Single J integral path Γ_i surrounding the crack tip through finite element integration points.

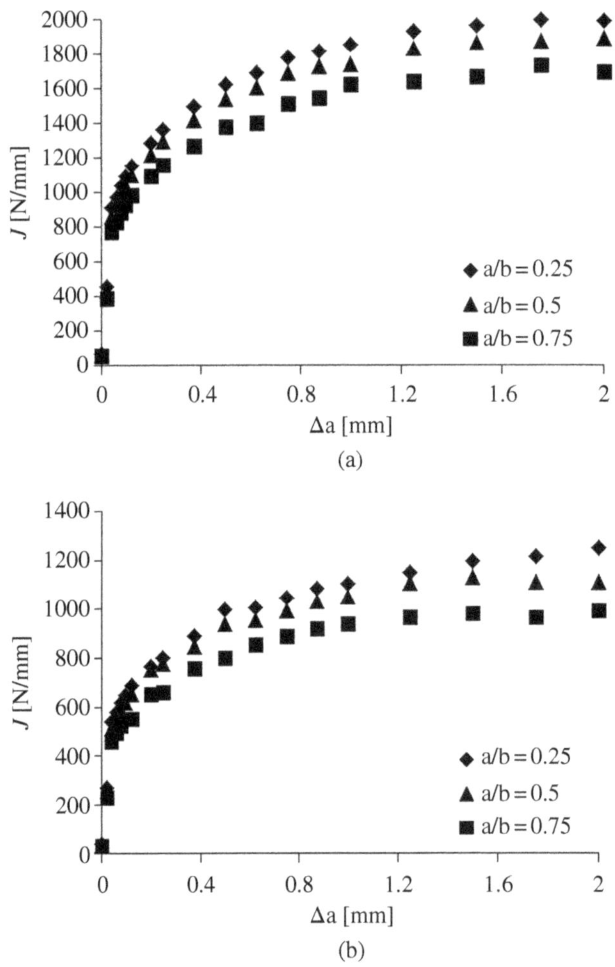

Figure 11.43 Numerically predicted J values using finite element models of a DCT specimen for steels: (a) 1.4122; (b) 1.0553.

11.5.3 Special Two-dimensional Finite Elements in Shear Stress Analysis

11.5.3.1 Introduction

For some often-used engineering elements subjected to torque, it is desirable to carry out an analysis of shear stresses in an easy way. As such, special finite elements, such as quadrilateral finite elements of arbitrary shape (general elements) can be used. Using a number of these general quadrilateral elements, the cross-sectional area of a considered engineering member can be discretized. Shear stress analysis on the considered cross-sectional area of the engineering element (let's say a beam or shaft) using these quadrilateral finite elements is based on their axial nodal displacements. The necessary details relating to these quadrilateral elements are outlined below, before an example is shown in Section 11.5.3.2. In section 11.5.3 stress and strain analyses relate to free warping.

Consider the quadrilateral finite element of arbitrary shape shown in Figure 11.44.

In accordance with the theory of elasticity, for an engineering member (let's consider a shaft) subjected to torque, the shear stresses are defined as follows (from Equation (6.28)):

$$\begin{aligned} \tau_{zx} &= G\gamma_{zx} = G\theta\left(-y+\frac{\partial \psi}{\partial x}\right) = \frac{\partial \Phi}{\partial y} = -G\theta y = G\left(\frac{\partial w}{\partial x} - \theta y\right) \\ \tau_{zy} &= G\gamma_{zy} = G\theta\left(x+\frac{\partial \psi}{\partial y}\right) = -\frac{\partial \Phi}{\partial x} = G\theta x = G\left(\frac{\partial w}{\partial y} + \theta x\right) \end{aligned} \tag{11.290}$$

As stated above, the cross-sectional area of the considered engineering member (shaft, beam) is discretized by a number of proposed general quadrilateral finite elements. For every finite element, displacement in the field of the finite element, in accordance with Saint Venant's theory, is described by Equations (6.26) and (6.27):

$$\left.\begin{aligned} u &= u(y,z) = -yz\theta \\ v &= v(x,z) = xz\theta \\ w &= w(x,y) = \theta\psi(x,y) \end{aligned}\right\} \tag{a}$$

where the twisting angle per unit length is (from Equation (6.20)):

$$\theta = \frac{d\varphi}{dz} = const \tag{b}$$

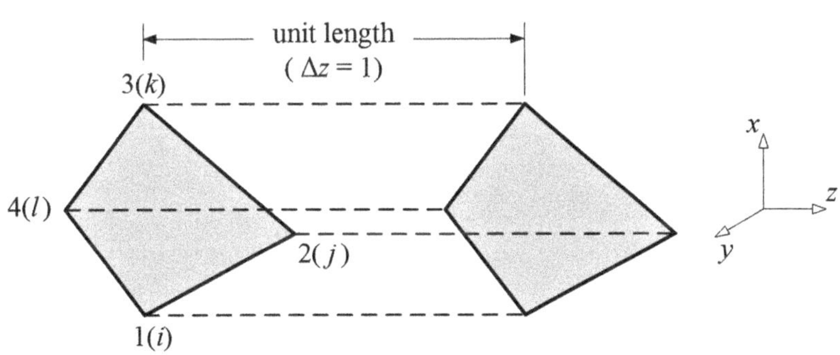

Figure 11.44 A quadrilateral finite element of arbitrary shape.

Displacement in the field of the finite element (displacement along the z axis due to torque) and displacements in the nodes of the finite element are (from Equation (11.3)):

$$w(x, y) = \mathbf{a}(x, y)\boldsymbol{\alpha}$$

and

$$u = \mathbf{a}\cdot\boldsymbol{\alpha} \rightarrow w(x, y) = \mathbf{a}(x, y)\boldsymbol{\alpha},\ \ \mathbf{u}^{\mathrm{e}} = \mathbf{a}^{\mathrm{b}}\boldsymbol{\alpha} \rightarrow \mathbf{w}^{\mathrm{e}} = \mathbf{a}^{\mathrm{b}}\boldsymbol{\alpha} \tag{11.291}$$

In this consideration, **a** is used to designate the polynomial matrix and w is used to designate the axial displacement. In accordance with previously explained theory, the relationship between the displacement in the field of the finite element and displacements in the nodes of the element, is (from Equation (11.6)) [25, 40]:

$$w(x,y) = \mathbf{a}(x,y)\mathbf{a}(x,y)^{\mathrm{b}-1}\mathbf{w}^{\mathrm{e}} = \mathbf{N}\mathbf{w}^{\mathrm{e}} \tag{11.292}$$

Using the following notation:

$$\mathbf{a}_x = \frac{\partial \mathbf{a}}{\partial x},\ \mathbf{a}_y = \frac{\partial \mathbf{a}}{\partial y} \tag{11.293}$$

the shear stress components are (from Equation (11.290)):

$$\tau_{zx} = G\left(\mathbf{a}_x\,\mathbf{a}^{\mathrm{b}^{-1}}\mathbf{w}^{\mathbf{e}} - \theta y\right),\ \tau_{zy} = G\left(\mathbf{a}_y\,\mathbf{a}^{\mathrm{b}^{-1}}\mathbf{w}^{\mathbf{e}} + \theta x\right), \tag{11.294}$$

where “b” means boundary.

The total potential (total potential energy) can be written in the form:

$$\Pi = \iint \frac{1}{2}G\left[\left(\frac{\partial w}{\partial x} - \theta y\right)^2 + \left(\frac{\partial w}{\partial y} + \theta x\right)^2\right]dx\,dy - M_t\theta. \tag{11.295}$$

Applying the necessary mathematical calculation:

$$\delta\Pi = 0 \rightarrow \delta\Pi_{(w,\theta)} = 0 \tag{11.296}$$

and taking (from Equation (6.27)):

$$w = \theta\psi(x,y) \tag{c}$$

we obtain:

$$\begin{aligned}\delta\Pi_{(w,\theta)} = &\iint G\theta\delta\theta\left[x^2 + y^2 + x\frac{\partial \psi}{\partial y} - y\frac{\partial \psi}{\partial x}\right]dx\,dy - M_t\delta\theta + \\ &+\iint G\left[\frac{\partial w}{\partial x}\frac{\partial \delta w}{\partial x} + \frac{\partial w}{\partial y}\frac{\partial \delta w}{\partial y} + \theta x\frac{\partial \delta w}{\partial y} - \theta y\frac{\partial \delta w}{\partial x}\right]dx\,dy = 0\end{aligned} \tag{11.297}$$

If the value of the torsional moment of inertia from Equation (6.28a) is considered, then Equation (11.297) can be separated into two equations, the first of which is the equilibrium equation:

$$1.\ \ (G\theta I_t - M_t)\delta\theta = 0, \tag{11.298a}$$

while comparing the second equation:

$$2. \quad \iint G\left[\frac{\partial w \partial}{\partial x} \frac{\delta w}{\partial x}+\frac{\partial w \partial}{\partial y} \frac{\delta w}{\partial y}+\theta x \frac{\partial}{\partial y} \frac{\delta w}{}-\theta y \frac{\partial}{\partial x} \frac{\delta w}{}\right] d x d y=0 \tag{11.298b}$$

with Equation (11.295), it is clear that:

$$\delta \Pi_{(w)}=\delta U_{(w)}=0 \tag{11.299}$$

The strain energy for a finite element of unit length $(\Delta z=1)$ can be written in the form:

$$U_{\mathrm{e}}=\iint \frac{1}{2} G\left[\left(\mathbf{a}_x \mathbf{a}^{\mathrm{b}^{-1}} \mathbf{w}^{\mathrm{e}}-\theta y\right)^2+\left(\mathbf{a}_y \mathbf{a}^{\mathrm{b}^{-1}} \mathbf{w}^{\mathrm{e}}+\theta x\right)^2\right] \mathrm{d} x \mathrm{d} y \tag{11.300}$$

and applying the condition in Equation (11.299) to Equation (11.300), it follows that:

$$\mathbf{k}^{\mathrm{e}} \mathbf{w}^{\mathrm{e}}=\mathbf{f}^{\mathrm{e}} \theta, \tag{11.301}$$

where the following values for the element stiffness matrix and the load vector are obtained:

$$\mathbf{k}_{\mathrm{e}}=G \mathbf{a}^{\mathrm{b}^{-\mathrm{T}}} \iint\left(\mathbf{a}_x^{\mathrm{T}} \mathbf{a}_x^{\mathrm{T}}+\mathbf{a}_y^{\mathrm{T}} \mathbf{a}_y^{\mathrm{T}}\right) \mathbf{a}^{\mathrm{b}^{-1}} \mathrm{d} x \mathrm{d} y \tag{11.302}$$

$$\mathbf{f}^{\mathrm{e}}=-G \mathbf{a}^{\mathrm{b}^{-\mathrm{T}}} \iint\left(\mathbf{a}_y^{\mathrm{T}} x-\mathbf{a}_x^{\mathrm{T}} y\right) \mathrm{d} x \mathrm{d} y \tag{11.303}$$

Let the polynomial matrix for the general quadrilateral finite element be prescribed in the form:

$$\mathbf{a}=\mathbf{a}(x, y)=\left[\begin{array}{llll}1 & x & y & x y\end{array}\right]. \tag{11.304}$$

The area of the finite element is:

$$A_e=\frac{1}{2}\left(x_i y_{j l}+x_j y_{k i}+x_k y_{l j}+x_l y_{i k}\right); \quad y_{r s}=y_r-y_s \tag{11.305}$$

On the basis of Equations (11.302) and (11.303), the finite element stiffness matrix and the load vector are:

$$\mathbf{k}^{\mathrm{e}}=G \mathbf{a}^{\mathrm{b}^{-\mathrm{T}}}\left[\begin{array}{cccc}0 & 0 & 0 & 0 \\ 0 & A_{\mathrm{e}} & 0 & S_x \\ 0 & 0 & A_{\mathrm{e}} & S_y \\ 0 & S_x & S_y & I_x+I_y\end{array}\right] \mathbf{a}^{\mathrm{b}^{-1}} \tag{11.306}$$

$$\mathbf{f}^{\mathrm{e}}=-G \mathbf{a}^{\mathrm{b}^{-\mathrm{T}}}\left[\begin{array}{c}0 \\ -S_x \\ S_y \\ I_y-I_x\end{array}\right] \tag{11.307}$$

In Equations (11.349) and (11.350)e: A_{e} represents the area of the finite element, S_x, S_y represent the first moments of the area and I_x, I_y represent the axial moments of inertia.

11.5.3.2 Application of General Quadrilateral Finite Elements

Example 11.16.

Consider the shaft with polygonal profile loaded by a torque of magnitude M_t (Nm), as shown in Figure 11.45 [40]. The cross-sectional area is modeled (discretized) by general quadrilateral finite elements (that is, by quadrilateral finite elements of arbitrary shape). The stress distribution, τ_{tot}, is presented in Figure 11.46.

This is the so-called three-sided configuration. It is suitable for high-torque and high-stress applications. The geometry can be selected in accordance with, for example, polygonal profile P3G, and the shear stress distribution, τ_{tot}, is shown in Figure 11.46.

Since the finite element method is applicable in various fields of engineering, it is always of practical significance to apply an appropriate procedure when problem solving [41].

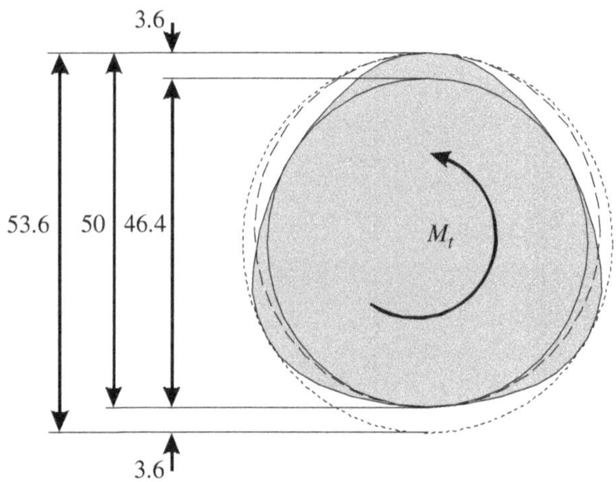

Figure 11.45 Geometry of a shaft with polygonal profile (P3) subjected to torque, M_t.

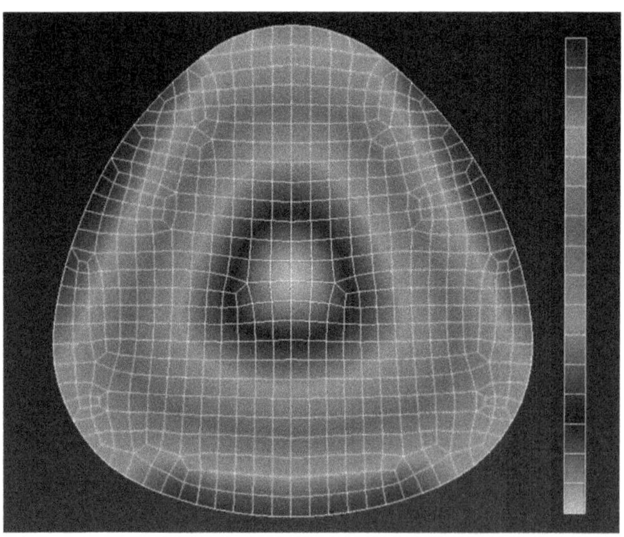

Figure 11.46 Shear stress distribution, τ_{tot}. Maximum shear stress on the three edges of the profile – at the points nearest to the center of the shear; minimum shear stress: at the very center of the shear.

References

1 Bathe, K. J. (1996) *Finite Element Procedures*, Prentice-Hall, Inc., New Jersey.
2 Bathe, K. J. (1982) *Finite Element Procedures in Engineering Analysis*, Prentice-Hall, New Jersey.
3 Gallagher, R. H. (1975) Finite Element Analysis *(Fundamentals)*, Prentice-Hall, New Jersey.
4 Reddy, J. N. (1993) *An Introduction to the Finite Element Method*, McGraw-Hill, Inc., New York.
5 Chrisfield, M. A. (1995) *Non-linear Finite Element Analysis of Solids and Structures*, Vol. 1, John Wiley & Sons, New York.
6 Zienkiewicz, O. C. and Taylor, R. L. (2000) *Finite Element Method*, Vol. 1, The Basis, 5th edition, Butterworth-Heinemann, London.
7 Zienkiewicz, O. C. and Taylor, R. L. (2000) *The Finite Element Method*, Vol. 1, Basic Formulation and Linear Problem, McGraw-Hill Book Company, New York.
8 Cook, R. D. (1995) *Finite Element Modeling for Stress Analysis*, John Wiley & Sons, USA.
9 Prelog, E. (1975) *Finite Element Method* (in Slovenian). Faculty for Architecture, Civil Engineering and Geodesy, Ljubljana.
10 Sorić, J. (2004) *Finite Element Method* (in Croatian), Golden Marketing-Tehnička knjiga, Zagreb.
11 Hughes, T. J. R. (2000) *The Finite Element Method, Linear Static and Dynamic Finite Element Analysis*, Dover Publications, Inc., New York.
12 Zienkiewicz, O. C. and Taylor, R. L. (1991) *The Finite Element Method*, Vol. 2, Solid and Fluid Mechanics, Dynamics and Non-Linearity, 4th edition, McGraw-Hill Book Company, Berkshire.
13 Zienkiewicz, O. C. and Taylor, R. L. (2000) *Finite Element Method*, Vol. 2, Solid Mechanics, 5th edition, Butterworth-Heinemann, New York.
14 Bathe, K. J. (1990) *Finite-Elemente Methoden*, Springer-Verlag, New York.
15 Brnić, J. and Čanađija, M. (2009) *Analysis of Deformable Bodies Using Finite Element Method* (in Croatian), Fintrade & Tours and University of Rijeka, Faculty of Engineering, Rijeka.
16 Smith, I. M. and Griffith, D. V. (1998) *Programming the Finite Element Method*, John Wiley & Sons, New York.
17 Kobayashi, S., Oh, S. I. and Altan, T. (1989) *Metal Forming and the Finite Element Method*, Oxford University Press, New York.
18 Sekulović, M. (1988) Finite Element Method *(in Serbian), IRO "Građevinska knjiga"*, Beograd.
19 Senjanović, I. (1986) *Finite Element Method in Analysis of Ship Structure* (in Croatian), *Faculty of Mechanical Engineering and Naval Architecture in* Zagreb, Zagreb.
20 Timoshenko, S. (1934) *Theory of Elasticity*, McGraw-Hill, New York.
21 Reisman, H. (1988) *Elastic Plates, Theory and Application*, John Wiley & Sons, New York.
22 Alfirević, I. (1999) *Strength of Materials II* (in Croatian), Golden Marketing, Zagreb.
23 Alfirević, I. (2006) *Linear Structural Analysis*, Pretei, d.o.o., Zagreb.
24 Brnić, J. and Turkalj, G. (2006) *Strength of Materials II* (in Croatian), Zigo, Rijeka.
25 Brnić, J. (1996) *Elastomechanics and Plastomechanics* (in Croatian), Školska knjiga, Zagreb.
26 Altan, T., Oh, S. I. and Gegel, H. L. (1983) *Metal Forming: Fundamentals and Applications*, American Society for Metals, Metals Parks, OH.

27 Dieter, G. R. (1997) Relationship between Materials Selection and Processing. In G. E. Dieter (Ed.) *ASM Handbook*, Vol. 20, Materials Selection and Design, ASM International, USA.
28 Zohdi, M. E., Webster, D. B. and Biles, W. E. (1998) Metal Forming, Shaping and Casting. In K. Mayer (Ed.) *Mechanical Engineers' Handbook*, 2nd edition, John Wiley & Sons, New York.
29 Al-Salehi, F. A. R., Firbank, T. C. and Lancaster, P. R. (1973) An experimental determination of the roll pressure distributions in cold rolling, *International Journal of Mechanical Sciences*, 15(9), 693.
30 Lenard, J. G. and Pietrzyk, M. (1992) Rolling Process Modelling. In P. Hartley, I. Pillinger and C. Sturgess (Eds) *Numerical Modelling of Material Deformation Processes*, Research, Development and Applications, Springer-Verlag, New York, pp. 274–302.
31 Roberts, W. L. (1983) *Hot Rolling of Steel*, Marcel Dekker, Inc., New York.
32 Štok, B. (1988) *Mechanics of Deformable Bodies, II*, Faculty of Mechanical Engineering, University of Ljubljana, Ljubljana.
33 Johnson, W. and Mellor, P. B. (1973) *Engineering Plasticity*, Van Nostrand Reinhold Company, London.
34 Brnic, J. (2013) *Basics of Optimization of Mechanical Structures* (in Croatian), Faculty of Engineering, University of Rijeka, Rijeka.
35 American *Society for Testing and Materials* (ASTM) (2005) *Standard test method for measurement of fracture toughness E1820*, Baltimore.
36 Kuna, M. (2013) *Finite Elements in Fracture Mechanics*, Springer, Berlin.
37 Brnic, J., Turkalj, G., Vukelic, G. and Brcic, M. (2012) Analysis of the Dependence of Material Properties on Temperature – Steel 1.4122, *High Temperature Materials and Processes*, 31(3), 259–266.
38 Brnic, J., Canadija, M., Turkalj, G. and Lanc, D. (2010) Behavior of S355JO Steel Subjected to Uniaxial Stress at Lowered and Elevated Temperatures and Creep, *Bulletin of Materials Science*, 33(4), 475–481.
39 Narasaiah, N., Tarafder, S. and Sivaprasad, S. (2010) Effect of Crack Depth on Fracture Toughness of 20MnMoNi55 Pressure Vessel Steel, *Materials. Science and Engineering A*, 527, 2408–2411.
40 Brnic, J., Turkalj, G. and Canadija, M. (2010) Shear stress analysis in engineering beams using deplanation field of special 2-D finite elements, *Meccanica*, 45(2), 227–235.
41 Khoei, A. R. (2015) *Extended Finite Element Method*, John Wiley & Sons, India.

Index

Analysis of Engineering Structures and Material Behavior, First Edition. Josip Brnić.

n

o

p

t